VIRUS AND VIRUS-LIKE
DISEASES OF POTATOES AND
PRODUCTION OF SEED-POTATOES

Virus and Virus-like Diseases of Potatoes and Production of Seed-Potatoes

Edited by

Gad Loebenstein
Philip H. Berger
Alan A. Brunt and
Roger H. Lawson

SPRINGER-SCIENCE+BUSINESS MEDIA, B.V.

A C.I.P. Catalogue record for this book is available from the Library of Congress.

ISBN 978-94-010-3736-5 ISBN 978-94-007-0842-6 (eBook)
DOI 10.1007/978-94-007-0842-6

Cover design: Dov Ronen
Potato plant with leafroll symptoms; Microtubers- Courtesy
by the late Dr A.Watad; Electron micrograph of Potato
leafroll virus - Courtesy Dr L.Salazar, CIP.

Contents

Preface

Potatoes, a major vegetatively-propagated crop, has been closely linked with plant virus research during the last 8 decades because, without their effective control potato viruses can cause considerable losses of crop quality and yield. Such research has resulted in marked advances in diagnosis, from relatively simple biological and serological tests to electron microscopy, sophisticated serological procedures and, more recently, the use of polymerase chain reaction (PCR) and nucleic acid hybridization methods. Associated tissue culture research during the past forty years or so has resulted in the successful production of virus-free plants from potato cultivars that were totally infected. Nevertheless, in many countries the high incidence of virus infection still causes considerable yield losses.

Because of their importance, potato viruses have also long been important subjects for research; much is thus now known about their intrinsic biological and physico-chemical properties, genomes, gene functions, virus-vector relationships (including specific sites of interaction between viral coat protein and the vector) and their potential as vehicles for transformation.

This book is a comprehensive up-to-date treatise that includes information on virus-, viroid- and phytoplasma-induced potato diseases and the detection, isolation, transmission and epidemiology of the pathogens, as well as their control by the rapid propagation of elite pathogen-tested clones, breeding of disease resistant genotypes by conventional and biotechnological methods and the implementation of certification schemes. Also included are chapters on the botanical characteristics and the economic importance of the potato, main cultivars, historical perspective of potato virology and emerging viruses in relation to current control procedures and regulations.

The contributing authors are internationally recognised specialists who were invited to contribute because of their specific knowledge and experience. Each has attempted to include all relevant recently published information; however, with the rapid advances in methodology, especially in diagnosis and production of transgenic resistant genotypes, valuable additional information will undoubtedly continue to be produced. Nevertheless, the principles presented here should long remain valid for the effective control of potato virus-, viroid- and phytoplasma-induced diseases.

We thank all the authors for their contributions and also the many colleagues who kindly provided illustrations, especially Drs. I. Ben Zeev, A. Gal-On, D-E. Lesemann, J. Cohen , L.Salazar and J.Valkonen. Our special thanks are also due to D.Ronen for preparing all the graphic and photographic illustrations.

We hope that this book will be of value not only in countries with high standards of potato agro-technology, but also in countries were potato yields are low and incidence of virus diseases is high, and thus be a contribution to further improvement of this important food crop.

G. Loebenstein, P. Berger, A.A. Brunt , R.H. Lawson

August 2000.

Contributors

Y. Alon
Phytodiagnostic Laboratory, Bactochem Ltd. Ness-Ziona 70400, Israel.

M. F. Askew
Alternative Crops & Biotechnology, Central Science Laboratory,Sand Hutton, York YO41 1LZ, U.K.

E.E. Banttari
University of Minnesota, Department of Plant Pathology, St. Paul, MN 55108, USA.

I. Bartolini
The International Potato Center (CIP), P.O.Box 1558, Lima 12, Peru.

I. Ben Zeev
Plant Protection and Inspection Service, Israel Ministry of Agriculture, Bet Dagan , P.O.B.78 50250 Israel.

P. Berger
Department of Plant, Soil and Entomological Sciences, University of Idaho, Moscow, ID 83844-2339 USA.

D. Bourdin
Institut National de la Recherche Agronomique, UMR 'Biologie des Organismes et des Populations appliquée à la Protection des Plantes' - Domaine de la Motte au Vicomte, B.P. 29; 35653 Le Rheu Cédex, France.

C. R. Brown
USDA Agricultural Research Service, Washington State University, Prosser, WA 99350 USA.

L.G. Brown
USDA/APHIS/PPQ- Center for Plant Health Science and Technology, Raleigh, North Carolina 27606 USA.

A.A. Brunt
Horticulture Research International, Wellesbourne, Warwiks CV35 9EF,U.K.

D. L. Corsini
USDA Agricultural Research Service, University of Idaho, Research and Extension Center, PO Box AA, Aberdeen, ID 83210 USA..

C.D. DiFonzo
Michigan State University, Department of Entomology, East Lansing, Michigan 44824 USA.

G. Faccioli
UCI S.T.A.A ,Istituto di Patologia vegetale, Universitá degli Studi di Bologna, Via Filippo Re 8, 40126 Bologna, Italy.

G.D. Franc
University of Wyoming, Department of Plant Sciences, P.O. Box 3354, Laramie, WY 82071-3354, USA

A. Gera
Department of Virology, Agricultural Research Organization, The Volcani Center, Bet Dagan 50-250 Israel.

T. German
Department of Plant Pathology, University of Wisconsin, Madison, WI 33706 USA.

O.A. Gutbrod
Department of Crop and Soil Science, Oregon State University, Corvallis, Oregon 97331-3001, USA.

A. Hurtado
The International Potato Center (CIP), P.O. Box 1558, Lima 12, Peru.

W.K. Kaniewski
Monsanto Co. St. Louis, MO 63198, USA.

M. Klein
Department of Entomology, Agricultural Research Organization, The Volcani Center, Bet Dagan, 50-250, Israel.

R.H. Lawson
National Program Staff, US Department of Agriculture, Agricultural Research Service, Beltsville, MD 20705-5137 USA..

R. Levin
Osmotek Ltd. P.O.B. 550, Rehovot, Israel.

D. Levy
The Volcani Center, Institute of Field Crops and Vegetables, Beit Dagan, Israel.

G. Loebenstein
Department of Virology, Agricultural Research Organization, The Volcani Center, Bet Dagan, 50-250 Israel.

S. Marco
Viran, Plant Disease Diagnostics, Neve Yarak 49045 Israel.

A.R. Mosley
Department of Crop and Soil Science, Oregon State University, Corvallis, Oregon 97331-3001 USA.

A. Nachmias
Potato R&D, P.O.B. 515, Ofaquim, Israel.

E.B. Radcliffe
University of Minnesota, Department of Entomology, St. Paul, Minnesota 55108, USA.

D.W. Ragsdale
University of Minnesota, Department of Entomology, St. Paul, Minnesota 55108, USA..

Y. Robert
Institut National de la Recherche Agronomique, UMR 'Biologie des Organismes et des Populations appliquée à la Protection des Plantes' - Domaine de la Motte au Vicomte, B.P. 29; 35653 Le Rheu Cédex, France.

L.F. Salazar
The International Potato Center (CIP), P.O. Box 1558, Lima 12, Peru.

C. Sluis
Tissue Grown Corp. Woodland, CA, USA.

R. Stace-Smith
Phyto Diagnostics Laboratory Ltd. 8801 East Saanich Rd, Sidney, British Columbia V8L 1H3, Canada.

P.E. Thomas
USDA , Agricultural Research Service, Prosser, WA 99350 USA.

A.A. Watad*
Department of Ornamental Horticulture, Agricultural Research Organization, The Volcani Center, POB 6, Bet Dagan, 50-250 Israel.
Dr. A. Watad passed away on March 15, 2000, after completing this chapter.

D.P. Weingartner
University of Florida, Institute of Food and Agricultural Sciences, Hastings Research and Education Center, Hastings, Florida 32145-0728, USA.

Colour Plates

Figure 2.1.1-2.1.4.

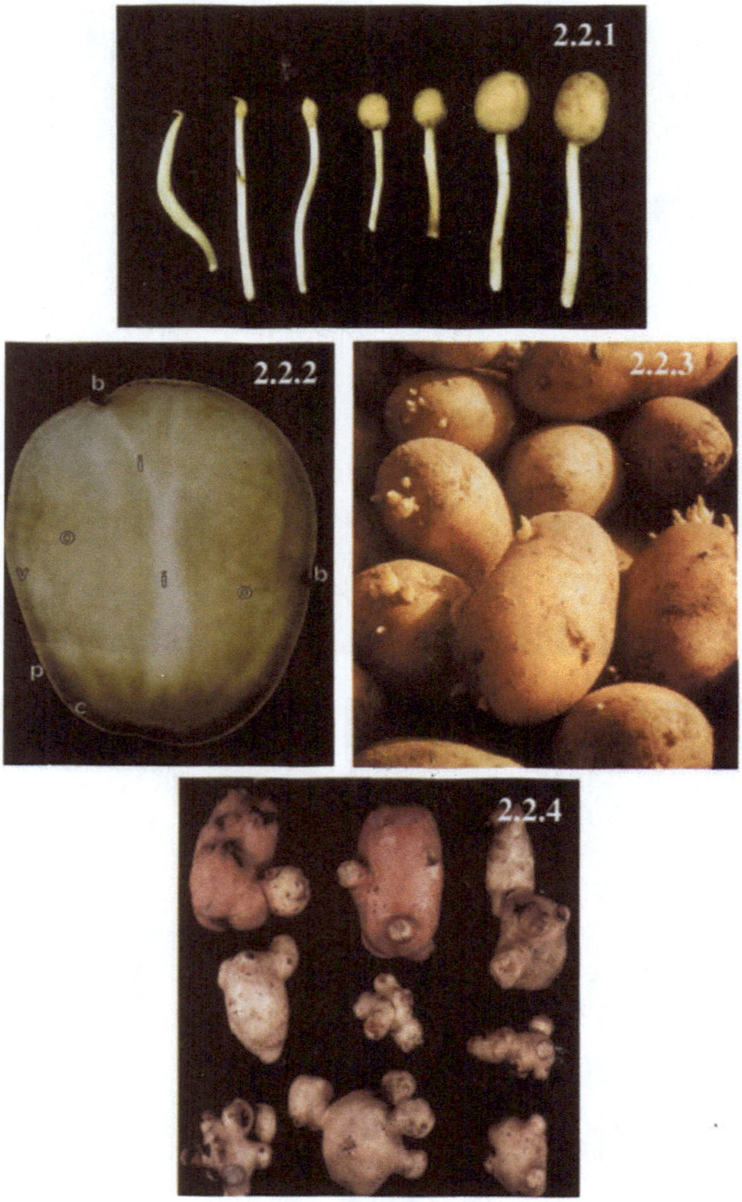

Figure 2.2.1-2.2.4

Figure 5.2.1-5.2.5

Figure 5.3.1-5.3.8

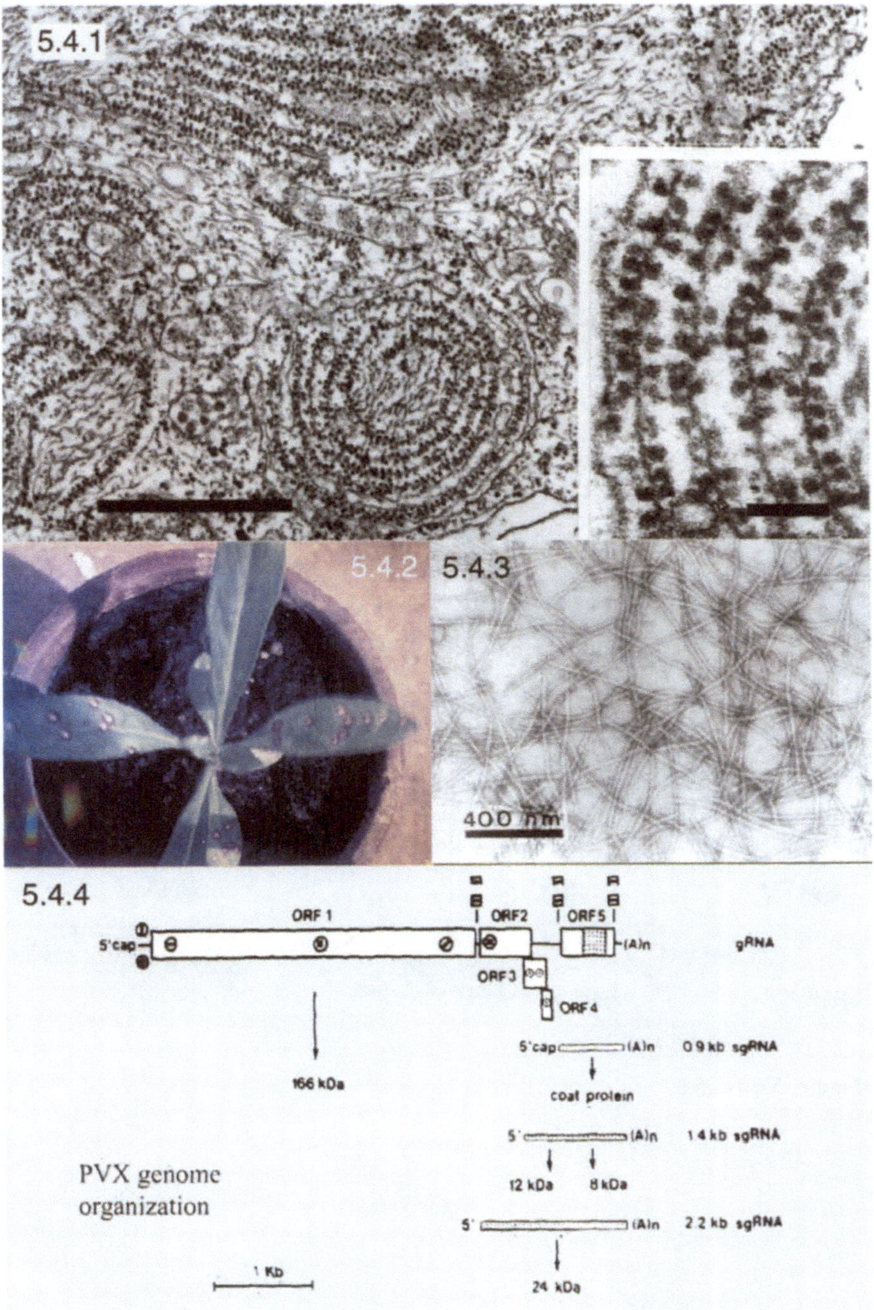

Figure 5.4.1-5.4.4

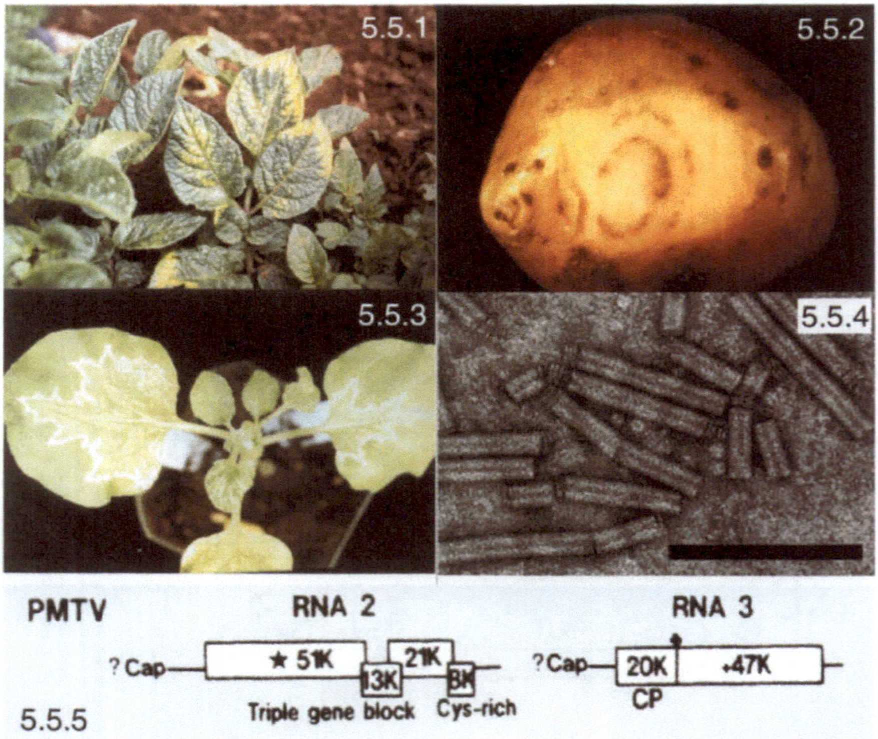

Figure 5.5.1-5.5.5

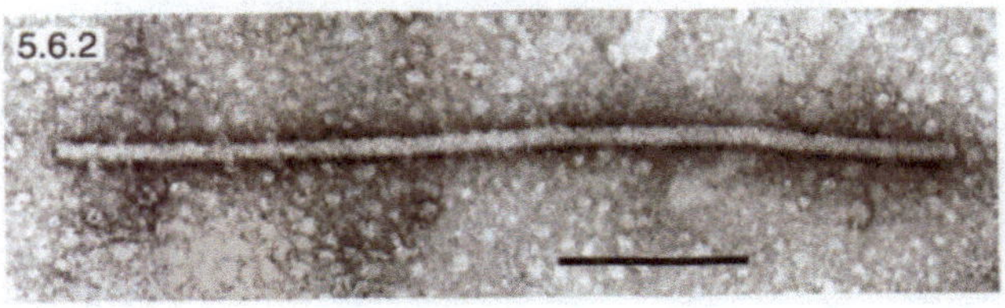

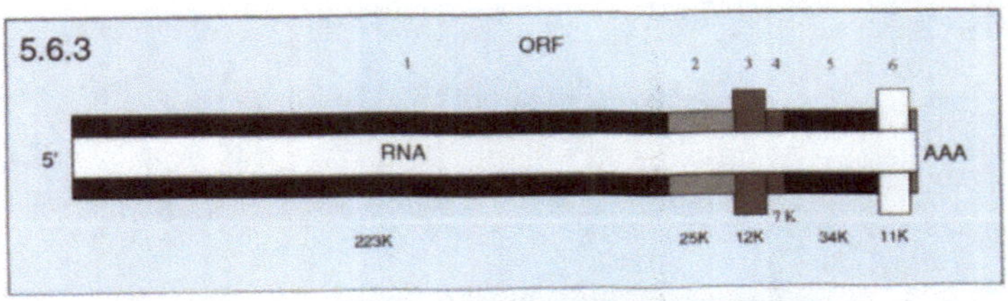

Figure 5.6.1-5.6.3 and Figure 5.7.1

Figure 7.2.A-7.2.H

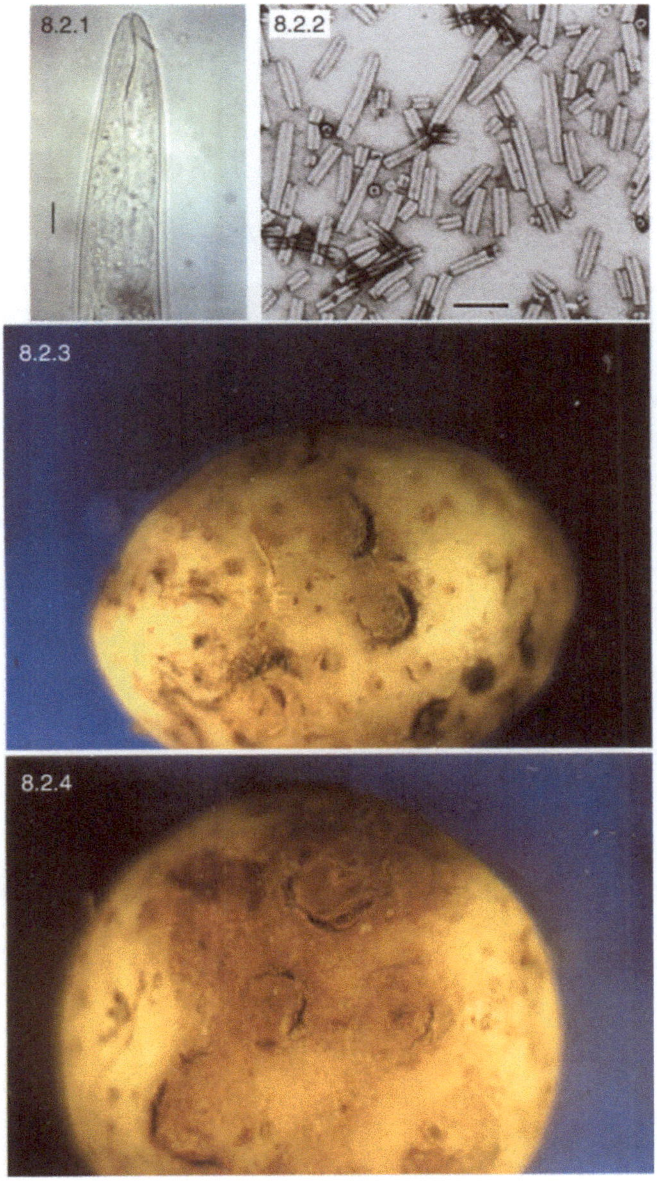

Figure 8.2.1.-8.2.4

Figure 8.4.1

Figure 9.1

Solar powered | Battery powered (12v)

Figure 9.2

Figure 12.1

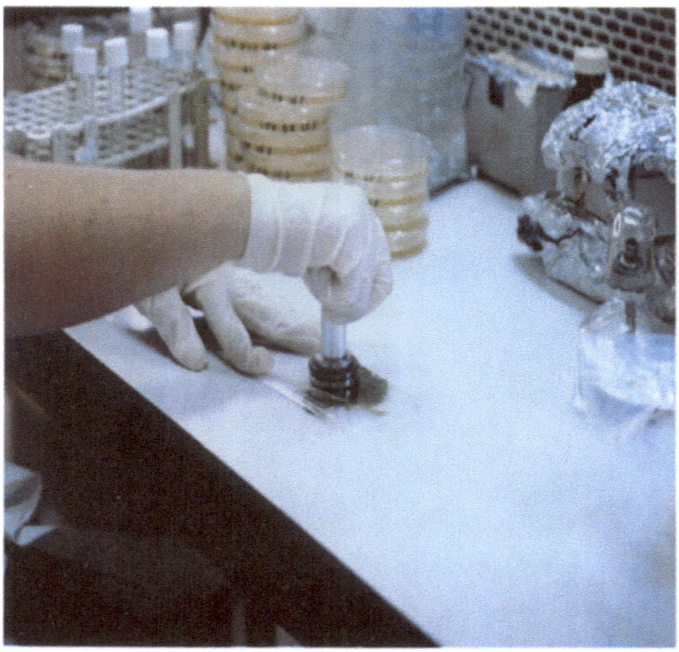

Figure 12.2

Figure 13.2.1

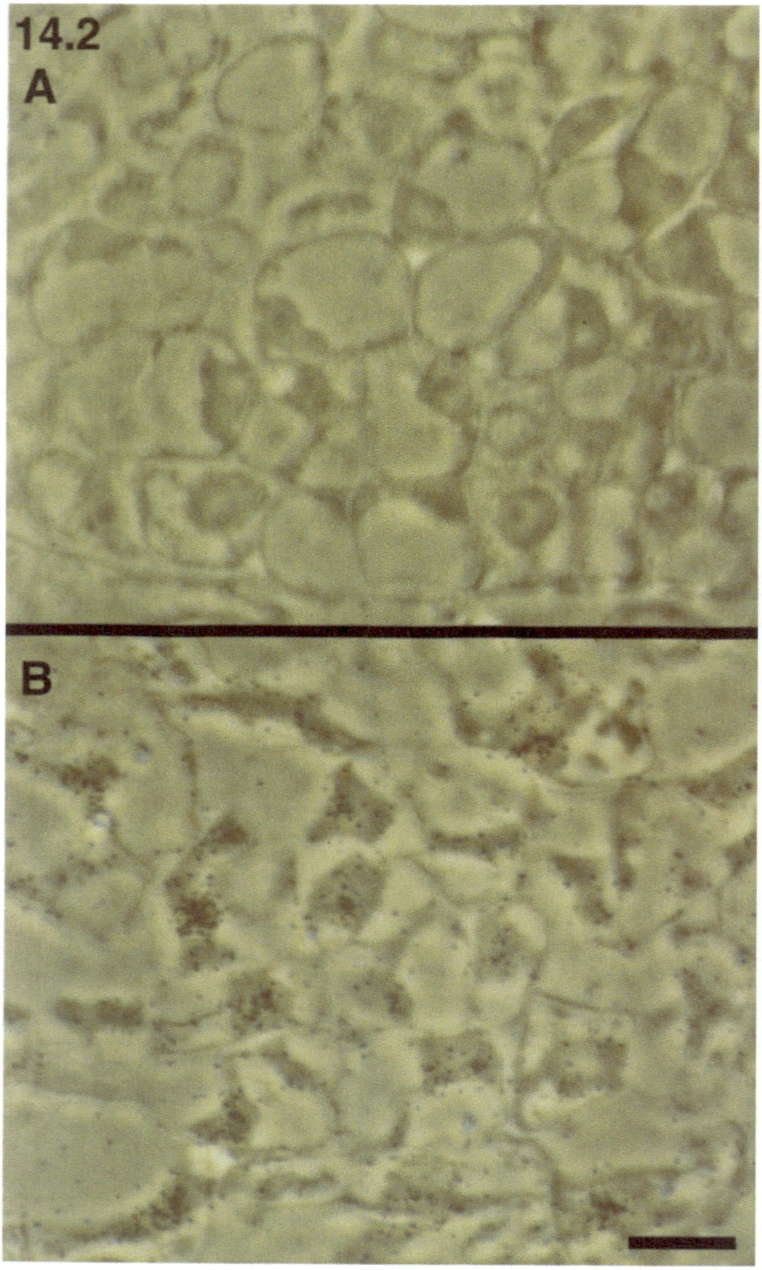

Figure 14.2

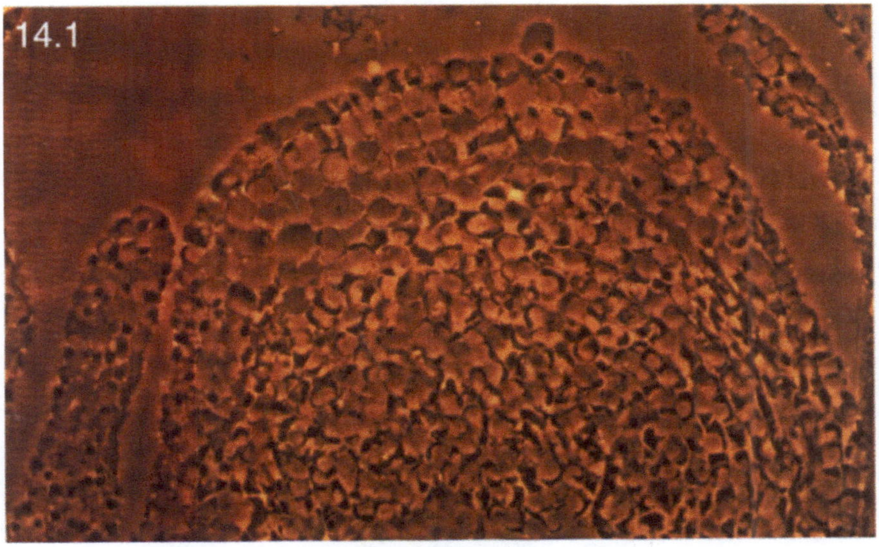

Figure 14.1

Figure 14.4

Figure 15.1

1. The Economic Importance of the Potato

Melvyn F. Askew
Alternative Crops & Biotechnology, Central Science Laboratory, Sand Hutton, York YO41 1LZ, U.K.

1.1 Introduction

The potato, *Solanum tubersum subsp. tuberosùm* and *andigena* and seven other related species, are cultivated, namely *S. ajanhuiri, S. chaucha, S. curtilobum, S. goniocalyx, S. juzepczukii, S. phureja and S. stenotomum.* Interestingly, clones of *S. phureja* are likely to form new potato cultivars for Europe. The potato is grown as a major crop in countries in different climatological zones, including temperate regions, the sub-tropics and tropics, under very different agro-ecological conditions, lowlands and highlands and in very different socio-economic environments. Currently, the crop is grown on a significant scale in about 130 countries.

A native of the remote high Andes and temperate Chilean lowlands, the potato was first seen by Europeans during 1532 and first formally recorded by López de Gómara in 1552 and Cieza de Leon during 1553 (Hawkes, 1992).

The precise time that the potato was first used in cultivation is unknown, but is thought to be some 7 to 10,000 years ago in the Lake Titicaca region of Peru/Bolivia.

Potatoes were probably first shipped into Europe during the period 1567 to 1574 and first multiplication of tuber number in European agriculture probably occurred towards the latter end of that time period (Hawkes, 1991).

G. Loebenstein et al. (eds.),
Virus and Virus-like Diseases of Potatoes and Production of Seed-Potatoes, 1–18.

However, interest in the potato crops as a foodstuff appears to have been restricted to the poorest strata of the population until about 200 years ago. The earliest known varieties according to J G Hawkes were Rough Purple Chile and a derivative, Early Rose.

The extension of the cultivated potato crop in Europe was northwards to England and Ireland. Considerable uptake and dependence occurred in Ireland, and the devastating outcome of that has been well documented (Salaman, 1949). Further extension of potato cultivation appears to have been via Scotland, Scandinavia, Germany and then the currently designated Central and Eastern European states.

Colonial influence had a major impact on dissemination of potatoes when colonists and Christian evangelists transferred tubers to North America, India and the Far East. According to Hawkes (1992), quoting Yen, potatoes were introduced to New Zealand by French explorers. The major portion of this expansion occurred during the 17th and 18th centuries.

At the current time, potatoes form the world's fourth largest crop after wheat, rice and maize and are the largest dicotyledonous food crop, based upon tonnage.

1.2 World Potato Production

It is important to note that whilst estimates of tonnage are quoted, there can be considerable variation in these according to data source.

Early trends in post World War II world potato production were calculated by Horton and Anderson (1992). They reported that average yields have increased by 25% in the period 1961 to 1987.

World potato production stands at an estimated 295 million tonnes for the year 1999. Of this, some 33¼ million tonnes was of seed tubers. Further data are given in Table 1.2 and 1.3.

Table 1.1 : The Potato Crop
Statistical data of regions and typical countries
(adapted from CIP 1998).

	Projected population in the year 2000 (millions)	Agricultural share of Gross Domestic Product (%)	Rank of importance of potato vs. other crops	Area (000 ha)	Production (000 t)	Yield (t/ha)
Asia						
China	1,225	20	5	3,489	47,777	14
India	1,022	27	3	1,116	18,627	17
Iran	68	-	2	152	3,182	21
Bangladesh	120	30	2	133	1,489	11
Africa						
Egypt	62	16	5	130	2,656	20
South Africa	46	5	3	56	1,539	28
Algeria	31	12	2	80	1,099	14
Malawi	12	36	2	51	379	7
Latin America						
Columbia	40	16	2	170	2,770	16
Brazil	175	14	7	182	2,701	15
Peru	26	7	1	240	2,355	10
Argentina	37	6	4	98	2,155	22
North America						
USA	275	-	4	556	21,200	38
Europe						
Poland	39	6	1	1,390	24,295	17
Germany	82	1	3	354	12,530	35
Netherlands	16	3	1	183	7,834	43
Eurasia						
Russian Federation	147	7	1	3,389	38,534	11
WORLD	6,123	-	4	18,381	295,118	16

Agricultural share of Gross Domestic Product is based on 1997 data. Ranking of crop's importance is based on fresh weight. Production and yield are based on fresh weight and are averages over the period 1995-1997. A dash means no reliable data available. Source: Struik and Wiersema, (1999).

Table 1.2 : States contributing more than 1% by weight of total world potato production during 1999 (tonnes)

Country	1998	1999	1999 as % 1998
Argentina	3,412,395	3,450,000	101.10%
Belarus	7,574,000	8,000,000	105.62%
Canada	4,329,000	4,204,000	97.11%
China (People's Republic)	60,739,000	55,354,163	91.13%
France	6,053,000	6,475,000	106.97%
Germany	11,711,720	11,419,908	97.51%
India	17,648,100	22,500,000	127.49%
Iran (Islamic Republic)	3,430,407	3,430,000	99.99%
Japan	3,074,000	3,400,000	110.61%
Netherlands	5,249,400	8,200,000	156.21%
Peru	2,589,670	3,049,000	117.74%
Poland	25,948,670	19,926,716	76.79%
Russian Federation	31,419,230	31,200,000	99.30%
Spain	3,246,800	3,311,900	102.01%
Turkey	5,315,000	5,315,000	100.00%
Ukraine	15,405,000	15,405,100	100.00%
UK	6,420,000	7,100,000	110.59%
USA	21,580,600	21,700,000	100.55%
World Total	293,377,360	294,332,000	100.33%

Source: FAO (1999)

Whilst tonnages of production are indicative of importance of the potato crop, they do not reflect well market sectors, nor necessarily the impact of novel technologies. This latter is particularly true for seed potato tubers where the impact of micropropagation and the development of micro and mini tubers has been considerable.

Van der Zaag (1990) in an overview paper, surveyed the use of micropropagation techniques and micro and mini tubers introduced in Europe. His primary findings are reported in Table 1.4.

Table 1.3 : States producing approximately 1% or more of seed potato tuber production in the world during 1999 (tonnes)

Country	1998	1999	1999 as % 1998
Belarus	1,644,000	1,844,000	112%
Brazil	315,000	315,000	100%
Canada	301,390	305,000	101%
China (People's Republic)	2,701,773	2,501,773	93%
France	373,000	373,000	100%
Germany	692,000	650,000	94%
India	2,100,000	2,100,000	100%
Kazakhstan	329,100	300,000	91%
Netherlands	410,000	410,000	100%
Peru	350,000	350,000	100%
Poland	3,100,000	3,100,000	100%
Romania	1,025,000	1,050,000	102%
Russian Federation	7,618,000	7,500,000	98%
Turkey	430,000	430,000	100%
Ukraine	4,650,000	4,650,000	100%
UK	478,000	478,000	100%
USA	1,260,000	1,260,000	100%

Source: FAO (1999)

Table 1.4 : Micropropagation Techniques in Europe

	% of Countries (rounded figures)
Pre-basic seed production	
Only traditional clonal selection	0
Clonal selection + stem cuttings	5
Clonal selection + MT	45
Only MT eventually in combination with clonal selection	50
Types of MT	
In vitro plantlets	90
Micro-tubers	0
Mini-tubers	60
Multiplication in	
Glass or screen-houses	75
Field	55
Micropropagated plants (MP) are equivalent to	
1st year clones	45
2nd year clones	30
3rd year clones	20
Which part of the basic seed is originating from MP?	
0-25%	35
15-95%	25
100%	40
Has quality improved by the use of MT?	
Yes	40
No reliable information	60
Why MT?	
Reducing cost	20
Improving quality concerning	
- virus diseases	65
- fungal diseases	20
- bacterial diseases	60

It is well proven that total biological yield of potato is a function of water availability, intercepted photosynthetically active radiation and the general nutrition and pest/pathogen complexes. Hence, comparing production or yield on an annual basis is likely to be misleading; longer term averages provide a more apt assessment. Such data is provided in Figures 1.1 and 1.2. These indicate that the genetic potential of most potato crops is far from fully exploited, even if maximum biological yield was not the absolute target and that considerable variations occur between some closely located countries with similar climatological conditions. In terms of the future, there is therefore considerable potential for enhancement of total level of production, given availability of inputs, or else a reduction of area and consolidation of production in more technically and economically efficient areas. This presumes more cross border trade in potatoes or potato products. However, the political, logistical and economic aspects of this option require elucidation.

As reported earlier potato yields have increased over time. Initially, in the period post World War II, improvements in agronomy, allied to the introduction of good quality inorganic fertilisers and crop protection products, gave considerable enhancement of yield. Thereafter, improved understanding of irrigation and improved machinery brought higher absolute and exploitable yields. However, with a changing focus in the market place, with much greater emphasis on quality as opposed to quantity and with some small but significant trends towards organic potato production, (i.e. production without use of synthetic materials) the trend is more likely to be towards fewer but more specialist growers with close focus on market specifications.

1.3 The Cost of Specialisation

The trend towards potatoes or their products being available throughout the year has enhanced the drive to improve potato tuber quality and to increase storage capacity. Unfortunately, the physical and physiological characteristics of the potato tuber do not lend themselves to simple and easily managed storage systems, as occur with cereals.

In potato, where tuber dry matter content may be from 18-28% depending upon cultivar and locality, consideration must be given to damage; pathogen loading; soil contamination; storage depth and pressure of tuber upon tuber, as well as storage temperature, this last item being governed by end market use. Hence design of potato stores needs to allow for all of these criteria and is therefore relatively expensive. For example in UK, the average cost of building one tonne capacity in a specialist potato store, would be approximately £150 (Author's own estimate) whereas in USA cost would be $80-$100 per tonne, depending on construction materials.

Additionally, the drive to enhanced potato quality in many western economies has caused major reductions in manual work with accompanying increases in investment in mechanisation. These changes in time have led to significant reductions in numbers of mainstream potato producers with accompanying increases in their individual potato production areas. This trend is typified by potato production trends in Great Britain and USA. In GB, producer numbers have declined by over 20% in the period 1992-1996, with no significant change in overall production. This trend is echoed in most developed countries.

Fig. 1.1

Potato Production and Yield in Africa — 1991–95 average

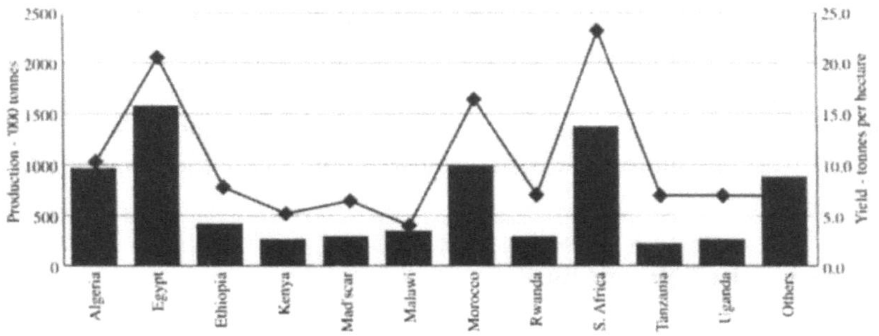

Potato Production and Yield in America — 1991–95 average

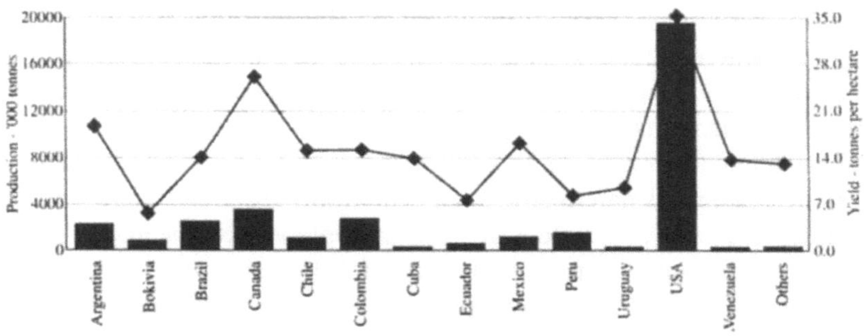

Potato Production and Yield in Asia — 1991–95 average

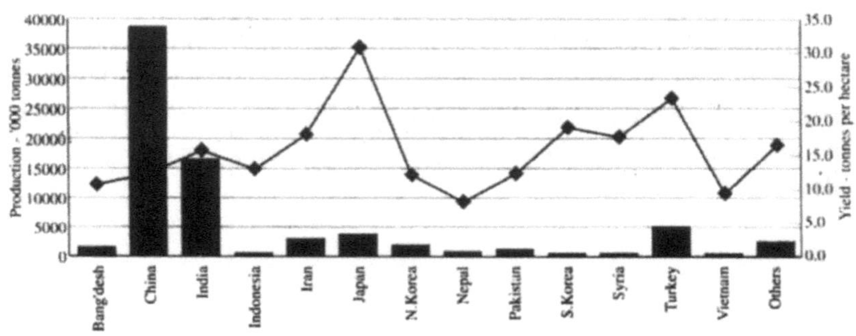

Fig.1.2

Potato Production and Yield in Non-EU Europe — 1991–95 average

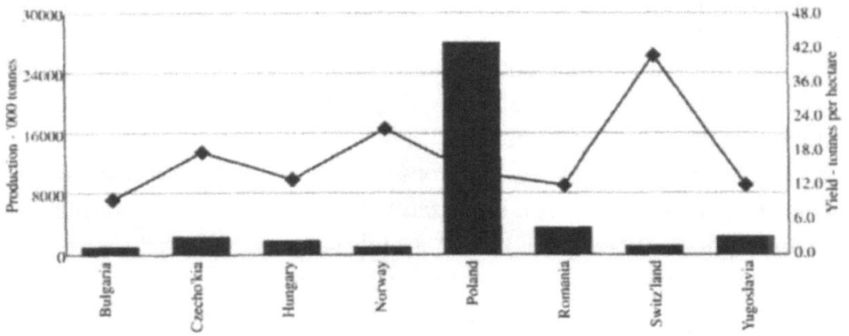

Potato Production and Yield in the European Union — 1991–95 average

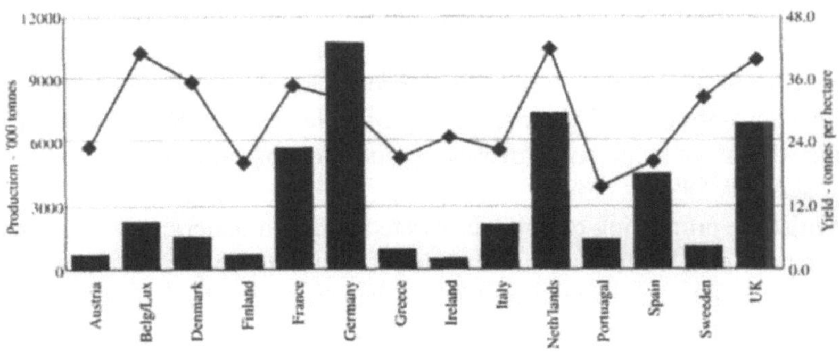

Potato Production and Yield in the Former USSR — 1991–95 average

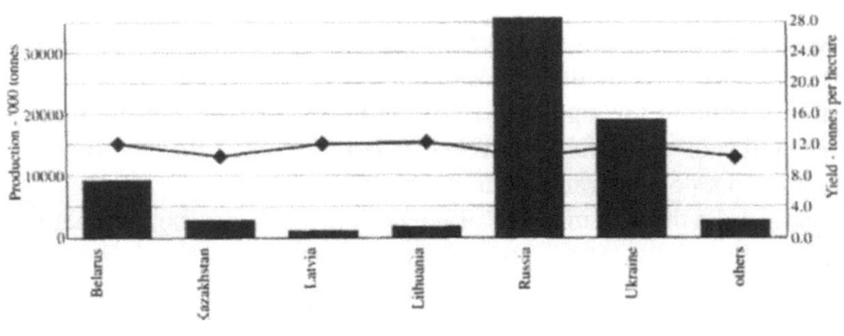

1.4 The Development of Market Segmentation

The split between seed tuber production and production of tubers for direct consumption or processing for food or non-food uses appears obvious, but is confounded by the use of once-grown home-produced tubers as seed tubers in some areas. Whilst a proportion of this use is well planned and intentional, it is important to note that opportunistic use of once-grown seed tubers occurs in years when potatoes are in surplus and therefore of low price, or, conversely when potato supplies are limited and purpose-grown seed tubers are perceived as expensive.

Market segmentation began in potatoes for direct consumption, usually because of time of harvest. In UK, or the Channel Islands for example, very early planted potato tubers give rise to a high value but limited tonnage and short duration market for so-called early, new or scraping potatoes, consumed as boiled potatoes having been scrubbed or scraped. Following this UK market example, later maturing or harvested crops would be bound for domestic or processing use, the former being characterised by "brown-bag" bulk (e.g. 50kg packs of mixed tuber size); pre-packs of various specification baking potatoes (usually >65mm and graded by weight and size) or free-flow potatoes sold loose by the greengrocer or supermarket.

The processing market has in it 2 broad components; processing for human consumption and the processing of potatoes for the production of starch, which could be used in a range of food or non-food products or processes.

Whilst the processing of potatoes existed in South America, where native Indians produced Chuno' (a "freeze" dried, desiccated potato) at least 1500 years ago and German experimenters attempted dehydration of potatoes surplus to requirements during the 18th century, the true processing of potatoes for specific consumer tastes did not develop until the 1950s and thereafter and was principally initiated in USA. In more developed economies, the processed potato sector is assuming greater significance and the final product, but not necessarily the primary potato feedstock, added value.

Expansion areas in this sector are to be found especially in ready purpose meals or snacks for immediate consumption after re-heating.

Table 1.5 shows the significance of the potato crop as a staple food in a number indicator countries. However, it is to be noted that the nutritional value of the potato crop varies with its mode of preparation.

Hence, whilst the consumption of potatoes in USA may be low relative to a number of other developed economies, the fact that much of the intake is as processed food indicates a much high calorific intake than would be the case with unprocessed potatoes.

1.4.1 Price and Value

The potato markets of the world are not well organised. Some are subjected to state controls or aids and standards are not uniform. Nevertheless, some stability has been introduced through contract arrangements, particularly in the processing sector. However, since

Table 1.5 : Utilisation (in % production) of the potato crop and per capita consumption in different regions and countries.

	Utilisation (%)					Per capita consumption (kg per year)
	Food	Processing	Feed	Seed	Other	
Asia						
China	36	22	31	6	5	14
India	73	0	0	10	17	14
Iran	85	0	0	5	10	39
Bangladesh	81	0	0	9	10	10
Africa						
Egypt	74	0	0	12	14	21
South Africa	75	6	10	5	5	26
Algeria	82	1	0	8	10	33
Malawi	76	0	3	11	10	29
Latin America						
Columbia	71	0	10	6	12	56
Brazil	79	1	0	12	8	14
Peru	68	1	1	11	20	63
Argentina	85	1	0	4	10	52
North America						
USA	32	60	1	7	0	61
Europe						
Poland	21	4	48	13	14	136
Germany	59	14	9	7	11	79
Netherlands	29	47	14	3	3	87
Eurasia						
Russian Federation	48	5	23	20	4	121
WORLD	54	8	19	12	8	28

Food is direct food only. Processing encompasses both food and non-food end products. All data are averages over 1994-1996, except for the USA data. A dash means no reliable data available. Adapted from CIP (1998). Source: Struik and Wiersema (1999).

Table 1.6 : Main Crop Potatoes : Production Costs and Gross Margins UK

Production Level	Low	Average	Hight
Yield : Tonnes per ha	32.5	42.5	52.5
Output	2590	3385	4180
Variable Costs:			
Seed:		570	
Fertiliser:		210	
Sprays:		425	
Casual Labour for harvesting and grading:		360	
Sundries (levy, sacks, etc):	310	360	410
Total Variable Costs	1875	1925	1975
Gross Margin per hectare	715	1460	2205

Source: Nix (1999)

Table 1.7 : Production Costs and Gross Margins for Specialist Seed Potato Production in UK

	Seed	*Ware*	*Waste/Stockfeed*
Yield t/ha	£25	£7	£3
Value £/t	£180	£60	£10
Output £/ha	£4,500	£420	£30

Total output £/ha	£4,950/ha

Variable costs	£880
Fertiliser	£147
Casual Labour:	
Lift	£89
Grade	£179
Contract	£120
Sprays	£290
Other crop costs	£169
Total Variable Costs	£1,874/ ha
Gross Margin (£/ha)	£3,076/ ha

Source: SAC

production is significantly affected by weather and harvesting conditions and since demand is fairly inelastic in economic terms, then surpluses can have depressing effects on prices and thereafter return, can occur.

Production costs of main crop potatoes and seed potatoes in UK are shown in Tables 1.6 and 1.7.

Clearly, any impact of reduction of end price will be reflected in attempts to reduce variable costs and with the high proportion of production costs attributable to seed tubers, be reflected in their demand and price.

In agricultural respects, even though the significance of the different market sector outlets is less marked than post farm gate, it has a noticeable effect on profitability. Examples are given for UK (Table 1.8).

Table 1.8 : Relative Value of Potatoes ex Farm in UK

	Index Value
Ordinary ware	100
Multiple sales outlets eg supermarkets	
'Freeflow'	140
Prepack	140
Bakers	210
Baby Roast	140
French fry	110
Oven ready	110
French fry	
Crisps (UK)	140
Dehydrate	70
Canned	250
Processing – damaged, peeled	20-40

(These data are calculated by the author)

1.5 Genetically modified potatoes

So-called genetically modified potatoes are those which have been subjected to novel transgenic technologies as a means to introduce novel traits or to prevent natural phenomena like production of reducing sugars. This latter process, where nothing is added but whereby existing genes are disabled is called antisense.

Whilst transgenic technologies offer the potential to move genes between species, genera and phyla, those same technologies could be employed as a means of transferring potato genes from one clone to another in a more controlled way than by traditional pedigree breeding methods. This could quicken breeding programmes and have positive impact on production costs.

Hence, genetically modified (GM) potatoes would not necessarily carry exotic genes.

Data presented at European Potato Industry Conference of February 1999, indicated that 19,800 ha of GM potatoes had been grown in USA during 1998 and the projections for 1999 were for an area in excess of 35,000 ha. The cultivars involved were of the 'New Leaf' brand introduced by Monsanto and had resistance to Colorado Beetle *Leptinotarsa decemlineata* or *Potato leafroll virus*. New cultivars with combined GM traits for Colorado Beetle resistance and resistance to Potato virus Y (PVY) were expected to be approved during 1999. However, in the recent past, marketing decisions by key potato processors in USA may have effectively stopped expansion of GM potatoes through closing down their marketplace. Whether or not this approach will change when GM potato cultivars with traits perceived as more consumer-orientated or consumer friendly are introduced, awaits to be seen.

Currently public opinion is somewhat divided on the desirability or acceptability of GM technology in its totality. Hence, comment here is restricted to major opportunities for GM potatoes, whether or not they are exploited in the longer term.

Potato production is faced with a number of key challenges for which GM technologies may provide solutions:

Potato virus Y (PVY*) and Potato Leaf Roll Virus* (PLRV)

The development of clones resistant to severe viruses would offer yield protection, cost savings and reduction in pesticide use, but could have a negative impact upon the specialist seed potato industry since supplies of new seed tubers may not be needed as frequently as they are at present.

Late Blight (*Phytophthora infestans*)

The impact of late blight is to reduce yield through early defoliation and to enhance storage problems which occur particularly where secondary pathogens infect through late blight lesions.

Reductions in risk of late blight infection, or varietal resistance, would enhance yield and quality of produce, enhance profitability of the enterprise and reduce use of fungicide and any environmental impact it may have.

Drought

Once critical soil moisture deficit has been reached, the yield potential of a potato crop is limited. Experimental evidence indicates that at least 1 tonne per hectare per millimetre water to be the benefit of irrigation in drought situations. This presumes that no other inputs are limiting. Availability of adequate irrigation equipment and irrigation water are practical, economic and logistical challenges which are often insuperable. Hence a genetic tolerance would have major economic benefits.

Cold Tolerance

There are a number of aspects to cold tolerance in potato. The canopy of the potato crop is relatively late to close relative to incident PAR in many temperate areas. Hence yield is limited. If the canopy was to be frost tolerant, earlier planting, earlier canopy development and closure and enhanced yields would accrue. Progress in developing these traits in potato has been made at University of Naples (Pooley, R. Pers. Com). Potato tubers are not frost tolerant. This offers partial control of volunteer tubers and is thus desirable. However, frost tolerance would offer safer storage of tubers in non-tropical areas and ease ambient air ventilation constraints. This would reduce storage, building and operational costs.

One key sector of the potato market is that of processing, where potato tubers with low reducing sugar concentrations are required. The development of tubers which were resistant to accumulation of reducing sugars when stored below 10°C, would be a particular benefit in terms of cost, value and chemical sprout suppressant use.

New Markets Through Chemistry Of The Potato Tuber

The ability to modify flavour and texture could be a primary GM target, especially where consumers prefer a particular taste or texture (eg waxy potatoes). Undoubtedly, several genes would be involved, so that this objective would not be achieved for some time. Starch content in the potato is composed of different starch types – amylose and amylopectin. These have quite different molecular structures. The introduction of clones with one starch type alone would be beneficial for end users and could enhance profitability in this sector.

The potato plant is relatively easy to grow, is not a significant out-pollinator and from a transgenic risk viewpoint, provides a relatively safe crop in which to produce exotic molecules. A number of opportunities exist but proteins for the pharmaceutical market in particular would provide high value opportunities on limited areas of production.

1.6 World Potato Trade

The physical bulk, high water content of the potato tuber and propensity to pathogen or physical damage do not predispose it to long distance transport. Nevertheless, with the bulk of the crop being produced from seed potato tubers rather than true potato seed (TPS), there is an obvious need for some movement.

Total world exports of potatoes for the calendar year 1998, amounted to approximately $1.6 billion and extended to 7.8 million tonnes (FAO). Exporting nations where exports exceeded $50 million in value, are listed in Table 1.9.

Table 1.9 : Countries exporting potatoes to value of $50 million or more during 1998

Country	Value ($000's)	Quantity (000's t)	Value Index
Belgium	168,969	1,179.8	143.2
Canada	128,942	631.5	204.2
France	255,210	1,248.9	204.3
Germany	102,577	930.5	110.2
Italy	90,686	278.7	325.4
Netherlands	354,734	1,420.9	249.7
Portugal	54,968	241.9	227.2
Spain	74,068	230.6	321.2
United Kingdom	76,167	171.1	445.2
USA	92,967	304.0	305.8

Source: FAO 1999

The value index expressed in Tables 1.9 and 1.10 is an indication of broad unit value of exports; lower values represent exports being primarily main crop potatoes for consumption or processing, whereas higher values represent seed potato tubers for planting, or, in examples like Italy, high value early market table potatoes.

Major importers of potatoes are shown in Table 1.10. It is to be noted that FAO data shows some discrepancy between world exports of potatoes at $1.8 billion and imports at $1.65 billion.

Table 1.10 : Major importers of potatoes by value above a threshold $50 million

Country	Value (000's)	Quantity (000's t)	Value Index
Belgium	146,715	903.1	162.5
Canada	65,981	239.9	175.0
France	94,616	360.6	262.4
Germany	181,039	685.2	225.5
Italy	101,873	451.7	94.8
Netherlands	126,911	1,339.4	227.2
Spain	139,124	595,8	233.5
United Kingdom	177,666	442.7	401.3
USA	106,867	481.3	220.0

Source: FAO 1999.

1.6.1 Future Potato Markets

The development of markets will occur in 3 areas:
The extension of mature markets (e.g. UK, USA) where the convenient food component and high added value post farm gate will be key features.

The reorganisation and re-focussing of markets in the Baltic States and Central and Eastern European Countries (CEEC). Key features will be continued restructuring and producer specialisation and extension of processor capacity, along with the development of fresh potato and processed potato consumption.

The development and extension of subsistence production as a staple food with barter and traded surpluses coming into the marketplace. Innovators in production and marketing will play key roles in rate and direction of development.

References

British Potato Council. Informal data from statistics department, British Potato Council, Oxford, UK.

FAO 1999. FAOSTAT Database results from FAO web site; http://appsl.fao.org/servlet/XteServlet.

Farmers Weekly, 19th November 1999.

Hawkes, J. G. 1993. The historical and social role of the potato. *Proceedings of 12th Triennial Conference and European Association for Potato Research*, Paris, France, pp75-84.

Hawkes, J. G. 1992. History of the Potato. In *'The Potato Crop'* ed. P. Harris. Chapman and Hall, London_ 909pp.

Nix, J. 1999. Farm Management Pocket Book. 30th Edition (2000), Sept 1999.

S.A.C. The Farm Management Handbook 1997/1998.

Salaman, R. N. 1949. The *History and Social Impact of Potato.* Reprinted with corrections by J. G. Hawkes, 1985. Cambridge University Press, UK.

Struik, P. and Wiersmema, S. G.; Seed Potato Technology. Netherlands, Wageningen, 1999.

USDA. 1999. *Potatoes and Sweet Potatoes.* Final estimates 1992-97. US Dept. Agriculture. National Agricultural Statistics Service. Statistical Bulletin 962. 56pp.

Van der Zaag, D. E. 1990. The implications of micropropagation for the future of seed potato production systems in Europe. *Proc. 11th Triennial Conf. Euro. Assoc. Potato Res.,* Edinburgh, Scotland. Pp28-45.

Walker, T.T.; Schmiediche, P.E. and Hijmans, R. J. 1991. World trends and patterns in the potato crop; an economic and geographic survey. *Proceeding of European Association of Potato Research,* Sorrento, Italy, 1998.

Woolfe, J. A. 1987. *The Potato in the Human Diet.* Cambridge University Press, Cambridge, UK.

Young, N.A. 1981. *Potato Products : Production and Markets in the European Communities.* Publ. Commission of the European Communities. Information on Agriculture. No 75, Luxembourg. 214pp.

Further information sources:

British Potato Council web site. www.potato.org.uk

CIP (International Potato Centre, Peru) web site. www.cipotato.org

Commission European – Directorate General VI, Agriculture V1/E.3. La pomme de terre en chiffres. [An international document ref JB/hve (Rev.1) dated 18th March 1999].

European Association for Potato Research web site. www.agro.wau.nl/eapr.

2. Biology and Physiology of the Potato

Haim D. Rabinowitch[1] and David Levy[2]

[1]*The Hebrew University of Jerusalem, Department of Field Crops, Vegetables and Genetics, Faculty of Agricultural, Food and Environmental Quality Sciences, Rehovot, Israel;* [2]*The Volcani Center, Institute of Field Crops and Vegetables, Beit Dagan, Israel.*

2.1 History and Economic Importance

The domestication of the potato (*Solanum tuberosum* L.) started millennia ago, around its center of evolution in South America. Many *Solanum* species, relatives of the domesticated potato grow wild in this area, mainly in the Andes of Peru and Bolivia. Archeological remains and archives of the early Spanish occupation provide evidence that the potato was domesticated by the native population between 5,000 and 2,000 years prior to the Spanish discovery.

Following domestication, the nourishing, highly productive plant spread rapidly from its centre of origin to neighboring areas, throughout the high Andes. Spaniards' accounts indicate that early in the 16th century, potato cultivation was common in the areas known today as Colombia, Ecuador, Peru, Bolivia and Chile (Hawkes, 1967). The first introduction in Europe was around the year 1570 when potatoes, probably the subspecies S.

G. Loebenstein et al. (eds.),
Virus and Virus-like Diseases of Potatoes and Production of Seed-Potatoes, 19–37.

andigena, were brought to Spain, and samples of this introduction were sent to Italy and thereafter to Austria, France and Belgium. Additional introductions were made into England in the eighties and nineties of the 16th centuries. The two successive introductions to Spain and later to England, were most probably the source for the potato spread to the rest of the world.

The first Andean potato introductions in Europe had a long period of stem growth. The plants produced many flowers and long branched stolons (underground stems) before producing a few small tubers, in the late fall, when days become short. The newly introduced plants were therefore first regarded as a botanical curiosity, and became an acceptable food crop only around the mid-18th century. This was made possible by breeding for early producing genotypes adapted to the long photoperiod common in western and northern Europe. The strong selection pressure resulted in the significant change from one subspecies (S. *andigena*) to the other (S. *tuberosum*) (Simmonds, 1966). Today, the potato is the fourth most important food crop worldwide after wheat, rice and maize, cultivated on about 20 million hectares and the yearly production is around 275 million tons (Anon., 1995; see Chapter 1).

A significant part of potato production is saved for reproduction and planting as seed tubers, amounting to over 35 million tons, over 13% of the world production. The quality of 'seed' tubers is of paramount importance to the subsequent yield and thus 'seed' production has become the specialization of certain regions, esp. The Netherlands, Scotland, Ireland, Canada, France and Argentina.

2.2 Classification

The genus *Solanum* belongs to the *Solanaceae* family, together with other important crops such as: tomatoes (*Lycopersicon esculentum*), pepper (*Capsicum annuum*), eggplant (S. *melongena*), tobacco (*Nicotiana tabacum)* and minor crops such as pepino dulce (S. *muricatum*). The genus *Solanum* includes about 1,000 species, out of which about six are cultivated and over 230 grow wild only near its center of origin, in South and Central America. Generally, all the tuber-bearing *Solanum* species are grouped in the *Petota* section. Yet, a number of related species that lack stolons and tubers have been included in this taxon due to other morphological traits. The latter are included in the series *Juglandifolia* and *Etuberosa*.

2.3 Cytology

With a base number of 12, most of the members of the *Petota* section are diploids, about 15% are tetraploids and the remaining 12% are tri-, penta- or hexaploids. All the diploids studied so far have shown regular pairing of homologous chromosomes at meiosis, and there are good indications that homologous chromosomes in polyploids also occur. Most of the diploids hybridize readily, and produce fertile F_1 plants, but the F_2 offspring are usually weak and mostly infertile.

The tetraploids, with the exception of the domesticated potato, behave like allotetraploids (producing 24 bivalents), whereas S. *tuberosum* is an autotetraploid (Swaminathan and Howard, 1953).

2.4. Botany

The potato, an herbaceous dicotyledonous annual plant, reproduces asexually by tubers, the only edible part of the plant. Tubers are formed at the end of underground stems, called stolons (Fig. 2.1). The plant also flowers and sets small green or purplish-green poisonous berries, 10 to 20 mm in diameter (Fig. 2.2). The true potato seeds (TPS) are used for breeding and in recent years also for propagation (Fig. 2.3). Floral development is usually inhibited in tuber-producing plants where reproductive buds and flowers drop. Hence, blooming, and fruit (and seed) set can be promoted by continuous removal of the newly produced tubers (Thijn, 1954); grafting potato scions on tomato rootstock (Ross, 1986); maintaining temperatures below 22°C, and exposing the plants to long photoperiods (Patterson, 1953).

2.4.1 Morphology and Development

True seed: Epigean germination begins with the radicle emergence from the micropylar end of the seed. The primary tap root soon forms numerous lateral roots. Following the elongation of the hypocotyl and the spread of cotyledons, the first true leaves develop on the main stem, and adventitious roots form on the hypo- and epicotyl. When the young seedling is only a few cm long, stolons develop in the cotyledonous node. These specialized side-shoots penetrate the soil and form tubers.

Roots: True seedlings first develop a tap root, and then adventitious roots form on the main stem immediately after the establishment of the plant. When grown from tubers, the sprouts develop adventitious roots at the nodes

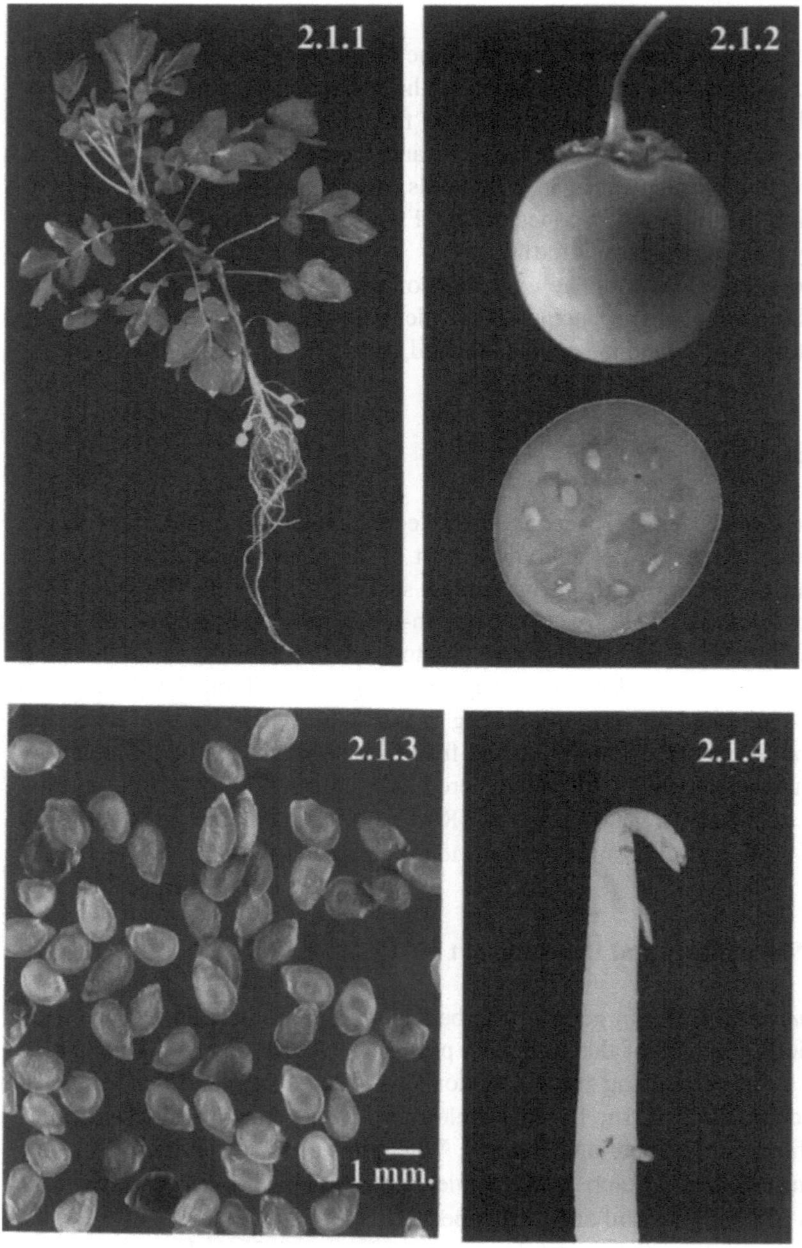

Fig. 2.1.1 - Potato plant with stolons, tubers and roots. The brown mass is the remains of the "seed" tuber.

Fig. 2.1.2 Fruit of potato. 1.5 cm in diameter.

Fig. 2.1..3 - True potato seed (TPS).

Fig. 2.1.4 - Young stolon with elongated internodes and hooked tip.

of the underground stems and stolons. These roots are generally shallow, and the main rhizosphere occupies the upper 40-50 cm soil layer.

Foliage: The herbaceous aerial stem with spiral phyllotaxis is considered to be a main stem, if it grows directly from the mother tuber ('seed' tuber). The main- and secondary-stems (a secondary stem is one that branches off near the basis of the main stem) show partial apical dominance, yet they may develop branches at several levels (Reestman and Schepers, 1971). The first level is made up of the lower part of the stem and leaves, up to the first flower. The second level includes the stem section between the first and second group of flowers, etc. The number of stem levels and the length of the stem at each level depend on the genotype and are greatly affected by the environment. Mature leaves consist of a petiole with a terminal leaflet, lateral leaflets, secondary- and sometimes tertiary-leaflets. The stolons are modified lateral shoots, usually from the basal nodes of the main and secondary stems, and later at progressively higher ones. Typically, these shoots have elongated internodes and hooked tips (Fig. 2.4). Environmental factors, such as temperature, nutrients and water availability and darkness have a pronounced effect on stolon development (Clark, 1921; Booth, 1959, 1963).

Tubers: The only edible part of the plant is the tuber, which is morphologically described as a fleshy, dwarfed stem developed from the end of the stolon tip. The subterranean tubers are adapted to starch-storage and reproduction (Fig. 2.5). Under certain conditions, however, the plant produces aerial tubers. This occurs when translocation of assimilates from photosynthetically active foliage is blocked by mechanical or biotic damage in the lower parts of the main stem. Each tuber produces apical- and lateral buds in the axils of leaf-scars (rudiments of scale leaves). The number of buds varies considerably with genotype, tuber size, and growth-conditions in the production field, and depends upon the degree of development of the tuber. For instance, medium size (45mm ∅) tubers of the Dutch cv. 'Bintje' commonly bear 12-15 buds. Each bud (both, apical and axillary buds), consists of a main bud with adventitious buds in its axil, thus capable of giving rise to a number of shoots.

A cross section of a tuber reveals the following tissues: outer periderm (corky skin), cortex (storage parenchyma), vascular ring, outer medulla and inner medulla (pith, branches to each bud) (Fig. 2.6).

Following tuber initiation, the cell layer immediately under the epidermis differentiates into a cork cambium (phellogen), which produces suberized phellem at the outer surface. The periderm consists of phellogen and phellem, and in mature tubers it consists of 5-15 layers of suberized cells. This skin is almost impermeable to chemicals, liquids and gases. It therefore, provides good protection against the penetration of micro-organisms and

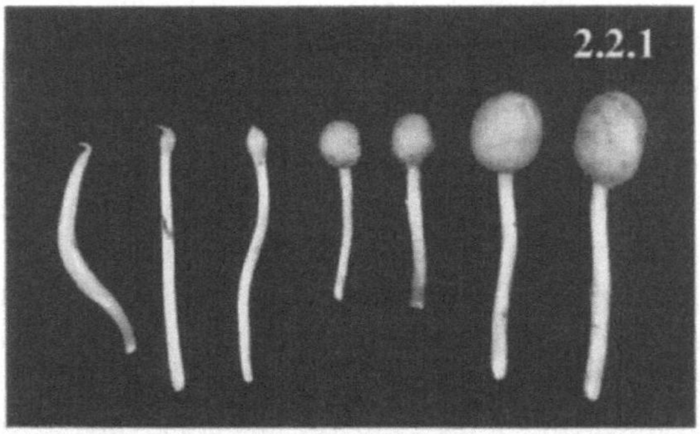

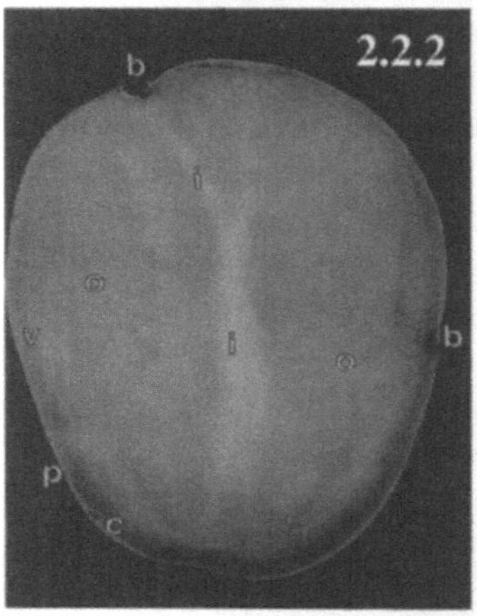

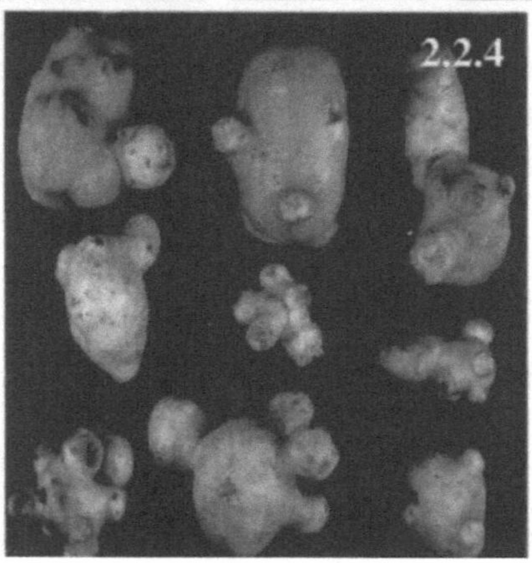

Fig. 2.2.1 - Developmental stages of a tuber: from stolon tip to a developed tuber.

Fig. 2.2.2 - Cross section of a tuber (B-bud; C-cortex; I-inner medulla; O-outer medulla; P-periderm; V-vascular ring).

Fig. 2.2.3 - Sprouted seed tuber.

Fig. 2.2.4 - Secondary growth and malformation of tubers.

slows down water loss. The main exchange of gases and water occurs through the numerous lenticels. When tubers develop under excessively wet conditions, the lenticels grow large and open up, thus facilitating infection with damaging microorganisms and loss of moisture. Wounded tubers are prone to rapid water loss and easy penetration of biotic and abiotic damaging agents and thus must be protected. Hence, the freshly cut surface undergoes fast healing processes, including: formation of phellogen (which generates new cells), and suberization of cell walls. Wound healing is faster in young than in old tubers, and the combination of intermediate temperatures of 10-20°C with relative humidity of ~90% and sufficient oxygen, provide optimal conditions for these processes (Appel, 1906; Wigginton, 1974). Similar environmental conditions are favorable for proper curing (i.e. skin setting and wound healing) of freshly harvested tubers.

Temperature and photoperiod effects on tuber formation: Although tubers develop from stolon tips, separate factors control the development of the two organs. Despite this, simultaneous formation of both is quite common. Low temperatures and short photoperiod promote tuberisation, and the process is significantly delayed under high temperatures, and/or long photoperiod. The latter conditions promote haulm growth (Scheme 2.1). Moderate temperatures, however, are dominant for tuberisation since no tubers are produced during two months of growth at 29°C (see: Bodlaender, 1963). Under short days tubers form 3-5 weeks earlier than under long days (Chapman, 1958; Ivins and Bremmer, 1964) and this day length stimulus can cross a graft union. These results indicate a quantitative effect of photoperiod in the potato. The induction is perceived by the plant haulm, including the terminal buds and the young expanding leaves (Kahn *et al.* 1983), and it takes about two weeks for the photoperiod induction to occur (Chapman, 1958).

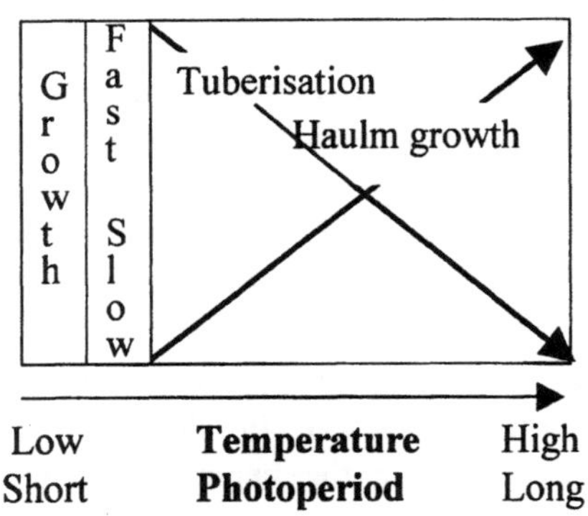

Scheme 2.1. Schematic presentation of the interaction between the environment and foliage and tuber development

Research has shown that stolon formation was promoted, and tuberisation was inhibited by gibberellic acid (GA), and that both were promoted by synthesis inhibitors of GA, such as cycocel (chloroethyl dimethyl ammonium; CCC) (Tizio, 1969). Auxins alone have no inductive effect, but tuber formation was greatly promoted by a combination of auxins and kinetin (such as benzyl adenine), and tuber size was larger than that of the control (Ahmed and Sagar, 1981). Under long photoperiod, absicic acid (ABA) had no inductive effect on tuber formation (Smith and Rappaport, 1969), and other experiments indicated that ABA may prevent tuberisation (Palmer and Smith, 1969), probably indirectly by suppressing the growth of the apical meristem (Wareing and Jennings, 1979). It is therefore likely that both cytokinins and ABA are involved in tuber formation. Some evidence indicates that ethylene may play a role in these processes, as shoot growth of ethylene-treated plants was inhibited and growing regions, including stolon tips, became swollen. The latter were very similar to that of normal potato tubers (Catchpole and Hillman, 1969) and some researchers reported that more starch accumulated in the ethylene-treated swollen sub-apical regions of the stolons than in the non-treated control (Minato *et al.* 1979).

Cutter (1992) proposed that in normal tuber formation, the following sequence of events may occur. The environmental induction results in GAs synthesis in the young leaves, which promote stolon formation and elongation. Auxins such as indole acetic acid (IAA) cause cell enlargement in the subapical region of the stolon, and ABA arrests its apical growth. In the young forming tuber, cytokinins induce cell division thus leading to the growth of the tuber. However, since these growth regulators affect very general processes, such as cell growth and cell division, it is of course possible that other substances or combination of substances could promote tuber formation (Cutter, 1992).

Tuber dormancy and rest: Under normal conditions, tubers at harvest do not sprout and cannot resume growth even under favorable environmental conditions of moisture and temperature (Scheme 2.2). This state of true dormancy, termed the rest period, may last from a few weeks to about three months. The length of dormancy is genotype-dependent and temperature has a significant effect. After the rest period, the buds of the tuber resume growth, and sprouting can be delayed by refrigeration for a few months, a period termed the dormancy period. In storage, tubers undergo aging processes. Exposure of tubers at the beginning of the dormancy period (young tubers) to optimal conditions results in slow sprouting of a single bud. At a later stage, sprouting of a number of buds occurs, and this stage is usually considered the preferred physiological stage for planting (Fig. 2.7).

Older tubers sprout autonomously even under suboptimal conditions of low temperatures and in the absence of an external source of moisture. Under similar environmental conditions, older tubers sprout faster than younger ones, growth vigor is reduced and lower yields are obtained.

Physiological age				
Young				Old
Rest	Dormancy			Senescence
No sprouting	Apical dominance	Multiple sprout growth	Branched sprouts	Hair thin sprouts and tuberisation

Scheme 2.2. Physiological aging of the potato tuber.

Emilsson (1949) and Schippers (1956) studied the differential length of dormancy and resting periods in tens of cultivars. They concluded that cultivars differ significantly in length of dormancy, that these traits are not related to the time of maturation (early or late crop) of the specific genotype, and that there is no correlation between dry matter content and length of dormancy. Growth conditions have a strong effect on dormancy, and thus the length of the dormancy period varies significantly with season and geographical region.

Dormancy and sprouting of the tuber are controlled by a balance of growth-inhibitors and growth-promoting substances (Bruinsma and Swart, 1970), the most important ones being ABA and GA (Wareing, 1982). During the dormant phase, even under refrigeration, the level of growth retardants (ABA and others) decreases, and when a certain threshold is reached, sprouting begins. At first, there is a strong apical dominance and therefore only the apical bud sprouts. The level of inhibitors further decreases with time, and there is a significant and continuous increase in the number of sprouting buds which develop into foliage shoots. Old tubers may sprout autonomously even in storage, and when planted they produce a large number of thin, branched shoots. Potato tubers do not keep well, even under refrigeration, from one crop year to another. Yet, there are big differences between genotypes in storage-ability, and both storage temperature and air-moisture have a dominant effect. Field conditions and post-harvest treatment can also result in reduced rest and dormancy periods of a given genotype. Foliage treatment with maleic hydrazide (commercial names: MH_{30}; Royal) prevents the tubers from sprouting altogether, probably by inhibiting proteins and nucleic acids synthesis.

2.5 Production of 'Seed' Tubers

Commonly, potatoes are propagated vegetatively from tubers. In the open field, however, plants and tubers are prone to infection by pests, viruses and micro-organisms. When infected tubers are used for propagation, the carried over diseases may cause severe losses of yield and quality. Hence, tissue culture procedures are widely used for the aseptic propagation, mainly through the reproduction of axillary cuttings of plants grown *in vitro* (plantlets). Rates of plantlets' propagation *in vitro* are very high. They amount to 8-84 fold per 40 days (Marinus, 1983), or 5 to 7 fold per 3-4 weeks (Levy *et al.* 1983), as compared to an annual (or seasonal) multiplication rate of 6-10 fold when multiplied from tubers in the field (See Chapter 15).

Production of 'seed' tubers is in many aspects similar to that of ware potatoes, only more caution is given to the genetic uniformity of the propagation material, to its characterization, and to the level of infestation with pathogens. The field is carefully selected in cool regions after long crop rotation to reduce contamination with soil-borne pests; plants are constantly monitored for pests and diseases; tight control measurements are taken to reduce contamination by virus transmitting aphids; and the tubers are handled gently after harvest to reduce mechanical damage, infections and rot. 'Seed' tubers are treated with protecting pesticides and use of maleic hydrazide or other sprouting inhibitors is prohibited. Storage conditions, especially temperatures, greatly affect the amount of sprout growth at re-planting, and consequently, the number and type of stems produced (Allen *et al.* 1979), foliage efficiency (O'Brien *et al.* 1983), the number of tubers and the yield. Hence, environmental factors during production and storage of tuber 'seed' can greatly influence the following ware crop, and in many ways the crop should be considered as biennial.

2.6 Production of Ware Potatoes

Potatoes for fresh consumption and for processing are produced from 'seed' tubers throughout the world. The quality of the 'seed' tubers is instrumental for obtaining good quality yields. Climate, namely temperature, day length, light intensity, water availability in rain fed areas, in conjunction with cultural practices, such as irrigation, fertilization, plant protection, weed control, soil management, all greatly influence yield and quality. Therefore, herbicides and pesticides are commonly used in addition to crop rotation, soil solarisation and below-ground drip irrigation, in order to control weeds and pathogens.

The cost of 'Seed' potatoes amounts to between 25 and 50% of the total growing costs, hence, it is important to use 'seed' tuber in an efficient manner. Desired 'seed' sizes range between 30-60 g/tuber, and 'seed' rates range between 1,800-3,000 kg/ha, depending on the distance between the rows (61-76 cm in many European countries, and 90 cm in the U.S. and other countries). 'Seed' spacing along the row ranges between 15 cm (smaller 'seed' tubers) to 33 cm with the larger 'seed' tubers.

The proper physiological age of the 'seed' for planting is when over 90% of the tubers are capable of developing 3-5 main stems per tuber. In many cultivars, this stage involves the presence of sprouts 2-6 mm long. Adequate water supply is essential for the production of high yields and good quality tubers, and variability in rainfall during growth has a major effect. It should be supplemented by irrigation. Assuming a rooting depth of ca. 50 cm for efficient water uptake, irrigation intervals and quantities are calculated to restore field capacity when about 50% of the available water has been depleted, taking into consideration the evapotranspiration and soil texture.

Tuber yield and quality is also dependent on the supply of nutrients. Potatoes require adequate supply of nitrogen, potassium, phosphorus, calcium, magnesium and sulfur, and microelements such as iron, manganese, zinc, boron, copper and molybdenum. Animal manure or green manure may supply the minerals, and the combination of organic and inorganic fertilization is beneficial since potatoes enjoy the presence of organic matter in soils.

2.7 Genetics and breeding

Potatoes of the *Andigena* taxon (S. *tuberosum* subsp. *andigena*) were first brought to Europe in the 16th century in limited quantity, and the genetic basis for crop improvement was therefore rather narrow. Plant breeders used this tetraploid material to develop the common potato, S. *tuberosum* var *tuberosum*. This was done in a relatively short period of time, since potato cultivation was already reported in Ireland in the 17th century (Salaman, 1949), and became the major food source in England by the middle of the 18th century (Glendinning, 1983). When compared to its ancestors, reduced vegetative growth, earlier tuberisation, a shorter growing cycle and higher tolerance to heat characterize the newly developed plant. More recent introductions from South America have been used all over the world in improvement programs, and many modern cultivars contain genes or chromosome fragments from those sources.

Most commercial potato breeding programs are aimed at the production of vegetatively propagated (from 'seed' tubers) clones. This nature of asexual propagation makes any F_1 combination a starting point of new potato clones. Once selected, tissue culturing and micro-propagation can accelerate the propagation, but special care must be taken with the field propagated 'seed' tubers to avoid infection by viruses, bacteria, fungi and pests.

The original *Solanum* plants consumed by man in the center of origin, were all diploids (2n=24). Polyploidization occurred later in the evolution, as a result of chromosome doubling and/or the production of unreduced gametes. Natural hybridization, within and between species, thus resulted in different ploidy levels. However, most modern potato cultivars (S. *tuberosum*) are tetraploid, and only a few diploids are currently cultivated (e.g., S. *phureja*). The former, however, are more prolific than the diploids.

At each locus, the four homologous chromosomes make the following five genotypic combinations possible: AAAA; AAAa; AAaa; Aaaa and aaaa. These genetic combinations give rise to three types of 2x gametes: AA; Aa and aa. A single dominant gene controls some traits of high economic importance, such as the resistance to race-specific blight (*Phytophthora infestans*), to potato viruses X and Y (PVX and PVY), and more.

The tetrasomic inheritance complicates any attempt to improve quantitative- and recessively controlled characters. Selection for traits controlled by recessive genes, or by polygenic mode, is much more difficult in the tetraploid than in the diploid level. In the former cases, desired genotypes with specific- or general combining-ability are hard to find, and screening for the desired traits (e.g., resistance) is rather difficult due to the presence of intermediates, which results in a gradual and continuous variation. (See chapter 13).

Many potato clones are sterile, and crossing barriers frequently prohibit gene introgression from wild *Solanum* species. Modern technologies, such as protoplast fusion and genetic engineering are currently being used to increase genetic variation in pre-breeding activities. In addition, some of the wild relatives are diploid and therefore cannot immediately be utilized in breeding programs of the tetraploid potato. Many breeders therefore generate diploid lines of S. *tuberosum* in order to facilitate the breeding work and to exploit the genetic variation potential of wild *Solanum* spp. for desired traits. Haploid gametes are generated in S. *tuberosum* by pollination with certain genotypes of S. *phureja* . The resulting parthenocarpic fruit bear diploid seed, which are being used for gene introgression, for fixation of desired traits in the homozygous state, and for the production of superior heterozygous hybrids. Due to the improved vigor and productivity at the tetraploid level, colchicine is applied to the diploid breeding lines prior to final selection and release of new potentially improved material. However, the resulting offspring still have to be crossed, evaluated and selected in the conventional way.

In traditional breeding programs up to 140,000 seedlings may be raised from true seed, but only about 1% of this population remains after the screening at the second clonal generation stage. In spite of the mass work and slow breeding procedures, numerous clones are introduced each year as potential new cultivars, yet only a few become commercially viable and these new introductions still face difficult competition from existing cultivars. For example, cv. 'Russet Burbank', released in 1876 in the U.S., and cv. 'Bintje', released in 1910 in the Netherlands, still dominate the U.S. and European markets respectively (Table 2.1). In North America, 86 cultivars are produced commercially, five of which were released in 1990. In the Netherlands, 'seed' of 213 cultivars were produced in 1996, but many of these new releases have never reached market acceptance.

Table 2.1. Major potato cultivars grown in North America (1997) and in the Netherlands (1996): Year of release and importance expressed as percentage of the total 'seed' production

Cultivar	Year released	% of 'seed' grown
	NORTH AMERICA	
Russet Burbank	1876	29.1
Shepody	1980	10.3
Russet Norkota	1987	8.3
Superior	1961	5.8
Atlantic	1976	5.0
Kennebec	1948	4.0
	THE NETHERLANDS	
Bintje*	1910	13.7
Desiree**	1962	6.7
Spunta**	1968	6.2
Diamant**	1982	4.1
Monalisa**	1982	3.1
Nicola**	1973	2.7

*** Grown mainly in Western and Northern Europe**

**** Exported to Europe, Mediterranean countries and South America**

2.7.1 Breeding procedures

Molecular mapping and recent developments of molecular markers have enhanced marker-assisted selection (MAS), especially by helping identify alleles closely linked to genes conferring resistance to pests and diseases. MAS is becoming helpful in breeding for nematode resistance in potatoes and the application of this procedure in the genetic analysis of quantitative traits (QTL) is envisioned. Another development is associated with gene manipulation within the potato which enables the control of starch composition in tubers, and the control of genes responsible for the activity of polyphenol oxidase thus reducing black coloration in tubers due to impact during handling. Potatoes are relatively amenable to various tissue culture techniques and manipulation, and thus are widely used to obtain somaclonal variation, as well as somatic hybrids, to overcome barriers to sexual crosses within S. *tuberosum* (the domesticated subspecies), between *Solanum* species, and the introgression of foreign genes from other organisms. Hence, incorporation of genes coding for virus coat-proteins provides novel forms of resistance to viruses. Introduction of gene coding for the *Bacillus thuringiensis* toxin confers resistance to the Colorado beetle. Similar procedures are being studied in an attempt to use the potato for producing valuable pharmaceuticals and other materials in the tubers, efficiently and at a lower cost.

2.8 Tuber quality

Many factors contribute to the quality of the tuber, the most important ones being the tuber structural and biochemical composition. The latter has a marked impact on the tuber processing value. Consumers also pay attention to the appearance of the produce, including shape, size, lack of defects, pests, disorders and mechanical injuries.

Nutritional value and main quality traits: Potato tubers are predominantly water (around 80%). The residual dry matter consists mainly of starch (60-80% of the dry matter), and also proteins (~10% of the dry weight), vitamin C (15-25 mg 100 g^{-1} fresh weight), iron and dietary fibers (~8% of the dry weight). The mean annual consumption figure for the European Community is estimated at ca 90 kg $person^{-1}$ and thus, in Europe, the current daily intake of potatoes provides 5-6% of the daily energy requirements (Elton, 1978).

In healthy plants, the dry matter content is mainly determined by genetic factors, i.e., by the cultivar in use, and it is well correlated with the starch content of the tuber. Other factors involved in the amount of stored reserves are the stage of tuber development and maturity, the level of fertilizers (mainly, N and K) and water in the growth medium, and the climatic conditions in the production field (for detailed reviews, see: Beukema and Van der Zaag, 1979, 1990).

Dry matter: Dry matter content is an important measure of the tuber processing quality, since the amount of product is directly dependent on its content in the raw material, and less energy is required to remove water from tubers with high, as compared to those with low dry matter. The high uniformity of the dry matter composition (~80% starch) enables the use of specific gravity as a quick, reliable, non-destructive method for the determination of dry matter content of batches of potatoes. This method is widely used to assess the quality of the raw material for processing.

All tubers contain a certain amount of soluble sugars, mainly sucrose, glucose and fructose. The level of sugars affects the taste of the potato. Moreover, during processing, tubers rich in reducing sugars are prone to develop a dark color due to the heat-sensitive Mailard reaction between tyrosine and glucose and fructose. The amount of sugars in the tuber depends mainly on the variety, on the stage of tuber development, and on storage temperature. Regardless of varietal differences and growth conditions, the reducing sugar content decreases with tuber maturation (Burton and Wislon, 1970), and supra-optimal growth conditions which promote excessive growth (e.g., high temperatures, high light intensity, high nitrogen, etc.) result in a high level of reducing sugars in the tubers. Storage temperatures have the most significant effect on sugar contents in the potato tuber. Respiration rate, and the conversions between sucrose and reducing sugars and between sugar and starch, are all temperature sensitive. Sugar accumulation occurs under low temperatures, whereas at high storage temperatures, most sugars are consumed by respiration and the remaining amounts are reconverted to starch. This 'reconditioning' process of potatoes from cold storage is commonly used by the processing industry to improve the quality of long stored raw material.

Glycoalkaloids: Potatoes, like many other *Solananceae* contain poisonous glycoalkaloids derived from the aglicone solanidine, which probably play a role in protection against pathogens (McKee, 1959; Allen and Kuc, 1968) and pests (Tingey, 1984). *Solanum* spp. differ greatly in solanidine glycosides, and thus fresh tubers of S. *acaule*, S. *juzepczukii* and S. *curtilobum* have been reported to contain 1,260, 470 and 290 ppm, respectively (Gregory, 1984). The domesticated potato contains mainly K-solanine and 2-chaconine which are concentrated mainly in the peel (Maga, 1980). Clones differ significantly in the content of glycoalkaloids in the tubers. Within a given clone, the stage of tuber development and light are the most important factors affecting the solanine and chaconine level in the tuber. The latter factor also promotes the synthesis of chlorophyll in the specialized stem, the tuber. However, the two processes (increase in glycoalkaloids and in chlorophyll) are independent (Gull and Isenberg, 1960). Mature tubers contain lower concentrations of glycoalkaloids than young developing ones, and even a short exposure to light significantly increases the content of solanine and chaconine (van Gelder, 1985). Conditions during growth (van Gelder, 1985; Dimenstein *et al.* 1997) such as low storage temperatures (Hilton, 1951) and tuber aging (Cronk *et al.* 1974), may promote the production of these solanidine glycosids.

Solanine and chaconine are heat-stable. They do not break down even at temperatures as high as 280°C and can thus withstand cooking or frying. They cause a bitter taste (Beukema and van der Zaag, 1990), and consumption of potatoes with glycoalkaloids at >200 mg kg^{-1} fresh weight, may result in gastroentritis and at higher concentrations, even in death. Therefore, it is recommended that freshly harvested ware potatoes should have less than 100 ppm glycoalkaloids, and most common European and American cultivars conform with this standard (Burton, 1989). Potato breeders routinely make use of wild plant species, and should therefore take this factor into consideration.

2.9 Climatic conditions

Growth of the potato plant and the formation and growth of tubers are greatly influenced by climatic conditions. The combination of day-length, light intensity, temperatures and season length are important factors that determine growth and yield, provided that good cultural practices are maintained.

Day-length affects haulm and tuber development and growth. Short photoperiods (about 12 hrs) reduce haulm growth and enhance the initiation of tubers, while long days (about 16 hrs) act in the opposite way (See also scheme 1). In many countries (e.g., Central and Northern Europe, USA, Canada, and more), potatoes are planted in early spring when day-length is relatively short and hence tuberisation is enhanced. Later in the season, the longer photoperiod promotes haulm growth and assimilation The additional photsynthates are partitioned between both haulm and tubers. Longer season and longer days at this phase result in high tuber yields provided the temperatures are favorable.

High light intensity during growth is important for efficient photosynthesis and sufficient production of photosynthate to maintain sufficient growth of both haulm and tuber, and for the accumulation of dry matter in the tubers.

Temperature regime is very important for potato productivity (Scheme 1). Plant and tuber development favor moderate temperatures around 20°C. However, high yields are obtained when low night temperatures (about 15°C) follow high temperatures during daytime (25-30°C). The high temperature during the day increases the rate of photosynthesis, while the lower temperature during the night enhances the transport of assimilates to the tubers and reduces respiration. Soil temperatures of 15-18°C are important for good tuber development and quality. High soil temperatures delay tuber formation and reduce tuber quality by reducing dry matter content and inducing secondary growth (Fig. 2.8). Secondary growth can also be induced by extreme changes in water and nitrogen supply, and varieties differ in their susceptibility to this tuber disorder.

High temperatures and sometimes also low temperatures, as well as calcium deficiency cause a severe tuber disorder known as 'internal brown spots', 'internal heat necrosis' or 'internal rust spot'. Varieties differ in their susceptibility to this disorder, cv. 'Desire' being relatively tolerant.

References

Ahmed, Ch. M. S. and Sagar, G. R. 1981. Effects of a mixture of NAA + BA on numbers and growth rates of tubers of *Solanum tuberosum* L. *Potato Res.* 24:267-278.

Allen, E. G., Bean, J. N., Griffith, R. L. and O'Brien, P. J. 1979. Effects of length of sprouting period on growth and yield of contrasting early potato varieties. *J. Agric. Sci. Cambridge* 92:151-163.

Allen, E. H. and Kuc, J. 1968. K-solanine and 2-chaconine as fungitoxic compounds in extracts of Irish potato tubers. *Phytopathology* 58:776-781.

Anon., 1995. *Potatoes in the 1990s. Situation and Prospects of the World Potato Economy.* FAO, Rome, 38 p.

Appel, O. 1906. Zur Kenntnis des Wundverschlusses bei Kartoffeln. *Berlin, Bot. Ges.* 26:118-122.

Beukema, H. P. and van der Zaag, D. E. 1979. *Potato Improvement. Some Factors and Facts.* International Agric. Ctr., Wageningen, Holland, 224 pp.

Beukema, H. P. and van der Zaag, D. E. 1990. *Introduction to Potato Production.* Pudoc, Wageningen, Holland, 208 pp.

Bodlaender, K.B.A. 1963. Influence of temperature, radiation and photoperiod on development and yield. In: *The Growth of the Potato.* (Eds.) J.D. Ivins and F.L. Milthorpe, pp. 199-210, Butterworths, London,.

Booth, A. 1959. Some factors concerned in the growth of stolons in the potato. *J. Linn. Soc. (Bot.).* 56:166-169.

Booth, A. 1963. The role of growth substances in the development of stolons. In: *The Growth of the Potato.* (eds) J. D. Ivins and F. L. Milthorpe, pp. 99-113. Butterworths, London.

Bruinsma, J. and Swart, J. 1970. Estimation of the course of dormancy of potato tubers during growth and storage, with the aid of gibberellic acid. *Potato Res.* 13:29-40.

Burton, W. G. 1989. *The Potato.* 3rd edition. Longman, London, 742 pp.

Burton, W. G. and Wislon, A. R. 1970. The apparent effect of the latitude of the place of cultivation upon the sugar content of potatoes grown in Great Britain. *Potato Res.* 13:269-283.

Catchpole, A. H. and Hillman, J. 1969. Effect of ethylene on tuber initiation in Solanum tuberosum. *Nature* 223:1387

Chapman, H. W. 1958. Tuberization in the potato plant. *Physiol. Plant.* 11:215-224.

Clark, C. F. 1921. *Development of tubers in the potato.* USDA Bull. 958, 27 p.

Cronk, T. C., Kuhn, G. O. and McArdle, F. J. 1974. The influence of stage of maturity, level of nitrogen fertilization and storage on the concentration of solanine in tubers of three potato cultivars. *Bull. Environ. Contamin. Toxicol.* 11:163-168.

Cutter, E. G. 1992. Structure and development of the potato plant. In: *The Potato Crop.* (ed.) P. M. Harris, pp. 65-161, 2nd edition. Chapman & Hall, London.

Dimenstein, L. Lisker, N., Kedar, N. and Levy, D. 1997 Changes in the content of steroidal glycoalkaloids in potato tubers grown in the field and in the greenhouse under different conditions of light, temperature and day length. *Physiol. Mol. Plant Pathol.* 50:391-402.

Elton, G. A. H. 1978. European diets in relation to standards of needs. In: *Diet of Man: Needs and Wants.* (ed.) Yudkin, J., pp.25-40, Applied Science Publishers, London.

Emilsson, B. 1949. Studies on the rest period and dormant period in the potato tuber. *Acta Agricultura Suecana* 3:189-284.

Gelder van, W. M. J. 1985. Nu veredelen voor voedingsveiligheid in de toekomst. *Zaadbelangen* 7:186-188.

Glendinning, D. R. 1983. Potato introductions and breeding up to the early 20th century. *New Phytol..* 94:479-505.

Gregory, P. 1984. Glycoalkaloid composition of potatoes. Diversity and biological implications. *Am. Potato J.* 61:115-122.

Gull, D. D. and Isenberg, F. M. 1960. Chlorophyll and solanine content and distribution in four varieties of potato tubers. *Proc. Am. Soc. Hort. Sci.* 75:545-556.

Hawkes, J. G. 1967. The history of the potato. J. Roy. Hort. Soc. 92:207-224.

Hilton, R. J. 1951. Factors in relation to tuber quality in potatoes. II. Preliminary trails on bitterness in Netted Gem potatoes. *Can. J. Agric. Sci.* 31:61-70.

Ivins, J. D. and Bremmer, P.M. 1964. Growth, development and yield in the potato. *Outlook Agric.* 4:211-217.

Kahn, B. A., Ewing, E. E. and Senesac, A. H. 1983. Effects of leaf age, leaf area, and other factors on tuberization of cuttings from induced potato (*Solanum tuberosum*) shoots. *Can. J. Bot.* 61:3193-3201.

Levy, D., Izhar, S., Fogelman, E., Itzhak, Y., Levy, Y., Ben Yehuda, R., Yanai, A. and Dinar, M. 1983. Multiplication rates of potato plantlets produced from tip culture and of tubers from this source grown in a screenhouse in the Golan. *Hassadeh* 53:934-938 (in Hebrew with English abstract).

Maga, J. A. 1980. Potato glycoalkaloids. *CRC Crit. Rev. Food. Sci. Nutr.* 12:371-405.

Marinus, J. 1983. Some aspects of the in vitro multiplication of potatoes. *Potato Res.* 26:85.

McKee, R. K. 1959. Factors affecting the toxicity of solanine and related compounds to *Fusarium caeruleum. J. General Microbiol.* 20:686-696.

Minato, T., Kikuta, Y. and Okazawa, Y. 1979. Effect of ethylene on sprout growth and endogenous growth substances of potato plants. *J. Fac. Agric. Hokkaido Univ.* 59:239-248.

O'Brien, P. J., Allen, E. J., Bean, J. N., Griffith, R. L., Jones, S. A. and Jones, J. L. 1983. Accumulated day-degrees as a measure of physiological age and the relationships with growth and yield in early potato varieties. *J. Agric. Sci,* Cambridge 101:613-631.

Palmer, C. E. and Smith, O. E. 1969. Effect of abscisic acid on elongation and kinetin-induced tuberization of isolated stolons of *Solanum tuberosum L. Plant Cell Physiol.* 10:657-664.

Patterson, C. F. 1953. A method of obtaining fruits in the potato variety Russet Burbank. *Am. Potato J.* 30:89-91.

Reestman, A. J. and. Schepers, A. 1971. Toepassingvan morphologische gewasanalysebij het toprol-onderzoek van aardappelen. Lelystad Res. Stn. Ann. Rept. for 1971. pp. 61-64.

Ross, H. 1986. *Potato Breeding - Problems and Perspectives*. Parey, Hamburg, Germany. 132p.

Salaman, R. N. 1949. *The History and Social Influence of the Potato.* Cambridge University Press, Cambridge.

Schippers, P.A., 1956. De invloed van de temperatuur op de duur van de rustperiode. Publikatie Aardappelbewaring (ATO), Wageningen, Ser. A, No. 108. 13pp.

Simmonds, N. W. 1966. Studies of the tetraploid potatoes. III. Progress in the experimental re-creation of the *Tuberosum* Group. *J. Linn. Soc. (Bot.),* 59:279-288.

Smith, O. E. and Rappaport, L. 1969. Gibberellins, inhibitors and tuber formation in the potato, *Solanum tuberosum*. *Am. Potato J.* 46:185-191.

Swaminathan, M. S. and Howard, H. W. 1953. The cytology and genetics of the potato (*Solanum tuberosum*) and related species. *Bibliografia Genetica* 16:1-192.

Thijn, G. A. 1954. Observations of flower induction with potatoes. *Euphytica* 3:28-34.

Tingey, W. M. 1984. Glycoalkaloids as pest resistance factor. *Am. Potato J.* 61:157-167.

Tizio, R. 1969. Action du CC (chlorure de (2-chloroethyl)-trimethyl ammonium) sur la tuberisation de la pomme de terre. *Eur. Potato J.* 12:3-7.

Wareing, P. F. 1982. The control of development of the potato plant by endogenous and exogeneous growth regulators. In: *Chemical Manipulation of Crop Growth and Development* (ed.) J. S. McLaren, pp. 129-138, Butterworths, London.

Wareing, P. F. and Jennings, A. M. V. 1979. The hormonal control of tuberisation in potato. In: *Plant Growth Substances* (ed.) F. Skoog. pp. 293-300. Springer-Verlag, Berlin,.

Wigginton, M. J. 1974. Effects of temperature, oxygen tension and relative humidity on the wound-healing process in potato tuber. *Potato Res.* 17:200-214.

3. Important Potato Cultivars

Dennis L. Corsini[1] and Charles R. Brown[2]
USDA Agricultural Research Service, [1]University of Idaho, Research and Extension Center, PO Box AA, Aberdeen, ID 83210 USA, and [2]Washington State University, 24106 N. Bunn Road, Prosser, WA 99350 USA

3.1 Introduction

The white potato became established in Europe and North America as an important crop during a 200 year period beginning in the mid 16th Century. The potato was first introduced to Europe from South America, either from the northern Andes as *Solanum tuberosum gp. andigena* (short day adapted), or from southern Chile as *Solanum tuberosum gp. tuberosum* (long day adapted) and the earliest descriptions are of what appear to be mixed genotypes (Salaman, 1949). From about 1550 to 1750, the potato was selected and adapted to the long day conditions of summer in the higher

G. Loebenstein et al. (eds.),
Virus and Virus-like Diseases of Potatoes and Production of Seed-Potatoes, 39–52.

latitudes of Europe. By the late 1700s, distinct cultivars were recognized, and attempts were being made to improve the crop through cultivar selection. Problems with *Potato leafroll virus* genus *Polerovirus*, (PLRV) and *Potato virus Y* genus *Potyvirus,* (PVY) infection called "curl" induced the early breeders to select new clones with less symptoms and higher yields. Even though they did not really understand the cause of curl, they were quite adept at applied breeding, growing small populations from botanical seed, and then selecting the best seedlings. Cultivars, like White Kidney, Manly, and Lumper, to name a few from Great Britain and Ireland, came to be grown on a large scale, and there was less cultivar diversity than is present in more recent years in Europe (Salaman, 1949).

After the first great late blight pandemics in 1845-1850, a new group of cultivars emerged. Introductions from Latin America, such as Rough Purple Chile, played an important role in developing some of these new cultivars. From Rough Purple Chile came Garnet Chile, which in turn gave rise to Early Rose, which in turn was a founding clone that is in the ancestry of a great many European and North American cultivars (Love, 1999; Salaman, 1949). In the British Isles, the cultivar Champion was developed along with several others that showed good late blight resistance. Other cultivars that originated in the late 1800s such as Up To Date, Majestic, and King Edward, still play an important role in potato production in the UK (Salaman, 1949; Scottish Agricultural College, 1996 and 1999). A single cultivar has often been marketed under many names. At one point earlier in this century, the cultivar Up-to-Date was being sold under more than 200 different aliases (Salaman, 1949). Today the situation is not quite so bad, but still a cultivar may be known by different names in different countries or in different regions of the same country. A few specialty cultivars that recently appeared in North America, such as Yellow Finn and German Butterball, seem to have originated from individuals bringing a European cultivar into the US and renaming it to suit their inclinations. Irish Cobbler has been the most important variety in Japan for nearly 100 years, and more recently in Korea as well, but is known there as Danshaku (Obhiro, 1999).

There are a very large number of potato cultivars now in use throughout the world. Although many of the most important cultivars have a relatively narrow genetic background, the potato is, in general, genetically diverse, incorporating wild germplasm for both disease resistance and improved quality (Bamberg, 1996; Love 1999; Swiezynski *et al.* 1997). Over 1,600 distinct cultivars are grown in 144 countries (Hamester and Hils, 1999). The great majority are locally adapted fresh market types with animal feed being an important secondary use. Cultivar use in important potato growing areas has been based on market needs rather than production efficiency. The relatively low production per hectare in many developing countries (10-15 mt/ha) (FAO, 1999) is, to some extent, a reflection of varietal selection based on market preferences rather than on resistance to late blight, viruses, and other disease and pest problems. The high level of production in North America and Western Europe (35-45 mt/ha) (FAO, 1999) is maintained by a combination of sophisticated and expensive seed programs to minimize virus impact, heavy fertilizer use, and the extensive use of pesticides and fumigants to control diseases and pests, especially late blight. There are certainly exceptions to this situation, particularly in localized growing areas where endemic disease problems are difficult to control with chemicals but are effectively managed with host resistance. *Tobacco rattle virus* genus *Tobravirus* (TRV) or cyst nematode, for example, may require that resistant cultivars be grown. Late blight has generally influenced which cultivars could be grown in more humid climates, but the transient nature of the vertical resistance in cultivars developed in the first half of the 20th century, and the availability of good fungicides to control late blight during the last 50 years, have allowed blight susceptible cultivars to continue in the forefront of production (Swiezynski *et al.* 1998; Wastie, 1991).

In areas such as the United States, Canada, Australia, and Western Europe, the requirements for processed potato products have had a major influence on the cultivars grown. Relatively few cultivars supply the needs of the frozen french fry and the chipping (crisping) industries (Table 3.1) (National Potato Council, 1999). Disease and pest resistance in general, and virus resistance in particular, have played a minor role in the importance of these cultivars. On the other hand, local and national preferences in skin color, flesh color, shape, and flavor determine which cultivars are grown for fresh market, and many of these show good disease resistance traits (Tables 3.5 and 3.7). So, although less than 10 cultivars account for most of the world production of frozen french fries, there are many hundreds of fresh market cultivars, most grown on very limited acreages. One country's preference in fresh market potatoes may be completely unacceptable to the people in another country. In North America, the long russet-skinned potato commands a premium in the fresh market, yet such types are unacceptable in the rest of the world. The large mealy textured baking potato so popular in the restaurant trade in the United States is in sharp contrast to the small waxy textured, yellow fleshed, boiling potato often preferred as table stock in many other potato growing regions.

In general, per capita fresh consumption of potatoes has been fairly constant for many years. Yet consumption of processed products has grown steadily (National Potato Council, 1999). Processed product requirements are much the same across national boundaries. The relatively few international fast food chains and snack food companies have standardized raw product quality needs. Processing industry requirements, therefore, are restricting cultivar diversity and will probably continue to do so in response to the rapid globalization that is now occurring. Unfortunately, such market needs will continue to pressure growers to put less emphasis on the use of cultivars that are more efficient to produce because they are resistant to

Table 3.1. The regions of potato production, area of potatoes harvested, the major producing countries, and some of the important cultivars grown in each region.

Region	Area Harvested (HA)	Major Producing Countries in the Region	Examples of Important Cultivars and Their Uses
Asia	5,719,690	China, India, Turkey, Kazakhstan, Iran, Japan, Nepal, Pakistan	Achirana – Fresh; Danshaku – Fresh; Desiree – Fresh; Granola – Fresh; Kufrii Joyti – Fresh
Russia, Ukraine, Belarus, Moldova	5,697,350	Russia, Ukraine, Belarus	Lugovska – Fresh; Nevsky – Fresh
Europe and Baltic States	3,728,250	Poland, Germany, Romania, Netherlands, France, Spain, United Kingdom, Lithuania	Agria – Fresh/Process; Aster – Fresh; Bintje – Fresh/Process; Desiree – Fresh; Saturna – Fresh/Process; Spunta – Fresh
Latin America and Carribean	1,049,695	Peru, Columbia, Brazil, Argentina, Bolivia, Chile, Ecuador, Mexico	Alpha – Fresh; Atlantic – Chip; Kennebec – Fresh/Process; Spunta – Fresh; Yungay – Fresh
Africa	815,350	Egypt, Kenya, Morocco, Uganda, South Africa	Desiree – Fresh; Spunta – Fresh
United States and Canada	706,440	United States, Canada	Atlantic – Chip; Russet Burbank – Fresh/Process; Russet Norkotah – Fresh; Shepody – Process
Oceania	54,230	Australia, New Zealand	Atlantic – Chip; Russet Burbank – Fry; Sebago – Fresh; Shepody – Fry
World Total	17,771,030		

diseases, pests, and environmental stresses, and more emphasis on cultivars with important quality traits – particularly storage quality. Potato breeding programs that in the past developed disease resistant cultivars for table stock are now shifting emphasis to processing quality.

3.2 The Most Important Cultivars

Table 3.2 lists the most important cultivars in the United States and Canada along with their market use (National Potato Council, 1999; Okeefe, et al., 1995). Three of the top five cultivars grown in North America, Russet Burbank, Shepody, and Ranger Russet are used more for frozen french fry production than for fresh market. They have certain characteristics demanded by the fry process industry, long tuber type, large average tuber size, medium to high solids content, mealy texture, and low sugar accumulation in storage at 7-10°C. Several more of the most important North American cultivars are chipping (crisping) types. These are Atlantic, Snowden, and a number of proprietary Frito Lay selections. Taken together, these few cultivars used for frozen french fries, chipping, and dehydration account for 55% of US production, whereas only 26% of potato production is consumed fresh in the US. The most important fresh market cultivars in the US and Canada are Russet Burbank, Russet Norkotah, Superior, Red Norland (all color variants), Yukon Gold, Red LaSoda, Chieftan, Russet Nugget, and GoldRush. There are about 60 others grown on a smaller scale. Seed needs are supplied by about 6% of the crop. Animal feed, starch, and storage losses account for the remaining 13% (National Potato Council, 1999).

Table 3.3 lists the most important potato cultivars in Western Europe based on seed acreages reported for 1997 (Netherlands Catalogue of Potato Varieties, 1997; NIVAA Facts, SAC online 1999; Scottish Agricultural College, 1996; 1999; Swiezynski *et al.* 1998). These also include major seed export cultivars that are important in Africa, the Middle East, the Pacific Rim, and Latin America. Their market uses are included. European processing uses are as fries (either frozen par- fries or fresh chips) and crisps (chips in the US and Canada). Dehydration is a major use of potatoes in both Europe and North America, but usually a cultivar is not grown specifically for dehydration. The dehydration industry utilizes potatoes from fry, chip (crisp), starch, and fresh market that often cannot meet the standards for their intended primary use.

In Western Europe, Bintje is still one of the most important cultivars, with more than twice the seed acreage of the next most important cultivars, Desiree and Spunta. Bintje is used for table stock and processing and Desiree for table stock in many countries, especially in Mediterranean and Middle Eastern countries. Agria, Saturna, Maris Piper, Pentland Dell, and a

Table 3.2. The most important potato cultivars in North America based on 1998 seed acreage.

Cultivar	Seed Acreage (ha)	Market Use
Russet Burbank	24,997	french fry/fresh
Russet Norkotah	9,305	fresh
Shepody	6,166	french fry
Atlantic	5,241	chipping
Ranger Russet	4,512	french fry/fresh
Snowden	4,000	chipping
Superior	3,745	fresh
Red Norland	3,649	fresh
Yukon Gold	2,096	fresh
Red LaSoda	1,869	fresh
Chieftain	1,806	fresh
Russet Nugget	1,647	fresh
Goldrush	1,264	fresh

Table 3.4. The virus reaction of the most important North American potato cultivars.

Cultivar	PVX	PVY	PVA	PLRV	TRV
Atlantic	VR	S	--	MS	S
Chieftain	MR	S	--	S	S
Goldrush	S	S	--	S	S
Ranger Russet	VR	MR	--	S	S
Red LaSoda	MS	S	--	S	S
Red Norland	S	S	--	S	S
Russet Burbank	S	S	S	S	S
Russet Norkotah	S	S	S	S	S
Russet Nugget	S	MS	--	S	S
Shepody	S	S	R	S	S
Snowden	--	--	--	S	S
Superior	S	S	--	S	R
Yukon Gold	R	S	--	MS	S

Table 3.3. The most important western European cultivars based on seed acreages in 1997.

Cultivar	Seed Acreage (ha)	Countries with Major Seed Acreages	Market Uses
Bintje	8886	NL, F, B, DK	fresh, process
Desiree	4846	NL, UK, F, B, DK	fresh
Spunta	4265	NL, F, B, DK	fresh
Agria	3196	NL, D	fresh, process
Saturna	2063	D, F, DK, NL, UK	fresh, process
Nicola	1796	NL, F, DK	fresh
Maris Piper	1770	UK	fresh
Kaptah	1687	F	starch
Diamant	1560	NL	fresh
Elkana	1546	D, NL, F	starch
Mona Lisa	1531	NL, F, DK	fresh
Kondor	1023	NL	fresh
Charlotte	982	F	fresh, process
Marfona	980	NL	fresh
Cilena	954	D	fresh
Kennebec	912	F, B, DK	fresh, process
Jaerla	879	NL	fresh
Secura	856	D	fresh
Quarta	835	D	fresh
Lady Rosetta	833	NL, UK	fresh, process
Pentland Dell	762	UK	fresh, process
Cara	685	UK	fresh
Hermes	682	UK	fresh, process
Premiere	673	NL	fresh, process
Karlena	654	D	fresh, process
Mondial	636	NL	fresh
Producent	629	D	fresh
Maris Bard	618	UK	fresh
Estima	597	UK	fresh
Asterix	580	NL	fresh

B (Belgium), D (Germany), DK (Denmark), F (France), NL (The Netherlands), UK (United Kingdom).

number of other popular cultivars are grown for both fresh market, and for processing as fries or crisps and dehydration.

Roughly 55% of the world potato acreage is in Russia, Ukraine, Belarus, China, and India (Table 3.1) (FAO, 1999). In these countries, pesticide supplies and costs, fertilizer availability, and seed production infrastructure limit overall production. There is a great diversity of cultivars grown, and growers place more reliance on partial resistance to viruses and late blight. These cultivars are grown mostly for fresh consumption and animal feed, but processing uses are growing (Obhiro, 1999; Wang, 1999; Personal Communications).

Poland has the largest potato production in Europe. The principal varieties are Aster, Maryna, and Ania developed by the Polish National Breeding Program. Russia, Ukraine, and Belarus, have about one third of the world's potato acreage. Wide ranges of cultivars developed in the region are grown. Nevskiy, Lugovska, and Belorusky 3 are a few examples of the many cultivars that have been developed by breeding institutes in the former Soviet Union (Hamester and Hils, 1999; Swiezynski *et al.* 1998; Personal Communications). The Peoples Republic of China, with 17% of the world potato acreage, grows many indigenous cultivars such as Mira and the Jinshu, and Zhengshu selections. Achirana and Serrana, developed by the Argentina National Program (INTA) in cooperation with the International Potato Center (CIP), have been widely used in China and Pacific Rim nations as well as in Latin America (Hamester and Hils, 1999; Personal Communications). These cultivars are mainly suitable for fresh consumption and animal feed uses. India, with 6% of the world production, grows cultivars developed by the Indian National Potato Breeding Program such as Kufri Jyoti and I-1085. These are particularly adapted to the distinct climate zones and preferences of the diverse population on the Indian subcontinent. Since processing is a minor use in all of these countries, local fresh market demand, seed availability, and adaptation to local growing conditions determine cultivar choice.

The Middle East and North African countries use mostly European cultivars with Desiree and Spunta being very important. South Africa and the tropical areas of Africa use many different cultivars, primarily for fresh consumption. Some of these have been developed by national programs, often in cooperation with CIP, and incorporate multiple disease resistance traits.

Latin America, the region from which the potato spread to the rest of the world, now has about 6% of the total world potato acreage, Table 3.1 (FAO, 1999). A large number of cultivars are grown; some are of European origin supplied by seed programs in Canada and Europe, while many others have been developed locally, often in cooperation with the International Potato Center. Cultivars like Agria, Bintje, and Atlantic are popular in Brazil, and Spunta in Argentina (Hamester and Hils, 1999; Personal Communications).

Table 3.5. The virus reaction of important western European potato cultivars.

Cultivar	PVX	PVY	PVA	PLRV	TRV
Agria	MR	R	R	MS	MR
Asterix	VR	MS	S	S	MR
Bintje	S	S	R	MR	R
Cara	VR	R	VR	MS	S
Charlotte	S	MR	--	S	MR
Cilena	--	R	R	MS	MR
Desiree	MS	R	MR	S	MR
Diamant	MR	MR	VR	MR	MR
Elkana	MR	R	VR	MS	R
Estima	MR	R	R	MS	MS-R
Hermes	--	MS	MR	MR	R
Jaerla	MR	MR	R	MS	MS-R
Kaptah	S	R	--	R	MR
Karlena	--	VR	--	R	R
Kennebec	S	R	VR	MS	S
Kondor	MS	MS	VR	MS	S-MR
Lady Rosetta	R	MR	R	MS	--
Marfona	MR	R	R	MR	R
Maris Bard	S	R	S	S	S
Maris Piper	VR	S	VR	MS	MS
Monalisa	R	R	VR	R	R
Mondial	R	MR	R	S	MS
Nicola	VR	S-R	VR	S	S
Pentland Dell	VR	S	VR	MS	S
Premiere	R	MS	R	MS	R
Producent	R	MR	R	MS	S
Quarta	R	MR	R	S	R
Saturna	MR	R	R	MS	MR
Secura	--	VR	VR	S	R
Spunta	MS	MR	R	MS	MR

Table 3.6. Some examples of other important cultivars in major potato growing regions and their market use.

Cultivar	Origin/Countries Grown	Use
Achirana INTA	Latin America, Pacific Rim	Fresh
Ania	Poland	Fresh
Aracy	Brazil	Fresh
Aster	Poland	Fresh
Belorusky 3	Belarus, Russia	Fresh
Granola	Germany, Pacific Rim	Fresh
I-1085	India, China	Fresh
Irish Cobbler	Japan, Korea	Fresh
Kufri Jyoti	India	Fresh
Maryna	Poland, Russia	Fresh
Lugovska	Russia, Ukraine	Fresh
Nevsky	Russia	Fresh
Serrana INTA	Latin America, Pacific Rim	Fresh
Temp	Belarus	Fresh/Starch
Yungay	Peru	Fresh

Table 3.7. Virus reactions of some other important cultivars.

Cultivar	PVX	PVY	PLRV	TRV
Achirana INTA	S	R	MR	MS
Ania	--	R	MS	R
Aracy	R	R	MR	MR
Aster	--	R	R	R
Belorusky 3	R	VR	R	MS
Granola	R	R	MR	S
I-1085	S	S	S	--
Irish Cobbler	S	S	S	S
Maryna	--	R	S	R
Lugovska	--	MR	MR	R
Nevskiy	--	R	R	MS
Serrana INTA	VR	MR	MR	R
Temp	--	MS	S	R
Yungay	S	S	S	--

Atlantic and Alpha are very important in Mexico along with cultivars developed by the Mexican National Program (INIFAP), such as Rosita and Nortena (Personal Communications). The cultivars Achirana and Serrana, developed by the Argentina National Program, and CIP are grown in Latin America and throughout the world, where their wide adaptation and consistent yields have made them popular with subsistence growers.

The International Potato Center, working with national cultivar development programs, has introduced many cultivars into non-traditional (sub-tropical and tropical) potato growing areas in Africa and the Far East. These are low input cultivars that incorporate multiple disease and pest resistance traits, particularly resistance to late blight and bacterial wilt, and have helped expand the potato as a staple crop in what had been considered to be marginal production areas (Collins and Pallais, 1998; Personal Communications). For example, Molinera (BR-63.65), with leafroll resistance, was developed for Peru, and Kenya Dhamana (BR-63.76), with PLRV and PVY resistance, was developed for Kenya, using CIP germplasm and resources. Breeding parents have also been selected that can be hybridized to produce botanical seed for use in areas where conventional seed programs are not feasible. The resulting progenies are grown as a homogeneous population rather than as a traditional single clone cultivar. Although these types of cultivars are only grown on modest acreages, their importance should not be underestimated. Breeding for virus resistance has often been given low priority because of the large number of other traits that must be incorporated into a new cultivar, particularly for processing use. Successful seed programs, combined with the right environments, have lessened the need for genetic resistance. Nevertheless, it is hoped that cultivars that require minimum inputs but that are adapted to specialized processing market needs can be developed in the future.

We have not attempted to be comprehensive in covering all cultivars in all major producing countries. We have selected those, which are the most important cultivars in Western Europe, and North America, including some of the dominant export cultivars from European and Canadian seed programs (Barclay and Scott, 1997; National Potato Council, 1999; NIVAA, 1997; SAC, 1996; SAC Online 1999; Seed Area in The Netherlands, 1999; Swiezynski *et al.* 1998).

3.3 Virus Reaction of Important Cultivars

Table 3.4 presents the reaction of important North American cultivars to the major viruses PVX, PVY, PVA, PLRV, and *Tobacco rattle virus* (TRV) (Collins and Pallais, 1998; deBokx, 1972; NIVAA, 1997; Okeefe *et al.* 1995; Stegemann and Schnick, 1985; Swiezynski *et al.* 1998; Personal Communications). Since the numerical scales often used to rate virus reaction can be confusing (9 is either the most resistant or the most susceptible depending on which programs' scale is used) we have described the virus reaction on a scale of very susceptible to very resistant. A field immune response is documented as very resistant. In the case of PVX, PVY, PVA, and PLRV, the resistance reported is to virus infection not symptom expression. In the case of TRV, the resistance is to symptom expression.

Table 3.5 lists the reaction of the most important European cultivars to the major viruses (Collins and Pallais, 1998; deBokx, 1972; NIVAA, 1997; Stegemann and Schnick, 1985; Swiezynski *et al.* 1998; Personal Communications). A number of sources were used for this information that did not always agree and that used different rating systems. We have done our best to assimilate all sources available and to resolve discrepancies. The primary references used for these tables were Stegemann and Schnick, 1995, and Swiezynski *et al.* 1998.

In Table 3.6 we have listed a number of other cultivars that are of importance in potato production regions outside of Western Europe and North America (Collins and Pallais, 1998; deBokx, 1972; NIVAA, 1997; Stegemann and Schnick, 1985; Swiezynski *et al.* 1998; Personal Communications). Some, like Achirana and Granola, are widely grown because of their general disease resistance characteristics, especially virus resistance and late blight resistance combined with culinary qualities. Others seem to be popular simply because of conservative consumer preferences. Table 3.7 lists the virus reactions of these cultivars as far as could be determined.

It is apparent from reviewing the three tables on virus reactions that there has been much more emphasis on virus resistance in European breeding programs and in developing countries when compared with the most popular cultivars developed in North America. The general level of virus resistance may in fact be improving in European breeding programs (Swiezynski and Domanski, 1998). There are efforts underway to correct problems with virus susceptibility in popular commercial cultivars through biotechnological approaches. Most national and private traditional breeding programs also continue to combine virus resistance with other market traits in their new cultivars.

Acknowledgments

We would like to acknowledge the assistance of the following individuals who contributed information regarding the relative importance of cultivars in diverse growing regions and the virus reactions of important cultivars.

Dao Huy Chien, Director, Biotechnology Research Institute, Potato Center, Hanoi, Vietnam.

George Mackay, Scottish Crop Research Institute, Dundee, Scotland.

Ilza Skrabule, Priekuli Plant Breeding Institute, Priekuli, Latvia.

D. Pete Weingartner, Agriculture Research Center, Hastings, Florida.

Jonathan Whitworth, Idaho Crop Improvement Association, Idaho Falls, Idaho.

Ewa Zimmoch-Gwzowska, Plant Breeding and Acclimatization Institute, Mlochow, Poland.

References

Bamberg, J.B., Martin, M.W., Schartner, and D.M. Spooner, D.M. 1996. Inventory of Tuber-Bearing Solanum Species. Catalogue of Potato Germplasm Potato Introduction Station NRSP-6 Project, Sturgeon Bay, WI.

Barclay, M. and Scott, P. 1997. *Potato Varieties in Canada*. New Brunswick Department of Agriculture and Rural Development, Florenceville NB, Canada.

British Atlas of Potato Varieties. 1965. Potato Marketing Board Knightsbridge, London.

Collins, W. and Pallais, N. 1998. Pathogen Tested Potato Germplasm for Distribution. Appendix A Status of Potato Varieties and Advanced Cultivars. International Potato Center, Lima Peru.

deBokx, J.A. 1972. Descriptions of potato varieties. In "*Viruses of Potatoes and Seed Potato Production*" ed. J.A.de Bokx. PUDOC, Wageningen, The Netherlands.

Food and Agriculture Organization of the United Nations, FAO, Statistical Databases 1999 FAOSTAT. http://www.fao.org

Hamester, W. and Hils, U. 1999. *World Catalogue of Potato Varieties*. Buch Edition Agrimedia GmbH Bergen/Dumme Germany.

Love, S. L. 1999. Founding clones major contributing ancestors and exotic progenitors of prominent North American potato cultivars. *Am. J. Potato Res.* 76: 263-272.

National Potato Council. 1999. *Potato Statistical Yearbook*. National Potato Council Publications, Englewood, CO, USA.

NIVAA, *Netherlands Catalogue of Potato Varieties*. 1997. NIVAA, The Hague.

NIVAAa Facts. *Seed area for classification in the Netherlands*. 1999. http://www.nivaa.nl/facts

Obhiro, K. I. 1999. Potato industry and marketing in Japan. *Proc. 1st Kangwon Int. Potato Symp.* pp. 13-20.

Okeefe, R.B., Sterrett, R., Chase. R. and Sieczka, J.B. 1995. *North American Potato Variety Handbook*, The Potato Association of America.

Salaman, R. N. 1949. *The History and Social Influence of the Potato*. Reprinted with corrections by J.G. Hawkes 1985. Cambridge University Press, Great Britain.

Scottish Agricultural College (SAC) and Gordon & Innes Ltd. 1996. *Scotland – The Natural Home of Seed Potatoes*. Scottish Seed Potato Development Council, Haddington.

Scottish Agricultural College Potatoes (SAC) Online Professionals Statistics. 1999. http://www.spud.co.uk

Stegemann, H. and Schnick, D. 1985. Index 1985 of European Potato Varieties. *Mitt. Biol. Bundesanstalt Land-und Forstwirtschaft*. Paul Parey, Berlin and Hamburg.

Swiezynski, K.M., Haynes, K.G., Hutten, R.C.B., Sieczka, M.T., Watts, P. and Zimnoch-Guzowska, E. 1997. Pedigree of European and North American potato varieties. *Plant Breed. Seed Science*. 41: 1-149 (Suppl.)

Swiezynski, K.M. and Domanski, L. 1998. Resistance to viruses and *Phytophthora infestans* in potato cultivars from Germany, The Netherlands, and Poland. *Plant Breed. Seed Science* 42:47-58.

Swiezynski, K.M., Sieczka, M.T., Stypa, I. and Zimnoch-Guzowska,E. 1998. Characteristics of major potato varieties from Europe and North America. *Plant Breed. Seed Science* 42: 1-44 (Suppl.).

Wastie, R.L. 1991. Breeding for resistance in "*Phytophthora infestans*, the cause of late blight of potato. *Adv. Plant Pathol.* 7: 193-224.

Wang, F. 1999. Potato Breeding in China. *Proc. 1st Kangwon Intern. Potato Symp*, pp. 210-214, Kangwon National University, Korea.

4. Historical Perspectives of Potato Virus Research

R.H.Lawson and R. Stace-Smith
National Program Staff, US Department of Agriculture, Agricultural Research Service, Beltsville, MD 20705-5137 USA, and Phyto Diagnostics Laboratory Ltd. 8801 East Saanich Rd, Sidney, British Columbia V8L 1H3, Canada

4. 1 Potato Degeneration

Today, several hundred pests and pathogens are recognized as causes of potato disease and decline. Some of them are probably indigenous to areas where the plant was introduced. In many instances, however, the disease agents and pests were carried along with the vegetative propagation material that was transported around the globe.

Many diseases of plants that are now recognized as caused by viruses undoubtedly existed for a long time. According to Bawden (1956), next to the tulip, the potato is the oldest sufferer of virus diseases on record. By 1775 potatoes were so severely attacked that farmers had to abandon cultivation. Prizes were offered to those who could identify the cause and develop a remedy for the destructive effects on the potato plant. Papers published at that time stressed the degenerative condition of the plants and the condition was known as "degeneration," "running out" and "senility" (Bawden, 1956). It was also noted that in general, new varieties were more vigorous than old ones. This was thought to be the result of continued propagation by asexual methods that were considered "unnatural."

Although there was little experimental work during this early period, Anderson noted in an essay written in 1778 and published in 1792 (mentioned in Bawdwn, 1956) that tubers from affected parents always produced degenerate plants. He also noted that potatoes grown in the south of Britain were more severely affected than those grown in the north. He concluded that the condition was infectious and compared it to smallpox and

G. Loebenstein et al. (eds.),
Virus and Virus-like Diseases of Potatoes and Production of Seed-Potatoes, 53–63.

that control might be achieved by destroying all affected plants as soon as they could be identified. Since the degeneration condition was apparently not known before 1770, it is likely that virus diseases were introduced into Europe about that time.

The degeneration that occurred in the middle of the Eighteenth Century in England was a combination of infectious diseases. The most striking symptoms of this degeneration were leaf curling and a severe reduction in yield. In some areas in England curl was so severe and so prevalent that farmers feared extinction of the potato as a field crop (Salaman, 1949; van der Want, 1972). Farmers soon realized that certain areas were more favorable for the production of "clean" seed potatoes. Although not recognized at the time, this was presumably due to the absence of insect vectors.

In France, the Netherlands and Germany, degeneration of potato crops was also prevalent in the Eighteenth and Nineteenth centuries. The science of plant virology was initiated nearly 100 years later when in 1886 Mayer named tobacco mosaic virus and showed that injecting sap from an infected plant into a healthy one could transmit it. Soon after came the pioneering work of Iwanowski in 1892 along with Beijerinck's discovery of the filterability of tobacco mosaic virus in 1898. Beijerinck concluded that TMV was different from a bacterium and was what he called a "contagium vivum fluidum."

4.2 Discovery of Potato Virus Diseases

In Germany, Appel (1906) first described potato leafroll and distinguished it from the "curl" complex. He stated that the disease had been known for more than a century. Appel described various forms of the disease, such as stunted growth, leaf curl and crinkled leaflets, and concluded that all are due to the same unidentified cause. He further observed that the disease is readily transmitted through the tubers and the virulence apparently increased from generation to generation.

Orton (1913) showed that degeneration of potato took more than one form. He described three types that he named leafroll, mosaic and streak. It was not, however, until later that leafroll was shown to be an infectious disease (Quanjer *et al.* 1916) and transmissibility by aphids (Botjes, 1920) was suspected. Aphids were later identified as leafroll vectors by Schultz and Folsom (1923a). The nature of the pathogen remained obscure for many years until Peters characterized the nature of the phloem limited virus by purification (Peters, 1970).

Little attempt was made by early investigators to distinguish between cause and effect of potato virus infections. The same name was usually used to refer to both the virus and the disease. Later it became clear that potatoes could be attacked by more than one virus and that similar symptoms may be caused by different viruses. The same virus might infect many different species of plants inducing different symptoms in different plants. These observations showed that the system of virus naming on the basis of host attacked and symptoms produced is inadequate. Attempts were made to compare some European and American virus diseases in the mosaic group to determine differences and similarities (Murphy and McKay, 1932b). Schultz and Folsom (1923a) recognized eight different potato virus diseases in the U.S. They included mild mosaic, crinkle mosaic, leaf-rolling mosaic, rugose mosaic, leafroll, streak, spindle tuber and unmottled curly dwarf. By 1929 nearly twenty different viruses were thought to infect potato (Johnson, 1929).Today we recognize mild mosaic as simple or common mosaic caused by PVX. Streak is caused by *Potato virus Y* genus *Potyvirus* (PVY) and spindle tuber is now recognized as a viroid disease.

Most potato mosaic diseases are complexes containing more than one mosaic virus (Smith, 1931b). In a landmark discovery Smith showed that needle and aphid inoculation from the same mosaic-affected potato source plant produced symptoms in tobacco characteristic of the mode of transmission. This was the first definite evidence of the composite nature of potato mosaic diseases. Smith designated the needle-transmitted virus X and the aphid-transmitted component he called Y. Smith (1931a) also found that the potato crinkle disease produced similar results in tobacco and that symptoms were characteristic of the method of infection. He noted that the symptoms produced by the aphid inoculation in tobacco were similar to the aphid inoculations from the potato mosaic affected potato. Symptoms produced by needle inoculation from the crinkle affected potato plants to tobacco were not identical to those produced from the mosaic potato.

In the U. S. the potato disease known as "rugose mosaic," probably the same as "crinkle" in England, reportedly contained two viruses (Valleau and Johnson, 1930). One of the viruses was referred to as "veinbanding." Smith (1931b) concluded that "veinbanding" was apparently identical to the disease he had produced in tobacco with aphid inoculation from mosaic and crinkle potato sources. Crinkle was further defined by Murphy and McKay (1932a) as a complex containing a previously undescribed virus designated virus A.

In his classic studies on potato viruses, Smith (1931b) further suggested three methods for separation of various mosaic viruses. They included transmission by the aphid *Myzus persicae*; the use of plant indicators and plant filters; taking advantage of differences in the lengths of the incubation periods of viruses in the complex within the tobacco plant and of the unequal rates of virus movement. These studies included a description of a selected range of *Solanaceous* hosts that the author hoped could be used as indicator plants whose reactions "may become a standard test for the identification and classification of potato mosaic viruses."

Latency was an important discovery in potato virus investigations made by Johnson (1925) who showed that sap from apparently healthy potato plants induced virus disease in certain species of plant. This observation demonstrated that the so-called "healthy" plant tolerated infection without showing symptoms. Furthermore, Smith (1931b) demonstrated that latent virus infections may only be detectable in other plant species. He showed that a strain of *Potato virus X* genus *Potexvirus* (PVX) that was nearly symptomless on potato caused symptoms on infected tomato plants. The discovery of latent virus infections has been of great significance in the production of seed potatoes certified free of known viruses.

4.3 Early Attempts of Identification and Classification

In the 1920's potato virologists from Europe came to the U. S. to visit potato production areas and Americans traveled in Europe to observe potato diseases. It was recognized that many of the descriptions of those diseases considered to be of virus origin differed greatly among investigators and countries where potatoes are grown. Johnson (1929) concluded that any additional information in the form of measurable characteristics should prove useful in the classification and identification of plant viruses. These features included thermal death-point, longevity in vitro and dilution endpoint and influence of chemicals on virus stability. Additional useful criteria used in virus classification were method of transmission; length of incubation period; transmission by a single method (grafting, insects, wounding); host range in different species or varieties; relation of source of inoculum to infection; influence of environmental conditions on symptom expression; variation in cytological and histological details; and filterability (Johnson, 1929).

In further attempts to investigate the similarities and/or differences among the reported potato virus diseases, Johnson (1929) studied the diseases described according to Schultz and Folsom's nomenclature. Special attention was given to the study of "spot necrosis" virus and its relation to "rugose mosaic" and to "mottle" from apparently healthy potatoes. Based on these studies Johnson (1929) concluded that rugose mosaic and Murphy and Quanjer's crinkle and spot necrosis were identical.

The influence of temperature on disease expression was also noted (Johnson, 1929). A tobacco plant showing a virulent form of spot necrosis on the lower leaves would outgrow the symptoms when placed in a higher temperature environment. The new mildly mottled or symptomless leaves yielded only the "mottle" form whereas the older leaves of the same plant with necrotic symptoms produced the spot necrosis form. If the plants infected with spot necrosis were exposed to different temperatures, or certain temperatures for different lengths of time, it was possible to isolate the virus in various degrees of virulence. Although this phenomenon was not explained, it provided important information on the movement and possible separation of different viruses or virus strains in the same plant.

In his classic 1929 paper, Johnson concluded that the agreements on description and nomenclature of a specific virus, "is the recognition of virus combinations and their separation into constituent entities, thereby reducing confusion by avoiding the publication of new descriptions of names of uncertain and unreliable significance." He went on to say "by the use of certain treatments of the extracts based on their respective properties, it is certainly feasible to separate certain combined viruses from each other." He suggested that the some potato varieties or other host species may be used in separating certain combinations of viruses.

4.4 Virus Collections

The need for a standard nomenclature was recognized beginning with the early descriptive work on potato viruses. In the U.S., Schultz and Folsom established a virus collection in 1916. The collection was based on their collaborative studies of the causes of degradation in the State of Maine (Webb, 1958). More than 20 different viruses and virus strains are represented in the collection. This unique collection has served as a source of reference cultures for investigators around the world for the last 80 years. Schultz and Folsom (1923a) conducted extensive studies on the transmission and symptomatology of several degeneration diseases.

4.5 How Many Potato Viruses ?

Potato is one of the plant species in which the largest number of viruses are found. The most economically important viruses are in the *Polero* (formerly *Luteo), Poty, Potex, and Carla* groups. According to Salazar (1989) nearly 30 different viruses have been reported infecting potato. This does not include the multiple strains of some of the viruses described. The principle potato viruses have been classified as 1) dependent on potato for survival and spread; 2) independent of potato for survival and spread and 3) those where these properties have not been determined.

Viruses that depend on the potato for survival and spread usually show a restricted or moderate host range, whereas those that do not depend on the potato for survival and spread have a wide host range and may cause economically important losses in other crops (Salazar, 1977). Some viruses are common in all areas where potatoes are grown whereas others are prevalent only in specific regions. The greatest variability in potato viruses is found in the Andean region, especially in those viruses that depend on potato for survival and dissemination. Salazar (1989) speculates that this is probably the result of their long association with the host. Some disease agents have been adapted to potato outside of the center of origin. For example, PSTVd has never been found in the Andean region in native cultivars of potato species. In contrast, *Potato virus M* genus *Carlavirus* (PVM) and *Potato virus A* genus *Potyvirus* (PVA), both prevalent in Europe and North America, are uncommon in Andean potatoes.

Spindle tuber was first described by Martin (1922). Schultz and Folsom (1923b) reported that the disease had been recognized by growers in Maine. They noted that the disease is characterized by spindliness of the potato plants and spindle-shaped tubers. Although the symptoms and epidemiological properties of the spindle tuber disease resembled those of conventional viruses, it was shown that the disease agent possessed properties of a free nucleic acid (Diener and Raymer, 1969). These investigators were the first to describe a new class of disease agent now known as a viroid.

4.6 Serology in Virus Identification

Following the discovery of PVX and PVY the separation of these viruses on the basis of transmission properties and host range, serology became important in establishing the identity of newly discovered potato viruses. In the Netherlands a new potato virus was first reported by DeBruyn Ouboter (1952).The virus was detected during an attempt to prepare antiserum against *Potato virus A*. The virus was named *Potato virus* S and antiserum was produced on a larger scale for disease diagnosis. In the Netherlands the virus S infected plants produced a crop of lower quality with an estimated reduction in value of 15 percent. Potato samples tested from different parts of the world indicated the virus was widespread (van Slogteren, 1955).

Potato virus M was also discovered through serological testing (Bagnall, *et al.* 1956). These workers demonstrated that the variety Irish

Cobbler with inverveinal mosaic harbored three distinct viruses. They included virus X, virus S and the newly described virus M. Virus M was described on certain differential test plants and the identity of the new virus was established by serological methods. Later potato virus M was found in the Netherlands in the variety Bintje (Rozendaal and van Slogteren, 1958).

4.7 Diagnosis by Serology and Electron Microscopy

Pioneering research on the application of serological methods for potato virus detection was first conducted in the Netherlands. It was in Holland that antiserum was first mass-produced and used in routine virus diagnosis. Immunization resulted in the production of enough antiserum to PVX and PVS for testing about one million potato plants annually in stocks intended for production of virus-free seed potatoes (van Slogteren and van Slogteren, 1957). One of the first procedures used was a simple agglutination method that did not require complicated equipment to perform the tests. The test method was further refined with the development of microprecipitin reactions (van Slogteren, 1955). This method was a significant improvement since a more dilute antiserum could be used in a droplet tested under paraffin oil. In efforts to obtain a more sensitive method for detection of potato viruses, Shepard (1972) applied diffusion-in-gel methods to the mass diagnosis of potato viruses X, S and M and described the reliability of the procedure. Radial immunodiffusion lies in the rapid diagnosis of virus infections by means of chemically dissociated capsid proteins. A mass indexing program for the presence of viruses in potato seed-stock was developed based on the use of virions degraded by pyrrolidine and pyridine.

A major advance in potato virus detection and certification programs occurred with the application of the enzyme-linked immunosorbent assay (ELISA) procedure. The extreme sensitivity of ELISA approaches that of radioimmunoassays and allows the detection of as little as 1-10 ng/ml virus (van Regenmortel, 1982). In addition to sensitivity and suitability for mass indexing, ELISA has the advantage of simplicity. (See Chapters 10 and 12) . Since the first reports of the applications of ELISA, the market for commercially available products to perform the tests expanded rapidly. Kits that include all the reagents needed to perform the tests have been offered since the 1980's. In recent years monoclonal antibodies to potato viruses have become widely used in mass screening programs. This source of antibodies has improved the specificity and sensitivity of detection that has been useful in virus strain identification. Dot-blot immunoassay and tissue printing have also been developed for detection of viruses on nylon or nitrocellulose membranes (Koenig and Burgermeister, 1986).

Seed potato certification programs in developing countries are constrained to apply the same serological procedures for disease diagnosis that are utilized in industrialized countries (Bravo-Almonacid *et al.* 1992). To overcome this problem, a squash-blot immunoassay (SBIA) was employed by these authors to detect PVX and PVY in samples obtained from potato leaf and tuber samples. Comparative assays using ELISA, squash-bot molecular hybridization and SBIA showed good agreement among the three methods in leaf and tuber sprout samples.

Another procedure applied to potato virus detection is the use of complementary nucleic acid probes. Sap-spot hybridization using ^{32}P labeled complementary DNA has been developed for PVS, PVY and PLRV (Boulton *et al.* 1986). Radioactive labeling has a significant disadvantage, however, since the use is limited to laboratories equipped for this purpose. Recently, a non-radioactive Digoxigenin labeled RNA probe enabled the detection of less than 1 pg/ml PLRV (Loebenstein *et al.* 1997) compared with 2 ng/ml by ELISA.

Electron microscopy was first utilized in the detection of plant viruses and disease diagnosis nearly 50 years ago (DeBruyn Ouboter *et al.* 1951). Particle morphology of elongated viruses was used in classification based on normal length (Brandes and Wetter, 1959). Gross morphology and serology were subsequently adopted as a basis for classification (Brandes and Bercks, 1965).

4.8 Meristem Cultures

In 1952 Morel and Martin obtained virus-free plants by isolating meristems from infected dahlias and culturing them *in vitro.* Tissue culture in combination with heat treatments has and is being used to recover virus-free potato clones (Stace-Smith and Mellor, 1968). Quak (1972) found that potato plants obtained from the meristem that included one leaf primordium were free of PLRV and 80% were free of PVA and PVY. Only one plant in 500 was free of PVX. Success depends on the nature of the virus, size of the bud and the method. Combining heat treatment with meristem culture resulted in renewal and replacement of infected stocks (See Chapter 14).

The recent history of potato virology includes new and improved serological and molecular methods. These procedures are now used in many countries where potatoes are commercially produced. Further advances in developing virus resistant transgenic plants have been made and are now commercialized. This topic will be included in another chapter.

Acknowlegments

I thank Ir. A.F.L.M. Derks, Bulb Research Centre in the Netherlands for the photograph and information on potato virus history and also Dr. D. Lesemann, Federal Biological Research Centre for Agriculture and Forestry in Germany for his contributions to this chapter.

References

.Appel, O. 1906. Neuere Untersuchungen über Kartoffel-und Tomaten Erkrankungen. *Jahresber. Vereinig. angewandten Bot.* 3 Jahrg. 1904/5, pp. 122-136.

Bagnall, R.H., Larson, R.H. and Walker, J.C., 1956. Potato viruses, M, S, and X in relation to interveinal mosaic of the Irish Cobbler variety. *Wis. Agr. Exp. Sta. Res. Bul.* 198:1-43.

Bawden, F.C. 1956. *Plant Viruses and Virus Diseases.* pp. 1-335. Chronica Botanica Company, Waltham, Mass.

Botjes, O.J 1920. Inaugural Dissertation, Wageningen In: Smith, K.M. 1931. *Biological Rev. 6, 302.*

Boulton, R.E., Jellis, G.J., Balcombe, D.C. and Squire, A.M. 1986. The application of complementary DNA probes to routine virus detection, with particular reference to potato viruses. In: *Developments and Applications in Virus Testing.(*eds) R.A.C. Jones and L. Torrance, pp.41-45. Lavenham Press, Lavenhaven, Sudbury, England.

Brandes, J. and Wetter, C. 1959. Classification of elongated plant viruses on the basis of particle morphology. *Virology* 8:99-115.

Brandes, J. and Bercks, R. 1965. Gross morphology and serology as a basis for classification of elongated plant viruses. *Adv. Virus Res.* 11:1-24.

Bravo-Almonacid, F., Haim, L., and Mentaberry, A. 1992. Rapid immunological detection of potato viruses in plant tissue squashes. *Plant Dis.* 76:574-578.

DeBruyn Ouboter, M.P. 1952. A new potato virus. *Proc. Conf. Potato Virus Diseases.* Wageningen-Lisse. 1951:83-84.

De Bruyn Ouboter, M.P., Beijer, J.J., and Slogteren, E. van. 1951. Diagnosis of plant diseases by electron-microscopy. *Antonie van Leeuwenhoek* 17:189-208.

Diener, T.O., and Raymer, W.B. 1969. Potato spindle tuber virus: A plant virus with properties of a free nucleic acid. II. Characterization and partial purification. *Virology* 37:351-366.

Johnson, J. 1925. Ttransmission of viruses from apparently healthy potatoes. *Wis. Agr. Exp. Sta. Res. Bul. 63, 12 p.*

Johnson, J., 1929. The classification of certain virus diseases of the potato. *Wis. Agr. Exp. Sta. Res. Bul.* 87:1-24.

Koenig, R., and Burgermeister, W. 1986. Applications of immunoblotting in plant virus diagnosis. In: *Developments and Applications in Virus Testing.* (eds) R.A.C. Jones and L. Torrance, pp.121-137. Lavenham Press, Lavenham, Sudbury, England.

Loebenstein, G. Akad, F., Filatov, V., Sadvakasova, G., Manadilova, A., Bakelman, H., Teverovsky, E., Lachmann, O. and David, A. 1997. Improved detection of potato leafroll luteovirus in leaves and tubers with a digoxigenin-labeled cRNA probe. *Plant Dis.* 81:489-491.

Martin, W.H., 1922. "Spindle tuber," a new potato trouble, Hints to Potato Growers. *New Jersey State Potato Assoc.* 3:8.

Morel, G. and Martin, C. 1952. Guérison de Dahlia attaints d'une maladie à virus. *C.R. Acad. Sci.* Paris 235: 1324- 1325.

Murphy, P.A. and McKay, R. 1932a. The compound nature of crinkle, and its production by means of a mixture of viruses. *Sci. Proc. Royal Dublin Soc.* 20:227-247.

Murphy, P.A and McKay, R. 1932b. A comparison of some European and American virus diseases of the potato. *Sci. Proc. Royal Dublin Soc.* 20:347-358.

Orton, W.A. 1913. Leaf roll, curly leaf and other new potato diseases. *Phytopathology* 3:69.

Peters, D. 1970. Potato leafroll virus. *C.M.I./A.A.B. Descriptions of Plant Viruses.* No. 36.

Quak, F. 1972. Therapy. In: *Viruses of Potato and Seed Potato Production. (*ed) J.A. de Bokx, pp.158-166. Centre Agr. Publ. and Doc. Wageningen, The Netherlands.

Quanjer, H.M. 1922. New work on leaf-curl and allied diseases in Holland. International Potato Conference convened in 1921. *Royal Hort. Soc.*, London. p.127-145.

Quanjer, H.M., van der Lek, H.A.A. and Oortwijn Botjes, J. 1916. On the nature, mode of dissemination and control of phloem-necrosis (leaf-roll) and related diseases. *Meded. R. Hoog. Land-Tuin en Boschboruwsch.* Wageningen 10:84-138.

Regenmortel, M.H.V. van. 1982. *Serology and Immunochemistry of Plant Viruses.* p.1-302. Academic Press, New York.

Rozendaal, A. and Slogteren, D.H.M. van. 1958. A potato virus identified with potato virus M and its relationship with potato virus S. *Proc. 3rd Conf. Potato Virus Diseases.* Lisse-Wageningen, 1957, pp.20-36.

Salaman, R. N. 1949. Some notes on the history of curl. Tijdschr. PlZiekt. 55:118-128.

Salazar, L.F. 1977. Studies on three viruses from South American potatoes. *Ph.D. thesis, Univ. of Dundee, Scotland,* 242 p.

Salazar, L.F. 1989. Main virus diseases of potato. In: Control of Virus and Virus-like Diseases of Potato and Sweet Potato. pp. 9-12. *Report of the III Planning Conf.,* Lima, Peru. International Potato Center.

Schultz, E.S., and Folsom, D. 1923a. Transmission, variation, and control of certain degeneration diseases of Irish potatoes. *J.. Agr. Res.* 25:43-147.

Schultz, E.S., and Folsom, D. 1923b. Spindling tuber and other degeneration diseases of Irish potatoes. *Phytopathology* 13:40.

Shepard, J.F. 1972. Gel diffusion methods for the serological detection of potato viruses X, S and M. pp.1-72. *Montana Agric. Exp. Stn. Bul.* 662.

Slogteren, D.H.M. van. 1955. Preparation of antisera against potato virus S with special reference to the preparation of non-toxic virus-suspensions for the immunization of rabbits from extracts of infected potato plants. *Proc. 2nd Conf. Potato Virus Diseases.* Lisse-Wageningen 1954, pp. 35-39.

Slogteren, E. van, and Slogteren, D.H.M. van. 1957. Serological identification of plant viruses and serological diagnosis of virus diseases of plants. *Ann. Rev. Microbiol.* 11, 149-164.

Smith, K.M. 1931a. Virus diseases of plants and their relationship with insect vectors. *Biol. Rev.* 6:302-344.

Smith, K.M. 1931b. On the composite nature of certain potato virus diseases of the mosaic group as revealed by the use of plant indicators and selective methods of transmission. *Proc. Royal Soc.* B. 109:251-267.

Stace-Smith, R. and Mellor, F.C. 1968. Eradication of potato X and S by thermotherapy and axillary bud culture. *Phytopathology* 58: 199-203.

Valleau, W.D. and Johnson, E.M. 1930. The relation of some tobacco viruses to potato degeneration. *Kentucky Agric. Exp. Sta. Bul.* No. 309:475-507.

Want, J.P.H. van der. 1972. Introduction of plant virology. In: *Viruses of Potato and Seed Potato Production.* (ed) J. A. de Bokx, pp. 19-25. Centre for Agricultural Publishing and Documentation. Wageningen, the Netherlands.
Webb, R.E. 1958. Schultz potato virus collection. *Am. Potato J.* 35:615-619.

Fig. 4.1 Prof. Dr. E. van Slogteren between Ir. D.H.M. van Slogteren (left) and Prof. Dr. T.H. Thung (right) at the First Conference on Potato Virus Diseases, August 1951, Wageningen , The Netherlands

5. The Main Viruses Infecting Potato Crops

5.1 Introduction

Alan A. Brunt
Horticulture Research International, Wellesbourne, Warwicks CV35 9EF, U.K.

At least 37 viruses naturally infect cultivated potatoes (e.g., Beemster and de Bokx, 1987; Salazar.1996; Jeffries, 1998); their natural modes of transmission, taxonomic status and accepted sigla are indicated in Table 5.1. Some of these viruses, notably *Potato leafroll virus* (PLRV), *Potato aucuba mosaic virus* (PAMV), *Potato mop-top virus* (PMTV), *Potato virus A* (PVA), *Potato virus M* (PVM), *Potato virus S* (PVS), *Potato virus X* (PVX) *and Potato virus Y* (PVY) occur worldwide in potato crops; others such as *Arracacha virus B* (AVB), *Beet curly top virus* (BCTV), *Eggplant mottled dwarf virus* (EMDV), *Potato black ringspot virus* (PBRV), *Potato virus U* (PVU) *and Potato T virus* (PVT) are important only in some geographical areas, and some such as *Alfalfa mosaic virus* (AlfMV), *Cucumber mosaic virus* (CMV), *Sowbane mosaic virus* (SMV), *Tobacco necrosis virus* (TNV) and *Tobacco streak viruses* (TSV) occur only rarely and sporadically in restricted geographical areas. Some of the latter, however, if they became more widespread would undoubtedly of greater importance. Although most of the viruses have long been known to infect potatoes, a few including *Potato latent virus* (PLV), *Potato rough dwarf virus* (PRDV), *Potato yellowing virus* (PYV) *and Potato yellow vein virus* (PYVV) have been recognised only during the past decade or so.

The viruses infecting potato are of two main types; some (such as PVY, PLV, PLRV, PVM, PVS, PAMV, PMTV, PVT, PYDV and PYVV) are primarily specific pathogens of potato, although they may occur also in other Solanaceous crops and weeds, and the remainder (e.g., AlfMV, *Tomato*

G. Loebenstein et al. (eds.),
***Virus and Virus-like Diseases of Potatoes and Production of Seed-Potatoes*, 65–67.**

Table 5.1. Viruses reported to infect potato

Vectors	Viruses	Siglum	Taxonomy	
			Genus	Family
Aphids	Alfalfa mosaic virus	AMV	Alfamovirus	Bromoviridae
	Cucumber mosaic virus	CMV	Cucumovirus	Bromoviridae
	Eggplant mottled dwarf virus	EMDV	Nucleorhabdovirus	Rhabdoviridae
	Potato latent virus	PLV	Carlavirus	
	Potato leafroll	PLRV	Polerovirus	Luteoviridae
	Potato rough dwarf virus	PRDV	Carlavirus	
	Potato virus A	PVA	Potyvirus	Potyviridae
	Potato virus M	PVM	Carlavirus	
	Potato virus P	PVP	?Carlavirus	
	Potato virus S	PVS	Carlavirus	
	Potato virus Y	PVY	Potyvirus	Potyviridae
	Potato yellowing virus	PYV	Alfamovirus	Bromoviridae
	(Wild potato mosaic virus)	WPMV	Potyvirus	Potyviridae
Beetles	Andean potato latent virus	APLV	Tymovirus	
	Andean potato mottle virus	APMV	Comovirus	Comoviridae
Thrips	Tobacco streak virus	TSV	Ilarvirus	Bromoviridae
	Tomato spotted wilt virus	TSWV	Tospovirus	Bunyaviridae
Leafhoppers	Beet curly top virus	BCTV	Curtovirus	Geminiviridae
	Potato yellow dwarf virus	PYDV	Nucleorhabdovirus	Rhabdoviridae
Whiteflies	Potato deforming mosaic virus	PDMV	?Begomovirus	Geminiviridae
	Potato yellow mosaic virus	PYMV	Begomovirus	Geminiviridae
	Potato yellow vein virus	PYV	? Crinivirus	Closteroviridae
Fungi	Potato mop-top virus	PMTV	Pomovirus	
	Tobacco necrosis virus	TNV	Necrovirus	Tombusviridae
Nematodes	Arracacha virus B	AVB	Nepovirus	Comoviridae
	Potato black ringspot virus	PBRV	Nepovirus	Comoviridae
	Potato virus U	PVU	Nepovirus	Comoviridae
	Tobacco rattle virus	TRV	Tobravirus	
	Tobacco ringspot virus	TRSV	Nepovirus	Comoviridae

	Tomato black ring virus	TBRV	Nepovirus	Comoviridae
Unknown	Potato aucuba mosaic virus* *	PAMV	Potexvirus	
	Potato virus T	PVT	Trichovirus	
	Potato 14R virus* *	P14R	?Tobamovirus	
	Potato virus X * *	PVX	Potexvirus	
	Solanum apical leafcurling virus	SALCV	?Begomovirus	Geminiviridae
	Sowbane mosaic virus**	SMV	Sobemovirus	
	Tobacco mosaic virus* *	TMV	Tobamovirus	
	Tomato mosaic virus**	ToMV	Tobamovirus	

?= possible or tentative species of the genus.

*= reported to occur naturally only in wild potatoes.

**= *these viruses are also transmitted mechanically from infected to healthy potatoes.*

spotted wilt virus and CMV) are general pathogens which occur naturally worldwide in a wide range of plant species.

Of these 37 viruses, some such as PVY, PLRV, PVM and PVS are very damaging pathogens, and others including PLV, PRDV, PYV and PYVV are potentially very important; by contrast, others that include TSV, EMDV, TNV, PVU, CMV and several tobamoviruses are relatively unimportant because they occur rarely and/or cause insignificant crop losses. The main viruses are reviewed below.

5.2 *Potato leafroll virus* (PLRV;Genus *Polerovirus*; Family *Luteoviridae)*

G. Loebenstein
Department of Virology, Agricultural Research Organization, The Volcani Center, Bet Dagan, 50-250 Israel

First described by Quanjer *et al.* (1916).

5.2.1 Symptoms and Damage in Potato

Symptoms of primary infection by PLRV consist of pallor and upward rolling of young leaves, especially at the base, with an upright habit. The edges of young leaflets of some cultivars may develop reddening. Secondary symptoms, in plants grown from infected tubers, are stunting of the shoots and upward rolling of leaflets, especially of basal leaves (Fig. 5.3.1), which become rigid and leathery. In some culivars of *Solanum tuberosum* ssp. *andigena* and other wild species grown in South America, leaf rolling is not a typical symptom of secondary infection. Such plants are stunted and their tip leaves develop marginal yellowing (enanismo amarillo) (Rodriguez and Jones, 1978).

Necrosis may develop in the phloem tissue of stems and petioles, and excessive callose occurs in the sieve tubes of stems and tubers. Callose staining with resorcin blue, the Igel-Lange test, (de Bokx, 1967) was often used for detecting PLRV in tubers before the introduction of ELISA. Carbohydrates accumulate in the leaves- sometimes 2-3 times more than in healthy leaves, with a corresponding reduction in the tubers - due to impaired phloem transport. This is probably due not only to necrosis in the phloem cells, but perhaps also because of blockage of photoassimilate movement from the chloroplast into the cytosol by the triose-phosphate translocator (TPT). Thus, in transgenic plants in which cDNA for TPT was expressed in reverse orientation, leaves accumulated five times as much

G. Loebenstein et al. (eds.),
Virus and Virus-like Diseases of Potatoes and Production of Seed-Potatoes, 69–75.

starch as leaves of control plants (Riesmeier *et al.* 1993). This could also be due to blockage of sucrose loading into the phloem of PLRV-infected plants by the sucrose transporter protein. Thus, in sucrose transporter antisense potato plants starch content increased up to 10-fold (Riesmeier *et al.* 1994). Recently it was suggested that the p28 (p0, see later) protein of PLRV may be involved in leaf symptom expression (Wilk *et al.* 1997) as healthy potato plants transformed with the PLRV p28 protein gene displayed an altered phenotype resembling virus-infected plants. In tobacco plants transformed with the PLRV MP17 movement protein, sugars and starch increased markedly in the leaves. It was suggested that the phloem tissue is the primary site of the PLRV MP17 protein action in altering the carbohydrate metabolism of the infected plant (Herbers *et al.* 1997).

Tuber symptoms are rare, but some cultivars such as Green Mountain and Russet Burbank have internal net necrosis (Fig.5.3.2), which can be seen when tubers are cut. Sometimes tubers from infected plants may develop thin sprouts - "spindling sprouts". This is an inconsistent symptom and depends on cultivar, virus strain and environmental conditions.

PLRV is of great economic importance. In plants grown from infected tubers (secondary infection), yields may be reduced by 33-50%. Tubers of infected plants are small to medium. Even greater losses are observed when PLRV occurs in complex with PVX or PVY.

5.2.2 Natural Host Range

This is mainly restricted to species of the Solanaceae, including:
Datura stramonium – interveinal necrosis
Solanum villosum – chlorosis, rolling and leatheriness of the leaves
Physalis floridana – stunting, chlorosis and rolling of the leaves
Lycopersicon esculentum – no distinct symptoms, athough some early reports from Australia describe yellow top strains that cause stunting of plants, marginal yellowing, curling of leaflets and death of flowers (Braithwaite and Blake, 1961)
Datura tatula, S. aculeatissimum, S. dulcamara S. erianthum, S. lycocarpum, S. nigrum, S. paniculatum, S. villosum, Nicotiana clevelandii and *Nicandra physalodes* are also alternative hosts . Non-solanaceous host species include *Amaranthus caudatus, Capsella bursa-pastoris, Celosia argentea, Gomphrena globosa, Montia perfoliata, Nolana lanceolatus* and *Sisymbrium altissimum* (Thomas, 1984; Tamada *et al.* 1984; Natti *et al.* 1953, Fox *et al.* 1993; Souza-Dias *et al.* 1993; Thomas, 1993; Ellis, 1992).

5.2.3 Geographical Distribution

Probably occurs worldwide in potato growing areas, especially in climatic regions that favour high vector populations and in agro-systems where use of virus-tested "seed" potatoes is not common.

5.2.4 Transmission

Several aphid species have been reported to transmit PLRV (Kennedy *et al.* 1962). *Myzus persicae* seems to be the most efficient and important vector. *Macrosiphum euphorbiae* transmits potato strains less efficiently, but is an efficient vector of Australian tomato yellow top isolates. In Brazil, *M. nicotianae* was found to be an efficient experimental vector (Cupertino *et al.* 1995).

PLRV is transmitted in a persistent manner, characterised by minimum access times for acquisition and inoculation of c. 1 h each. There is a latency period between acquisition and transmission of the virus, and the minimum time for transmission is c. 12 h. Transmission frequency increases with an increase in the access-feeding period of up to 2 days or more. Both larvae and adults can transmit the virus. Aphids remain infective after moulting and remain viriferous for life. PLRV is circulative in the vector aphids and can be found in the haemolymph of *M. persicae.*

Recently it was shown that endosymbiontic bacteria (*Buchnera* sp.) are involved in the persistent transmission of PLRV by *M. persicae* (Hogenhout *et al.* 1996, 1998). PLRV displayed a strong affinity for symbionin, the major protein synthesized and released into the aphid's haemolymph by the bacterial endosymbiont. This protein had a high degree of homology with *Escherichia coli* heat shock protein GroEl. PLRV binding sites of *Buchnera* GroEl are located in the equatorial domain of the protein. Mutants lacking this domain lost the ability to bind PLRV. Absence of *Bucnera* GroEl in the haemolymph of aphids treated with antibiotics led to the rapid degradation of the viral capsid protein and a concomitant loss of infectivity. Also *Azadirachta indica* metabolites, which interfered with the host-endosymbiont relationship inhibited transmission of PLRV by *M. persicae* (Heuvel *et al.* 1998). Availability of virus for acquisition by aphids generally declines with increases in plant age and symptom severity (Heuvel and Peters, 1990).

PLRV is not transmitted by mechanical inoculation, by seed or pollen; it is transmitted experimentally by grafting.

5.2.5 Strains

Isolates have been detected that, although they are serologically indistinguishable, induce symptoms of different severity on potato cultivars or *Physalis pubescens (syn. P.floridana)*, or by their ease of transmission by *M. persicae* (Tamada *et al.* 1984). Several avirulent strains protected *P. pubescens* plants from virulent strains (Webb *et al.* 1952).

Tomato yellow top strains (Thomas, 1984) that cause yellow edge symptoms in tomato leaves cause inconspicuous symptoms or symptomless infection in potato, but are readily transmitted by *Macrosiphum euphorbiae* (Braithwaite and Blake, 1961).

5.2.6 Experimental Hosts

5.2.6.1 Diagnostic Species

Datura stramonium - systemic interveinal yellowing

Physalis pubescens (syn P.floridana) – severe stunting when young seedlings become infected (Fig.5.3.3); systemically infected leaves develop mild interveinal necrosis, older leaves become slightly rolled.

Physalis sp. close to *P. heterophylla* – stunting, interveinal necrosis and epinasty become evident within 8-12 days (Souza-Dias *et al.* 1991) of infection.

Solanum tuberosum ssp. tuberosum (potato) – stunting and leafroll (Fig.5.2.1).

Brassica campestris var. pekinensis (Chinese cabbage), *Raphanus sativus* (radish) and *Vicia faba* (broad bean) are non-hosts.

5.2.6.2 Propagation Species

*Physalis pubescens (syn. P.floridana), Datura stramonium and Solanum tuberosum (*especially for maintenance by storing dormant infected tubers at 4° C).

5.2.7 Purification

From *P. pubescens* or *D. stramonium* as described by Takanami and Kubo (1979). See also Chapter 11.

5.2.8 Particle Morphology and Properties

Virions are isometric *c.* 24 nm in diameter, and not enveloped (Fig. 5.3.4). Purified preparations contain one sedimenting component of 115 S (Takanami and Kubo, 1979). A260/ A280: 1.78; A260/ A240: 1.43.
Specific absorbance at 260:8.6.
Density in CsCl: 1.39 g/cm³ (Rowhani and Stace-Smith, 1979).
Density in Cs_2SO_4: 1.34 g/cm³ (Thomas, 1984).
The virus contains 30% nucleic acid and 70% protein.
Mol.wt of the coat protein 26.3 kDa (Rowhani and Stace-Smith,1979).

5.2.9 Nucleic Acid Component

The genome consists of a single-stranded, linear messenger sense RNA molecule of 5.88-5.99 kb (mol. wt. 2.0 x10^6), which is covalently linked to a small protein (VPg) of 32 amino acids at it's N-terminal (Rowhani and Stace-Smith, 1979). The RNA does not contain a Poly A sequence at it's 3'terminus, and is infectious when used to inoculate tobacco mesophyll protoplasts (Mayo *et al.* 1982). Sequence database accession codes include: D00530 Em(40)_vi:PLVGR Gb(84)_vi:PLVGR 5,987bp, D00733 Em(40)_vi:PLRVA (Australian isolate) 5,882bp, D00734 Em(40)_vi:PLVRC (Canadian isolate) 5,883bp, X13906 Em(40)_vi:PLRVCOAT Gb(84) (PLRV coat protein gene) 5,987bp.

The genomic sequence of PLRV consists of 8 open reading frames (ORFs) (Fig. 5.3.5) (Martin *et al.* 1990; Mayo and Ziegler-Graff, 1996; Ashoub *et al.* 1998). The six major ORFs are separated by a small intergenic region into two gene clusters – ORFs 0, 1 and 2, and ORFs 3, 4 and 5. ORF 0 encodes a factor involved in symptom development. ORF 1 and 2, with motifs characteristic of helicases (ORF 1) and polymerases (ORF 2), form part of the viral replicase. The other ORFs are located in the 3' half of the genome. ORF 3 encodes the capsid protein. Initiation of an internally located AUG codon within the CP gene, but in a different reading frame, codes for the movement protein (ORF 4). Suppression of the CP amber stop codon results in formation of an ORF 3/ ORF 5 read-through protein, which is supposedly involved in aphid transmission (Ashoub *et al.* 1998).

ORF 0, ORF 1 and ORF 2 are translated from genomic RNA. In addition, two subgenomic RNAs have been observed - sgRNA 1 (~2.3 kb) (Smith and Harris, 1990) and sgRNA 2 (~0.8 kb) (Ashoub *et al.* 1998). sgRNA 1 serves as mRNA for ORF 3, ORF 3/5 and ORF 4. sgRNA 2 may code for two viral proteins of 7.1 kDa (ORF 6) and 14 kDa (ORF 7), respectively. The VPg sequence has been mapped to position 400-431 of ORF 1, downstream of the putative protease domain and in front of the RNA-dependent RNA polymerase (Wilk *et al.* 1997). The structural organization of the PLRV

Fig. 5.2.1 Potato cv. Desirée showing potato leafroll symptoms.

Fig. 5.2.2 Internal net necrosis in PLRV-infected tuber. Courtesy Dr. L. Salazar , CIP.

Fig. 5.2.3 Stunting of *Physalis pubescens (syn. P.floridana;* (left) after aphid inoculation with PLRV (right: healthy).

Fig. 5.2.4 Purified virus particles stained with uranyl acetate. Bar represents 100 nm. Courtesy Dr. L. Salazar, CIP.

Fig. 5.2.5 Genomic organization of potato leafroll virus.

genome in the coat protein region is similar to that of beet western yellows virus and barley yellow dwarf virus (Tacke *et al.* 1989).

A full-length cDNA copy of the virus has been expressed in protoplasts of *Chenopodium quinoa,* and intransgenic potato plants that accumulated the 17 kDa movement protein and had symptoms typical of PLRV (Prüfer *et al.* 1997).

5.2.10 Detection Methods

PLRV can be detected in potato leaves by serology, using commercial ELISA kits (see Chapters 10 and 12). However, the concentration of PLRV varies and, in plants grown at temperatures of c. 30°C or in older plants, ELISA may not always detect infection. It is also difficult to detect PLRV by ELISA in unsprouted tubers (Hill and Jackson, 1984).

Various methods based on PCR or nucleic acid probes are now being developed and evaluated. Thus, PLRV could be detected in tubers within one day by immunocapture and a fluorogenic 5'nuclease RT-PCR assay (Schoen *et al.* 1996; Russo *et al.* 1999). Using a digoxigenin-labeled cRNA probe, PLRV was easily detected in dormant tubers (Loebenstein *et al.* 1997). The limit of PLRV detection with this probe was 1 pg/ml compared with 2 ng/ml by ELISA. Such methods, if adopted by testing laboratories, may become very important in inspection schemes by eliminating the present necessity of sprouting dormant tubers.

5.2.11 Therapy of Individual Plants

PLRV can be eliminated by keeping tubers for c. 25 days at 37.5°C in a humid atmosphere (Kassanis, 1950; Lizárraga *et al.* 1991) (See Chapter 14). Heat treatment causes tubers to deteriorate and is therefore not used for commercial stocks, but is mainly employed to obtain virus-free plants of small special potato lines which can then be propagated. In certain regions in India, an increase in PLRV was observed when cold storage replaced open-air storage; apparently the high ambiant temperatures eliminated PLRV from the tubers (Quak, 1987). Tissue culture, with or without thermotherapy, is the most common practice to retrieve virus-free potatoes in general and PLRV-free plants in particular (Lizárraga *et al.* 1991) (See Chapter 14). It should, however, be emphasised that following the therapy step reliable testing for viruses, preferably over two growing periods and evaluation of trueness-to-type of the cured lines are essential.

5.3 Potyviruses

Alan A Brunt

Horticulture Research International, Wellesbourne, Warwicks CV35 9EF, U.K.

Three potyviruses (*Potato viruses A, Y and V*) are economically-important in potato crops. All three viruses are transmitted from infected to healthy plants in the non-persistent manner by aphids, induce in infected plants cylindrical cytoplasmic inclusions ("pinwheels") composed of a protein of c. 74 kDa, have slightly flexuous filamentous particles mostly measuring c.750 x 11 nm each with a coat protein of c. 34 kDa encapsidating a single molecule of ss positive-sense RNA of c. 10 kb.

As *Wild potato mosaic virus* has been found only in wild potatoes and a few Solanaceous species (Jones & Fribourg, 1979; Salazar, 1996), it will not be discussed further.

5.3.1 *Potato Virus A* (PVA)

This was first described by Murphy & McKay (1932) and was earlier also known as potato mild mosaic or potato virus P (Bartels, 1971).

5.3.1.1 Symptoms and Damage in Potatoes

The severity of symptoms depends on virus isolate, potato cultivar and environmental conditions, but infection can reduce tuber yield by up to 40% (e.g., Dedic, 1975; Hooker, 1981). Leaves of potatoes infected with PVA alone may remain symptomless or develop a mild mosaic, rugosity and marginal distortion; hypersensitive cultivars develop tip necrosis. In complex with PVX or PVY, PVA induces leaf crinkling and significant yield losses.

5.3.1.2 Natural Hosts

Its only known natural hosts are potato and *Solanum betacea* (tamarillo).

5.3.1.3 Geographical Distribution

PVA is widely distributed in potato growing areas in Europe and N.America, but does not occur in the Andean region of S.America. It occurs less commonly than PVY in tropical countries.

G. Loebenstein et al. (eds.),
Virus and Virus-like Diseases of Potatoes and Production of Seed-Potatoes, 77–86.

5.3.1.4 Transmission

At least ten aphid species are its vectors, but the more important are *Aphis frangulae, Aulocorthum solani, Macrosiphum euphorbiae* and *Myzus persicae.* Like other potyviruses, the transmission of PVA is dependent on the virus-encoded helper component proteinase and the presence of an aspartic acid-alanine- glycine (DAG) sequence in the coat protein. PVA is not seed-transmitted.

5.3.1.5 Experimental Hosts

PVA has a very restricted experimental host range that is limited to the Solanaceae. The most useful of these are:

Nicotiana tabacum cvs Samsun and White Burley - vein clearing and dark-green vein-banding of systemically infected leaves.

Nicandra physalodes - systemic leaf mottling, severe necrosis, rugosity and palnt stunting.

Solanum demissum x S.tuberosum cv. Aquila (=A6) and *S.demissum* "SdA" – systemic vein necrosis (Fig. 5.3.1).

5.3.1.6 Strains

Of the five strains groups generally recognised by the hypersensitive response of potato cultivars (Valkonen *et al.*1995; Oruetxebarria *et al.* 2000), three from potato are recognised by their differential interactions with potato cv. King Edward. Another strain consists of potato isolates which, after repeated experimental propagation in Nicotiana spp., fail to infect potato systemically or to be aphid transmissible (Rajamaki *et al.*1998). A strain from *Solanum betacea* (tamarillo) is distinguishable by experimental host reactions; like the strains from potato and *Nicotiana* spp., its genomic sequence has been determined (Puurand *et al.* 1994; Kekarainen *et al.* 1999).

5.3.1.7 Particle Morphology and Properties

The filamentous particles are mostly measure 730 x 11 nm.

5.3.1.8 Nucleic Acid Component

The genomic RNA of 9565 nucleotides has been completely sequenced and has at its 3' terminus an untranslated region of 227 or 229 nucleotides (the number is dependent on the strain) which like that of other potyviruses, ends in a polyadenylated tract (Puurand *et al.* 1994; Kekarainen *et al.* 1999). The arrangement of the genomic RNA is generally similar to that of PVY; the genomic RNA is translated into a polyprotein (3,059 amino acids) that is subsequently processed into up to 10 functional proteins, including the coat protein of 269 amino acids.

5.3.1.9 Detection Methods

The test plant species 'A6' is a very sensitive indicator host for PVA

PVA is more conveniently detected and identified in leaves and tubers by serological tests, of which DAS-ELISA (Gugerli, 1979; Vetten *et al.*1983; Singh and Somerville, 1992), dot-ELISA (Weidemann, 1988) and direct tissue blotting (Samson *et al.* 1993) are now commonly used; the use of MAbs in ELISA has also proved reliable for mass testing (Browning *et al.*1995).

5.3.2 *Potato Virus V* (PVV)

PVV was first recognised as a distinct potyvirus by Fribourg and Nakashima (1984b). It is serologically closely related to PVY^{c}, but unrelated to other PVY strains, and was earlier variously described by as a deviatiating strain (designated G1) of PVY^{c} (Rozendaal *et al.* 1971), PVY^{c}-ab (*Calvert et al.* 1980), and PVY-UF (Fribourg and Nakashima, 1984b). This confusion was due to the hypersensitive reaction that PVV induced in potato cultivars containing the gene (Nc) which causes the hypersensitive reaction to PVY^{C}; these cultivars were later shown also to contain another hypersensitive gene (Nv) specific for PVV. The virus was then characterised by Fribourg and Nakashima (1984b). Because PVV is also distantly related to *Peru tomato mosaic virus* (PTV), it has been suggested that it should be considered a strain of PTV (Fernandez-Northcote, 1986); however, phylogenetic analyses show the viruses to be distinct (Oruetxebarria *et al.* 2000).

5.3.2.1 Symptoms

Naturally infected plants of most susceptible potato cultivars in Europe remain symptomless although some have slightly pallid, smaller and distorted leaves; those of a few cultivars such as cv. Wauseon develop mild mosaic and crinkling and plants of cv. Foxton develop a systemic necrosis. PVV also causes severe symptoms of some native cultivars in Bolivia.

5.3.2.2 Natural Hosts

The only known natural hosts of PVV are potato and tomato; the virus is widespread in only a few potato cultivars.

5.3.2.3 Geographical Distribution

PVV has been reported only from the Andean Region of S.America (Peru and Bolivia) and N.Europe (France, Germany, Norway, The Netherlands and the U.K.) (Fribourg and Nakashima, 1984b; Rozendaal *et al.* 1971; Calvert *et al.* 1980; Jones and Fuller, 1984).

5.3.2.4 Transmission

PVV is transmitted in the non-persistent manner by aphids including *Brachycaudus helichrysi, Macrosiphum euphorbiae, Myzus persicae and Rhopalosiphoninus latysiphon (*Jones and Fribourg, 1986).

5.3.2.5 Experimental Hosts

PVV is mechanically transmissible, but its experimental hosts are restricted to species of the Solanaceae and Chenopodiaceae (Fribourg and Nakashima, 1984b; Jones and Fuller, 1984). Of these, the following are the most useful:-

Nicotiana clevelandii, N.glutinosa, N.occidentalis, N.tabacum - vein clearing, chlorotic vein banding and mosaic patterns of systemically infected leaves.

Nicotiana debneyi - local diffuse chlorotic spots, followed by vein clearing, vein banding, mosaic, chlorotic spots and rings.

Solanum demissum x *S.tuberosum* 'Aquila', A6 - systemic infection (this species is hypersensitive to PVY and PVA).

Solanum tuberosum - faint systemic leaf mosaic in cv.Estima, systemic mottle in in cv.Desiree, and local necrotic spots in cvs Maris Piper and Pentland Crown.

5.3.2.6 Strains

None differentiated by biological and serological properties, or by RNA sequence comparisons.

5.3.2.7 Particle Morphology and Properties

The virus particles are mostly c.760 x 11 nm. Purified virus has an absorption ratio (A260/A280) of 1.31-1.35.

5.3.2.8 Nucleic Acid Component

PVV particles each contain a single molecule of ss-messenger polarity RNA of 9851 nucleotides (isolate Dv42); it contains a 5' terminal UTR (204 nt), a single ORF (9201 nt that encodes a protein of 3067 amino acids), an unusually long 3' UTR (446 nt), and a poly-A tail. The arrangement and functions of the genes are similar to those of PVY. Six other isolates have been partially sequenced. Phylogenetic analysis has shown that, depending on the sequences compared, isolates of PVV are 95-100% homologous, and that PVV has the closest relationship to PVY and *Pepper mottle virus* (Oruetxebarria *et al.*2000).

5.3.2.9 Detection Methods

Symptoms of PVV in some test plant species are difficult to distinguish from those of PVY^{o} and PVY^{c}. However, it is readily detected, and differentiated from PVY strains by ELISA using polyclonal antisera (Gugerli and Fries, 1983).

5.3.3. *Potato Virus Y* (PVY)

PVY, the type member of the *Potyvirus* genus (*Potyviridae* family), is the most important of the three potyviruses infecting potatoes. It was first described by Smith (1931).

5.3.3.1 Symptoms and Damage in Potatoes

In many cultivars, the primary symptoms of PVY^{O} and PVY^{C} are leaf mottling and necrosis, and secondary symptoms are leaf mottling and crinkling, and plant dwarfing (Figs. 5.3.2 and 5.3.3). PVY^{N} causes mosaic primary symptoms, and, usually, leaf mottling and mosaic secondary symptoms. Infected plants are more severely affected if PVY occurs in complex with other viruses, especially PVX. Yield losses of 10-80% have been reported (de Bokx and Huttinga, 1981).

PVY^{NTN} causes a damaging disease, designated potato tuber necrotic ringspot disease, in which tubers develop superficial rings that initially are raised but later are sunken and necrotic (Fig. 5.3.4), and often become more conspicuous during storage; such symptoms can develop in as much as 90% of crops of susceptible cultivars (Le Romancer and Nedellec,1997).

The cylindrical cytoplasmic inclusions of PVY isolates are of Edwardson's Division IV.

5.3.3.2 Natural Hosts

PVY has a wide natural host range; it occurs naturally in at least 41 species of four families. Various strains occur in potato worldwide and, in some tropical and sub-tropical countries, in pepper, tobacco, tomato and several other solanaceous species.

5.3.3.3 Geographical Distribution

PVY^{O} occurs worldwide. PVY^{N} occurs in Europe, Africa, New Zealand, S.America (Argentina, Chile, Colombia, and Peru) and, in isolated outbreaks, in N.America. PVY^{C} occurs in Australia, Ecuador, Europe N.America, S.Africa and New Zealand. PVY^{NTN} is reported from Europe (Austria, Belgium, Czech Republic, Denmark, France, Great Britain, Hungary, Italy, The Netherlands, the former Yugoslavia), Israel and Lebanon (Jeffries, 1998; Boonham *et al.* 1999).

5.3.3.4 Transmission

More than 40 aphid species are known to transmit PVY (Kennedy *et al.* 1962; Sigvald, 1984), but they vary in efficiency. The more important vectors are probably those that also colonise potatoes such as *Macrosiphon euphorbiae, Aphis fabae,* Myzus *persicae and Rhopalosiphoninus latysiphon*; however, others such as *M. (Nectarosiphon) certus, M. (Phorodon) humuli and Rhopalosiphon insertum* are also important natural vectors (Kennedy *et al.* 1962). Aphid transmissibility is dependent on virus-encoded helper component proteinase and a DAG triplet in the coat protein. There is an unconfirmed report that PVY may also occasionally be transmitted mechanically by plant contact (Banttari *et al.* 1993 in Jeffries, 1998).

5.3.3.5 Experimental Hosts

Strains of PVY are mechanically-transmissible to more than 400 plant species, most of which (almost 300) are included in the Solanaceae (Edwardson and Christie, 1997). Of these, the following are important test species:

Nicotiana glutinosa - mild to severe systemic mottling.

N.tabacum - systemic vein clearing (Fig. 5.3.5) followed by mottling by all strains except PVY^{N} which induces veinal necrosis (Fig. 5.3.6).

Solanum tuberosum cv.Duke of York - local lesions induced by PVY^{C} but not by PVY^{N} or PVY^{O}.

Tinantia erecta - severe systemic mottling (but immune to *PAMV, PVS* and *PVM).*

Solanum demissum Y - local lesions.

Datura stramonium and *Solanum demissum* "A" are immune to PVY strains.

5.3.3.6 Strains

Three strains, the common or ordinary strain (PVY^{O}), the stipple streak strain (PVY^{C}) and the tobacco veinal necrosis strain (PVY^{N}) have long been differentiated by their differential serological reactions and symptoms in a range of test plant species including *Nicotiana tabacum, Physalis pubescens (syn.P.floridana)* and potato cultivars such as Duke of York; PVY^{O} and PVY^{C} are also differentiated, respectively, by the strain-specific interaction with the genes Ny and Nc for hypersensitive resistance in potato, and PVY^{N} by the vein necrosis it causes in tobacco. During the last decade, four additional unusual isolates or pathotypes have been described: PVY^{NTN} causes potato tuber necrotic ringspot disease and, since first described in Hungary (Beczner *et al.* 1984), has been reported in most Eurpean potato-growing countries (Le Romancer and Kerlan, 1991; Weidemann and Maiss, 1996; Tomassoli and Lumia, 1998); this is now recognised as a sub-group of PVY^{N} (van den Heuvel *et al.* 1994; Glais *et al.* 1996). The isolate originally described as PVY^{N}-Wi in Poland but known also to occur in Canada and Spain induces vein necrosis typical of PVY^{N} isolates in tobacco but its RNA has a high nucleotide sequence homology with that of PVY^{O} and PVY^{NTN}(Chrzanowska, 1994). PVY^{Z}, reported from the UK, overcame genes (Ny/thr and Nc) which confer resistance to PVY^{O} and PVY^{C} (Jones, 1990). Another pathotype reported from Spain is designated PVY^{ZE} (Kerlan *et al.*,2000). Attempts to further characterise these unusual isolates are now in progress.

5.3.3.7 Particle Morphology and Properties

The filamentous particles are mostly 730 x 11 nm which sediment as a single component with an sedimentation coefficient of c.150 S. They consist of a coat protein of c.33 kDa and ss-RNA of c.10 kb. The coat protein consists of 267 amino acids which corresponds to a mol.wt of 30 kDa. The sequences of about 30 PVY isolates have been reported, and these have homologies of 93-99% (van der Vlugt et al. 1993).

5.3.1
5.3.2
5.3.3
5.3.4
5.3.5
5.3.6
5.3.7
VPg
P1
HC-Pro
P3
6K1
CI
6K2
VPg-NIa
NIb
CP
poly(A)
Viral encoded protease -
P1
HC-Pro
NIa
polyprotein
Viral proteins
5.3.8
VPg - NIa
CI
6K
NIb
CP
VPg
P3
HC-Pro
P1

Fig. 5.3.1 Systemic vein necrosis induced by *Potato virus A* on graft inoculated *Solanum demissum x S.tuberosum* cv. Aquila (=A6). Courtesy Dr. J. Valkonen.

Fig. 5.3.2 Mosaic leaf symptoms on cv. Sieglinde (tuber borne PVY^{O} infection), left; healthy plant , right. Courtesy of Dr. J. Valkonen.

Fig. 5.3.3 Leaf drop symptoms induced by PVY^{O} in cv. Pinto. Courtesy of Dr. J. Valkonen.

Fig. 5.3.4 Necrotic rings on potato tuber caused by PVY^{NTN} . Courtesy of Dr. C. Kerlan, INRA, Le Rheu, France.

Fig. 5.3.5 Vein clearing induced by PVY^{O} in *Nicotiana tabacum* cv. Samsun nn. Courtesy of Dr.J. Valkonen.

Fig. 5.3.6 Vein necrosis symptoms induced by PVY^{N} in *Nicotiana tabacum* cv. Samsun nn. Courtesy of Dr. J. Valkonen.

Fig. 5,3.7 Electon micrograph of *Potato virus Y* (PVY). Bar=0.5μ. Courtesy J. Cohen, ARO.

Fig. 5.3.8 Genomic organization of PVY. Courtesy of Dr. A. Gal-On, ARO.

5.3.3.8 Nucleic Acid Component

The genomic RNAs of all three strains have been fully sequenced and shown to contain 9698-9704 nt. The arrangement of the genomic RNA and gene products are shown in Fig. 5.3.7.

5.3.3.9 Detection Methods

An indication of the strain can be obtained by the reactions induced in a range of test plant species that include *Solanum demissum x S.tuberosum* A6, *S.demissum* Y and *S.chacoense* "TEL". Serological techniques, however, have long been used for rapid detection and identification; although immuno-diffusion, microprecipitin, latex agglutination and immunosorbent electron microscopy were used earlier, various formats of ELISA have been used routinely during the past two decades for bulk sampling (e.g., Gugerli and Gehringer, 1980; Vetten *et al.* 1983; Weidemann, 1988; Barker *et al.* 1993; Singh *et al.* 1993). Dot immuno-binding assay has also great potential for mass screening (Singh *et al.*1993). Serological differentiation of strains has been improved by the use of monoclonal antibodies (Fernandez-Northcote and Gugerli, 1988; Singh *et al.*1993; Ellis *et al.*1996; Cerovska, 1998). The strains can also be detected and differentiated by nucleic acid spot hybridisation using radioactive and non-radioactive probes (LeClerc *et al.*1992; Baulcombe and Fernandez-Northcote, 1988), RT- and IC-PCR (e.g., Barker *et al.*1993; Weidemann and Maiss, 1996). PVY^{NTN} can be distinguished from other PVY strains by differences in the electrophoretic mobility of its RNA transcripts (Rosner and Maslenin, 1999).

5.3.3.10 Control

The virus is best controlled by the production, propagation and distribution of elite virus-free plants, nuclear stocks of which are retained under stringent conditions to prevent infection.

5.4 *Potato X Virus* (PVX; Genus *Potexvirus*)

G.Loebenstein
Department of Virology, Agricultural Research Organization, The Volcani Center, Bet Dagan, 50-250, Israel.

First described by Smith (1931).

5.4.1 Symptoms and Damage in Potatoes

Symptoms of PVX are variable and depend on cultivar, virus strain, environmental conditions and synergism in mixed infections. Most isolates cause only mild leaf mosaic or symptomless infections, especially at higher temperatures. For example, a strain that in cv. Alpha causes interveinal mosaic when plants are grown below 22° C, remained symptomless in those grown above 22° C (Beemster and de Boxk, 1987). The yield reductions in such symptomless plants are generally less than 10%, even though they may contain high virus titers. These plants remain carriers of the virus and are important sources of infection. Other strains induce necrotic streaks, severe mosaic, crinkling and rugosity of leaves and may cause significant yield losses. In potato cultivars with genes Nx and Nb, some strains of PVX induce a hypersensitive severe top necrosis (Cockerham, 1955; Jones, 1985). In mixed infections with PVY, yields are reduced by up to 50%.

PVX induces a unique type of cellular inclusion, described as an aggregate of laminated inclusions or beaded sheets (Kozar and Sheludko, 1969; Shalla and Shepard, 1972). The sheets are c. 3 nm thick, and may or may not be studded on both sides with beads of c. 11-14 nm in diameter (Fig. 5.4.1). The sheets and beads are antigenically unrelated to the viral coat protein.

5.4.2. Natural Host Range

This is limited mainly restricted to solanaceous species including *Solanum tuberosum, Datura stramonium, Nicotiana* spp., *Lycopersicon esculentum, Petunia hybrida, Solanum nigrum* and *Cyphomandra betacea.* PVX also infects *Brassica campestris ssp.rapa* plants causing stunting, mild mosaic and leaf distortion.

5.4.3 Geographical Distribution

Distributed worldwide in potato growing areas.

G. Loebenstein et al. (eds.),
Virus and Virus-like Diseases of Potatoes and Production of Seed-Potatoes, 87–94.

5.4.4 Transmission

PVX is transmitted by mechanical inoculation. It is spread naturally by contact between infected and healthy plants, on contaminated farm equipment and workers or by animals that have been in contact with diseased plants. Healthy "seed" potato tubers may become infected during storage by contact of their sprouts with those of infected tubers. Contact between roots may result in infection in the field, but is probably unimportant.

Although the concentration of PVX in plants is very high, the virus is not transmitted by aphids. Planthoppers and other chewing insects can transmit the virus due to mechanical contact. Transmission by zoospores of the fungus *Synchytrium endobioticum* has also been reported (Nienhaus and Stille, 1965), and infection of *N. debneyi* and *N. benthamiana* by PVX was frequently obtained when they were planted in soil from the Andes infested by *S. endobioticum* (Salazar, 1966). PVX can be transmitted experimentally by *Cuscuta campestris* (Ladeburg *et al.* 1950).

PVX is not transmitted through true seed or pollen.

5.4.5 Strains

Strains of PVX have been divided into three groups on the basis of their thermal inactivation point (Köhler, 1962), into 4 groups according to their serological properties (Matthews, 1949) and into 4 groups based on their ability to overcome resistance conferred by two dominant resistance genes Nx and Nb in *Solanum tuberosum* (Cockerham, 1955). Group 1 strains induce a hypersensitive (HR) reaction in potatoes carrying either the Nx or Nb gene; group 2 strains induce HR only in Nb potatoes, whereas group 3 strains induce HR only in Nx potatoes; group 4 strains overcome both Nx and Nb-mediated resistance (Table 5.4.1). The extreme resistance gene (Rx) (see Chapter 13.1) can be overcome by strain PVXHB found in 7% of Bolivian clones of *S. tuberosum* ssp. andigena (Jones, 1985; Moreira *et al.* 1980).

Table 5.4.1. Genes for hypersensitivity in potatoes and strain groups of PVX (Cockerham, 1955)

Cultivar	Genotype		Strain Group			
			1	2	3	4
Arran Banner	nx	nb	s	s	s	s
Epicure	Nx	nb	R	s	s	s
Arran Victory	nx	Nb	R	R	s	s
Craigs Defiance	Nx	Nb	R	R	R	s

R- hypersensitive; s- susceptible

In the UK, PVX strain group 3 predominates, but in Australia strain group 1 is widespread (Wilson and Jones, 1995).

For PVX strains able to overcome resistance conferred by both the Nx and Rx1 genes, the viral coat protein is the principal determinant (Kavanagh *et al.* 1992). The ability to overcome Rx1 resistance by strain PVXHB is conferred by a lysine residue at position 121 of the coat protein, while PVX strains avirulent to Rx1 potatoes have a threonine residue at this position (Goulden *et al.* 1993). Evaluating different potato genotypes expressing extreme resistance to PVX (Rx) confirmed that elicitation of resistance was determined by the amino acid 121 of the viral coat protein (Querci *et al.* 1995). However, strain PVXHB does not induce local lesions on *Gomphrena globosa.* By analysing the nucleotide sequence of the coat protein gene of different PXV strains, it was shown that all that were avirulent on potato cultivars carrying the Nx resistance gene had type X protein, whereas strains capable of overcoming the Nx resistance had type B coat protein (Santa Cruz and Baulcombe, 1995). The strains with type X proteins showed 97-100% identity at the amino acid level and were all 237 amino acids in length. Type B proteins were less closely related and could be divided into two subgroups on the basis of protein length, i.e., 248 (Bi) and 236 (Bii) amino acids in length, respectively. Strains with subtype Bi coat protein seem to be common in Europe, whereas subtype Bii coat protein was observed in strains from North and South America.

PVX DX strains induce a hypersensitive reaction in potato cultivars carrying the Nx resistance gene. Strain PVX DX4, a natural mutant, overcomes Nx-mediated resistance (Jones, 1982). Sequence comparison with PVX DX showed that a substitution of glutamine 78 for proline in the coat protein of DX4 was sufficient to avoid the hypersensitive reaction in Nx potato cultivars (Goulden *et al.* 1993). However, in the coat protein of PVX MS, an Argentine isolate, which like PVXHB is able to break the Rx gene-mediated resistance and the hypersensitive reaction conferred by genes Nx and Nb, no substitute to proline 78 was determined. Strain MS, which is similar to strain PVXHB, has a glycine 78 (Feigelstock *et al.* 1995). It may be, therefore, that in addition to coat sequences other parts of the viral genome are involved in pathogenicity. A similar suggestion was made by Watts *et al.* (1997) when Canadian isolates of PVX were analysed for coat protein differences.

Recently, the Rx DNA from the potato cv. Cara has been cloned (Kanyuka *et al.* 1999). Transferring the Rx DNA as a transgene to *Nicotiana benthamiana* induced resistance against PVX (Bendahmane *et al.* 1999).

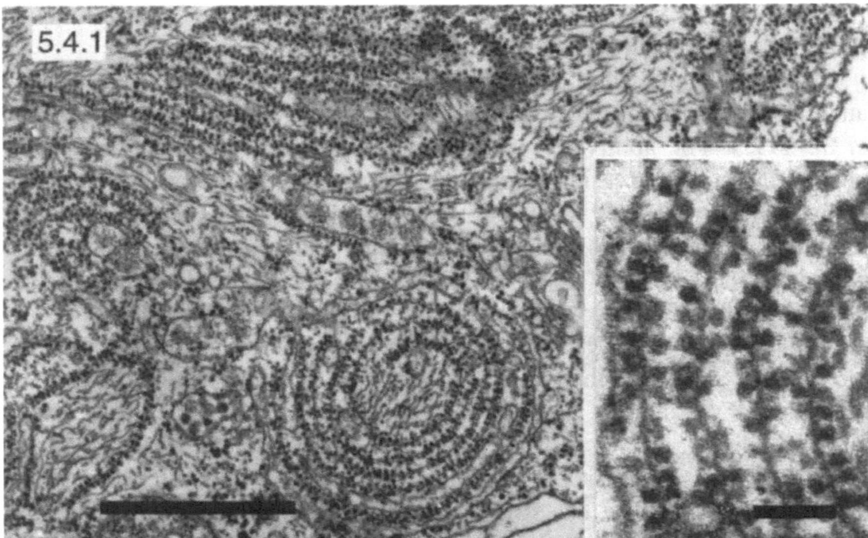
5.4.1

5.4.2

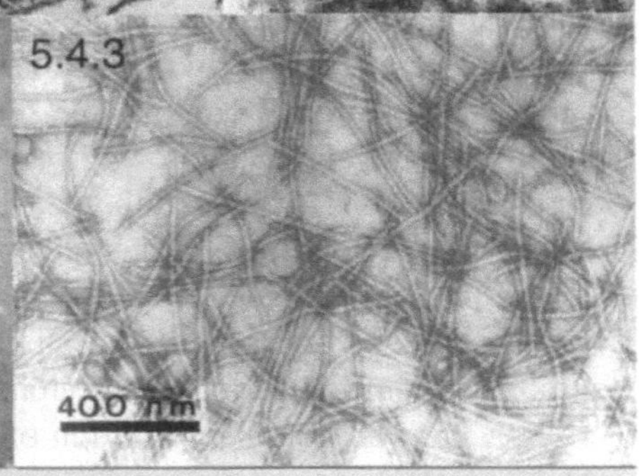
5.4.3
400 nm

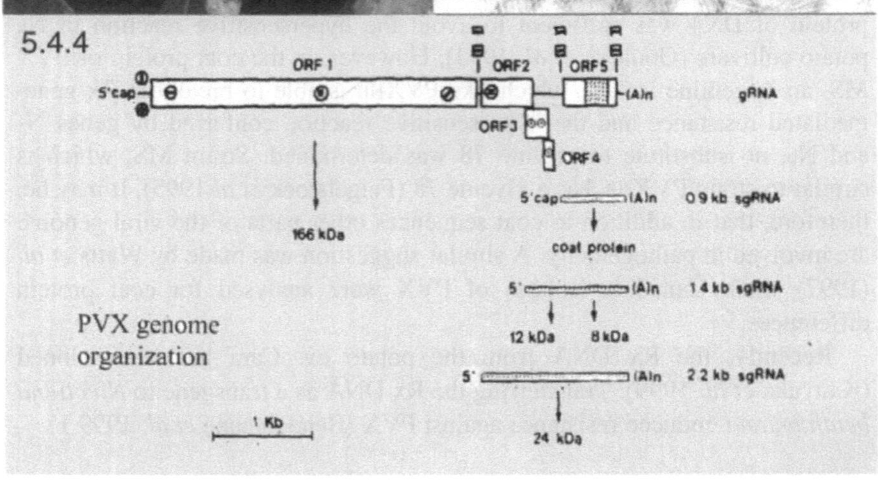
5.4.4
ORF 1
ORF2
ORF5
ORF3
ORF4
5'cap
(A)n
gRNA
166 kDa
5'cap
(A)n
0.9 kb sgRNA
coat protein
5'
(A)n
1.4 kb sgRNA
12 kDa
8 kDa
5'
(A)n
2.2 kb sgRNA
24 kDa
1 Kb
PVX genome
organization

Fig.5.4.1 Laminated inclusion components induced by PVX in *Chenopodium quinoa.* Insert shows details of cross-sectioned beaded sheets. Bars, 500 nm and (insert) 100 nm. (Courtesy of Dr. D.-E. Lesemann , BBA, Braunschweig, Germany).

Fig. 5.4.2 Local lesions induced by potato X virus on *Gomphrena globosa.*

Fig. 5.4.3 Purified virus particles stained with uranyl acetate. Bar represents 100 nm.

Fig. 5.4.4 Genome organisation and expression strategy of potato virus X.

5.4.6 Experimental Hosts

5.4.6.1 Diagnostic species

Datura stramonium –systemic chlorotic rings, mosaic and mottling; symptoms differ according to strain. As this host is immune to PVY it may be used to separate PVX from the latter when the two viruses occur in complex. with PVY.
Gomphrena globosa – all strains of PVX except PVX_{HB} induce local lesions. Four to five days after inoculation small grey spots appear, which later increase in size and develop a reddish halo (Fig. 5.4.2). The middle leaves of *G. globosa* plants with 8-10 leaves on the main shoot are the most susceptible.
Nicotiana tabacum cvs Samsun, Samsun NN, White Burley, etc. – systemic veinclearing, mottling and ringspots.
Other hosts include *Lycopersicon, esculentum* (systemic), *N. glutinosa* (systemic) and *Chenopodium amaranticolor* (local lesions).

5.4.6.2 Propagation species

Nicotiana tabacum, N. glutinosa (to avoid contamination with TMV), other *Nicotiana* species.

5.3.7 Stability in Sap

Thermal inactivation point (10 min): 68-76° C; dilution end point $\log_{10}$ minus 5-6; infectivity is retained at 20° C for several weeks, and in the presence of glycerol for more than one year.

5.4.8 Purification

According to Francki and McLean (1968), but simplified by Sadvakasova *et al.* (1996). (See Chapter 11).

5.4.9 Particle Morphology and Properties

Virions not enveloped. Nucleopcapsids filamentous, usually flexuous, 470-580 nm long, with a modal length of 515 nm and 13 nm in diameter (Brandes, 1964) (Fig.5.3.3). Helical symmetry. Axial canal 3.4 nm in diameter. Basic helix obvious. Pitch of helix 3.4 nm (Varma *et al.* 1968).
8 7/8 subunits per turn (Tollin *et al.* 1980). Purified preparations contain one sedimenting component.

Sedimentation coefficient ($s_{20,w}$) : 115-130 S.
Isoelectric point: pH 4.4.
A_{260} / A_{280}:1.20. Specific absorbance at 260: 2.97 (Paul, 1959).
Density in CsCl: 1.31 g /cm^3.
Mol. wt: 3.5 x 10^6, particles stable.
The virus contains 6% RNA (Knight, 1963) and 94% protein.

Mol. wt. of the coat protein: c. 25 kDa, as deduced from the amino acid sequence obtained from the nt sequence of the coat protein gene. In SDS PAGE intact protein appears to have an apparent mol. wt of c. 30 kDa, but for preparations which have been kept at room temperature faster moving bands are observed, with calculated mol. wt of c. 22-24 kDa (Koenig *et al.* 1970). The intact PVX coat protein shows an anomalous behaviour in SDS PAGE, depending on the gel concentration (Koenig, 1972). The coat protein is an O-glycoprotein, based on periodate oxidation, digestion with glycosidases and trifluoromethanesulfonic acid treatment of several PVX strains (Tozzini *et al.* 1994). The amino acid composition of the coat protein was determined by Shaw *et al.* (1962) and Miki and Knight (1968).

Infectivity is retained when PVX is deproteinized with proteases, phenol or detergent.

5.4.10 Nucleic Acid Component

One molecule of linear positive-sense ssRNA (mol. wt 2.1 x 10^6; 6,435 nt). Base composition: 22% G, 32% A, 24% C and 22% U. The 5'terminus has a methylated nucleotide cap m7 G5 pppGpA. Poly (A) at the 3'terminus. Nucleotide sequence deposited at EMBL/GenBank under the following accession numbers M63141; M28049. Sequence database accession codes include: D00344 Em(40)_vi:PVXX3 Gb(84)_vi:PVXX3 PVX genomic RNA 6,435bp. M31541 Em(40)_vi:PVXKPA Potato virus X 166K RNA, complete cds; 24K RNA, complete cds; 12K RNA, complete cds; 8K RNA. M38655 Em(40)_vi:PVXCT23 Gb(84)_vi:PVXCT23 Potato virus X coat protein gene, complete cds, 861bp. X72214 Em(40)_vi:PVXHBRNA Gb(84)_vi:PVXHBRNA Potato virus X strain HB RNA. 6,432bp. For additional accession codes see Brunt *et al.* (1996) or http://life.anu.edu.au/viruses/CTVdB/56010001.htm.

The genomic RNA of PVX contains 5 ORFs (Fig. 5.4.4). ORF 1 codes for a 166 K polypeptide that presumably possesses an RNA-dependent RNA polymerase activity (Mentaberry and Orman, 1995; Morozov *et al.* 1990). Products of ORFs 2, 3 and 4 (the "triple block") are involved in cell-to-cell transport of PVX (Lough *et al.* 1998), together with a functional coat protein (Santa Cruz *et al.* 1998). The polypeptide encoded by ORF 2 contains a second NTPase/helicase motif similar to that found in the product of ORF 1. Products of ORF 3 and 4- polypeptides of 12 and 8 kDa, respectively, contain sequences of uncharged amino acids, resembling membrane-spanning domains (Morozov *et al.* 1987). ORF 5 encodes the coat protein. The gRNA serves as a messenger for the 5'proximal ORF, while several 3'co-terminal subgenomic (sg) RNAs direct the synthesis of the other proteins (Fig.5.4.4). Thus, the 0.9 kb sgRNA serves as a messenger for coat protein. It was suggested that the 1.4 kb sgRNA acts as a bicistronic mRNA for the products of ORFs 3 and 4. The 25 K protein is expressed as a single translation product of the 2.1 kb sgRNA (Morozov *et al.* 1991). It is generally assumed that the transcription of the sgRNAs is controlled by promoter regions located in the minus-strand RNA (Mentaberry and Orman, 1995).

5.4.11 Detection Methods

PVX can be detected easily in potato leaves and dormant and sprouted tubers by serology, using commercial ELISA kits (see Chapters 10 and 12), or by immunoelectron microscopy (Milne and Luisoni, 1977; Koenig, 1988). Diffusion tests in agar can be used both for detection and determining strain relationships, but PVX virions have to be dissociated into soluble antigens, which migrate more readily into the agar medium. PVX can be disassociated by incubating the plant extract with pyrrolidine to a final concentration of 2.5% (Shepard, 1970).

5.4.12 Therapy of Individual Plants

PVX–free plants can be obtained by culturing 0.1 mm meristems, with or without a leaf primordium (Kassanis and Varma, 1967), although only a small percentage of plantlets was regenerated from the meristems. Combining heat-treatment before the excision of the meristem tip enhanced the proportion of virus-free plants obtained (Stace-Smith and Mellor, 1968). Incorporation of Virazole (ribavirin) into the culture medium yielded a higher percentage of virus-free progeny then comparable cultures without Virazole (Cassells and Long, 1982; Faccioli and Marani, 1998; Quak, 1987; see also Chapter 14). It is essential that following therapy, progeny should be thoroughly tested for viruses, preferably over two growing periods, and evaluated for trueness-to-type.

Acknowledgement

Thanks are due to Dr. D-E.Lesemann, Biologische Bundesanstalt für Land- und Forstwirtschaft (BBA), Braunschweig, Germany, for providing figure 5.4.1 and to Prof. Renate Koenig (BBA) for critically reading this chapter and for valuable suggestions.

5.5 *Potato Mop-Top Virus* (PMTV; Genus *Pomovirus*)

G. Loebenstein
Department of Virology, Agricultural Research Organization, The Volcani Center, Bet Dagan, 50-250, Israel.

First described by Calvert and Harrison (1966). PMTV, previously included in the genus *Furovirus,* is now recognised as the type member of the genus *Pomovirus* (Torrance and Mayo, 1997; Pringle, 1998).

5.5.1 Symptoms and Damage in Potatoes

Symptoms of PMTV vary according to potato cultivar and season. Leaves develop aucuba-yellow patches (Fig 5.5.1.) and also mop-head or "mop-top" – as the bunched foliage on shortened stems resembles a household mop. Shoot symptoms develop best in cool conditions (5-15° C). In the year of infection from soil, tubers of some cultivars, such as Arran Pilot, develop internal brown arcs ("spraing"), seen as brown rings on the tuber surface (Fig. 5.5.2). Secondary symptoms may consist of cracks leading to malformations. In cv. Arran Pilot brown blotches develop on the tuber surface, and necrotic spots, rings and arcs occur in the tuber flesh. Secondary symptoms in tubers are often found in cultivars that develop mop-top symptoms on their stems (e.g., cvs Alpha and Patrones). Symptoms are often more conspicuous when tubers are kept at 9° C than 4° C, and inoculation during flowering resulted in high rates of "spraing" (Sandgren, 1995). It should, however, be mentioned that "spraing" symptoms may also be caused by *Tobacco rattle virus* (TRV).

"Spraing" in the tubers may cause severe economic damage, mainly in cultivars such as Saturna, used for crisp production (Sandgren, 1995). In Finland, damage was greater in fields in which potatoes were grown for the starch industry than in those for production of table potatoes (Kurppa, 1989). In Sweden severe symptoms occurred in cvs Evergood, Lady Rosette and Saturna, while cvs Bintje, King Edward and Redbad exhibited few or no symptoms (Ryden *et al.* 1989). Latent infection seems to be common (Sandgren, 1995). In cv. Saturna, a decrease in "spraing" symptoms is sometimes observed during storage (Nielsen and Molgaard. 1997). Apparently, the tubers are able to re-absorb the brown colour of "spraing". In other observations a distinct increase in "spraing" in cv. Saturna was noted (Kurppa, 1989; Ryden *et al.* 1989). Soil type seems to influence disease incidence, and "spraing" symptoms were more than three times

G. Loebenstein et al. (eds.),
Virus and Virus-like Diseases of Potatoes and Production of Seed-Potatoes, 95–100.

higher on coarse-grained sandy soils than on fine-grained sandy soil or loam (Nielsen and Molgaard, 1997). This, however, could be due to a long tradition of potato growing on the light soils, with a concomitant build-up of infection sources. Clusters of microtubule-like elements occur in the cytoplasm of PMTV–infected leaves (Fraser, 1976).

No effects of mop-top on total yields or dry matter content were observed in Denmark (Nielsen and Molgaard, 1997), although in Scotland yields of "spraing"-infected plants were significantly lower than those of healthy plants (McNaughton *et al.* 1974).

When partially-infected stocks were grown in the absence of re-infection, the virus was self-eliminating (Cooper *et al.* 1976), apparently because of slow and erratic movement of the virus in the plant.

5.5.2 Natural Host Range

Solanum tuberosum and *Chenopodium album* (Foxe, 1980). PMTV experimentally infects species in the Solanaceae (including *Datura stramonium, Lycopersicon chilense, Solanum muricatum* and *Physalis peruviana*) and in the Chenopodiaceae (*Chenopodium quinoa, Spinacia oleracea* and *Beta vulgaris*). These species are probably natural hosts in the Andean region of South America where they occur in weeds in potato and adjacent fields (Jones, 1988).

5.5.3 Geographical Distribution

PMTV is common in Scandinavian countries, UK (Scotland and Northern Ireland) and in the Andean region of South America (Jones, 1975; Salazar and Jones, 1975), and is present in Japan and Pakistan.

5.5.4 Transmission

PMTV is transmitted by the primary zoospores of *Spongospora subterranea*, the potato powdery scab fungus (Jones and Harrison, 1969). The fungus is able to acquire the virus from the roots of infected plants and transmit it to healthy plants (Arif *et al.* 1995), and PMTV particles occur within resting spores of the fungus (Merz, 1995). The long-lived resting spores (cystosori) of *S.subterranea* are able to survive in the field for more than a decade (Jones and Harrison, 1972).

PMTV isolates differ in their ability to be transmitted by *S. subterranea*. A monofungal culture was unable to acquire and transmit the T-isolate which had been maintained by manual transmission, whereas it readily transmitted a field isolate (PMTV-S). Sequence analysis of the readthrough

(RT) protein-coding region of PMTV-S showed the presence of an additional 543 nt in the 3'half of the coding region relative to that of PMTV-T. These additional nucleotides inserted 181 amino acids into the RT protein (Reavy *et al.* 1998; see also Chapter 8.4). There seem to be structural similarities between the capsid readthrough product of PMTV and the furovirus and bymovirus proteins implicated in transmission by *Polymyxa* spp. (Dessens and Meyer, 1996).

PTMV can be transmitted by mechanical inoculation and by grafting, but is not seed transmitted.

5.5.5 Strains

Isolates differing in virulence (Harrison and Jones, 1970), and strains differing in their coat protein (CP) sequence, affecting transmission by *S. subterranea* (Reavy *et al.* 1998) have been reported. In general, however, CP genes of PMTV isolates are highly conserved (Reavy *et al.* 1997; Mayo *et al.* 1996). Thus, CP sequences of Scandinavian, Scottish and Peruvian isolates were very similar; Scandinavian and Scottish isolates differed by a maximum of three amino acids and less than 2% change in CP nucleotide sequence (Reavy *et al.* 1997).

5.5.6 Experimental Hosts

5.5.6.1 Diagnostic species

Chenopodium amaranticolor – concentric necrotic etched-ring lesions, appearing c. 7-10 days after inoculation. A single lesion may spread to cover half a leaf. Not systemic.

Nicotiana debneyi – necrotic spots or necrotic or chlorotic ringspots on inoculated leaves. First systemically infected leaves show chlorotic or necrotic "thistle leaf" line pattern.

N. tabacum cvs Xanthi and Samsun NN – necrotic or chlorotic ringspots on inoculated leaves, when plants are kept below 20° C. Usually no symptoms at higher temperatures. Systemic "thistle leaf" patterns during winter months (Fig.5.5.3).

PMTV from potatoes can be detected more readily if inoculated first to tobacco, and then 2 weeks later to *C. amaranticolor.* To test the infectivity of soil, it is dried at 20° C for 2 weeks to kill nematodes, then moistened and baited with *N. debneyi* seedlings. *N. debneyi* reacts with necrotic symptoms 5-6 weeks later.

5.5.6.2 Propagation Species

N. tabacum cv. Samsun NN (inoculated leaves), *N. benthamiana, N. debneyi* and *N. clevelandii.*

5.5.7 Stability in Sap

Thermal inactivation point (10 min): 75-80° C. Most of the infectivity in tobacco sap kept at 20° C is lost in one day, but a little is retained for 10 weeks. Dilution end point log 10 minus 2 - minus 4.

5.5.8 Purification

According to Kassanis *et al.* (1972), Koenig *et al.* (1984), Kurppa (1990), and Torrance *et al.* (1993); see also Chapter 11.

5.5.9 Particle Morphology and Properties

Fragile rod-shaped particles with two predominant lengths of 100-150 and 250-300 nm (Fig. 5.5.4). They are helically constructed with a pitch of 2.4-2.5 nm, 18-20 nm wide, a hollow core and not enveloped (Harrison and Jones, 1970; Kassanis *et al.* 1972; Roberts and Harrison, 1979).

Possibly three sedimenting components in purified virus preparations. Sedimentation coefficient ($s_{20,w}$): 236 (largest most infective), 171 and 126 S.

Mol. wt. of coat protein: 18.5-20 kDa.

5.5.10 Nucleic Acid Component

The genome consists of linear ssRNAs. Particles of isolate PMTV/T contain three RNA species: 6.4, 3.0 and 2.5 kb, none of which is derived from another (Kallender *et al.* 1990). Nucleotide sequence accession codes: D16193 Em (44)n: Pxm67kp Gb(90)n: Pxm67kp PMTV RNA 3 for coat protein and 67K readthrough protein. D30753 Em(43)_vi:Pxmtodd Gb(89)_vi:Pxmtodd PMTV triple-gene for 51K protein, 13K protein, 21K protein and 8K protein. For additional accession codes see http://life.anu.edu.au/viruses/CTVdB/27010004.htm .

RNA 2 (2962 nt) has 4 ORF at the 5' end, which encodes proteins of 51 kDa, 13 kDa, 21 kDa and 8 kDa. The first three of these have sequence similarities to the triple-block proteins of other viruses which are probably involved in cell-to-cell movement (Scott *et al.* 1994). RNA 3 (2315 nt) contains one ORF for the coat protein (20 kDa), followed by an in-phase

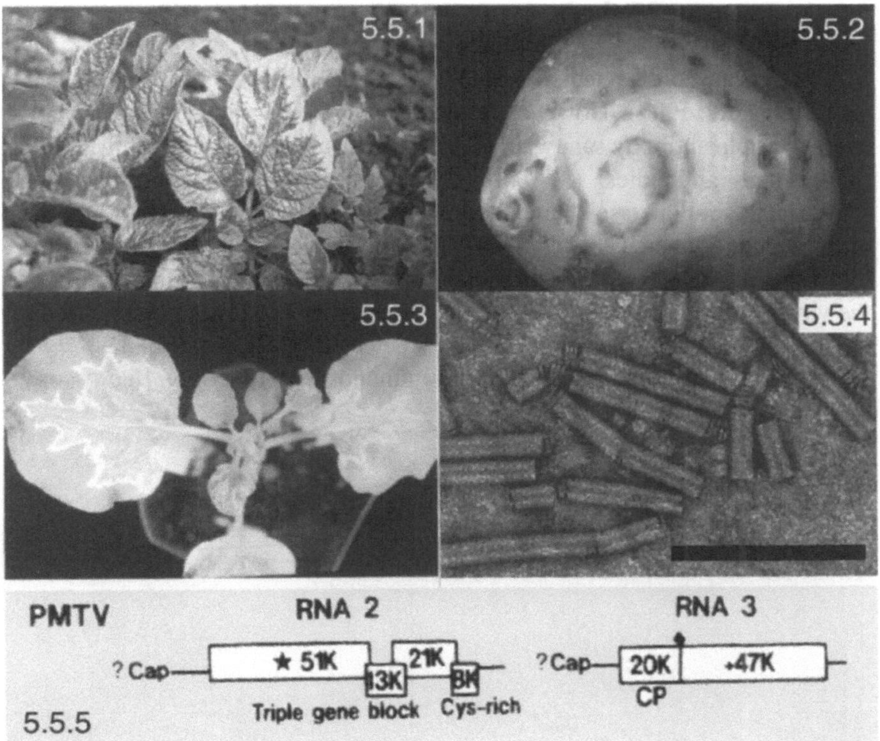

Fig. 5.5.1 Aucuba yellow patches on potato leaves induced by *Potato mop-top virus*. Courtesy Dr. L. Salazar (CIP).

Fig. 5.5.2. Symptoms of "spraing" in a potato tuber. Courtesy Dr. L. Salazar (CIP).

Fig. 5.5.3. "Thistle-leaf" line patterns in leaf of tobacco cv. Xanthi infected with *Potato mop-top virus*. Courtesy Dr. l. Salazar (CIP).

Fig. 5.5.4. PMTV particles, bar - 150 nm.

Fig 5.5.5. Genomic organisation of PMTV RNA 2 and RNA 3 (Adapted from Kashiwazaki *et al.* 1995).

coding region for an additional 47K protein (Kashiwazaki *et al.* 1995). (Fig. 5.5.5).

RNA 1 of broad bean necrosis virus, a furovirus closely related to PMTV, contains methyltransferase, helicase and RNA-dependent RNA polymerase motifs (Lu *et al.* 1998).

5.5.11 Detection Methods

PMTV can be detected in potato leaves and tubers by ELISA (Nielsen and Molgaard, 1997). However, effective polyclonal antisera are in short supply. Triple antibody sandwich (TAS) ELISA using a panel of anti-PMTV monoclonal antibodies (MAbs) (Torrance *et al.* 1993) and reverse transcriptase-polymerase chain reaction (RT-PCR) improved detection of the virus in potato tubers (Arif *et al.* 1994). More infected tubers were detected by ELISA than by RT-PCR, probably because of the uneven distribution of the virus. Storing of tubers at 20° C for 4 weeks prior to testing increased the number of samples in which PMTV was detected by ELISA (Sokmen *et al.* 1998). PMTV can also be detected by immunosorbent electron microscopy (Roberts and Harrison, 1979).

5.6 *Potato Virus M* (PVM; Genus *Carlavirus*)

Alan A.Brunt
Horticulture Research International. Wellesbourne, Warwicks CV35 9EF, U.K.

Symptoms of the virus first designated *Potato virus M* (PVM; Bagnall *et al.* 1956) were observed in potatoes in the U.S.A. at least 33 years previously (Schultz and Folsom, 1923). Before its designation as PVM, it was variously known as potato leaf rolling mosaic virus (Schultz and Folsom, 1923), potato interveinal mosaic virus (McKay and Dykstra, 1932; Bagnall *et al.* 1956), potato leaf rolling virus (Köhler, 1935), potato paracrinkle virus (Salaman and Le Pelley, 1930; Kassanis, 1960), potato virus E (Dykstra, 1939), potato virus K (Köhler, 1942), and Fortuna virus (Wetter and Brandes, 1956).

5.6.1 Symptoms and Damage in Potato

Most iolates of the virus alone symptomlessly infect many potato cultivars (e.g., cvs Green Mountain, Irish Cobbler, Sebago and Up-to-Date), but some may induce very mild leaf chlorosis and leaf distortion in other cultivars (e.g., cvs Climax, Fortuna, Katahdin, King Edward and White Rose) or, in a few cultivars (e.g., Arran Victory), severe leaf mottling, chlorosis, crinkling and rolling ("paracrinkle") (Fig. 5.6.1) and stunting. The severity of symptoms and importance of the virus are dependent on the virulence of the virus strain, the tolerance of the cultivar, the incidence of infection and, possibly, environmental factors (e.g., Beemster and Rozendaal, 1972). However, even isolates that induce only mild symptoms can reduce tuber yield by 10-18% (e.g., 1961; Bawden and Kassanis 1965; Gladysink, 1977; Beemster and de Bokx, 1987). Thus, as the virus is geographically widespread and prevalent in many cultivars, on a global scale even mild strains are very important. Infection of intolerant cultivars by more severe isolates, which are especially common in some eastern European countries, can result in tuber yield losses of 40-75% (Chrzanowska, 1977). PVM frequently occurs in complex with other viruses (especially PVS), and symptoms are usually more severe. Although it frequently occurs together with *Potato virus S* in potato plants, the two viruses can be separated from such complexes by inoculation of differential host species (see Experimental Hosts).

G. Loebenstein et al. (eds.),
Virus and Virus-like Diseases of Potatoes and Production of Seed-Potatoes, 101–107.

Infected cells often contain irregularly-shaped inclusions consisting of endoplasmic reticulum, ribosomes, virus particles and vesicles (Edwardson and Christie, 1997). PVM particles also occur in infected plants either scattered throughout cytoplasm of immature parenchyma cells and sieve tubes, or as aggregates forming paracrystals or banded bodies (Tu and Hiruki, 1970).

5.6.2 Natural and Experimental Host Ranges

Potato is the major natural host of PVM, although a strain of the virus occurs in tomato (*Lycopersicon esculentum*) in Italy, usually in complex with other viruses (Grieco *et al.* 1997); there is an unconfirmed report that *Capsicum annuum* and *C.frutescens* are also natural hosts in India (Mishra *et al.* 1979). The virus has been detected also occasionally in weed species including *Datura metel* in Egypt (Habib, 1980), *Fumaria officinalis, Galium aparine, Solanum dulcamara* in Europe (Kaczmarek, 1985) and *Solanum fernandezianum* in S. America (Valkonen *et al.* 1992b).

5.6.3 Geographical Distribution

Because the virus alone induces inconspicuous or symptomless infection in many potato cultivars, it has in the past probably been inadvertently disseminated in infected seed stocks; it thus probably now occurs in many countries (Wetter, 1972), but is especially common in those of eastern Europe and the former-USSR. However, it does not occur in the Andean Region of S.America (Salazar, 1996), although it is present in Argentina (Butzonitch *et al.* 1994).

5.6.4 Transmission

The virus is transmitted readily in the non-persistent manner by aphid species including *Myzus persicae* and *Aphis nasturtii* and, less efficiently, by *Aulacorthum solani*, *Macrosiphum euphorbiae* and those of the *Aphis gossypii-frangulae* complex (Wetter and Volk, 1960; Kassanis, 1961; Bode and Weidemann, 1971; MacKinnon, 1974). Some isolates, however, are transmitted very inefficiently (Bode and Weidemann, 1971) and others are not aphid-transmissible (Kassanis, 1961;Horio *et al.* 1969).

The virus is not seed-borne. Although it is experimentally transmissible by mechanical inoculation of sap to potato and test plant species (e.g., Bawden *et al.*1950), it is insufficiently infectious to be transmitted to a significant level by plant contact (e.g., Rasocha and Nohejl, 1993).

The epidemiology of PVM has been reviewed (Gabriel, 1988; Weidemann, 1986; Salazar, 1996).

5.6.5 Strains

Numerous strains, variants or isolates of differing virulence occur in potato crops worldwide; a serologically-distinct strain (PVM-ID) has been described recently (Cavileer *et al.*1998). Strains that differ in virulence are best distinguished by their symptomatology in potato cultivars such as Kennebec and Prinslander.

5.6.6 Experimental Hosts

PVM is experimentally transmissible to potato and to another 122 species in 9 genera of the Solanaceae and to 47 species in the Amaranthaceae, Caryophyllaceae, Chenopodiaceae, Compositae, Cucurbitaceae, Fabaceae, and Rubiaceae (e.g., Bagnall *et al.*1956, 1959; Kowalska and Was, 1976; Slack, 1983; Edwardson and Christie, 1997). Of these, the following are useful indicator plant species:-

Chenopodium amaranticolor, C. quinoa, Phaseolus vulgaris, Vigna unguiculata - chlorotic or necrotic lesions in inoculated leaves, but no systemic infection.

Chenopodium album, Nicotiana glutinosa, N.tabacum, Physalis pubescens (syn. P.floridana) – no infection.

Datura metel - local chlorotic and necrotic spotting and systemic chlorotic mottling and rugosity.

Lycopersicon esculentum, Nicotiana occidentalis, N. rosulata - symptomless systemic infection. Several cultivars of *Lycopersicon esculentum* (e.g., cv. Marmande) are immune to potato virus S (PVS) but can be infected systemically with PVM; such cultivars can be used to separate PVM from PVS with which it often occurs in complex in potatoes (e.g., Bagnall *et al.* 1959).

Nicotiana hesperis, N. rotundifolia - lesions in inoculated leaves followed by chlorosis of systemically infected leaves.

Nicotiana debneyi - brown ring-like local lesions and, rarely, symptomless systemic infection

Solanum rostratum, S. demissum x S. tuberosum cv. Aquila clone 6 - chlorosis of systemically infected leaves.

5.6.7 Particle Morphology and Properties

The virus has particles mostly measuring 650 x 12 nm (Brandes *et al.* 1959) (Fig. 5.6.2). Purified virus preparations have a u.v.absorbance spectrum typical of a nucleoprotein with maximum and minimum absorbance at c.260 and c.244 nm, respectively, an A260/280 ratio of 1.2, an extinction coefficient of 2.88, a sedimentation coefficient of 157 S and a buoyant density in caesium chloride of 1.32 g/cm^3 (Tavantzis, 1984).

The particles contain c. 6% positive-sense single-stranded RNA and c. 94% protein. The genomic RNA is encapsidated in a coat protein composed of c.1,800 identical polypeptides each of 33 kDa. The polypeptide subunits are arranged in a primary helix having a pitch of c.3.4 nm and with c.10 subunits per turn of the helix (Koenig, 1982).

PVM is serologically related to *Carnation latent, Chrysanthemum B, Elderberry latent, Helenium S, Hop latent, Hop mosaic, Lily symptomless, Passiflora latent and Potato S carlaviruses* (Wetter, 1972; Koenig, 1982; Edwardson and Christie, 1997).

5.6.8 Nucleic Acid Component

Each particle contains a single molecule of RNA of 8,534 nucleotides (M_r 2.5 x 10^6), the full sequence of which has been reported (Zavriev *et al.* 1991). The genomic RNA (Genbank Accession number X53062) contains six large open reading frames (ORFs) and non-coding sequences of 75 nucleotides at the 5' terminus, 70 nucleotides followed by a polyadenylated tail at the 3' terminus and 38 and 21 nucleotides between three large blocks of coding sequences (Fig. 5.6.3). The ORFs encode polypeptides (from 5' to 3') of 223, 25, 12, 7, 34 and 11 kDa. The 223 Da protein is probably the viral RNA replicase, those encoded by the triple gene block (25,12 and 7 kDa) may facilitate cell-to-cell movement of virus, and the 34 kDa protein is the capsid protein. The function of the 11 kDa protein has yet to be determined but it possibly facilitates aphid transmissibility or is involved in host gene transcription and/or viral replication.

Infected plants contain a double-stranded form of the genomic RNA (M_r 5.0 x 10^6) (Tavantzis, 1984).

5.6.9 Detection Methods

Symptoms induced in mechanically infected indicator plant species (see Experimental Hosts) can indicate the occurrence of PVM. The presence of paracrystalline, banded-body and irregularly-shaped inclusions which are detectable in suitably stained potato tissue by light microscopy, although not specific for PMV, also indicates infection by a carlavirus (Christie and Edwardson, 1977).

However, as the virus induces no conspicuous symptoms in many potato cultivars, and often occurs in complex with other viruses, it can be detected reliably only by the use of one or, preferably, more of the following assay procedures:-

Serological methods. Precipitation and micro-precipitation tests were earlier used for virus detection (e.g., Shepard, 1972). However, latex agglutination, a procedure significantly more sensitive than the microprecipitation test, is still a very useful and simple method for detecting PVM (Fribourg and Nakashima, 1984a).

ELISA is now used routinely for the detection of PVM in potato leaves and and tubers (e.g., de Bokx *et al.* 1980; Schiessendoppler and Forschum, 1990; Dedic, 1995). The double antibody sandwich type of ELISA can detect as little as 5 ng virus /ml of plant sap; this is equivalent to the detection of virus in infective potato sap diluted to c.1/30,000 (e.g., Faccioli and Colombarini, 1996). ELISA on nitrocellulose membranes (dot-ELISA) is also now often used (e.g., Schiessendoppler and Forschum, 1990; Dedic, 1995) and allows virus to be detected in potato sap diluted to 1/64,000, the equivalent to a concentration of 0.005 ng of virus/ml of sap (Faccioli and Colombarini, 1996). Plain paper (Heide and Lange, 1988) or PVC foil (Mierzwa et al, 1993) can also be used as solid substrates for dot ELISA (Heide and Lange, 1988). Diagnosis by direct tissue blotting is reported to be more sensitive, faster and less expensive than ELISA (Samson *et al.* 1993). Time-resolved fluoroimmunoassay, especially using monoclonal antibodies, is 10-20 times more sensitive than DAS-ELISA (Jarvekulg *et al.* 1989).

The virus can be readily detected by the polymerase chain reaction (e.g., Badge *et al.* 1996).

5.6.10 Therapy of Individual Plants

The virus is best controlled by the production, propagation, distribution and planting of elite virus-tested nuclear stocks. Such stocks are obtainable by meristem tip culture (e.g., Kassanis, 1957); however, because the virus occurs in potato meristem tips (Rubies-Autonell *et al.* 1989), meristem tip culture is more effective if preceded by or combined with thermotherapy (35°C) and/or chemotherapy using Ribavirin and 2,4-dioxo-hexahydro-1,3,5-triazine (Cassells and Long,1982; Faccioli and Colombarini, 1996). Production schemes utilising elite virus-tested stocks, now used in many countries, are fully reviewed by Salazar (1996).

Fig. 5.6.1 Mild mosaic, crinkling and rolling of leaves (paracrinkle) in potato plant infected with *Potato virus M* (PVM). Courtesy of Dr. H.- Weidemann, BBA.

Fig. 5.6.2 Electron mocrograph of *Potato virus M* (PVM) particle. Bar= 100nm.Courtesy Dr. D.-E. Lesemann, BBA.

Fig. 5.6.3 Genomic organization of *Potato virus M* (PVM) (modified from Zavriev *et al.* 1991). Courtesy of Association of Applied Biologists.

Fig. 5.7.1 Potato plant with mild leaf rugosity infected with *Potato virus S*. Courtesy of Dr. H.-L. Weidemann, BBA.

5.6.1

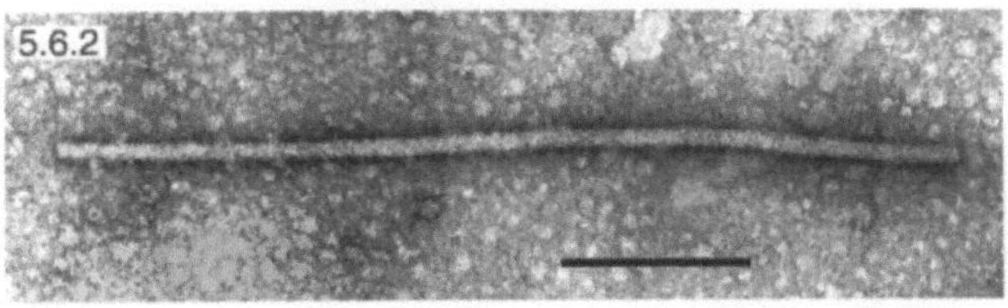
5.6.2

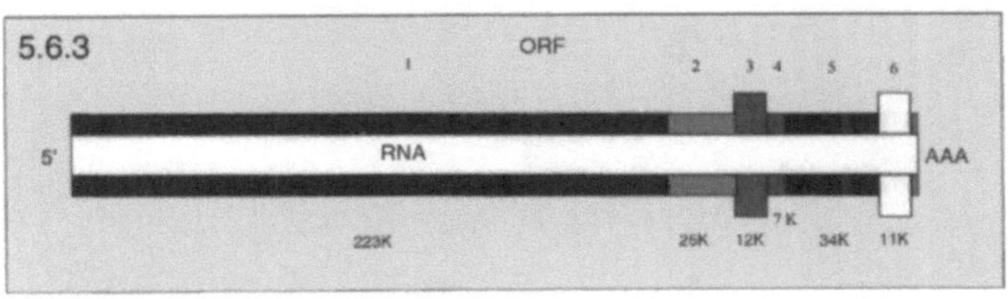
5.6.3
ORF
1
2
3 4
5
6
5'
RNA
AAA
7 K
223K
25K
12K
34K
11K

5.7.1

5.7 *Potato Vvirus S* (PVS; Genus *Carlavirus*)

Alan A. Brunt
Horticulture Research International, Wellesbourne, Warwicks CV35 9EF, U.K.

PVS was first first reported in potato crops in the Netherlands almost 50 years ago (de Bruyn Ouboter, 1952), but it probably also occurred previously in other countries where potatoes were grown. Two major (PVS and, PVS) and several minor virus strains are recognised.

5.7.1 Symptoms and Damage in Potatoes

Many isolates of PVS^{O} alone induce no conspicuous symptoms in many potato cultivars. However, in susceptible cultivars some isolates cause undulation of leaf margins and some rugosity of leaf surfaces (Fig. 5.7.1). The importance of PVS is mainly dependent upon its local incidence, the virulance of the isolate, tolerance of the potato cultivar and environmental conditions. However, most isolates reduce tuber yield by 3-20% (Wetter, 1972; Beukema and van der Zaag, 1979). The virus, however, often occurs in complex with other viruses and may then exacerbate the severity of infection.

Infected plant cells contain aggregated virus particles, seen as paracrystalline inclusions or banded bodies, and irregular inclusions consisting of virus particles, ribosomes and proliferated endoplasmic reticulum (Edwardson and Christie, 1977).

5.7.2 Natural Hosts

Potato is the major host of PVS^{O} and PVS^{A}, although the latter also occurs in pepino (*Solanum muricatum; formerly Cyphamandra muricatum)* in New Zealand and the Netherlands. PVSA has also been reported to occur occasionally in some weed species including *S. brevidens, S.chacoense, S. dulcamara, S. spegazzini* and *Geranium dissectum* (Valkonen *et al.*1992b).

G. Loebenstein et al. (eds.),
Virus and Virus-like Diseases of Potatoes and Production of Seed-Potatoes, 109–112.

5.7.3 Geographical Distribution

PVS^{O} occurs worldwide in potato cultivars (Wetter, 1972). PVS^{A} has been reported in potatoes in some S.American countries (Hinostroza-Orihuela, 1973; Santillan *et al.* 1980), and has been detected also in some breeders potato selections and cultivars in Germany, the Netherlands, the U.K. and the U.S.A. (Slack, 1983; Dolby and Jones, 1987; Weidemann and Koenig, 1990), and in *S.muricatum* in New Zealand (Dolby and Jones, 1988) and the Netherlands (Verhoeven and Roenhorst, 1995).

5.7.4 Transmission

Some isolates are not aphid-transmissible but others are readily transmitted by several aphid species including *Aphis fabae, A. nasturtii, Myzus persicae* and *Rhopalosiphum padi* (Wetter and Volk, 1960; Bode and Weidemann, 1971; MacKinnon, 1974; Weidemann, 1986). PVS is not seed-transmitted but is sufficiently infectious to be transmitted mechanically from infected to healthy field-grown potato plants (Franc and Banttari, 1984; Salazar, 1996).

5.7.5 Experimental Hosts

Although the virus has a very restricted natural host range, it is experimentally transmissible by mechanical inoculation to at least 56 other solanaceous species and to 33 species in 12 other families (Delhey,1981; Valkonen *et al.*1992b; Edwardson and Christie,1997). Useful indicator plant species are:-

Chenopodium album, C.amaranticolor, C.quinoa - chlorotic local lesions, often with a green halo on older leaves. PSV^{A}, but not PVS^{O}, infects *C.quinoa* systemically and induces chlorotic spotting.

Cyamopsis tetragonoloba - small brown necrotic lesions in inoculated cotyledons, but no subsequent systemic infection.

Lycopersicon esculentum - symptomless systemic infection by PVS^{A} but immune to PVS^{O}.

Nicotiana clevelandii - conspicuous chlorosis of systemically infected leaves. This is a very useful host for the propagation and maintenance of virus cultures.

Nicotiana debneyi - symptomless local infection but vein-clearing, mottling and necrosis of systemically infected leaves. This species is immune to PVM, and so allows PVS to be separated from complexes of the two viruses.

Solanum rostratum - numerous small necrotic lesions in inoculated and systemically infected leaves.

5.7.6 Strains

It was long considered that only minor variants of the virus occurred in potato cultivars worldwide (Wetter, 1972). However, unlike PVS isolates described previously, some variants from a few S. American countries were later shown to induce systemic leaf mottling in *Chenopodium amaranticolor* and *C .quinoa*, and to be more readily transmissible by aphids and plant contact (Hinostroza-Orihuela,1973; Santillan *et al.* 1980); these isolates were soon after recognised by Jones *et al.* (1981) as those of a distinct strain which was designated the Andean strain (PVS_A) to distinguish it from the more widely distributed, commonly occurring ordinary strain (PVS_O). *Potato virus P was* previously thought to be a strain of PVS in Brazil (Salazar, 1996), but it is now recognised as a new virus (Daniels *et al.* 1993).

5.7.7 Particle Morphology and Properties

PVS has particles mostly measuring c. 650 x 12 nm which contain c. 5% single-stranded RNA and 95% protein. The protein subunits have a mol. wt of 33-34 kDa.

5.7.8 Nucleic Acid Component

The genomic RNA contains c.7.500 nucleotides, the sequence of which at the 3' terminus have been sequenced (Foster and Mills, 1992); the RNA has six ORFs which encode proteins of 10, 33, 7, 11, 25 and 41 kDa. The arrangement and functions of the ORFs are similar to those of *PV M.*

5.7.9 Detection Methods

Serological precipitation and gel diffusion tests were used successfully for many years to detect infection (e.g., Shepard, 1970, 1972), but these have been superseded by more sensitive methods such as latex agglutination (Fribourg and Nakashima, 1984a). Immunosorbent electron microscopy (e.g., Milne and Luisoni, 1977) and ELISA are now also used routinely for the detection of the virus in tubers (de Bokx *et al.* 1980; Banttari and Franc, 1982) and leaves (e.g., Banttari and Franc, 1982; Singh and Sommerville, 1983; Goodwin and Banttari, 1984; Schiessendoppler and Forschum, 1990), by immunoblotting (dot blotting), and ELISA on nitrocellulose (Banttari and Goodwin, 1985) or paper (Heide and Lange, 1988). Direct tissue blotting is reported to be more sensitive, cheaper and faster than conventional ELISA

(Samson *et al.* 1993). Monoclonal antibodies have been produced for the specific identification of PVS (Cerovska and Filigarova, 1995).
Nucleic hybridisation is a sensitive and reliable procedure for the detection of PVS using radioactive (Weidemann and Koenig, 1990; Foster and Mills, 1990) or non- radioactive probes (Audy *et al.*1991). More recently, PCR has been shown to be c. 1,000-fold more sensitive than ELISA as a diagnostic procedure (e.g., Badge *et al.* 1996).

5.7.10 Therapy

Virus-induced yield losses are best minimised by the production, large-scale propagation and distribution of virus-free stocks. These are obtainable by meristem tip culture (e.g., Cassells and Long, 1982; Faccioli and Colombarini, 1996), especially if preceded by thermotherapy (Stace-Smith and Mellor, 1968; Aruta and Funtealba, 1977) or if antiviral agents such as Virazole or Ribavirin are included in the culture media (Cassells and Long, 1982; Faccioli and Colombarini,1996). Although the use of elite virus-tested stocks in production schemes is well established in many countries, virus-resistant cultivars are also utilised. The use of transgenic resistant plants also has great potential (MacKenzie and Tremaine, 1990), but they have not yet been used commercially.

5.8 *Potato Latent Virus* (PLV; Genus *Carlavirus*)

Alan A.Brunt
Horticulture Research International, Wellesbourne, Warwicks CV35 9EF, U.K.

PLV was first detected in 1992 in potato plants of cv. Red LaSoda imported into Scotland from the USA (Brattey *et al.*1995). Although it was then tentatively designated Red La Soda virus, it is now recognised as a distinct carlavirus (Brattey *et al.*1998; Goth *et al.*1999).

5.8.1 Symptoms and Damage in Potatoes

Potato plants are usually infected symptomlessly.

5.8.2 Natural Host

Potato is the only known natural host of PLV; it has been detected in only a few cultivars, including cvs Denali, High Plains, Kanona, Pembina Chipper, Platte, Pungo, Purple Chief, Red LaSoda, Red Pontiac, and Tejon (Goth *et al.*1999). The virus may occur in other untested cultivars in N.America and elsewhere.

5.8.3 Geographical Distribution

PLV has been detected in the USA (California, Maine, Minnesota), Canada (British Columbia) and in imported germplasm in Scotland (Brattey *et al.*1995, 1998; Goth and Goins, 1998; Goth *et al.*1999).

5.8.4 Transmission

PLV is transmitted by *Myzus persicae*, the only aphid species so far tested.

5.8.5 Experimental Hosts

PLV is transmissible by mechanical inoculation, and induces mild chlorosis of systemically leaves of *Nicotiana biglovii, N.benthamiana,*

G. Loebenstein et al. (eds.),
Virus and Virus-like Diseases of Potatoes and Production of Seed-Potatoes, 113–114.

N.debneyi, N.glutinosa, N.megalosiphon, N.occidentalis-P1, N.rustica, N.tabacum and Physalis pubescens (syn. P. floridana), and local lesions only in *Chenopodium murale* (Brattey *et al.* 1995; Jeffries, 1998; Goth *et al.* 1999).

5.8.6 Strains

None reported.

5.8.7 Particle Morphology and Properties

PLV has slightly flexuous filamentous particles mostly measuring c.690 nm (Goth *et al.* 1999).

5.8.8 Nucleic Acid Component

No information.

5.8.9 Detection Methods

Unlike commonly occurring isolates of PVS and PVM, PLV induces systemic infection in some mechanically-inoculated Solanaceous species including *N.tabacum*, *N.rustica* and *P. pubescens*. Conventional ELISA can be used for detecting PLV but Indirect TAS-ELISA using specific monoclonal antibodies (MAb) can be used both for its rapid identification and its differentiation from the other carlaviruses infecting potatoes; one MAb is specific for PLV and three others differentiate PLV from PVS and PVM (Goth *et al.* 1999).

5.8.10 Control

The virus is best controlled by the detection and removal of infected potatoes (Goth *et al.* 1999).

5.9 Other Possible Carlaviruses

Alan A Brunt
Horticulture Research International, Wellesbourne, Warwicks CV35 9EF, U.K.

Two possible carlaviruses, *Potato P and Potato rough dwarf viruses*, also infect potato. The latter two viruses, like the definite carlaviruses, have slightly flexuous particles mostly c. 12 x 650 nm which are transmitted from infected to health plants by aphids in the non-persistent manner.

5.9.1 *Potato rough dwarf virus* (PRDV)

Although tentatively considered to be a possible carlavirus, the taxonomic status of PRDV is uncertain as it is serologically related to *Potato virus P* but resembles PVS^{A} in inducing systemic infection in *Chenopodium* (Jeffries, 1998).

Symptoms. Some cultivars (e.g., Atlantic, Spunta) are infected symptomlessly, others (e.g., Bintje) develop mild symptoms and infected plants of some (e.g., Kennebec) have thickening and roughening of mature leaves and plants are stunted.

Natural Hosts. Potato is its only known natural host.

Geographical Distribution. The virus is reported only from Argentina and Uruguay (Jeffries, 1998).

Transmission. PRDV is transmitted by *Myzus persicae* and possibly other aphid species.

Experimental Hosts. Its host range is generally similar to that of PVS^{O} but, unlike the latter, it induces symptomless systemic infection in *N.benthamiana, N.biglovii, N.edwardsonii* and *P.hybrida*, occasional local lesions in *Impatiens balsamina* and vein clearing and leaf rugosity in *N.megalosiphon*.

Strains. None reported.

Particle Morphology and Properties. Particles are mostly c.638 nm in length.

Nucleic Acid Component. No information.

Detection Methods. Local chlorotic lesions are induced in clone "A6" and *C. quinoa*, but inoculation of these herbaceous host species is unreliable. The virus is best identified by PCR and by ELISA.

G. Loebenstein et al. (eds.),
Virus and Virus-like Diseases of Potatoes and Production of Seed-Potatoes, 115–116.

5.9.2 *Potato virus P* (PVP)

PVP, which occurs in Brazil, was previously thought to be an isolate of PVS^{A} as the two are serologically related (Salazar, 1996). However, it is now considered to be a distinct possible carlavirus, although the extent of its relationship to PVS and other carlaviruses (especially PRDV) has yet to be determined (Jeffries, 1998).

Symptoms. Although the virus induces no conspicuous symptoms in potato, it reduces tuber yield of cvs Baronesa and Macaca by an average of 20% but up to 84%.

Natural Hosts. Potato is its only known natural host.

Geographical Distribution. The virus is reported only from Brazil (Daniels *et al.*1993).

Transmission. PVP is transmitted by *Myzus persicae* and possibly other aphid species.

Experimental Hosts. PVP infects symptomlessly some solanaceous species (*Datura metel, D.stramonium, Lycopersicon esculentum* and *Nicotiana glutinosa*).

Particle Morphology and Properties. Particles have a length of c. 640 nm.

Nucleic acid component. No information.

Detection. The virus is best detected and identified by ELISA.

5.10 *Potato Aucuba Mosaic Virus (PAMV;* Genus *Potexvirus)*

G.Loebenstein
Department of Virology, Agricultural Research Organization, The Volcani Center, Bet Dagan, 50-250, Israel.

First described by Quanjer (1921)
Synonyms: potato virus F (PVF), potato virus G

5.10.1 Symptoms and Damage in Potatoes

Leaf symptoms of PAMV differ depending on virus strain and potato cultivar. Thus, several virus isolates cause yellow leaf flecking in cvs Triumf and Gloria, whereas other isolates cause deformation and stunting of cv. Bintje. Some cultivars develop necrosis in the tubers or sunken patches on the tuber surface; such symptoms occur mainly when tubers are stored at high (20°-21°) temperatures (Hooker, 1981; Munro, 1960). The symptoms resemble those of *Potato mop-top virus* (PMTV); both cause necrosis on the tuber surface, but those of PAMV are less clearly patterned than the necrotic rings caused by PMTV.

5.10.2 Natural and Experimental Host Ranges

Its natural host is *Solanum tuberosum,* and experimental hosts in the Solanaceae include *Capsicum annuum, C. frutescens, Datura stramonium, Lycopersicon esculentum and Nicotiana spp.* (Kollmer and Larson, 1960).

5.10.3 Geographical Distribution

Probably distributed worldwide, but not common.

G. Loebenstein et al. (eds.),
Virus and Virus-like Diseases of Potatoes and Production of Seed-Potatoes, 117–119.

5.10.4 Transmission

PAMV is transmitted mechanically and by *Myzus persicae* in a non-persistent manner; aphid transmission requires a helper potyvirus such as PVA or PVY (Kassanis and Govier, 1971). Nucleotide sequence analysis of the coat protein gene of PAMV indicated that amino acid residues 14 to 16 from the N-terminus have the DAG sequence (Baulcombe *et al.* 1993). This sequence, also found in the coat of potyviruses, is required for aphid transmissibility.

5.10.5 Diagnostic and Propagation Species

Capsicum annuum, C. frutescens – necrotic local lesions, then epinasty, systemic mosaic and necrosis; young plants are killed.
Nicotiana glutinosa – light green mottle with dark veinbanding.

Propagation species – *Nicotiana tabacum* cv. Xanthi-nc.

5.10.6 Purification

As described by Govier and Kassanis (1972).

5.10.7 Particle Morphology and Properties

Virions not enveloped, filamentous ca. 580 x 11-12 nm. Purified preparations contain one sedimenting component; sedimentation coefficient ($s_{20,w}$) : 130S.

A_{260}/A_{280}: 1.1. Specific absorbance at 260: 2.6 .

Virions contain 5% nucleic acid and 95% protein. The mol. wt of the coat protein: 26 kDa.

Temperature inactivation point (TIP) 60° - 70° C.

5.10.8 Nucleic Acid Component

The genome consists of single-stranded RNA with 7057 nucleotides. Base composition: 25% G, 30% A, 24% C and 21% U. Sequence data accession code: S73580_Gb(89)_un: S73580. The genome contains five major putative open reading frames (ORFs), designated from the 5' terminus as encoding putative proteins with M_r of 187,379 (187 K), 26,326 (26 K), 12,101 (12 K), 8,183 (8 K) and 27,114 (27 K) (Xu *et al.* 1994). The sizes and composition of the proteins encoded by the PAMV ORFs are generally similar to those found in other potexviruses.

5.10..9 Detection Methods

PAMV can be detected by ELISA using polyclonal (Gnutova *et al.* 1989) and monoclonal (Engel'gardt *et al.* 1984) antibodies. PAMV, but not PMTV, infects *Capsicum annuum* systemically.

5.10.10 Therapy of Individual Plants

PAMV can be eliminated from potato plants by heat treatment and meristem tip culture (Dhingra *et al.* 1982.).

5.11 *Tobamo- and Tobamo-like Viruses*

Alan A Brunt
Horticulture Research International, Wellesbourne, Warwicks CV35 9EF, U.K.

Tobacco mosaic virus, although it occurs rarely, has been reported in potato crops in China, India (Phatak & Verma, 1965; Khurana & Singh, 1983), Saudi Arabia (Al-Shahwan *et al.* 1997) and, in native cultivars only, in the Peruvian Andes (L.Salazar in Jeffries, 1998); it was earlier reported to occur naturally in *S.commersonii* subsp. *malmeanum* in Argentina (Hansen, 1960). Symptoms in naturally infected potato plants included interveinal yellowing, mild mottling, chlorotic chevron patterns and severe plant stunting.

Tomato mosaic virus has been reported to occur in potato only in cv. Astilla in Hungary (Juretic *et al.* 1977; Horvath *et al.* 1978) and in glasshouse-grown *S.bulbocastanum* in the Netherlands (J.Verhoeven in Jeffries, 1998).

A possible tobamovirus virus designated *Potato 14R* was isolated from cv. Aleli, a native variety in Peru, with a yellowish leaf mosaic; the plants, however, also contained PVX. The virus has particles mostly measuring 300 x 18 nm which have a coat protein of 19.4 kDa and a sedimention coefficient of 180 S. This virus was not transmitted experimentally to other potato cultivars (Salazar, 1996).

Unidentified tobamo-like viruses infecting potatoes have been reported also from Sumatra (Thung and Hadiwidjaja, 1958) and Chile (Accatino, 1966).

References.

Accatino, J.A. 1966. Virus desconocido en dos variedades de papa (Solanum tuberosum) autóctonas de Chile. *Agric. Tec.* 26:85-86.

Agur, M.1987. Susceptibility of tomato to infection with potato viruses M and S. *Bioloogia* 36: 306-312.

G. Loebenstein et al. (eds.),
Virus and Virus-like Diseases of Potatoes and Production of Seed-Potatoes, 121–134.

Al-Shahwan,I.M., Abdalla,O.A. and Al-Saeh,M.A.,1997. Viruses in the northern potato-producing regions of Saudi Arabia. *Plant Pathology* 46:91-94.

Arif, M., Torrance, L. and Reavy, B. 1994. Improved efficiency of detection of potato mop-top furovirus in potato tubers and in the roots and leaves of soil-bait plants. *Potato Res.* 37: 373-381

Arif, M.,Torrance, L. and Reavy, B. 1995. Acquisition and transmission of potato mop- top furovirus by a culture of *Spongospora subterranea f.sp. subterranea* derived from a single cystosorus. *Ann. appl. Biol.* 126: 493-503.

Audy, P., Parent, J.G. and Asselin, A. 1991. A note on four non-radioactive labeling systems for dot hybridization of potato viruses. *Phytoprotection* 72:81-86.

Aruta, M. and Fuentealba, A.J.1977. Virus free potato (Solanum tuberosum L.) obtained by heat treatment and meristem tip culture. *Phyton* 3:53-59.

Ashoub, A., Rohde, W. and Prüfer, D. 1998. *In planta* transcription of a second subgenomic RNA increases the complexity of the subgroup 2 luteovirus genome. *Nucleic Acids Res.* 26: 420-426.

Badge, J.,Brunt, A.,Carson, R.,Dagless, E.,Karamagioli, M.,Phillips, S., Seal, S.,Turner, R. and Foster, G.D.1996. A carlavirus-specific PCR primer and partial nucleotide sequence provides further evidence for the recognition of cowpea mild mottle virus as a whitefly-transmitted carlavirus. *Eur.. J. Plant Pathology* 102:305-310.

Bagnall, R.H.,Larson, R.H. and Walker, J.C.1956.Potato viruses M, S and X in relation to interveinal mosaic of the Irish Cobbler variety. *Bull. Wisconsin Agric. Res. Stn 198*, 45 pp.

Bagnall, R.H.,Wetter, C. and Larson, R.H.1959. Differential host and serological relationships of potato virus M, potato virus S and carnation latent virus. *Phytopathology* 49:435-442.

Banttari, E.E. and Franc, G.D. 1982. ELISA with single or combined antisera for viruses S and X in potato tubers and plants. Amer.Pot. J. 59: 375-387.

Banttari, E.E. and Goodwin, P.H. 1985. Detection of potato viruses S, X and Y by enzyme-linked immunosorbent assay on nitrocellulose membranes (Dot-ELISA). *Plant Dis.* 69:202-205.

Barker,H, Websyer,K.D. and Reavy,B. 1993. Detection of potato virus Y in potato tubers: a comparison of polymerase chain reaction and enzyme-linked immunosorbent assay. *Potato. Res.*36:13-20.

Bartels,R. 1971. Potato virus A. *CMI/AAB Descr.Pl.Viruses* No.54, 4 pp.

Baulcombe,D.C. and Fernandez-Northcote,E.N. 1988. Detection of strains of potato virus X and of a broad spectrum of potato virus Y isolates by nucleic acid spot hybridization (NASH). *Plant Dis.* 72: 307-309.

Baulcombe, D.C., Lloyd, J., Manoussopoulos, I.N., Roberts, I.M. and Harrison, B.D. 1993. Signal for potyvirus-dependent aphid transmission of potato aucuba mosaic virus and the effect of its transfer to potato virus X. *J. gen. Virol.* 74: 1245- 1253.

Bawden,F.C. and Kassanis, B.1965. *Rep. Rothamsted Exp. Stn 1964*, p.282.

Bawden, F.C.,Kassanis, B. and Nixon,H.L.1950. The mechanical transmission and some properties of potato paracrinkle virus. *J.gen. Microbiol.* 4:210-219.

Beczner, L., Horvath, J., Romhanyi, I. and Forster, H.1984. Studies on the aetiology of tuber necrotic ringspot disease in potato. *Potato Res.* 27:339-352.

Beemster, A.B.R. and de Bokx, J.A.1987. Survey of properties and symptoms. In: de Bokx JA and van der Want JPH (eds) Viruses of Potato and Seed Potato Production, pp.84-113. *Pudoc, Wageningen, The Netherlands..*

Beemster, A.B.R. and Rozandaal,A.1972. Potato viruses; properties and symptoms. In: Bokx, J.A. de (ed.), Viruses of Potatoes and Seed-Potato Production, pp.115-143. Pudoc, *Wageningen, The Netherlands.*

Bendahmane, A., Kanyuka, K. and Baulcombe, D.C. 1999. The Rx gene from potato controls separate virus resistance and cell death responses. Rep. *John Innes Centre & Sainsbury Lab. 1998/99*, p.55.

Beukema, H.P. and van der Zaag, D.E. 1979. Potato Improvement. Some Factors and Facts, 224 pp. International Agricultural Centre, Wageningen, The Netherlands.

Blanco-Urgoiti,B., Sanchez, F., Perez de San Roman,C., Dopazo,J. and Ponz,F.1998.

Bode, O. and Weidemann, H.L.1971.Untersuchungen zur Blattlausübertragbarkeit von Kartoffel-M und S-Virus. *Potato Res.*14:119-129.

Boonham, N., Hims, M., Barker,I. and Spence,N. 1999. Potato virus Y from Petunia can cause symptoms of potato tuber necrotic ringspot disease (PTNRD). Eur.J.Plant Path. 105:617-621.

Braithwaite, B.M. and Blake, C.D. 1961. Tomato yellow-top virus: its distribution, characteristics and its transmission by the aphid *Macrosiphum euphorbiae* (Thom.). *Austr. J. agric. Res.* 12:1100-1107

Brandes, R. 1964. Identifizierung von gestrekten pflanzenpathogenen Viren auf morphologischer Grundlage. *Mitt. Biol. Bundesanstalt Land-forstwirtsch.* Berlin-Dahlem 110: 1-130.

Brandes, J., Wetter, C., Bagnall, R.H. and Larson, R.H.1959. Size and shape of the particles of potato virus S, potato virus M and carnation latent virus. *Phytopathology* 49:443-446.

Brattey, C., Jeffries,C., Goodfellow,S., Rose,G., Burns,R. and George,E. *1995*. Some characteristics of a virus infecting Solanum tuberosum cv.Red LaSoda imported from the USA. *Proc.9th EAPR*, pp.65-69. Bled.

Brattey,C., George,E., Goodfellow,H., Jeffries,C., McDonald,J., Badge,J.& Foster,G. 1998. A newly described carlavirus infecting potato. *Abstr. Int.Congr.Pl.Path.*, 1.11.33, *Edinburgh*, U.K.

Browning,I.A., Burns,R., George,E.L. and Darling,M.1995. Development and evaluation of ELISA assays incorporating monoclonal antibodies for the detection of potato virus A. *EPPO Bull.* 25:259-268.

Brunt, A.A., Crabtree, K., Dallwitz, M.J., Gibbs, A.J. and Watson, L. 1996. *Viruses of Plants.* CAB International, Wallingford, UK.

Butzonitch, I.P.,Colavita, M.L. and Lucarini, O.1994.Potato virus M in Argentina. *Boletim Tecnico Estacion Experimental Agropecuaria, Balcarce* 127:1-11.

Calvert, E.L. and Harrison. B. D. 1966. Potato mop-top, a soil-borne virus. *Plant Pathology* 15: 134-139.

Calvert,E.L., Cooper,P. and McClure,J. 1980. An aphid-transmitted strain of PVY^{C} recorded in potatoes in Northern Ireland. *Rec.Agric.Res.Dept Agric. Northern Ireland* 28:63-74.

Cassells, A.C. and Long, R.D. 1982. The elimination of potato viruses X, Y, S and M in meristem and explant cultures of potato in the presence of Virazole. *Potato Res.* 25: 165-173.

Cavileer,T.D., Clark,R.C., Corsini,D.L. and Berger,P.H. 1998. A new strain of potato carlavirus M. *Plant Dis.*82:98-102.

Cerovska,N. 1998. Production of monoclonal antibodies to potato virus Y^{NTN} strain and their use for strain differentiation. *Plant Path.*4: 505-509.

Cerovska,N. and Filigarova,M.1995. Specific detection of the Andean strain of potato virus S by monoclonal antibodies. *Ann.appl.Biol.*127: 87-93.

Christie, R.G. and Edwardson, J.R.1977. Light and electron microscopy of plant virus inclusions. *Monogr.Florida Agric. Exp. Stn* 9, 155 pp.

Chrzanowska, M.1977. Relationship to yield and its structure and the intensity of disease symptoms caused by virus M on potato variety Uran. *Biuletyn Instytutu Ziemniaka* 19:27-33.

Chrzanowska,M.1994. Differentiation of potato virus Y (PVY) isolates. *Phytopath. Polonica* 8:15-20.

Cockerham, G. 1955. Strains of potato virus X. *Proc. 2nd Conf. Potato Virus Diseases.* Lisse-Wageningen 1954, pp. 89 - 92.

Cooper, J.L., Jones, R.A.C. and Harrison, B. D. 1976. Field and glasshouse experiments on the control of potato mop-top virus. *Ann. appl. Biol.* 83: 215-230

Cupertino, F.P., Silva, A.M.R. da, Dusi, A.N., Avila, A.C.de, Costa, C.L. and Buso, J.A.1995. Transmission of potato leafroll virus by the red tobacco aphid, *Myzus persicae. Fitopat. Bras.* 20: 440-443 (in Portuguese).

Daniels, J., Carvalho, T.C.P. and Bisognin, D.A. 1993. Um novo virus latente de batata encontrada no Rio Grande do Sul. Fitopat. Bras. 18: 287.

De Bokx, J.A. 1967. The callose test for the detection of potato leafroll virus in potato tubers. *Eur. Potato J.* 10: 221-234.

De Bokx, J.A. and Huttinga, H. 1981. Potato virus Y. C.M.I./A.A.B. Descr. Pl. Viruses 242, 6 pp.

De Bokx, J.A.,Piron, P.G.M. and Cother, E.1980. Enzyme-linked immunosorbent assay (ELISA) for the detection of potato viruses S and M in potato tubers.Neth. *J. Plant Path.* 86:285-290.

De Bruyn Ouboter, 1952. A new potato virus. Proceedings of the Conference on Potato Virus Diseases, Wageningen-Lisse 1951, pp.83-84.

Dedic,P.1975. The effect of potato virus A (PVA) on yield in some potato varieties. Ochr. *Rostl.* 11:127-133.

Dedic,P. 1995. The reliability of potato virus M detection in the progeny of primarily infected potato plants. *Ochr. Rostl.* 31: 277-285.

Delhey, R.1981.Incidencia de los virus y M en los cultivos de papa en la Argentina.*Fitopatologia* 16:1-5.

Dhingra, M.H., Sangar, R.B.S. and Singh, R.A. 1982. Elimination of potato aucuba mosaic virus by heat treatment and meristem tip culture. *J. Indian Potato Assoc. Simla* 9: 150-152.

Dolby, C.A. and Jones,R.A.C. 1987. Occurrence of the Andean strain of potato virus S in imported plant material and its effects on potato cultivars. Plant Pathology 36: 381-388.

Dolby, C.A. and Jones, R.A.C. 1988. The relationship between the Andean strain of potato virus S and pepino latent virus. Ann. Appl. Biol. 112:231-234.

Dykstra, T.P.1939.A study of viruses infecting European and American varieties of potato, Solanum tuberosum. *Phytopathology* 29:40-67.

Edwardson, J.R. and Christie, R.G.1997. Viruses infecting peppers and other solanaceous crops; Monogr. Agric.Exp.Stn, Univ. Florida 18, Vol.1:106-123.

Ellis, P.J. 1992. Weed hosts of beet western yellows virus and potato leafroll virus in British Columbia. *Plant Dis.*76: 1137-1139.

Ellis,P., Stace-Smith,R.,Bowler,G. and MacKenzie,D.J. 1996. Production of monoclonal antibodies for detection and identification of strains of potato virus Y. *Can.J.Plant Path.* 18:64-70.

Engel'gardt, V.A., Vengerov, Yu. Yu., Volkov, D.V., Sveshnikov, P.G., Menelova, L.V., Semenov, T.E., Nisanov, V.Ya., Popova, A.B. and Trofimets, L.N. 1984. Utilization of monoclonal antibodies for immunoenzyme diagnosis of the potato aucuba mosaic virus. (tranlated from Russian). *Sov. Agric. Sci. Allerton Press N.Y.* 7: 18-21.

Faccioli, G. and Colombarini, A.1996. Correlation of potato virus S and potato virus M contents of potato meristem tips with the percentage of virus-free plantlets produced in vitro. *Potato Res.* 39:129-140.

Faccioli, G. and Marani, F. 1998. Virus elimination by meristem tip culture and tip micrografting. In: *Plant Virus Disease Control.* (eds) A. Hadidi, R.K. Khetarpal and H. Koganezawa, pp. 346- 380. APS Press, St. Paul, Minnesota.

Feigelstock, D.A., Tozzini, A.C. and Hopp, H.E. 1995. Coat protein sequence of a resistance-breaking strain of potato virus X isolated in Argentina. *Virus Genes* 10: 289-292.

Fernandez-Northcote, E.N.1986. Peru tomato virus. ISHS Vegetable Virus Working Group *Newsl.* 14,13-14.

Fernandez-Northcote,E.N. and Gugerli,P. 1988. Reaction of a broad spectrum of potato virus Y isolates to monoclonal antibodies in ELISA. *Fitopat.* 22:33-36.

Foster,G.D.,1992.Molecular variation between the ordinary and Andean strains of potato virus S. *Virology* 142:413-416.

Foster,G.D. and Mills,PR.,1990.Detection of strains of potato virus S by nucleic acid spot hybridisation. *Potato Res.*33:487-495.

Foster, G.D. and Mills, P.R. 1992. The 3'-nucleotide sequence of an ordinary strain of potato virus S. Virus Genes 6:213-220.

Fox, L., Biever, K.D., Toba, H.H., Duffus, J.E. and Thomas, P.E. 1993. Overwintering and monitoring potato leafroll virus in some wild crucifers. *Amer. Potato J.* 70: 505-515.

Foxe, M.J. (1980). An investigation of the distribution of potato mop-top virus in County Donegal. *J. Life Sci. Royal Dublin Soc.* 1: 149-155.

Franc, G.D. and Banttari, E.E. 1984. The transmission of potato virus S by the cutting knife and retention time of infectious PVS on common surfaces. Amer. Potato J. 61: 253-260.

Francki, R.I.B. and McLean, G.D. 1968. Purification of potato virus X and preparation of infectious ribonucleic acid by degradation with lithium chloride. *Austr. J. biol. Sci.* 21: 1311-1318.

Fraser, T.W. 1976. Mop-top tubules. The ultrastructure of unusual tubular elements associated with two different leaf symptoms of potato mop-top virus-infected potato plants. *Protoplasma* 90: 15-31.

Fribourg, C.E. and Nakashima, J.1984a.An improved latex agglutination test for routine detection of potato viruses. *Potato Res.* 27:237-249.

Fribourg,C.E. & Nakashima,J.1984b. Characterization of a new potyvirus from potato. *Phytopathology* 74:1363-1369.

Gabriel W,1988. The carlaviruses of potatoes, M and S viruses; epidemiology and control. *Potato Res.*31: 667-680.

Gladysiak, S.1977. Effect of M infection on yield of potato tubers. *Roczniki AR Poznan* 97:67-77.

Glais, L., Kerlan,C., Tribodet,M., Astier-Manifacier,S. and Robaglia,C. 1996. Molecular characterization of potato virus Y isolates by PCR-RFLP. *Eur.J.Pl. Path.*102: 655-662.

Goodwin,P.H. and Banttari, E.E. 1984. Increased sensitivity of ELISA for potato viruses S, X and Y by polystyrene pretreatments, addiditives and a moodified assay procedure. Plant Dis. 68: 944-947.

Goth,R.W. and Goins,E.W.1998. Distribution of carlavirus Red LaSoda (RLSV). *Amer.J.Pot.Res.*75: 277.

Goth,R..W., Ellis,P.J., Villiers,G., Goins,E.W. and Wright,N.S.1999 Characteristics and distribution of potato latent carlavirus (Red LaSoda virus) in North America. *Plant Dis.*83: 751-753.

Goulden, M.G., Köhm, B.A., Santa-Cruz, S., Kavanagh, T.A. and Baulcombe, D.C. 1993. A feature of the coat protein of potato virus X affects both induced virus resistance in potato and viral fitness. *Virology* 197: 293 -302.

Govier, D.A. and Kassanis, B. 1972. Potato aucuba mosaic virus. *CMI / AAB Descr.Plant Viruses,* no. 98, 4 pp.

Grieco,F., Franco,A. di and Gallitelli,D. 1997. Potato virus M in tomato crops in southern Italy. *J. Plant Path.*78:45-49.

Gugerli,P.,1979. Potato virus A and potato leafroll virus: Purification, antiserum production and serological detection in potato and test plants by enzyme-linked immunosorbent assay (ELISA). *Phytopath.Z.*96: 97-107.

Gugerli,P. and Fries,P.1983. Characterisation of antibodies to potato virus Y and their use for virus detection. *J.gen.Virol.*64: 2471-2477.

Gugerli,P. and Gehringer,W.1980. Enzyme-linked immunosorbent assay (ELISA) for the detection of PLRV and PVY in potatoes after artificial break of dormancy. *Potato Res.*23:353-359.

Habib, H.M.1980. Natural infection of Datura metel by potato virus M in Egypt. *Egyptian J. Bot.* 23:163 -172.

Hansen, H.P. 1960. Tobacco mosaic virus carried in potato tybers. Amer.Potato J. 37: 95-101.

Harrison, B.D. and Jones, R.A.C. 1970. Host range and some properties of potato mop-top virus. *Ann. appl. Bio*l. 65: 393-402.

Heide, M. and Lange, L.1988. Detection of potato leaf roll and potato viruses M, S, X and Y by dot immunobinding on plain paper. *Potato Res.* 31:367-373.

Herbers, K., Tacke, E., Hazirezaei, M., Krause, K., Melzer, M., Rohde, W. and Sonnewald, U. 1997. Expression of luteoviral movement protein in transgenic plants leads to carbohydrate accumulation and reduced photosynthetic capacity in source leaves. *Plant J.* 12: 1045-1056.

Heuvel, J.F.M. van den and Peters, D. 1990. Transmission of potato leafroll virus in relation to the honey dew excretion of *Myzus persicae. Ann. appl. Biol.* 116: 403-502.

Heuvel, J.F.J. M. van den, Hogenhout, S.A., Verbeek, M. and Wilk,F. van der. 1998. *Azadirachta indica* metabolites interfere with the host-endosymbiont relationship and inhibit the transmission of potato leafroll virus by *Myzus persicae. Entomol. Experiment. Applic.* 86: 253-260.

Hill, S.A. and Jackson, E.A. 1984. An investigation on the reliability of ELISA as a practical test for detecting potato leafroll virus and potato virus Y in tubers. *Plant Pathology* 33: 21-26.

Hinostroza-Orihuela, A.M. 1973. Some properties of potato virus S isolated from Peruvian potato varieties. Potato Res. 33: 244-250.

Hogenhout, S. A., Verbeek, M., Hans, F., Houterman, P.M., Fortass, M., Wilk, F. van der, Huttinga, H. and Heuvel, J. F. J. M. van den. 1996. Molecular bases of the interactions between luteoviruses and aphids. *Agronomie* 16: 167- 173.

Hogenhout, S.A., Wilk, F. van der, Verbeek, M., Goldbach, R.W. and Heuvel, J.F.M. van den. 1998. Potato leafroll virus binds to the equatorial domain of the aphid endosymbiotic GroEl homolog. *J. Virol.* 72: 358-365

Hooker, W.J.1981. Compendium of plant virus diseases (ed.), 125 pp. APS Press, St Paul, MN, U.S.A.

Hooker,W.J. and Salazar,L.F. 1984. A new potato virus from the high jungle of the E.Andes; Solanum apical leaf curling virus (SALCV). *Ann.appl.Biol.*103: 449-454.

Hooker,W.J., Salazar,L.F. and Brown,C.R. 1985. Field infection of potato by the Solanum apical leaf curling virus (SALCV). *Amer. Potato J.* 62:263-272.

Horio, H.,Yano, I. and Esumi, K.1969.Potato virus M found in Japan. *Ann. Phytopath. Soc. Japan* 35:47-54.

Horvath,J., Horvath,A., Lonhard,M.and Besada,W.H.,1978. Natural occurrence of a strain of tomato mosaic virus on potato in Hungary. *Acta Phytopath.Acad.Sci.Hung.*13: 299-305.

Jarvekulg, L., Sober, J., Sinijaru, R., Toots, I. and Saarma, M.1989.Time-resolved fluoroimmunoassay of potato virus M with monoclonal antibodies. *Ann. appl. Biol.* 114: 279-291.

Jeffries,C.J.1998. Potato. *FAO/IPGRI Technical Guidelines for the Safe Movement of Germplasm,* No.19, 177 pp. FAO & IPGRI, Rome.

Jones, R A.C. 1988. Epidemiology and control of potato mop-top virus. In: *Developments in Applied Biology,* II. *Viruses with Fungal Vectors.* (eds.) J.I. Cooper and M. J. C. Asher . pp. 255- 270. Association of Applied Biologists, Wellesbourne, U.K.

Jones, R.A.C. 1975. Presence of potato mop-top in Bolivia. *Pl. Dis. Reptr* 59: 193.

Jones, R.A.C. 1982. Breakdown of potato virus X resistance genes Nx : selection of a group four strain from strain group three. *Pl. Pathol.* 31: 325 -331.

Jones, R.A.C. 1985. Further studies on resistance-breaking strains of potato virus X. *Plant Pathology* 34: 182 -189.

Jones,R.A.C.1990. Strain group specific and virus specific hypersensitive reactions to infection with potyviruses in potato. *Ann.appl.Biol.*117: 93-105.

Jones, R.A.C. and Harrison, B.D. 1969. The behaviour of potato mop-top virus in soil, and evidence for its transmission by *Spongospora subterranea* (Wallr.) Lagerh. *Ann. appl. Biol.* 63: 1-17.

Jones, R.A.C. and Harrison, B.D. 1972. Ecological studies on potato mop-top virus in Scotland. *Ann. appl. Biol.* 71: 47-57.

Jones,R.A.C. and Fribourg,C.E. 1979. Host plant reactions, some properties, and serology of wild potato mosaic virus. *Phytopathology* 69: 446-449.

Jones,R.A.C. and Fribourg,C.E.1986. Potato virus V. *CMI/AAB Descr.Pl. Viruses* No.316, 4 pp.

Jones,R.A.C. and Fuller,N.J. 1984. Incidence of potato virus V in potato stocks in England and Wales. *Plant Pathology* 33:595-597.

Jones, R.A.C., Fribourg, C.E. and Slack, S. A. 1981. Potato viruses and virus-like diseases. In: *Plant Virus Slide Series,* No.2 (eds O.W. Barnett and S.A.Tollin). Clemson University, U.S.A.

Juretic,N., Horvath,J., Besada,W.H.,Horvath,A. and Lonhard,M.,1977. Serological relationships of tomato mosaic virus isolated from potato to two members of tobamovirus. *J.Indian Potato Assoc.*4: 44-65.

Kaczmarek, U.1985. Weeds as sources of potato viruses. *Biuletyn Instytutu Ziemniak* 1985: 69-91.

Kallender, H., Buck, K.W. and Brunt, A.A. 1990. Association of three RNA molecules with potato mop-top virus. *Neth. J. Plant Pathol.* 96: 47-50.

Kanyuka, K., Bendhamane, A., Voort, J.N.A.M.R. van der, Vossen, E.A.G. van der and Baulcombe, D.C. 1999. Mapping of intra-locus duplications and introgressed DNA: aids to map-based cloning of genes from complex genomes illustrated by physical analysis of the Rx locus in tetraploid potato. *Theor. Appl. Genetics* 98: 679 – 689.

Kashiwazaki, S., Scott, K.P., Reavy, B. and Harrison, B. 1995. Sequence analysis and gene content of potato mop-top virus RNA 3: further evidence of heterogeneity in the genome organization of furoviruses. *Virology* 206: 701-706.

Kassanis, B. 1950. Heat inactivation of leaf-roll virus in potato tubers. *Ann. appl. Biol.* 37:339-341.

Kassanis, B.1957. The use of tissue cultures to produce virus-free clones from infected potato varieties. *Ann. appl. Biol.* 45:422-427.

Kassanis, B.1960. Potato virus M and paracrinkle.*Nature (London)* 188:688.

Kassanis, B.1961. Potato paracrinkle virus. *Eur. Potato J.* 4:13-24.

Kassanis, B. and Govier, D.A. 1971. The role of the helper virus in aphid transmission of potato aucuba mosaic virus and potato virus C. *J. gen. Virol.* 13: 221- 228.
Kassanis, B. and Varma, A. 1967. The production of virus-free clones of some British potato varieties. *Ann. appl. Biol.* 59: 447-450.
Kassanis, B., Woods, R.D. and White, R.F. 1972. Some properties of potato mop-top virus and its serological relationship to tobacco mosaic virus. *J. gen. Virol.* 14: 123-132.
Kavanagh, T., Goulden, M., Santa Cruz, S., Chapman, S., Barker, I. and Baulcombe, D. 1992. Molecular analysis of a resistance-breaking strain of potato virus X. *Virology* 189: 609-617.
Kekarainen,T., Metits,A., Oruetxebarria,I., Rajamaki,M.-L. and Valkonen,J.P.T. 1999. Comparison of the complete sequences of five different isolates of potato virus A (PVA). *Arch.Virol.*144: 1-12.
Kennedy, J.S., Day, M.F. and Eastop, V.F. 1962. *A Conspectus of Aphids as Vectors of Plant Viruses. Commonwealth Institute of Entomology, London,* 114 pp.
Kerlan, C., Tribodet,M., Glais,L. and Guillet,M.2000. Variability of potato virus Y in potato crops in France. *J. Phytopathology* (in press).
Khurana, S.M.P. and Singh, M.N.1980.Occurrence of potato virus M in Simla Hills. *Indian J. Mycol. & Plant Path.*10:179-181.
Knight, C.A. 1963. Chemistry of viruses. *Protoplasmatologia* 4: 1-177.
Koenig, R. 1972. Anomalous behaviour of the coat proteins of potato virus X and cactus virus X during electrophoresis in dodecylsulfate containing polyacrylamide gels. *Virology* 50: 263 -266.
Koenig, R.1982. Carlavirus Group. *CMI/AAB Descr.Plant Viruses* No.259, 4 pp.
Koenig, R. 1988. Serology and immonochemistry. In: *The Plant Viruses.* R.G. Milne (ed.) Vol. 4 pp. 111-158. Plenum Press, New York.
Koenig, R., Lesemann, D.-E. and Burgermaster, W. 1984. Beet necrotic yellow vein virus: Purification, preparation of antisera and detection by means of ELISA, immunosorbent electronmicroscopy and electro-blot immunoassay. *Phytopath. Z.*111: 244-250.
Koenig, R., Stegemann, H., Francksen, H. and Paul, H. 1970. Protein subunits in the potato virus X group. Determination of the molecular weights by polyacrylamide electrophoresis. *Biochim. Biophys. Acta* 207: 184 – 189.
Köhler, E.1935. Erfahrungen beim feldmassigen Anbau von kunstlich blattrollinfezierten Kartoffeln (Sorte Kl.-Sp.Wohltmann). *Arb. Biol. Anst.* 21:517-529.
Köhler, E.1942. Untersuchungen uber das "K-Virus" der Kartoffel.1. *Mitt. Angew.Botanik* 24:118-130.
Köhler, E. 1962. Über die unterschiedlichen Temperaturempfindlichkeit von Typen und Stämmen des Kartoffel-X-Virus. *Phytopathol. Z.* 44: 189 -199.
Kollmer, G.F. and Larson, R.H. 1960. Potato virus F in relation to host range, properties, reaction of American potato varieties and resistance. *Res. Bull. agric. Exp. Stn Univ. Wisc.* 223, 40 pp.
Kowalska, A. and Was, M.1976. Detection of potato virus M and potato virus S on test plants. *Potato Res.* 19:131-139.
Kozar, F. and Sheludko, Y.H. 1969. Ultrastructure of potato and *Datura stramonium* plant cells infected by potato virus X. *Virology* 38: 220-229.
Kurppa, A. 1989. The distribution and incidence of potato mop-top virus in Finland as determined in 1987 and on the variation of disease symptoms in infected potatoes. *Ann. Agric. Fenniae* 28: 285-295.
Kurppa,A. 1990. Potato mop-top virus: purification, preparation of antisera and detection by means of ELISA. *Annal. Agric. Fenniae* 29: 9-17.

Ladeburg, R.C., Larson, R.H. and Walker, J.C. 1950. Origin, interrelation and properties of ringspot strains of virus X in American potato varieties. *Res. Bull. Wisc. Agr. Exp. Stn*, 165, 47 pp.

Le Romancer, M. and Kerlan,C. 1991. La maladie des nécroses annulaires superficielles des tubercules; une infection de la pomme de terre, due á virus Y. *Agronomie* 11:889-900.

Le Romancer,M. and Nedellec,1997). Effect of plant genotype, virus isolate and temperature on the expression of potato tuber necrotic ringspot disease (PTNRD). Plant Pathology 46: 104-111.

LeClerc,D., Eweida,M., Singh,R.P. and AbouHaidar,M.G. 1992. Biotinylated probes for detecting virus Y and aucuba mosaic virus in leaves and dormant tubers of potato. *Potato Res.*35: 173-182.

Lihnell, D., Bang, D., Kwist, K. and Nilsson, C.1975. *Meddelanden Stations Vaxtskyddanstalt* 16:149-213.

Lizárraga, R., Panta, A., Jayasinghe, U. and Dodds, J. 1991. Tissue culture for elimination of pathogens. CIP Research Guide 3. International Potato Center (CIP), Lima, Peru, 21pp.

Loebenstein, G., Akad, F., Filatov, V., Sadvakasova, G., Manadilova, A., Bakelman, H., Teverovsky, E., Lachman,O. and David, A. 1997. Improved detection of potato leafroll luteovirus in leaves and tubers with a digoxigenin-labeled cRNA probe. *Plant Dis.* 81:489-491.

Lough, T.J., Shash,K., XoconostleCazares, B., Hofstra, K.R., Beck, D.L., Balmori, E., Forster, R.L.S. and Lucas, W.J. 1998. Molecular dissection of the mechanism by which potexvirus triple gene block proteins mediate cell-to-cell transport of infectious RNA. *Molec. Pl. Microbe Interact.* 11: 801- 814.

Lu, X., Yamamoto, S., Tanaka, M., Hibi, T. and Namba, S. 1998. The genome organization of the broad bean necrosis virus (BBNV). *Arch. Virol.* 143: 1335-1348.

MacKenzie, D.J. and Tremaine, J.H. 1990. Transgenic Nicotiana debneyi expressing coat protein are resistant to potato virus S infection. J.gen.Virol. 71: 2167-2170.

MacKinnon, J.P.1974. Detection, spread and aphid transmission of potato virus S. *Can. J. Bot.* 52:461-465.

Martin, R.R., Keese, P.K., Young, M.J., Waterhouse, P.M. and Gerlach, W.L. 1990. Evolution and molecular biology of luteoviruses. *Ann. Rev. Phytopathol.* 28: 341-363.

Matthews, R.E.F. 1949. Studies on potato virus X. *Ann. appl. Biol.* 36: 449-474.

Maury, Y., Quemener, J. and Spire, D.1965.Expérience de dissémination du virus de la pomme de terre. *Annls Epiphytie* 16:171-175.

Mayo, M.A. and Ziegler-Graff, V. 1996. Molecular biology of luteoviruses. *Adv. Virus Res.* 46: 413-460.

Mayo, M.A., Barker, H., Robinson, D.J., Tamada, T. and Harrison, B.D. 1982. Evidence that potato leafroll virus RNA is positive-stranded, is linked to a small protein and does not contain polyadenylate. *J. gen. Virol.* 59: 63-67.

Mayo, M.A., Torrance, L., Cowan, G., Jolly, C.A., Macintosh, S.M., Orrega, R., Barrera, C. and Salazar, L.F. 1996. Conservation of coat protein sequence among isolates of potato mop-top virus from Scotland and Peru. *Arch. Virol.* 141: 1115-1121.

McKay, M.B. and Dykstra, T.P.1932. Potato virus diseases. *Bull. Oregon Agric. Res. Stn* 294:5-40.

McNaughton, I.H., Malcolmson, J.F., Davidson, T.M.S., Stewart, H.E. and Rogers, W.G. 1974. *Rep. Scott. Plant Breeding Stn 1973-1974.* Roslin, U.K.

Mentaberry, A. and Orman,B. 1995. Potexviruses. In: *Pathogenesis and Host Specificity in Plant Diseases. Vol. 3. Viruses & Viroids*, pp. 19-33 (eds) R.P. Singh, U.S. Singh and K Kohmoto, Pergamon, Elsevier Science UK.

Merz, U. 1995. PMTV-like particles inside resting spores of *Spongospora subterranea. J. Phytopath.* 143: 731-733.

Mierzwa, Z., Zurck, H. and Marciniak-Dziopa, M.1993. The use of PVC foil produced in Poland as a carrier in ultramicro-ELISA test for detection of PVM and PVY. *Biuletyn Instytutu Ziemniaka* 42:23-27.

Miki, T. and Knight, C.A. 1968. The protein subunit of potato virus X. *Virology* 36: 168-173.

Milne, R.G. and Luisoni, E. 1977. Rapid immune electron microscopy of virus preparations. In: *Methods in Virology* (eds) K. Maramorosch and H. Koprowsky. Vol. 6, pp. 264-281, Academic Press, New York.

Mishra, A., Choudhary, D.P., Mishra, P.K. and Jha, A. 1979. Electron microscopical and serological investigations on Indian chilli mosaic virus. *Z. Pflanzenkr. Pflanzensch.* 86:39-42.

Moreira, A., Jones, R.A.C. and Fribourg, C.E. 1980. Properties of a resistance-breaking strain of potato virus X. *Ann. appl. Biol.* 95: 93-101.

Morozov, S. Y., Kanyuka, K.V., Levay, K.E. and Zavriev, S.K. 1990. The putative RNA replicase of potato virus M : obvious sequence similarities with potex- and tymoviruses. *Virology* 179: 911-914.

Morozov, S.Y., Lukasheva, L.I., Chernov, B.K., Skryabin, K.G. and Atabekov, J.G. 1987. Nucleotide sequence of the open reading frames adjacent to the coat protein cistron in potato virus X genome. *FEBS Lett.* 213: 438-442.

Morozov, S.Y., Miroshnichenko, N.A., Solovyev, A.G., Fedorkin, O.N., Zelenina, D.A., Lukasheva, L.I., Karasev, A.V., Dolja, V.V. and Atabekov, J.G. 1991. Expression strategy of the potato virus X triple gene block. *J. gen. Virol.* 72: 2039-2042.

Munro, J. 1960. The reaction of some potato varieties and seedlings to potato virus F. *Amer. Potato J.* 37:249-256.

Murphy, P.A and McKay, R. 1932. A comparison of some European and American virus diseases in potato. Scientific Proceedings, Royal Dublin Society (New Series) 20: 347-358.

Natti, J. J., Kirkpatrick, H.C. and Ross, A.F. 1953. Host range of potato leaf-roll virus. *Amer. Potato J.* 30: 55-64.

Nielsen, S.L. and Molgaard, J.P. 1997. Incidence, appearance and development of potato mop-top furovirus-induced spraing in potato cultivars and the influence on yield, distribution in Denmark and detection of the virus in tubers by ELISA. *Potato Res.* 40: 101-110.

Nienhaus, F. and Stille, B. 1965. Übertragung des Kartoffel-X-Virus durch Zoosporen von *Synchytrium endobioticum. Phytopahol. Z.* 54: 335 -337.

Oruetxebarria,I., Kekarainen,T., Spetz,C. and Valkonen,J.P.T.,2000. Molecular characterisation of Potato virus V genomes from Europe indicates limited spatio-temporal strain differentiation. *Phytopathology* (in press).

Paul, H.L. 1959. Spektralphotometrische Untersuchungen am Kartoffel-X-Virus. *Arch. Microbiol.* 32: 416- 422.

Pringle, C.R. 1998. The universal system of virus taxonomy of the International Committee on Virus Taxonomy (ICTV), including new proposals ratified since publication of the Sixth ICTV Report 1995. *Arch. Virol.* 143; 203-210.

Prüfer, D., Schmitz, L., Tacke, E., Kull, B. and Rohde,W. 1997. In vivo expression of a full-length cDNA copy of potato leafroll virus (PLRV) in protoplasts and in transgenic plants. *Mol. Gen. Genetics* 253: 609-614.

Puurand,U., Makinen,K., Paulin,L. and Saama,M.1994. The nucleotide sequence of potato virus A genomic RNA and its sequence similarities with other potyviruses. *J.gen.Virol.*75: 457-461.

Quak, F. 1987. Therapy of individual plants. In: *Viruses of Potatoes and Seed-Potato Production.* (eds) J.A. de Bokx and J.P.H. van der Want, pp. 151-161. Pudoc, Wageningen, The Netherlands.

Quanjer, H.M. 1921. New work on leaf-curl and allied diseases in Holland. *Roy. hort. Soc., London, Rep. Int. Potato Conf.* pp. 127-145.

Quanjer, H.M., H.A. van der Lek and Oortwijn Botjes J.G. 1916. Aard, verspreidingswijze en bestrijding van phloememnecrose (bladrol) en verwante ziekten, o.a. *Meded. LandbHogesch. Wageningen* 10: 1-138.

Querci, M., Baulcombe, D.C., Goldbach, R.W. and Salazar, L.F. 1995. Analysis of the resistance-breaking determinants of potato virus X (PVX) strain HB on different potato genotypes expressing extreme resistance to PVX. *Phytopatology* 85: 1003-1010.

Rajamaki,M., Merits,A., Rabenstein,F,Andrejeva,J., Paulin,L., Kekarainen,T., Kreuze,J.F., Forster,R.L.S. and Valkonen,J.P.T. 1998. Biological, serological and molecular differences among isolates of potato A potyvirus. *Phytopathology* 88:311-321.

Reavy, B., Arif, M., Cowan, G.H. and Torrance, L. 1998. Association of sequences in the coat protein/readthrough domain of potato mop-top virus with transmission by *Spongospora subterranea. J. gen. Virol.* 79: 2343-2347.

Reavy, B., Sandgren, M., Barker, H., Heino, P. and Oxelfelt, P. 1997. A coat protein transgene from a Scottish isolate of potato mop-top virus mediates strong resistance against Scandinavian isolates which have a similar coat protein gene. *Eur. J. Plant Pathol.* 103: 829-834.

Riesmeier, J.W., Willmitzer, L., and Frommer, W.B. 1994. Evidence for an essential role of the sucrose transporter in phloem loading and assimilate partitioning. *EMBO J.* 13: 1-7.

Riesmeier, J.W., Flügge, U.I., Schulz, B., Heineke, D., Heldt, H.W., Willmitzer, L. and Frommer, W.B. 1993. Antisense repression of the chloroplast triose phosphate translocator affects carbon partitioning in transgenic potato plants. *Proc. Natl Acad. Sci.* USA 90: 6160-6164.

Roberts, I.M. and Harrison, B.D. 1979. Detection of potato leafroll and potato mop-top viruses by immunosorbent electron microscopy. *Ann. appl. Biol.* 93: 289-297.

Rodriguez, A. and R.A.C. Jones. 1978. Enanismo amarillo disease of *Solanum andigena* potatoes is caused by potato leafroll virus. *Phytopathology* 68:39-43.

Rowhani, A. and Stace-Smith, R. 1979. Purification and characterization of potato leafroll virus. *Virology* 98: 45-54.

Rozendaal,A., van Binsbergen,J., Anema,B., van Slogteren,D.H.M. and Bunt,M.H. 1971. Serology of a deviating potato virus Y strain in the potato variety Gladblaadje. *Potato Res.*14: 141.

Rubies-Autonell, C., Faccioli, G. and Colombarini, A.1989. Detection of potato virus M in potato meristem tips and preliminary results on its eradication. *Potato Res.* 33, 42.

Rosner,A. and Maslenin,L. 1999. Transcript conformation polymorphism: a novel approach for differentiating PVY^{NTN} . *J.Phytopathol.* 147: 661-664.

Russo,P., Miller,L., Singh,R.P. and Slack, S.A. 1999. Comparison of potato leafroll virus and potato virus Y detection in seed potato samples tested by Florida winter field inspection and RT-PCR . *Amer. Potato J.* 76: 313-316.

Ryden, K., Sandgren, M. and Lovgren, L. 1989. Potato mop-top virus (PMTV) – studies on occurrence in soil samples and symptom development in some potato cultivars. Vaxtskyddsnotiser 53: 124-129 (in Swedish with English summary).

Sadvaksova, G., Tam, Y. and Loebenstein, G. 1996. A simple method for purification of potato virus X. *Appl. Biochem. Microbiol.* 32: 393-396 (in Russian).

Salaman, R.N. and Le Pelley, R.H.1930. Paracrinkle: a potato disease of the virus group.*Proc. Royal Soc.*,B106: 140-175.

Salazar, L.F. 1996. *Potato Viruses and their Control, 214 pp.* Intern. Potato Center, Lima, Peru.

Salazar, L.F. and Jones, R.A.C. 1975. Some studies on the distribution and incidence of potato mop-top virus in Peru. *Amer. Potato J.* 52: 143-150.

Samson, R.G., Allen, T.C. and Whitworth, J.L.1993. Evaluation of direct tissue blotting to detect potato viruses. *Amer. Potato J.* 70:257-265.

Sandgren, M. 1995. Potato mop-top virus (PMTV): distribution in Sweden, development of symptoms during storage and cultivar trials in field and glasshouse. *Potato Res.* 38: 379-389.

Santa Cruz, S. and Baulcombe, D. 1995. Analysis of potato virus X coat protein genes in relation to resistance conferred by the genes *Nx, Nb* and *Rx1* of potato. *J. gen. Virol.* 76: 2057-2061.

Santa Cruz, S., Roberts, A.G., Prior, D.A.M., Chapman, S. and Oparka, K.J. 1998. Cell-to-cell and phloem-mediated transport of potato virus X: the role of virions. *Plant Cell* 10: 495-510.

Santillan, F.W., Fribourg, C.E. and Jones R.A.C. 1980. Estudo comparitavo de once aislamiento de virus S der la papa de la region Andiana. Fitopat. 15: 42-43.

Schiessendoppler, E. and Forschum, H.1990. Detection of potato viruses Y, X, A, M, S and potato leaf roll virus by dot ELISA. *Pflanzenberichte* 51:17-24.

Schoen C.D., Knorr,D. and Leone,G. 1996. Detection of potato leafroll virus in dormant potato tubers by immunocapture and a fluorogenic 5' nuclease RT-PCR assay. *Phytopathology* 86: 993-999.

Schultz, E.S. and Folsom, D.1923. Transmission, variation, and control of certain degenerative disease of Irish potatoes. *J. Agric. Res.*25: 43-117.

Scott, K.P., Kashiwazaki, S., Reavy, B. and Harrison, B. 1994. The nucleotide sequence of potato mop-top virus RNA 2: a novel type of genome organization for a furovirus. *J. gen Virol.* 75:3561-3568.

Shalla, T.A. and Shepard, J.F. 1972. The structure and antigenic analysis of amorphous inclusion bodies induced by potato virus X. *Virology* 49: 654-66.

Shaw, J.G., Reichmann, M.E. and Hatt, D.L. 1962. Amino acid analysis of the proteins of two strains of potato virus X. *Virology* 18: 79-88.

Shepard, J.F. 1970. A radial-immunodiffusion test for the simultaneous diagnosis of potato viruses S and X. *Phytopathology* 60: 1669-1671.

Shepard, J.R.1972.Gel-diffusion methods for the serological detection of potato viruses X,S and M. *Bull. Montana Agric. Res. Stn* 662,72 pp.

Sigvald,R. 1984. The relative efficiency of some aphid species as vectors of potato virus Y. *Potato Res.* 27:285-290.

Singh,R.P. and Somerville,T.H. 1992. Evaluation of the enzyme-amplified ELISA for the detection of potato virus A,M, S, X, Y and leafroll . *Amer. Potato J.* 69:21-30.

Singh,R.P., Boucher,A., Somerville,T.H. and Dhar,A.K. 1993. Selection of a monoclonal antibody to detect PVY/N and its use in ELISA and DIBA. *Can.J.Plant Path.*15: 293-300.

Slack, S.A.1983. Identification of an isolate of the Andean strain of potato virus S in North America. *Plant Dis.* 67:786-789.

Smith, K.M. 1931. On the composite nature of certain potato virus diseases of the mosaic group as revealed by the use of plant indicators. *Proc. Royal Soc. London B* 109; 251- 267.

Smith, O.P. and Harris, K.F. 1990. Potato leafroll virus 3' genome organization: sequence of the coat protein gene and identification of a viral subgenomic RNA. *Phytopathology* 80: 609-614.

Sokmen, M.A., Barker, H. and Torrance, L. 1998. Factors affecting the detection of potato mop-top virus in potato tubers and improvement of test procedures for more reliable assays. *Ann. appl. Biol.* 133: 55-63.

Souza-Dias, J.A.C. de, Costa, A.S. and Nardin, A.M. 1993. Potato leafroll virus in solanaceous weeds in Brazil explains severe outbreaks of the disease in absence of known potato donor sources. *Summa Phytopatho.* 19: 80-85.

Stace-Smith, R. and Mellor, F.C. 1968. Eradication of potato viruses X and S by thermotherapy and axillary bud culture. *Phytopathology* 58:199-203.

Tacke, E., Sarkar, S. and Rohde, W. 1989. Cloning of the gene for the capsid protein of potato leafroll virus. *Arch. Virol.* 105:153-163

Takanami, Y. and Kubo, S. 1979. Enzyme-assisted purification of two phloem-limited plant viruses: tobacco necrotic dwarf and potato leafroll. *J. gen. Virol.* 44:153-159.

Tamada,, T., Harrison, B.D. and Roberts, J..M. 1984. Variation among British isolates of potato leafroll virus. *Ann. appl. Biol.* 104: 107-116.

Tavantzis, S.M.1984. Physicochemical properties of potato virus M. *Virology* 133:427-430.

Thomas, J.E. 1984. Characterisation of an Australian isolate of tomato yellow top virus. *Ann. appl. Biol.* 104: 79-86.

Thomas, J.E. 1993. Alternative hosts and the epidemiology of potato leafroll virus in Queensland. *Austr. J. agric. Res.* 44: 1905-1916.

Tollin, P., Wilson, H.R. and Bancroft, J.B. 1980. Further observations on the structure of potato virus X. *J. gen. Virol.* 49: 407-410.

Tomassoli,L. and Lumia,V. 1998. Occurrence of potato tuber necrotic ringspot disease (PTNRD) in Italy. *Plant Dis.*82:350.

Torrance, L. and Mayo, M.A. 1997. Proposed re-classification of furoviruses. Virology Division News. *Arch. Virol.* 142: 435-439.

Torrance, L., Cowan, G.H. and Pereira, L.G. 1993. Monoclonal antibodies specific for mop-top virus, and some properties of the coat protein. *Ann. appl. Biol.* 122: 311-322.

Tozzini, A,C., Ek, B., Palva, E.T. and Hopp, H.E. 1994. Potato virus X coat protein is a glycoprotein. *Virology* 202: 651-658.

Tu, J.C. and Hiruki, C.1970. Ultrastructure of potato infected with potato virus M. *Virology* 42:238-242.

Valkonen,J.P.T., Contreras,A., Pehu,E. and Salazar,L.F. (1992). Naturally occurring viral infections in Solanum brevidens and S.fernandezianum. *Potato Res.* 35:411-417.

Valkonen,J.P.T., Puurand,U., Slack,S.A., Makinen,K. and Saarma,M.1995. Three strains of potato A potyvirus based on hypersensitive responses in potato, serological properties and coat protein sequences. *Plant Dis.*79: 748-753.

Van den Heuvel, J.F.J.M., van der Vlugt,R.A.A., Verbeek,M., Haan,P.T. and Huttinga,H.1994. Characteristics of a resistance breaking isolate of potato virus Y causing tuber necrotic ringspot disease. *Eur.J.Pl.Path.*100: 347-356.

Van der Vlugt, R.A.A., Leunissen,J. and Goldbach,R.1993. Taxonomic relationships between distinct potato virus Y isolates based on detailed comparison of the viral coat proteins and 3'-nontranslated regions. *Arch.Virol.*131: 361-375.

Varma, A., Gibbs, A.J., Woods, R.D. and Finch, J.T. 1968. Some observations on the structure of the filamentous particles of several plant viruses. *J. gen. Virol.* 2: 107-114.

Verhoeven,J.Th.J. and Roenhorst,J.W. 1995. Virological risks accompanying the increased interest in pepino (Solanum muricatum). Proc. 8th Conf. Virus Dis. Vegetables, Prague, Czech Republic, 9-15 July, 1995, pp.109-112.

Vetten,H.J., Ehlers,U. and Paul,H.L.,1983. Detection of potato virus Y and A in tubers by enzyme-linked immunosorbent assay after natural and artificial break of dormancy. *Phytopath.Z.*108: 41-53.

Watts, N.R., Singh, M. and Singh, R.P. 1997. Potato virus X isolates from potato collected in eastern Canada with different symptoms in tobacco differ in their coat proteins. *Amer. Potato J.* 74: 245 – 253.

Webb, R.E., Larson, R.H. and Walker, J.C. 1952. Relationships of potato leaf-roll virus strains. *Res. Bull. Univ. Wisc., Madison,* no. 178.

Weidemann, H.L. 1986. Die Ausbreitung der Kartoffel Viren S und M unter Feldbedingungen. *Potato Res.* 29:109.

Weidemann,H,L.1988. Rapid detection of potato viruses by dot-ELISA. *Potato Res.*31:485-492.

Weidemann, H.L. and Koenig, R. 1990. Differentiation of isolates of potato virus S which infect Chenopodium quinoa systemically by means of quantitative cDNA hybridization tests. Plant Dis. & Prot. 97: 323-327.

Weidemann,H.L. and Maiss,E.1996. Detection of the potato necrotic ringspot strain of potato virus Y (PVY^{NTN}) by reverse transcription and immunocapture polymerase chain reaction. *J.Plant.Dis.Prot.*103: 337-345.

Wetter, C. and Brandes, J.1956. Untersuchungen über das Kartoffel-S-Virus. *Phytopath. Z.* 26:81-92.

Wetter, C. and Völk, J.1960.Versuche zur übertragung der Kartoffel-viren M- und S-Viren durch Myzus persicae Sulz. *Eur. Potato J.* 3:158-163.

Wetter, C.1972. Potato virus M. C.M.I./A.A.B. Descr.Plant Viruses No. 87, 4 pp.

Wilk, F. van der, Verbeek, M., Dullemans, A.M. and Heuvel, J.F.J..M. van den. 1997. The genome-linked protein of potato leafroll virus is located downstream of the putative protease domain of the ORF1 product. *Virology* 234: 300-303.

Wilson, C.R. and Jones, R.A.C. 1995. Occurrence of potato virus X strain group 1 in seed stocks of potato cultivars lacking resistance genes. *Ann. appl. Biol.* 127: 479-48.

Xu, H., Leclerc, D., Leung, B. and Abouhaidar, M.G. 1994. The entire nucleotide sequence and genomic organization of potato aucuba mosaic potexvirus. *Arch. Virol.* 135: 461-469.

Zavriev, S.K.,Kanyuka, K.V. and Leavy, K.E.1991. The genomic organisation of potato virus M RNA. *J. gen. Virol.* 72: 9-14.

6. Viroids

L.F. Salazar, I. Bartolini and A. Hurtado
The International Potato Center (CIP), P.O. Box 1558, Lima 12, Peru

6.1 Introduction

Viroids are the smallest known infectious agents of several plant species and consist of a unique nucleic acid molecule. Viroid nucleic acid is a single-stranded, covalently closed, circular RNA molecule with a rod-like secondary structure, characterised by an alternated arrangement of double-helical regions and single-stranded loops (Gross, 1985). Viroid nucleic acids range from 250 up to 375 nucleotides (Symons, 1981; Flores, 1995). In potatoes, the potato spindle tuber viroid (PSTVd) is the only viroid reported, though the disease it causes has been known since 1922 (Martin, 1922). PSTVd was considered a virus with unusual properties until 1967, when Diener and Raymer (1967) determined that its properties suggested a free nucleic acid nature (Diener and Raymer, 1967). Several strains of PSTVd, with 356 to 360 nucleotides, have been reported (Puchta *et al.* 1990; Gross *et al.* 1978; Owens *et al.* 1992; Behjatnia *et al.* 1996). These strains can be grouped as mild, intermediate, and severe according to the symptoms they produce in potato and tomato.

G. Loebenstein et al. (eds.),
Virus and Virus-like Diseases of Potatoes and Production of Seed-Potatoes, 135–144.

6.2 Symptoms and Damage

Symptoms caused by PSTVd in the foliage of infected potato plants range from symptomless to severe.. Symptom expression is influenced by the type of potato cultivar (Werner, 1926; Pfannenstiel and Slack, 1980), the strain of PSTVd (Singh *et al.* 1971), and the environmental conditions (Goss, 1930). The tubers may be very spindly, usually showing growth cracks, a decrease in eye depth, and raised bud scales or "eyebrows" (Fig. 6.1).

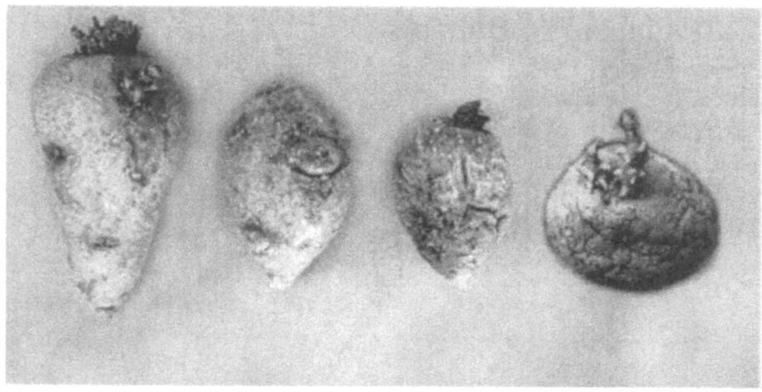

Fig. 6.1. Typical symptoms by PSTVd in tubers of a hybrid Clone (*S. tuberosum x S. phureya).* Tuber at right is healthy control.

In addition to loss of market quality of the tubers, yield losses of 17 to 24% with mild strains and up to 64% with severe strains have been reported (Singh *et al.* 1971). Pfannenstiel and Slack (1980) showed that the decrease in tuber weight depended on the potato cultivar and age of infection. A decrease of tuber weight with a severe strain varied from insignificant to 46% in the first year of infection, up to 87% in the second year, and 97% in the third year. The authors found that mild strains did not reduce tuber weight, but in some cultivars tuber symptoms were more pronounced, market quality was affected.

6.3 Viroid Replication and Movement

Replication takes place in the nucleus, and most viroids are found in the nucleoli. There are two components involved in nucleic acid replication: the template, and the enzymes that catalyse the process. It is well known that viroid RNA does not code for any protein and that no viroid-specific DNA is present in the infected plant. Viroid replication apparently involves direct RNA to RNA copying. PSTVd infection must activate host enzyme, such as RNA polymerase, RNase, or RNA ligase, that recognise and replicate monomeric PSTVd units.

Two rolling-circle models for viroid replication have been proposed. The most accepted is the asymmetric model, in which the plus circular RNA unit of PSTVd serves as a template to synthesise multimeric minus strands. These serve directly as templates for multimeric plus strands, which are cleaved by an RNase to obtain monomers. The monomers are circularized by an RNA ligase, producing copies of the original plus circular RNA PSTVd units that start the replication cycle. The minus linear multimeric strands were detected in PSTVd infected tissue by Branch *et al.* (1981), who later isolated the replication complex of template-complementary transcript strands with a circular plus and monomeric template (Branch *et al.* 1988). No monomeric circular minus PSTVd RNA could be detected in naturally infected plants (Feldstein *et al.* 1998).

Specific inhibitors were used to demonstrate that RNA polymerase II is the enzyme that amplifies viroid RNA (Mulhbach and Sanger, 1979; Flores, 1989; Flores, 1982). It is unclear, however, if one RNase catalyses both cleavage and ligation (Tsagris *et al.* 1987), or if these activities require both an RNase and an RNA ligase (Tabler *et al.* 1992).

Systemic infection by viroids must include nuclear import, export, cell-to-cell and phloem movement. Woo *et al.* (1999) suggest that nuclear import of the viroid results from the interaction of a viroid sequence or structural motif(s) with cellular factor(s). This cellular factor may be either cytosolic or associated with the nuclear pore complex (NPC). Viroids may diffuse from the cytoplasm to the NPC and are then recognised via special motifs and imported.

6.4 Pathogenesis and Variability

Despite their extreme simplicity, viroids cause syndromes in plants that are almost as varied as those caused by plant viruses. Because viroids are not translated, their effects on plants must be a consequence of direct interaction of the viroid RNA with host constituents. There is probably more than one mechanism responsible for viroid pathogenesis.

Analysis of molecular chimeras between viroids of different pathogenicity levels has revealed that the severity of symptoms is the result of complex interactions among three of the five viroid domains (Owens *et al.* 1991; Hammond, 1992).

Mutations in the pathogenicity domain of PSTVd determines symptom expression, and there is a strong correlation between symptom severity and the relative orientation of three short helical regions within the overall rod-like native structure (Owens *et al.* 1995; Owens *et al.* 1996).

Recent evidence suggests that one pathway is due to viroid RNA activating a plant RNA–activated protein kinase, or PKR. Protein synthesis is reduced, and this causes pathogenic effects. In the case of PSTVd, there is a good correlation between the pathogenicity of the strain and its ability to activate PKR in vitro (Diener *et al.* 1993).

6.5 Transmission

PSTVd is easily transmitted by contact (Diener, 1979). Farm machinery plays an important role in this transmission (Manzer and Merriam, 1961). Transmission through the botanical seed of its most important hosts in nature, potato and tomato is well known (Benson and Singh, 1964; Hunter *et al.* 1969). The rate of transmission through seed is highly variable, ranging from negligible to 100% (Fernow *et al.* 1970). Recently, two pathways of potato seedling infection have been confirmed (Lazarte *et al.* 1999). In "true infection", PSTVd infects one or both parents and invades the embryo and seed coats. Therefore, the seedlings carry the viroid. The other, called "contamination", PSTVd appears to be carried only on the outside of the seed coat, but seedlings may become infected during the germination process. This contamination probably occurs during extraction of seeds from the berries (from external sources of infection such as contaminated tools or labourers' hands). In this case, PSTVd can be effectively eliminated by treating the seeds for 30 min with 2% calcium hypochlorite before sowing.

Reports on the transmission of PSTVd by chewing insects (Goss, 1928) have not been confirmed. Of particular interest, however, are the recent reports on transmission by aphids. De Bokx and Pirone (1981) reported low rates of transmission of PSTVd by *Macrosiphum euphorbiae*, but not by *Myzus persicae* or *Aulacorthum solani*. More recently, however, high rates (up to 100%) of PSTVd transmission aided by potato leafroll virus (PLRV) were found (Salazar *et al.* 1995). When PSTVd and PLRV infect a plant simultaneously, the viroid can be encapsidated in PLRV particles and can be transmitted by the aphid *M. persicae* in a persistent manner (Querci *et al.* 1997). This type of transmission may play an important role in long-distance dissemination of the viroid under field conditions. A strong correlation between incidence of PSTVd and PLRV was found in China (Querci *et al.* 1997), suggesting that PLRV facilitates transmission of the viroid.

6.6 Detection

6.6.1 Indicator Hosts

Tomato (*Lycopersicon esculentum*) was the only host used to detect PSTVd for several years. Tomato plants mechanically inoculated with PSTVd develop symptoms in two to four weeks when grown under high temperature (>30C) and high light intensity (>1500 f-c) (Fig. 6.2).

Fig. 6.2. Systemic symptoms caused by PSTVd in *Lycopersicon esculentum* cv. Rutgers (left), uninoculated control (right).

The severity of the symptoms depends on the viroid concentration in the sample, the growing conditions, and the type of viroid strain. Severe strains cause severe bunchiness, rugosity, epinasty, and necrosis of the midveins of the upper leaves and stems. Symptoms caused by mild strains are sometimes indistinguishable from uninoculated tomatoes. Mild strains may be detected using a "challenge" test developed by Fernow (1967). In this test, a group of tomatoes are inoculated with a sample and about 15 days later some of the plants are reinoculated with the severe strain. If no symptoms of the severe strain develop in the reinoculated tomatoes, the sample is assumed to be infected with a mild strain that protected the plants from infection with the severe strain. The second approach, developed by Yang and Hooker (1977), relies on albinism produced by mild or severe strains of PSTVd in tomatoes inoculated and maintained under continuous fluorescent light (>1500 f-c).

Other susceptible hosts such as *Scopolia sinensis* or *Solanum berthaultii* have also been reported (Singh, 1971; 1984) but their use was limited, mostly due to the requirement of specific growing conditions for reliable development of symptoms. In general, the biological indicator tests, though sensitive, became impractical for large-scale detection of the viroid because their use requires large space in heated greenhouses and because of the long time needed for symptom development.

6.6.2 Nucleic Acid Based Methods.

Electrophoresis in polyacrylamide gels of low molecular weight nucleic acid extracted from infected plants was first developed by Morris and Wright (1975) as an alternative to diagnosis using indicator hosts. Some modifications to the original procedure reduced the time needed to complete the test to only a few hours (Pfannenstiel *et al.* 1980). However, sensitivity was lower than in the tomato test. Procedures that combined inoculation of tomatoes and subsequent electrophoresis were reportedly adequate for testing composite samples. This procedure was especially effective for testing true potato seed for PSTVd infection (Salazar, 1989). Electrophoresis under denaturing conditions was also reported to increase the sensitivity of the original procedures and allowed distinction of mild and severe strains (Singh and Boucher, 1987). Electrophoresis was used by several laboratories until the development of nucleic acid spot hybridisation (NASH) by Owens and Diener (1981). In this procedure, the PSTVd nucleic acid is immobilised on a nitrocellulose membrane, which is later hybridised with a radiolabeled (^{32}P) probe, and the reaction detected by autoradiography. Several modifications have been made regarding the type and preparation of the probe (Salazar *et al.* 1988) and the type of label used (Querci and Salazar, 1998). Non-radioactive labels have been developed that offer a good alternative to radioactive testing. The most recent non-radioactive procedure adopted for detecting PSTVd is based on the incorporation of fluorescein-11–dUTP into a DNA probe during a random-prime-labelling reaction catalysed by the Klenow fragment of DNA polymerase I. After the hybridisation step, detection is carried out by an anti-fluorescein antibody conjugated with alkaline phosphatase; and the reaction of a dioxyetane substrate that produces a chemiluminescent signal that can be captured on film (Fig. 6.3) (Agrawal *et al.* 1986). Fig. 6.4 shows the comparison of PSTVd detection by radioactive and non-radioactive procedures.

Detection of PSTVd by RT-PCR appears to be highly sensitive and has been used for the detection of the viroid in dormant potato seed and tubers, using primers constructed for the nucleotide sequence of the upper central conserved region and its adjacent segment (Shamloul *et al.* 1995).

6.7 Control of PSTVd

Because potato tubers and botanical seed are the most important natural pathways for viroid dissemination, assuring their freedom of PSTVd in any established seed program, germplasm collection, or breeding program is fundamental for its control. The technology currently used to detect the viroid is sufficiently sensitive and can be applied to test large number of samples with minimum effort and low costs. The recent finding of dissemination of PSTVd by aphids due to encapsidation of PLRV particles (Querci *et al.* 1997) reinforces the need for application of detection technology. Application of

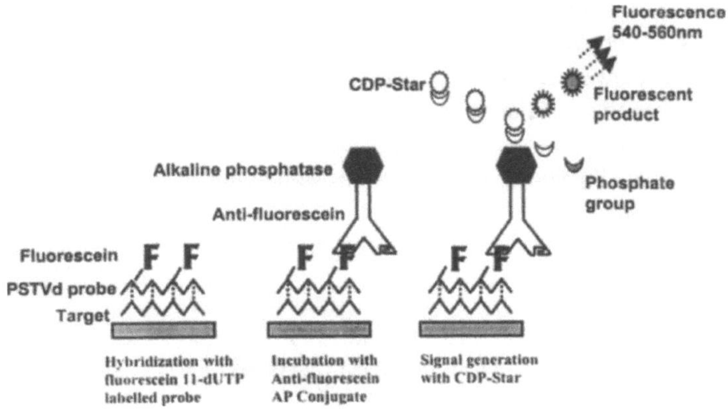

Fig. 6.3. Detection of PSTVd using fluorescein 11-dUTP labelled probes. CDP-Star = Disodium 2-Chloro-5-(4-methoxyspiro [1,2-dioxetane-3,2-(5'-chloro)-tricyclo [3,3,1,1]decan]-4yl) Phenyl phosphate. AP = Alkaline phosphatase.

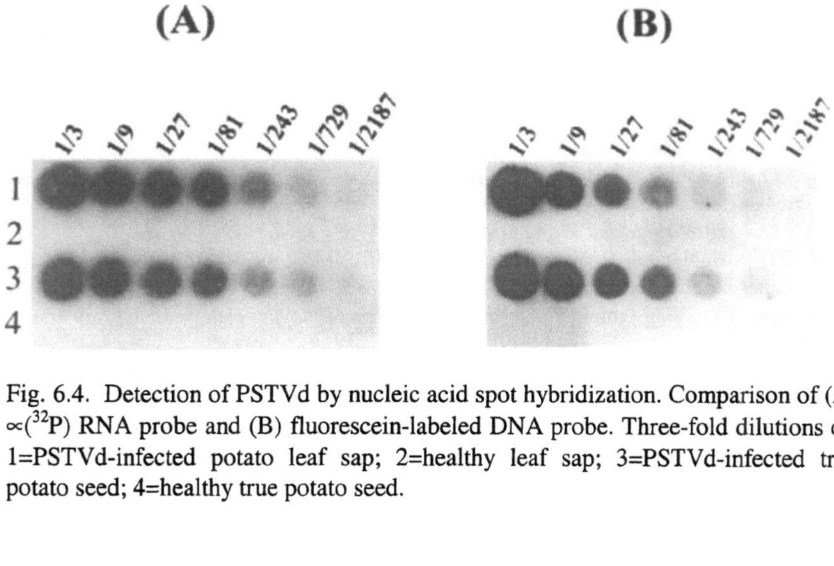

Fig. 6.4. Detection of PSTVd by nucleic acid spot hybridization. Comparison of (A) $\propto(^{32}P)$ RNA probe and (B) fluorescein-labeled DNA probe. Three-fold dilutions of: 1=PSTVd-infected potato leaf sap; 2=healthy leaf sap; 3=PSTVd-infected true potato seed; 4=healthy true potato seed.

detection methods to assure freedom of tubers and seeds from PLRV should also provide a means for controlling the viroid itself.

Hygienic measures such as disinfecting tools with calcium hypochlorite have proved effective in reducing PSTVd contamination at the International Potato Center (CIP) for a number of years. Treatment of seeds with this chemical is also used routinely to eliminate contamination with PSTVd carried on the seed coats.

Several of the modern cultivars are usually asymptomatic to infections with PSTVd, probably because of unintentional selection during cultivar development. However, no practical efforts have been made to search for and incorporate genetic resistance to PSTVd. Resistance to mechanical infection but not by grafting or "agroinfection" with the viroid has been confirmed in *Solanum acaule* (Salazar *et al.* 1988). Resistance has also been reported in *S. berthaultii* (Singh, 1985).

References

Agrawal, S., Christodoulou, C. and Gait, M.J. 1986. Efficient methods for attaching nonradioactive labels to the 5'ends of synthetic digodeoxyribonucleotides. *Nucleic Acids Res.* 14: 6227–6245.

Behjatnia, S.A.A., Dry, I.B., Krate, L.R., Conde, B.D., Connelly, M.O., Randles, J.W. and Rezaian, M.A. 1996. New potato spindle tuber viroid and tomato leaf curl geminivirus strains from a wild *Solanum* sp. *Phytopathology,* 86 :880–886.

Benson, A.P. and Singh, R.P. 1964. Seed transmission of potato spindle tuber virus in tomato. *Am. Pot. J.* 41: 294.

Branch, A.D., Robertson, H.D. and Dickson, E. 1981. Longer than unit viroid minus strands are present in RNA from infected plants. *Proc. Natl. Acad. Sci.* U.S.A. 78: 6381–6385.

Branch, A.D., Benenfeld, B.J. and Robertson, H.D. 1988. Evidence for a single rolling circle in the replication of potato spindle tuber viroid. *Proc. Natl. Acad. Sci.* U.S.A. 85: 9128–9132.

De Bokx, J.A, and Piron, P.G.M. 1981. Transmission of potato spindle tuber viroid by aphids. *Neth. J. Plant Pathol.* 87: 31–34

Diener, T.O, and Raymer, W.B. 1967. Potato spindle tuber virus: a plant virus with properties of a free nucleic acid. *Science* 158: 378–381.

Diener, T.O. 1979. *Viroids and Viroid Diseases*. John Wiley & Sons, New York.

Diener, T.O., Hammond, R.W., Black. T. and Katze, M.G. 1993. Mechanism of viroid pathogenesis-differential activation of the interferon-induced, double stranded RNA-activated, m(r) 68000 protein-kinase by viroid strains of varying pathogenicity. *Biochimie* 75: 533–538.

Feldstein, P.A.; Hu, Y.; Owens, R.A. 1998. Precisely full length, circularizable, complementary RNA: An infectious of potato spindle tuber viroid. Proc. Natl. Acad. Sci. USA, Vol 95: 6560-6565.

Fernow, K. H. 1967. Tomato as a test plant for detecting mild strains of potato spindle tuber virus. *Phytopathology* 57:1347–1352.

Fernow, K.H., Peterson, L.C and Plaisted, R.L. 1970. Spindle tuber virus in seeds and pollen of infected potato plants. *Am. Pot. J.* 47: 75-80.

Flores, R.; and Semancik, J.S. 1982. Properties of a cell-free system for synthesis of citrus exocortis viroid. *Proc. Natl. Acad. Sci.* U.S.A. 79: 6285–6288.

Flores, R. 1989. Synthesis of RNA specific to citrus exocortis viroid by a fraction rich in nuclei from infected *Gynura aurantiaca:* Examination of the nature of the products and solubilization of the polymerase-template complex. *J. Gen. Virol.* 70: 2695–2703.

Flores, R. 1995. Subviral Agents: Viroids. In: Murphy, F.A., Fauquet, C.M., Bishop, D.H.L., Ghabrial, S.A., Jarvis, A.W., Martelli, G.P., Mayo, M.A., and Summers M.D. (eds.). *Viru Taxonomy, Classification, and Nomenclature of Viruses*: *Sixth Report of the International*

Committee on Taxonomy of Viruses. Springer-Verlag, Wien, New York. pp. 495–497.

Goss, RW. 1928. Transmission of potato spindle tuber by grasshoppers (Locutidae). *Phytopathology* 18: 445.

Goss, RW. 1930 The symptoms of spindle tuber and unmottled curly dwarf on the potato. *Nebr. Agric. Exp. Stn. Res. Bull.* 47: 39.

Gross, H.J., Domdey, H., Lossow, D., Jank, P., Raba, M., Alberty, H. and Sanger, H.L., 1978. Nucleotide sequence and secondary structure of potato spindle tuber viroid. *Nature* 273: 203–208.

Gross, H.J. 1985.Viroids: Their structure and possible origin. In: Maramorosch, K. and J.J. MacKelvey, Jr (eds.) *Subviral Pathogens of Plants and Animals: Viroids and Prions*. Academic Press, New York, pp. 165–182.

Hammond. R.W. 1992. Analysis of virulence modulating region of potato spindle tuber viroid (PSTVd) by site-directed mutagenesis. *Virology* 187: 654– 662.

Hunter, D.E., Darling, D.H, and Beale, W.L. 1969. Seed transmission of potato spindle tuber virus. *Am. Pot. J.* 46: 247.

Lazarte, V.; Bartolini, I.; Querci, M. and Salazar, L.F. 1999. Pertpetuación y diseminación del Viroide ahusado de la papa (PSTVd) en la semilla botánica. Fitopatologia 34 (4) : 177-178 .

Manzer, F.E. and Merriam, D. 1961. Field transmission of the potato spindle tuber virus and virus X by cultivating and hilling equipment. *Am Pot. J.* 38 : 346-352.

Martin, W.H. 1922. "Spindle tuber," a new potato trouble. Hints to Potato Growers, *New Jersey State Potato Association*, 38 p.

Morris, T.J. and Wright, N.S. 1975. Detection on polyacrylamide gel of a diagnostic nucleic acid from tissue infected with potato spindle tuber viroid. *Am. Pot. J. 52*:57-63.

Muhlbach, H.P. and Sanger. H.L. 1979. Viroid replication is inhibited by o-amanitin. *Nature* 278:185–188.

Owens, R.A. and Diener, T.O. 1981. Sensitive and rapid diagnosis of potato spindle tuber viroid disease by nucleic acid hybridization. *Science* 213: 670–672.

Owens, R.A., Thompson, S.M. and Steger, G. 1991. Effects of random mutagenesis upon potato spindle tuber viroid replication and symptom expression. *Virology* 185: 18–31.

Owens, R.A., Khurana, S.M. P., Smith, D.R., Singh, M.N. and Garg, I.D. 1992. A new mild strain of potato spindle tuber viroid isolated from wild *Solanum* spp. in India. *Plant Dis.*, 76: 527– 529.

Owens, R.A., Chen, W., Hu, Y. and Hsy, Y.H. 1995. Suppression of potato spindle tuber viroid replication and symptom expression by mutations which stabilize the pathogenicity domain. *Virology* 208: 554–564.

Owens, R.A., Steger, G., Hu, Y., Fels, A., Hammond, R.W. and Riesner, D. 1996. RNA structural features responsible for potato spindle tuber viroid pathogenicity. *Virology* 222: 144–158.

Pfannenstiel, M.A. and Slack, S.A. 1980. Response of potato cultivars to infection by the potato spindle tuber viroid. *Phytopathology* 70: 922–926.

Pfannenstiel, M.A.; Slack, S.A.; Lane, L.C. 1980. Detection of potato spindle tuber viroid in field grown potatoes by an improved electrophoretic assay. Phytopathology 70:101-1018.

Putcha, H., Herold, T., Verhoeven, K., Roenhorst, A., Ramm, K., Schmidt-Puchta, W. and Sanger, H.L. 1990. A new strain of potato spindle tuber viroid (PSTVd-N) exhibits major sequence differences as compared to all other PSTVd strains sequenced so far. Plant *Mol. Biol.* 15: 509–511.

Querci, M., Owens, R.A., Bartolini, I., Lazarte, V. and Salazar, L.F. 1997. Evidence for heterologous encapsidation of potato spindle tuber viroid by potato leafroll virus. *J. gen. Virol.* 78: 1207–1211.

Querci, M. and Salazar, L.F. 1998. Molecular Probes for Detection of Viruses/Viroids in Potato. In: *Comprehensive Potato Biotechnology* (Khurana, S.M.P.; Chandra, R.; and Upadhya, M.D., Eds.) pp. 187–206. Malhotra Publishing House, New Delhi, India.

Salazar, L.F., Hammond, R.W., Diener, T.O. and Owens, R.A.1988. Analysis of viroid replication following *Agro*ted inoculation of non-host species with potato spindle tuber viroid cDNA. *J. gen. Virol.* 69: 879–889.

Salazar, L.F. 1989. Potato spindle tuber. In: *Plant Protection and Quarantine*. Volume II. Selected Pests and Pathogens of Quarantine Significance [edited by Kahn, RP]. Boca Raton, Florida, USA; CRC Press, Inc., 155–167 pp.

Salazar, L.F., Querci, M., Bartolini, I. and Lazarte, V. 1995. Aphid transmission of potato spindle tuber viroid assisted by potato leafroll virus. *Fitopatologia* 30: 56–58.

Shamloul, A.M., Hadidi, A. and Singh RP. 1995. Detection of potato spindle tuber viroid from infected tissue using polymerase chain reaction technology. *Phytopathology* 85: 1181.

Singh, R.P. 1971. A local lesion host for potato spindle tuber virus. *Phytopathology* 61:1034–1035.

Singh, R.P. 1984. *Solanum berthaultii*, a sensitive host for indexing potato spindle tuber viroid from dormant tubers. *Potato Res.* 27:163– 172.

Singh, R.P. 1985. Clones of *Solanum berthaultii* resistant to potato spindle tuber viroid. *Phytopathology* 75: 1432–1434.

Singh, R.P. and Boucher, A. 1987. Electrophoretic separation of a severe from mild strains of potato spindle tuber viroid. *Phytopathology* 77: 1588–1591.

Symons, R.H. 1981. Avocado sunblotch viroid—primary sequence and proposed secondary structure. *Nucleic. Acid Res.* 9: 6527–6537.

Tabler, M., Tzortzakaki, S. and Tsagris, M. 1992. Processing of linear longer-than-unit-length potato spindle tuber viroid RNAs into infectious monomeric circular molecules by a G-specific endoribonuclease. *Virology* 190: 746–753.

Tsagris, M., Tabler, M., Muhlbach, H.P. and Sanger, H.L. 1987. Linear oligomeric potato spindle tuber viroid (PSTVd) RNAs are accurately processed in vitro to the monomeric circular viroid proper when incubated with a nuclear extract from healthy potato cells. *EMBO J.* 6: 2173–2183.

Werner, H.O. 1926. The spindle tuber disease as a factor in seed potato production, *Nebr. Agric. Exp. Stn. Res. Bull.*, 32, 128 p.

Woo, Y., Itaya, A., Owens, R.A., Tang, L., Hammond, R.A., Chou, H., Lai, M. and Ding, B. 1999. Characterization of nuclear import of potato spindle tuber viroid RNA in permeabilized protoplasts. *Plant J.* 17: 627–635.

Yang, T.C. and Hooker, W.J. 1977. Albinism of potato spindle tuber viroid-infected Rutgers tomato in continuous light. *Am. Potato J.* 54: 519–530.

7. Phytoplasma Diseases

M. Klein
Department of Entomology, Agricultural Research Organization, The Volcani Center, Bet Dagan 50 250, Israel

Phytoplasmas (formerly Mycoplasma-like organisms, or MLOs) are considered to be the smallest free-living prokaryotes, but unlike mycoplasmas, they have not so far been isolated from their plant hosts or their insect vector and cultivated in artificial media. Phytoplasmas have been seen only in phloem tissues of infected plants and observed in thin sections by transmission electron microscopy as pleomorphic bodies: rounded bodies, 200–1100 nm in diameter, and filamentous ones, 50–400 nm in diameter and 200–3000 nm in length (Fig. 7.1). Phytoplasmas contain DNA, RNA, ribosomes and DNA strands. Extra chromosomal DNA can also be isolated from these organisms. The overall concentration of the phytoplasmas in plants is low and they are distributed unevenly throughout the plants. Molecular techniques are employed for diagnosis and for finer analysis of the interrelationships of strains. For identification of phytoplasmas in infected plants or in the leafhopper vector, the universal polymerase chain reaction (PCR) was the first technique to be used; it has been used in detecting animal mycoplasmas since 1989 (Bernet *et al.* 1989). Ahrens and Seemüller (1992) were the first to use this method for MLOs. The 16S rRNA gene is highly conserved within all prokaryotes and is used as a template DNA for the detection of phytoplasmas. Sequencing the 16S rRNA of the *Oenothera* phytoplasma revealed that plant-pathogenic phytoplasmas are

G. Loebenstein et al. (eds.),
Virus and Virus-like Diseases of Potatoes and Production of Seed-Potatoes, 145–158.

distinct from animal mycoplasmas (Lim and Sears, 1989). Two oligonucleotide primers were designed from the 16S rRNA of *Oenothera* phytoplasma, and were shown to be fully homologous with corresponding sequences of all other phytoplasmas, including animal and saprophytic mycoplasmas (Ahrens and Seemüller, 1992). Use of the restriction fragment length polymorphism (RFLP) technique enabled further diagnosis of phytoplasmas into cluster groups (Lee *et al.* 1998). The amplified sequences are digested by one of the two restriction endonucleases Alu I or Hink II. The profiles of the DNA fragments were distinct for each cluster group of phytoplasmas (Lee *et al.* 1998).

Cloned phytoplasma-specific DNA fragments from infected plants or leafhopper vectors which were used as probes in nucleic acid hybridizations also facilitated detection of phytoplasma strains (Lee and Davis, 1988). A series of phytoplasma 16S rRNA group-specific primer pairs were designed accordingly and were nested within the regions amplified by the universal primer pairs. They are now employed for the relatively easy detection of target phytoplasma-infected plants and vectors, and also to identify infection of more then one phytoplasma in infected plants or vectors (Lee *et al.* 1994). These assays have been improved during recent years. The most recent classification of phytoplasmas is based on RFLP analysis of the 16S rRNA gene and ribosomal protein (rp) gene sequence (Lee *et al.* 1998).

Most phytoplasmas are maintained in the laboratory by grafting diseased material onto periwinkle plants (*Catharanthus roseus*). Figures 7.2 A and B show periwinkle plants which have been inoculated with two isolates of aster yellows (AY), one isolate causing phyllody (7.2.B) and the second yellowing (7.1 A) but not phyllody.

Not all phytoplasmas of potatoes have been genetically diagnosed, therefore, it could be that we attribute several different identities to the same phytoplasma. In contrast to most plant viruses, phytoplasmas are not mechanically transmissible. They are probably not transmitted through seeds of infected plants. They can be transmitted by vegetative organs (*e.g.*, bulbs, tubers, stolones, etc.), by grafting of infected material and by dodders. In nature phytoplasmas are transmitted mostly by *Auchenorrhynchan* insects and only a few by psyllids. They are transmitted by the circulating propagative process. An insect vector, once infested by a phytoplasma remains infective for long periods, usually for the duration of its life. Some phytoplasmas can be vectored by many species of leafhoppers; for example, the aster yellows (AY) phytoplasma is transmitted by at least 24 known species of leafhoppers (Lee *et al.* 1998). Leafhoppers, such as the aster leafhopper *Macrosteles fascifrons*, may transmit the phytoplasma to a long list of plant species which belong to various botanical families (McCoy *et al.* 1989).

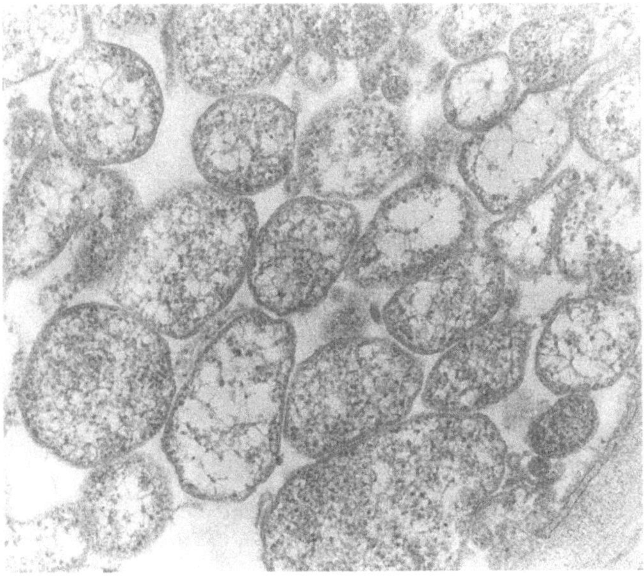

Fig.7.1. Safflower phyllody phytoplasma bodies in phloem cells of a diseased periwinkle plant (*Catharanthus roseus*). Courtesy of the Dept. of Virology, ARO. (Bar = 0.5 μ).

7.1 Aster Yellows (AY)

Phytoplasma group 16S rRNA I, subgroup B (western AY). Symptoms first described by Folsom (1926). First identified experimentally by Severin and Haasis (1934).

7.1.1 Symptoms and Damage

Growth of AY-infected plants is characterized by purple slender sprouts and purple aerial tubers which appear from the axils of the leaves. Small and narrow leaves are frequently developed from these abnormal tubers. The margins of the leaves usually roll inward. Petioles are very often curved downwards. Early effect of the pathogen on the leaves can be observed in Figures 7.2 C and B. Old leaves turn yellow and wilt comparatively early in the season. Internodes are short and many shoots develop from the axils of the leaves and frequently become purple: a phenomenon which mostly characterizes the infection of red varieties of potato. Aerial tubers grow on

Fig.7.2.A. Periwinkle plant (*Catharanthus roseus)* infected with a phytoplasma by the leafhopper *Orosius albicintus.* Yellowing of leaves and thin shoots are observed , but no flower disintegration.Fig.7.2 B. Beginning of phyllody in the flowers of an aster yellows infected periwinkle plant (*Catharanthus roseus)* (left), and a healthy plant (right).Figs. 7.2 C and D. Healthy (right) and aster yellows-infected (left) potato leaves. Fig.7.2.D – Asymmetry of upper leaflet due to the infection. Fig.7.2 C – Thickening of veins.Fig. 7.2 E. Tomato plant with tomato big bud symptoms – rolling of leaves and big buds.Figs. 7.2 F and G. Stolbur-infected potatoes. F- Aerial tubers; G- rolling and partial flavescence. (Courtesy of Dr. E. Tanne, Israel).Fig. 7.2 H. *Orosius albicintus,* the leafhopper vector of witches' broom, purple top roll and marginal flavescence in India, and of an aster yellow isolate in Israel . (Courtesy of Prof. E. Swirski, Israel

the secondary shoots also (Severin and Haasis, 1934). Tubers are not usually produced in AY-infected plants. Raymer and Milbrath (1960) described symptoms of a disease in potatoes called "the late breaking disease"; these symptoms are essentially identical to those described for the AY disease, and the pathogen is transmitted by the same leafhopper vector. These two diseases are probably caused by the same pathogen or by two closely strains of AY. The AY disease is an important limiting factor in potato production in certain sections of Canada, Mexico and the US (Macleod, 1954).

7.1.2 Natural Host Range

The western AY is transmitted by vectors to many hosts plants belonging to various botanical families (McCoy *et al.* 1989).

7.1.3 Geographical Distribution

The Western AY is distributed worldwide, infecting various host plants (*e.g.*, mulberry in Asia, cabbage in Europe, gladiolus in Italy, etc.) (Lee *et al.* 1998). Probably some of the known potato diseases of phytoplasmal etiology are caused by the western AY or by a closely related strain.

7.1.4 Strains

Fifteen phytoplasmal strains in the AY cluster have been identified by molecular techniques (Lee *et al.* 1998); some of them are transmitted by the aster leafhopper and therefore may include strains of the AY pathogen infecting potatoes. One of the strains seems to be the purple top wilt described in 7.2.

7.1.5 Transmission

The western AY pathogen is transmitted by the six-spotted leafhopper (= the aster leafhopper) *Macrosteles quadrilineatus* which was mistakenly identified in the past as *M. fascifrons* (Hamilton, 1983). Severin and Haasis (1934) infected many potato plants in laboratory with *M. fascifrons,* but found only one naturally infected plant in the field, whereas in the fifties the AY disease became a big problem in potato production in western North America (MacLeod, 1954). Other AY vectors were probably not tested on potatoes in North America.

7.1.6 Source Plants

Potatoes do not serve as source plants for the vector. Clover plants with phyllody symptoms, which were found near potato fields, could serve as a source of AY infection in potato (Raymer and Milbrath, 1960).

7.2 Purple Top Wilt (PTW) of Potatoes

The same group and subgroup as AY, in the USA (Lee *et al.* 1998); subgroup unidentified in other countries (Davis and Sinclair, 1998; Lee *et al.* 1998). Synonym: tomato big bud (Norris, 1954).

7.2.1 Symptoms and Damage

Symptoms of PTW in potatoes include stunting of the whole plant, leaf rolling and development of a purple or yellow pigmentation (depending on the potato variety) in the upper young leaves. Besides these symptoms, auxiliary buds proliferate heavily. PTW-infected plants wilt very soon after the symptoms become clear. The disease starts with cortical necrosis of the lower stems and discoloration of the vascular tissues. It seems that wilting of the shoots is caused by secondary invaders, rather than by the PTW organisms. Tubers are usually absent from infected plants, and those which are produced either do not sprout, or develop into hairy sprouts without tubers (Norris, 1954). The symptoms described for PTW in potatoes in Australia (Norris, 1954) show various similarities to those attributed to aster yellows in the US (MacLeod, 1954; Wright *et al.* 1981). PTW or aster yellows cause severe problems in the chip industry because of the browning of the vascular ring from the small amount of viable tubers produced on infected plants (Banttari *et al.* 1990). The damage caused by PTW in Australia ranges from trace to over 50% of the yield (Harding and Teakle, 1985); in the USA it is normally low but in some years it may be very high (Banttari *et al.* 1990).

7.2.2 Natural Host Range

The PTW is probably caused by AY everywhere. In the USA AY phytoplasma is transmitted to at least 190 plant species belonging to 42 botanical families (McCoy *et al.* 1989). Therefore, it seems that PTW can be hosted by plants all over the world, and that the disease etiology depends more on the vector than on a natural host range.

7.2.3 Geographical Distribution

PTW was described in Australia from grafts done in the early 1930s taken from tomato big bud-infected tomatoes (Samuel *et al.* 1933). Fig. 7.2 E shows a tomato plant infected with tomato big bud. A similar disease was recorded in Japan (Murayama *et al.* 1967). In the USA this disease (Leach and Bishop, 1946) was later proven to be caused by the AY agent (Li and McCrum, 1973). A disease with similar symptoms was identified in Israel (Zimmerman-Gries *et al.* 1972) and was proven to be caused by an isolate of AY (Klein, unpublished).

7.2.4 Strains

No typical strains of PTW have yet been identified. However, such strains may exist as we know that at least six different strains of the western AY have already been identified (Lee *et al.* 1992).

7.2.5 Transmission

Infected tubers fail to produce normal plants which might develop new tubers. Insect transmission – Grylls (1979) suggested that the leafhopper, *Orosius argentatus* is the vector of PTW in Australia. The vector in the USA is the aster leafhopper *M. fascifrons* (Li and McCrum, 1973) but it cannot be transmitted from a potato source (Banttari *et al.* 1990).

7.3 Stolbur Disease of Potatoes (=Tomatoes) (PS)

16S rRNA XII group (Stolbur group of phytoplasma), subgroup A (Davis and Sinclair, 1998; Lee *et al.* 1998).

7.3.1 Symptoms and Damage in Potatoes

First symptoms in young seedlings appear when plants assume their vertical growth. Leaves start rolling and the youngest ones become flavescent. In a later stage of growth shoots become increasingly stunted, relatively thick, and start producing aerial tubers (Figures 7.2.F and G) and aerial stolons from the leaf axis. All symptom-showing shoots wilt very soon. Infected plants do not usually produce normal tubers: the tubers, if present, remain very small, and sometimes flaccid, and they produce spindly sprouts. Plants developing from infected tubers are usually dwarfed, with thin shoots and tiny simple leaflets (Valenta *et al.* 1961). In some areas the rate of disease incidence may reach about 90% (Çitir, 1985).

7.3.2 Natural Host Range

The PS disease has been identified in nature in several other hosts such as *Solanum nigrum, Datura stramonium* and *Convolvulus arvensis* in addition to potato and tomato.

7.3.3 Geographical Distribution

The PS disease of potatoes is known, so far, only from eastern and southern Europe (Blattny *et al.* 1954; Cousin and Moreau, 1977) and from the Middle East (Çitir, 1985; Zimmerman-Gries, 1970).

7.3.4 Strains

Various strains of stolbur known in potato differ from one another in their virulence or in the symptoms induced in some hosts.

7.3.5 Transmission

The pathogen is easily transmitted by dodders and by grafting. Insect transmission – the planthopper species *Hyalesthes obsoletus* was established by Sukhov and Vovk (1945) as the first known and efficient vector of the PS pathogen. A closely related species, *H. molokosiewiczii,* had been reported as the vector of PS in the Georgian S.S.R. (Valenta *et al.* 1961). Leafhopper species *Aphrodes bicinctus, Euscelis plebejus* and *Macrosteles laevis* were also identified as vectors to the PS pathogen (Musil and Valenta, 1958; Valenta *et al.* 1961).

7.3.6 Natural Hosts of Stolbur

Bindweed (*C. arvensis*) is considered in Europe as the principle reservoir of the PS phytoplasma (Savulescu and Ploaie, 1967).

7.4 Potato "Witches' Broom" (PWB)

Phytoplasma group: Clover proliferation 16Sr RNA VI subgroup: A (Lee *et al.* 1998).

7.4.1 Symptoms and Damage

First symptoms of infection with PWB are very similar to those of the AY (western strain): upward rolling, occasionally marginal chlorosis of the upper leaflets. Differences in the progressive stage of infection: stem growing in a cylindrical shape rather than angular, bulbous nodes. Numerous fine stems with simple leaves develop around the bases of infected plants. Many small pointed tubers are usually produced, in contrast to the very few tubers yielded by AY-infected plants. These tubers have spindling sprouts appearing from most of the buds. After some time these become very branched with extremely numerous, fine cylindrical stems with the typical witches' broom feature (Wright, 1952). Sometimes there are also various degrees of chlorosis or of leafrolling. From time to time, aerial tubers may be found. Flowers usually degenerate with small downward curling phylloid sepals (Nagaich, 1979). In North America it is a minor disease and the economic importance elsewhere is also low. In India incidence sometimes exceeds 30%.

7.4.2 Natural Host Range

There was no information about natural hosts in western North America. Additional hosts of PWB were found experimentally in India. They include various species of the Solanaceae, one species each of Compositae and Leguminosae, and *C. roseus* (periwinkle) (Nagaich *et al.* 1978).

7.4.3 Geographical Distribution

In spite of its low importance, the disease is widespread. PWB can be found in North America (Wright, 1952), in Japan, where in fact, the first microscopic observation of phytoplasma particles was made in PWB-infected material (Doi *et al.* 1967), in Europe (Valenta, 1969), in India (Nagaich *et al.* 1974), in China (Tu, 1939) and in Australia (Anonymous, 1943).

7.4.4 Strains

Variations occur in the degree of chlorosis, the amount of leafrolling and attainment of the advanced stage of infection. There were minor differences between the plants, in the symptoms appearing when they were grafted onto tomato and tree tomato (*Cyphomandra betacea*), but this does not indicate the involvement of different strains (Wright, 1952). On the other hand, five distinct strains of PWB were identified in India according to their reactions on a set of solanaceous differentials, *i.e.*, several varieties of tree-tomato, brinjal and several varieties of tomato (Khurana *et al.* 1979). Two distinct strains have been isolated in Europe (Valenta, 1969).

7.4.5 Transmission

PWB is readily transmitted with tubers (Khurana *et al.* 1979; Wright, 1952) and the pathogen may be transmitted by several dodder species (Khurana *et al.* 1979). Insect transmission – transmission by the leafhopper vector has not yet been investigated in North America. The pathogen of PWB is transmitted in India by two leafhopper species: *Alebroides dravidanus* and *Orosius albicintus* (Fig. 7.2 H) (Khurana *et al.* 1979). It is transmitted in Japan by the leafhopper *Ophiola* (= *Scleroracus*) *flavopicta* (Fukushi *et al.* 1955). In Europe no vector has yet been identified.

7.5 Purple Top Roll (PTR or PT) of Potatoes

16S rRNA I group (aster yellows group, subgroup unidentified (Davis and Sinclair, 1998; Lee *et al.* 1998). PTR in potatoes is also referred to as purple top, purple top wilt, purple dwarf, haywire, bunchy top and several other terms (MacLeod, 1954).

7.5.1 Symptoms and Damage

First symptoms are rolling of the basal parts of the leaflets and reduced size of the appearing young leaves. These symptoms are accompanied in most potato varieties by a color change to purple. Infected plants do not flower even under suitable conditions (short days). Tubers produced by infected plants remain small. Tubers from infected plants and the plants themselves may show symptom recovery when kept at 36–40°C for 4 h daily (Nagaich and Giri, 1973). In France the PTR disease induces aerial tubers, and the ground tubers develop grassy shoots (Cousin, 1975). In India the disease may cause tonnage losses of about 50–75%, through severe and early infection (Nagaich, 1979).

7.5.2 Natural Host Range of PTR

PTR is observed on potato, tomato, *Datura*, several species of *Nicotiana, Calendula officinalis, Physalis floridana Trifolium repens* and periwinkle plants (*C. roseus*), of which the last three species showed only mild symptoms (Nagaich *et al.* 1978).

7.5.3 Geographical Distribution

PTR has been reported from several localities in India (Nagaich and Giri, 1973) and from France (Cousin, 1975).

7.5.4 Strains

Two distinct strains of PTR have been isolated in India. These two strains differed in symptoms and virulence on potato varieties and had great effects on yield (Nagaich, 1979).

7.5.5 Transmission

The PTR disease in India was transmitted by tubers from infected plants, but infectivity is lost in the second generation. Three different leafhopper species, *O. albicintus, A. dravidanus* and *Seriana equata*, transmitted the pathogen in India (Nagaich and Giri, 1973). The vector of the PT pathogen in France is still unknown.

7.5.6 Source Plants for Infection

Infected clover plants (*T. repens*) are suspected to be the natural reservoir of the PTR pathogen in India (Nagaich and Giri, 1973).

7.6 Potato Phyllody (PP), Hairy Sprout (HS), and Marginal Flavescence (MF) in India

7.6.1 Symptoms and Damage

The four diseases PWB, PP, HS and MF found in India exhibit their characteristic symptoms only under appropriate environmental condition and at the specific growth stages of the infected plants for which each disease is typical. In the early stages of infection it is sometimes difficult to distinguish MF from PTR and PWB from PP.

Potato phyllody – Initially, large phylloid flowers can be seen on infected plants but at plant maturity a number of rudimentary flowers develop, in the form of stunted anthers fused on the small sepals. These late flowers give rise to aerial roots and pigmented tubers. The plants gradually become severely chlorotic, with elongated internodes; the stems are enlarged with wavy wings and margins; and the compound leaves are intensely chlorotic and hairy with elongated petioles (Nagaich *et al.* 1978).

Hairy spout – HS is probably a reaction to one of the four other phytoplasmal diseases (PWB, PTR, PP and MF) found in India, or to all of these diseases (Nagaich *et al.* 1973). The plants which emerged from HS tubers had weak, spindly stems with small and simple leaves.

Marginal flavescence – The young leaves of potato plants with MF disease become chlorotic. The plants remain stunted with short bushy internodes. The color of the leaves in infected plants is light yellow, and several of them show clear flavescence symptoms along their margins. Infected plants produce very few and very small tubers. Incidence of MF can reach 35% in some areas in India.

7.6.2 Host Range

Eighty-three species/varieties of plants belonging to nine botanical families have been tested, mainly through grafting, for PTR, MF, PWB and PP. Of 82, 73, 83, 57 of these species tested as hosts 45, 32, 40 and 28 have been confirmed to be new hosts of PTR, MF, PWB and PP, respectively (Nagaich *et al.* 1978).

7.6.3 Transmission

No vector for the PP and the HS pathogen has yet been reported.

The MF pathogen was transmitted by tubers and by two species of leafhoppers: *O. albicintus* and *S. equata* (Nagaich *et al.* 1978).

References

Ahrens, U. and Seemüller, E. 1992. Detection of DNA of plant pathogenic mycoplasmalike organisms by a polymerase chain reaction that amplifies a sequence of the 16S rRNA gene. *Phytopathology* 82:828-832.

Anonymous 1943. Virus diseases of potatoes. *Agr. Gaz. N.S. Wales* 54: 359-362.

Banttari, E.E., Orr, P.H. and Preston, D.A. 1990. Purple top as a cause of potato chip discoloration. *Trans. ASAE* 33: 221-226.

Bernet, C., Garret, M., de Barbeyrac, B., Bebear, C. and Bonnet, J. 1989. Detection of *M. pneumoniae* by using the polymerase chain reaction. *J. Clin. Microbiol.* 27: 2492-2496.

Blattný, C., Brčàk, I., Pozděna, J., Dlabola, J., Limberk, J. und Bojňanský, V. 1954. Die ubertragung des Slolburvirus bei Tabak und Tomaten und seine virogeographischen Beziehungen. *Phytopathol. Z.* 22: 381-416.

Brčàk, I., Králik, O., Limberk, J. and Ulrychova, M. 1969. Mycoplasma-like bodies in plants infected with potato witches' broom disease and the response of plants to tetracycline treatment. *Biol. Plantarum* 11: 470-476.

Çitir, A. 1985. Preliminary investigation of potato diseases caused by mycoplasmalike organisms (MLO) in Erzurum region. *J. Turk. Phytopathol.* 14: 53-53.

Cousin, M.-T. 1975. L'Enroulement violac de la pomme de terre. *Ann. Phytopathol.* 7: 167-173.

Cousin, M.-T. and Moreau, J.-P. 1977. Les Stolbur des Solanaceés. *Phytoma* 292 (November): 15-19.

Davis, R.E. and Sinclair, W.A. 1998. Phytoplasma identity and disease etiology. *Phytopathology* 88: 1372-1276.

Doi, Y., Teranaka, M., Yora, K. and Asuyama, H. 1967. Mycoplasma-like or PLT group-like microorganisms found in the phloem elements of plants infected with mulberry dwarf, potato witches broom, aster yellows, or paulownia witches broom. *Ann. Phytopath. Soc. Japan* 33: 259-267.

Folsom, D. 1926. Virus diseases of the potato. *Que. Soc. Prot. Plants Rept.* 1925-1926: 14-29.

Fukushi, T., Shikata, E., Shioda, H., Sekiyama, E., Tanaka, L., Oshima, N. and Nishia, Y. 1955. Insect transmission of potato witches broom in Japan. *Proc. Jap. Acad.* 31: 234-235.

Grylls, N.E. 1979. Leafhopper vectors in Australia. In: "*Leafhopper Vectors and Plant Disease Agents.*" (ed.) K. Maramorosch and K.F. Harris, pp. 179-214. Academic Press, New York.

Hamilton, K.G.A. 1983. Introduction and native leafhoppers common to the old and new worlds (Rhynchota: Homoptera: Cicadellidae). *Can. Entomol.* 115: 473-511.

Harding, R.M. and Teakle, D.S. 1985. Mycoplasma-like organisms as causal agents of potato purple wilt in Queensland. *Aust. J. Agric. Res.* 36: 443-449.

Khurana, S.M.P., Singh, S. and Nagaich, B.B. 1979. Indentification of five strains of the potato witches' broom pathogen in India. *JIPA.* 6: 119-129.

Leach, J.G. and Bishop, C.F. 1946. Purple top wilt (blue stem) of potatoes. *Univ. W. Va. Agric. Exp. Stn. Bull.* No. 326.

Lee, I.-M. and Davis, R.E. 1988. Detection and investigation of genetic relatedness among aster yellows and other mycoplasmalike organisms by using cloned DNA and RNA probes. *Mol. Plant Microbe Interact.* 1: 303-310.

Lee, I.-M, Davis, R.E., Chen, T.-A., Chiykowski, L.N., Fletcher, J., Hiruki, C. and Schaff, D.A. 1992. A genotype-based system for identification and classification of mycoplasmalike organisms (MLOs) in aster yellows MLO strain cluster. *Phytopathology* 82: 977-986.

Lee, I.-M., Gundersen-Rindal, D.E. and Bertaccini, A. 1998. Phytoplasma: Ecology and genetic diversity. *Phytopathology* 88: 1359-1366.

Lee, I.-M., Gundersen-Rindal, D.E. and Davis, R.E. 1994. Use of mycoplasmalike organism (MLO) group specific oligonucleotide primers for nested-PCR assays to detect mixed-MLO infections in a single host plant. *Phytopathology* 84: 559-566.

Li, T.C.C. and McCrum, R.C. 1973. Electron microscopy of purple-top wilt of potato in relation to the suspected mycoplasmalike pathogen. *Phytopathology* 63: 204 (Abstr.).

Lim, P.O. and Sears, B.B. 1989. 16S rRNA sequence indicates that plant-pathogenic mycoplasmalike organisms are evolutionarily distinct from animal mycoplasmas. *J. Bacteriol.* 171: 5901-5906.

MacLeod, D.J 1954. Aster yellows (purple-top) of potatoes. *Am. Potato J.* 31: 119-127.

McCoy, R.E., Caudwell, A., Chang, C.J., Chen, T.A., Chiykowski, L.N., Cousin, M.-T., Dale, J.L., de Leeuw, G.T.N., Golino, D.A., Hackett, K.J., Kirkpatrick, B.C., Marwitz, R., Petzold, H., Sinha, R.H., Sugiura, M., Whitcomb, R.F. Yang. I.L., Zhu, B.M. and Seemüller, E. 1989. Plant diseases associated with mycoplasmalike organisms. In: "*The*

Mycoplasma". (ed.) R.F. Whitcomb and J.G. Tully. Vol. 5 pp. 545-560. Academic Press, New York.

Musil, M. and Valenta, V. 1958. Pernos stolbura a pribuzných virusov pomocou niektorých cikád (Transmission of the stolbur and related viruses by some leafhoppers). *Biológia* (Bratislava) 13: 133-136.

Nagaich, B.B. 1979. The yellows diseases of potatoes. *JIPA* 6(1): 1-11.

Nagaich, B.B. and Giri, B.K. 1973. Purple top roll disease of ptoato. *Potato J.* 50: 79-85.

Nagaich, BB., Lal, S.B. and Upreti, G.C. 1973. Hairy sprout disease of potatoes. *Ind. Potato* J. 5: 71-75.

Nagaich, B.B., Khurana, S.M.P., Singh, S. and Singh, V. 1978. Studies on the potato witches' broom and phyllody disease. *Proc. 7th Triennial Conf. EAPR*, 26th June-1st July, Warsaw, Poland, pp.289-290 (Abstr.).

Norris, D.O. 1954. Purple top wilt, a disease of potato caused by tomato big-bud virus. *Aust. J. Agric. Res.* 5: 1-9.

Raymer, W.B. and Milbrath, J.A. 1960. The identity and host relations of the potato late-breaking virus. *Phytopathology* 50: 3212-314.

Samuel, G., Bold, J.G. and Fardley, C.M. 1933. ''Big bud", a virus of the tomato. *Phytopathology* 23: 641-653.

Savulescu, A. and Ploaie, P.G. 1967. Virogeographische Studien über das kleeberlaubungs Virus und seine Vektoren. *Phytopathol.* Z. 58: 315-322.

Severin, H.H.P. and Haasis, F.A. 1934. Transmission of California aster yellows to potato by *Cicadula divisa. Hilgardia* 8: 328-335.

Suknov, K.S. and Vovk, A.M. 1945. Cikada *Hyalesthes obsoletus* Sign. perenoschik stolbura paslyonovykh (The leafhopper *Hyalesthes obsoletus* Sign. a vector of stolbur of the Solanaceae) *Doklady Acad. Nauk, S.S.S.R.* 53: 153-156.

Tu, T.F. 1939. A list of plant viruses observed in China. *Phytopathology* 29: 459-461.

Valenta, V. 1969. Vergleich eines aus Niedersachsen stammenden Kartofelhexenbesen-Virus mit anderen aus Europa bekhannten Viren dieser Gruppe. *Zentr. Bakteriol. Parasiten. Infekt. Hygiene* 123: 352-357.

Valenta, V., Musil, M. and Misiga, S. 1961. Investigation on European yellows-type viruses. I. The stolbur virus. *Phytopathol.* Z. 42: 1-38.

Wright, N.S. (1952). Studies on the witches broom virus disease of potatoes in British Columbia. *Can. J. Bot.* 735-742.

Wright, N.S., Raine, J. and Valenta, V. 1981. Mycoplasmas. In: "*Compendium of Potato Diseases*" (ed.) W.J. Hooker. pp. 91-98, The American Phytopathology Society, St. Paul.

Zimmerman-Gries, S. 1970. "Stolbur" a new potato disease in Israel. *Potato Res.* 13(2): 146-150.

Zimmerman-Gries, S., Klein, M. and Sneh, B. 1972. Incidence of yellows diseases in potato fields grown for certified seeds in Israel. *6th Triennial Conf. EAPR*, 15-19 Sept. 1975, Wageningen, The Netherlands, pp. 196-197 (Abstr.).

8. Transmission of Viruses

8.1 Mechanically Transmissable Viruses of Potato

Gary D. Franc[1] and Ernest E. Banttari[2]
[1]University of Wyoming, Department of Plant Sciences, POB 3354, Laramie, WY 82071-3354 (francg@uwyo.edu) and [2]University of Minnesota, Department of Plant Pathology, 495 Borlaug Hall, St. Paul, MN 55108. (ernestb@puccini.crl.umn.edu), USA

8.1.1 Introduction

In the early 1900's farmers, potato breeders, and plant pathologists described problems in potatoes now believed to have been caused by viruses. Terms such as "running out" and "degeneration" were coined to describe the general "loss of vigor," "loss of stamina," and "loss of yielding power" associated with tubers harvested from potato cultivars affected by viruses (Krantz and Bisby, 1921; Whipple, 1919). Schulz *et al.* (1919) and Schulz and Folsom (1920) were able to mechanically transmit a "mosaic dwarf agent," probably *Potato virus Y* genus *Potyvirus* (PVY), but perhaps including *Potato virus X* genus *Potexvirus* (PVX), to potato that caused running out. The discovery that several North-American potato cultivars that appeared to be healthy, but were infected with one or more mechanically transmitted viruses, was made by Johnson (1925). He transmitted a virus or viruses latent in potato plants, to several other *Solanaceous* spp. including tobacco, tomato, and eggplant, which subsequently developed various symptoms including mild mottle, spot-necrosis or ringspot. The virus or one of the viruses he transmitted mechanically eventually came to be known as PVX. Since then, numerous viruses mechanically transmissible to potato

G. Loebenstein et al. (eds.),
Virus and Virus-like Diseases of Potatoes and Production of Seed-Potatoes, 159–175.

have been characterized. Some also have vectors that serve as the principal means of transmission (Table 8.1).

Virus-infected seed tubers serve as an important source of primary inoculum for mechanically transmitted viruses. Depending on the virus, secondary spread from infected seed tubers occurs during seed handling and, following tuber sprouting and plant growth, via plant-to-plant contact. The degree to which the spread of mechanically transmissible viruses occurs is dependent on many factors, including the host, environment, and foliage contact by animals, equipment, and man. The detection and elimination of infected seed tubers effectively reduces the carryover and geographical movement of virus inoculum, and, thus forms the basis for maintaining healthy potato seed stocks. Factors influencing the movement of viruses to tubers and the secondary spread of inoculum are discussed below.

8.1.2 Penetration and Host Infection by Mechanically Transmitted Viruses

Mechanically transmissible viruses have no intrinsic means for entering through the surface of intact plants. Wounds that permit virus entry into cells, yet are not severe enough to preclude infection by killing cells, are avenues for virus entry. Inoculum efficiency for mechanically transmissible viruses is increased in the presence of abrasive powders on foliage during mechanical inoculation (Matthews, 1970). Increased wounding caused by the abrasive increases the frequency of infectable wounds created during mechanical inoculations. Under natural conditions in the field, sublethal wounding of plant cells readily occurs during normal leaf abrasion, cultivation practices, and through other types of foliar contact, including that by various animals and man.

Matthews (1970) estimated that mechanical inoculation requires 10^4 to 10^5 virus particles applied to leaf surface for each cell infected. Therefore, it follows that the spread of mechanically transmissible viruses to healthy plants is favored by high concentrations of the virus in epidermal cells and if the host plants are easily wounded during abrasion. This scenario places the greatest concentration of viral inoculum at the infection court. Indeed, inclusion bodies containing virions of various mechanically transmitted viruses are present in plant cells associated with the leaf surface, and all epidermal cell types including leaf hairs and guard cells, are capable of becoming infected during mechanical inoculation (Matthews, 1970). The presence of numerous ectodesmata, the plasmodesmata that occur in the outer cell wall of epidermal cells, also may increase host susceptibility during mechanical inoculation by providing pathways for virus entry (Brants, 1964; Thomas and Fulton, 1968). Virus particles directly in contact with protoplasmic membranes exposed during wounding are probably assimilated into host cells by pinocytosis (Mandahar and Garg, 1990). Once the virus enters the host cell, uncoating and exposure of viral nucleic acid is necessary for infection to proceed. Viral nucleic acid is subsequently translated and/or replicated by viral-encoded replicases and host cell machinery is utilized for the eventual production of new virions.

Once infection of the cell occurs, the virus apparently utilizes the same trafficking system for cell-to-cell movement that also is used to move plant-derived macromolecules from cell to cell (Gilbertson and Lucas, 1996). Because viruses appear to be able to replicate in cells from a wide range of non-host species, it is hypothesized that both non-host and host resistance stems from the virus' inability to move from the initial site of infection and/or the ability of the host to interfere with such movement (Gilbertson and Lucas, 1996). Although few studies have been done with mechanically transmitted viruses of potato, a 25 kDa "movement protein" has been identified for PVX (Angell *et al.* 1996; Malcuit *et al.* 1999). In a compatible host reaction, this movement protein facilitates cell-to-cell movement by interacting with plasmodesmata to increase the molecular size exclusion limit (SEL), thus, facilitating viral movement among cells. In resistant potato cultivars, the PVX movement protein triggers a hypersensitive response suggesting that resistance for this particular virus-host interaction prevents extensive cell-to-cell virus movement (Malcuit *et al.* 1999). The infection front of mechanically transmitted viruses, characterized as localized and slowly spreading from cell to cell, progresses rapidly throughout the plant once virus enters the vascular tissue. Mechanisms utilized for cell to cell movement apparently differ from those utilized for long distance movement of the virus within vascular tissue (Gilbertson and Lucas, 1996).

Mechanically transmissible viruses must utilize the vascular system for movement throughout the plant and eventual translocation to tubers. Therefore, systemic movement of the virus following primary infection as well as systemic secondary infection, characterized by virus movement from infected tubers to shoots of the developing plant, is dependent upon eventual virus entry into and exit from the host vascular system. Natural selection will favor those viruses that move most rapidly to tubers, thus, providing a means for inoculum survival between growing seasons. Indeed, numerous independent studies demonstrated that reduced virus translocation to tubers is correlated with increased natural resistance to virus infection (Beemster, 1958, 1961, and 1969; De Bokx, 1968; Dedic, 1979; Franc and Banttari, 1996).

One method to measure virus translocation rates in potato is by determining the incidence of tubers infected at various time intervals following foliar inoculation. Studies that measured virus movement from mechanically inoculated leaves to daughter tubers generally revealed that fewer tubers became infected as increasingly older plants were inoculated (Beemster, 1958, 1961, and 1969; De Bokx, 1968; Dedic, 1979). Thus, "mature plant resistance" was the term coined to describe this relationship. The degree of mature plant resistance observed is influenced by cultivar, virus isolate, and, no doubt, by other factors. Several examples detailing the systemic movement of mechanically transmissible viruses in potato are described below.

The rate of *Potato virus A* genus *Potyvirus* (PVA) movement from inoculated leaves to tubers was greater in susceptible cultivars and less dependent on age than it was for a cultivar known to be more resistant (Dedic, 1979). In the susceptible cultivar Jara, 6 to 7 days were required for PVA movement from inoculated leaflets to tubers. The incidence of infected tubers increased as the time between inoculation and tuber assay increased. Weak mature plant resistance was detected in plants 8 and 10 weeks old at the time of inoculation, with 84 and 56 percent of the tubers becoming infected, respectively. However, for the resistant cultivar Krasava, it was increasingly difficult to detect virus movement out of inoculated leaves to tubers, and plants 8 weeks old at the time of inoculation failed to develop infected tubers.

Potato virus S genus *Carlavirus* (PVS) moved out of mechanically inoculated leaves within 24 hour (Franc and Banttari, 1996). However, 13 and 20 days were required before primary infection was detected by enzyme-linked immunosorbent assay (ELISA) in foliage sampled above and below the inoculated leaf, respectively. Translocation of PVS from inoculated leaves to daughter tubers occurred within 13 days for cultivars Russet Burbank and Norland and within 20 days for the more resistant Kennebec. Inoculation of European potato cultivars revealed that PVS could be translocated to tubers within 14 days following inoculation of a single leaf (De Bokx, 1968; De Bokx and Waterreus, 1967). Mature plant resistance to European PVS isolates was readily detected 4 weeks after planting and became more pronounced at 6 weeks (De Bokx, 1968). The degree of mature plant resistance varied with cultivar and with PVS isolate. In contrast, Franc and Banttari (1996) were unable to detect mature plant resistance to a Minnesota isolate of PVS. They hypothesized that the lack of mature plant resistance explained, at least partially, the rapid natural reinfection of healthy potatoes observed in Minnesota. The lack of mature plant resistance would contribute to tuber reinfection by enabling PVS spread to occur throughout the entire growing season, especially late in the growing season when vines are larger and foliage contact is most likely. In contrast, spread of European PVS isolates was limited to the early portion of the growing season because older plants become increasingly resistant. Thus, mature plant resistance would reduce reinfection of seedlots in European potato production areas, even when inoculum is present.

PVY is mechanically transmissible and also is transmitted by aphids in a non-persistent manner. Following mechanical inoculation of single leaves of cultivar Bintje, the earliest PVY translocation to tubers required longer than 2 days, but less than 4 days (Beemster, 1961). Mature plant resistance was detected because, as plants became increasingly older, it took more time for PVY to translocate to tubers. Assays of tubers from inoculated plants revealed that, if a plant yielded infected tubers, it did not imply that all tubers were infected from that plant. Furthermore, infection was not necessarily uniform in a given tuber, with some tubers possessing both infected and healthy eyes. A relative comparison of PVY translocation rates to that of the necrotic strain (PVY^N), revealed that PVY^N was translocated more rapidly to tubers following foliar inoculation (Beemster, 1976). Although mature plant resistance was demonstrated for both viruses, it was more evident for PVY than for PVY^N.

Leaflets of several European cultivars were mechanically inoculated with PVX and the percentage of tubers that became infected was measured over time (Beemster, 1958). The earliest PVX translocation to tubers required longer than 2 days, but less than 4 days. In general, as plant age increased at the time of inoculation, the percentage of tubers infected with PVX decreased. The degree of mature plant resistance was cultivar dependent and became evident for cultivar Bintje when plants were at least 9 weeks old. However, even old plants yielded infected tubers. Beemster concluded that old leaves were readily infected by inoculation and that, perhaps, the vascular system becomes increasingly "less penetrable for the virus," thus accounting for the inhibited translocation of PVX to tubers following the inoculation of older plants.

The presence of multiple stems on a single plant also was found to influence the pattern of virus distribution in the plant following mechanical inoculation (Beemster, 1969). Inoculation of a leaf on one stem of two-stemmed potato plants (cv. Bintje) favored virus movement to tubers produced on the inoculated stem rather than tubers produced by noninoculated stems on the same plants. Tubers recovered from noninoculated stems were infected less frequently than tubers produced by inoculated stems. This effect was more obvious for PVX, whereas inoculation with the necrotic strain of PVY (PVY^N) had a greater propensity to infect tubers from both inoculated and noninoculated stems. In this experiment, at least 14 days after inoculation was allowed for virus movement to tubers.

8.1.3 Rates of Reinfection and Cultural Practices affecting Mechanical Spread of Potato Viruses.

8.1.3.1 Reinfection Rates

In experiments performed in Scotland, Cockerman (1958) found that reinfection of cultivar Majestic seed potatoes by PVX, starting from a virus incidence of approximately 1 percent infection, increased over the course of a three-year study to a final incidence of 4.8 percent. These seed potatoes were planted in isolated plots to prevent the introduction of PVX from inoculum sources outside of the plot. In contrast, plots of the same parent stock surrounded by PVX-infected potatoes, experienced a reinfection of 82.9 percent over the same period. It was postulated that the spread of virus was mainly, although not entirely, by plant to plant contact, and that the close proximity of PVX-infected fields to healthy potato seed plots provided an important source of inoculum that accelerated reinfection. Implements used in various cultural operations also were implicated in the movement of virus and the rapid reinfection of healthy potatoes by PVX.

Aphid-transmissible isolates of PVS may influence reinfection rates measured for some production areas (Bode and Weidemann, 1971; Kostiw, 1975; Wardrop *et al.* 1989). In a trial program, healthy potato seed stocks derived from tissue-culture, were increased in Minnesota on four farms that were part of the Foundation Certified Seed Potato program. Reinfection of five potato cultivars with PVS and PVX was evaluated in 1976-1978 (Banttari *et al.* 1978). After three generations of increase in the field, the overall means for the percentage of tubers reinfected with PVS/PVX were: Norland 24/1, Norchip 16/Trace, Kennebec 15/1, Norgold Russet 31/Trace and Russet Burbank 32/6. Isolation from other potatoes and differing cultural practices may have been important factors influencing reintroduction of these viruses. Although reinfection with PVX, PVY or aphid-vectored *Potato leafroll virus* genus *Polerovirus* (PLRV) was slight during the 3-year trial period, PVS reinfected these stocks more rapidly.

Hahm *et al.* (1981) reported reinfection of cultivars Norgold Russet and Ontario with PVS, and LaChipper, Norchip, and Norgold Russet with PVX. They attributed differences in reinfection rates to the relative susceptibility of the cultivars to each virus and to differences in cultural practices among growers. When composite samples were considered, 59 and 70 percent were PVS-positive, and 19 and 55 percent were PVX-positive after 1 and 3 generations in the field, respectively. Manzer *et al.* (1978) also found that PVS reinfected seed stocks more rapidly than did PVX.

Franc and Banttari (1996) observed rapid reinfection of healthy stocks from point source inoculations with PVS. This PVS isolate was not aphid-transmissible. These studies were conducted at two field sites over a two year period with Russet Burbank, Norland, and Kennebec. Reinfection rates for Norland (71.8%) and Russet Burbank (73.0%) did not differ significantly, but were significantly greater than that for the more resistant Kennebec (29.5%; $P \leq 0.01$). Khalil and Shalla (1982) detected plant-to-plant spread of PVS following leaf contact among potatoes in the greenhouse and field. The virus was not an aphid-transmissible isolate. After two months in contact with PVS-infected plants in an insect-free greenhouse, five of 30 previously healthy plants became infected. In field plots, 11 of 108 plants previously free of PVS became infected. De Bokx (1972) established field plots which contained approximately 10 percent of the plants with secondary PVS infections. After one growing season, increased primary infections of 56 to 76 percent were measured for cultivar Eersteling and two to 28 percent were measured for cultivar Alpha. The pattern of infected plants indicated that under Dutch field conditions PVS spread was probably by foliar contact.

Healthy potato seed stocks are often derived from plantlets increased in tissue culture. Plantlets are subsequently transplanted into greenhouses or screenhouses to reduce chances of virus re-introduction and to protect the minituber crop. However, plantlets are sometimes transplanted directly into field plots for production of seed. McDonald (1987) compared PVY and PVS reinfection of plantlets transplanted in the field versus plants grown from tubers. Comparative leaf assays for two years revealed significantly greater infection with both viruses in transplanted plantlets compared to reinfection of plants propagated from tubers ($P \leq 0.05$). Plantlets flowered approximately 1 to 2 weeks later than did tuber-generated plants and, thus, later maturity may have accounted for the differences in virus infection. He concluded that in addition to the inherent resistance of cultivars, mature plant resistance may have further reduced susceptibility of the tuber-generated plants. Therefore, cultural practices that hasten maturity of plantlets are recommended to help reduce movement of virus to tubers.

8.1.3.2 Yield Effects

Teri *et al.* (1977) found no significant differences in total tuber yield for three cultivars infected with a mild strain of PVX at Ithaca and Mount Pleasant, NY. However, at Riverhead, NY, the yield of larger tubers was significantly reduced ($P \leq 0.05$) for all three cultivars. Wright (1970) compared the yields of healthy and PVX plus PVS-infected seed stocks for cultivars Netted Gem and White Rose. Seed stocks freed from these viruses by thermotherapy and planted in British Columbia, Oregon, and California yielded 10 to 25 percent more than did the same seed stocks that carried these viruses. Healthy seed stocks of these cultivars planted in California and Oregon produced 15 to 32 percent more tubers than did PVS plus PVX-infected stocks. In a field trial in British Columbia, healthy Netted Gem produced 10 to 12 percent more tubers and yields exceeded those of regular foundation seed stock by 11,655 kg per ha with a marketable increase of almost 30 percent. Wright (1977) also reported that mottle and latent strains of PVX and PVS did not affect specific gravity and numbers of tubers of Netted Gem potatoes. However, the mottle strain or the latent strains of PVX reduced tuber size and yields significantly ($P \leq 0.05$) when plants were co-infected with PVS.

Manzer *et al.* (1978) found that overall yields for healthy seed stocks of cultivars Katahdin, Kennebec and Russet Burbank were 3 percent more than those of PVS-infected stocks and 5 percent more than yields for stocks co-infected with PVS plus a mild strain of PVX. When seed stocks were co-infected with a moderate PVX strain plus PVS, losses were an additional 5 percent. Infections were mostly latent, and mosaic symptoms were observed only in plants co-infected with a moderate strain of PVX plus PVS.

8.1.3.3 Virus Sources and Methods of Spread

Wright (1974) tested the retention of infectivity of PVX by applying sap from infected potato leaves to various materials that commonly contact potato tubers during harvesting, grading, and storage and materials that contact foliage during plant growth. The virus retained infectivity for 3 hr on unpainted wood, iron, rubber and human skin; 6 hr on painted wood, jute, and cotton; and 24 hr on soil. Longevity *in vitro* of PVX was at least five days, so virus infectivity may be longer in other test environments. Franc and Banttari (1984) reported that infectious PVS was recovered from various surfaces contacted by tubers or potato plants during seed cutting, planting, or cultural practices conducted in the field. Infectious PVS persisted for at least 25 and 120 hr on soil and burlap, respectively. If materials were kept in a moist environment at 4°C, infectious PVS was recovered for up to 59 and 24 hr on metal and rubber, respectively.

Contaminated cutting knives transmitted PVS to healthy tubers of cultivars Norland, Kennebec and Russet Burbank (Franc and Banttari, 1984). Greenhouse studies revealed that 15.5, 4.2 and 24.5 percent of Norland, Kennebec and Russet Burbank plants, respectively, grown from seedpieces became PVS-infected when the knife was contaminated by cutting PVS-source tubers, avoiding eyes or sprouts. When PVS-source tubers were cut through sprouts, 24.7, 54.2 and 45.2 percent of the plants for Norland, Kennebec and Russet Burbank plants, respectively, became infected. Banttari, Gyenis and DiFonzo (unpublished) were unable to demonstrate significant reinfection of three potato cultivars by PVY using three different types of commercial seedpiece cutters. However, Draper and Gudmestad (1992) reported that cutting tubers with hand-held cutting knives resulted in PVY transmission. A ringspot strain of PVX also was transmitted with greater frequency by the cutting knife when virus-source tubers were cut through eyes (52 percent) compared with cuts made through source tubers avoiding eyes (24 percent) (Larson, 1950). Because mechanically transmissible viruses are potentially transmitted among tubers during seed cutting, hand-held cutting knives should be disinfested between tuber cuts. Even low levels of mechanical virus transmission to seedpieces in early generation seedlots will result in widely dispersed infection centers from which secondary spread occurs.

Tillage or spray equipment that moves through fields also has been implicated in virus spread. Although there are few quantitative studies, virus movement and transmission by equipment may account for the difference in reinfection rates experienced by growers. Experiments conducted by Banttari, Gyenis and DiFonzo (unpublished) evaluated the role of tillage equipment in the reinfection of healthy potato seed stocks with PVY. Although interpretation of results was complicated by the aphid transmission of PVY, significant spread of the virus by tillage equipment was demonstrated during one year of the two-year study. Todd (1958) examined factors involved in the spread of PVX among seed potatoes in field plots. He demonstrated that the virus can be spread in the field simply by walking through PVX-infected plots in which there was foliage contact with clothing or boots. Clothing that came in contact with infected plants retained infectious virus for at least three weeks. Rabbits and dogs permitted to run among PVX-infected potato plants carried infectious virus for at least 10 days on the back, muzzle and paws of rabbits and for 24 hr on the back, legs and belly of dogs. The virus was readily acquired and transmitted when infected and healthy potato foliage was handled.

In some production areas, where winters are sufficiently mild or where abundant snow cover protects tubers from freezing, volunteer potato plants may serve as an inoculum source. Wright and Bishop (1981) described one field that had approximately 30,300 volunteer potato stems per hectare, and 65 percent were infected with PVX. Even if tillage removed one-half of the volunteers prior to planting a potato crop, one infected potato stem would remain for each 1.1 m of row and would serve as a potent source of inoculum for the new crop. Thomas (1983) measured the persistence PVX and PVS in tubers that overwintered in the Columbia Basin of the northwestern United States. The frequency of virus infection in volunteer plants in the spring was similar to that measured for tubers the previous fall.

Weeds may be a potential source of virus inoculum in potato fields. Allen and Davis (1982) found that 12 of 28 weed species could harbor *Tobacco rattle virus* genus *Tobravirus* (TRV) and that PVX occurred in 11 of 28 species examined. Hairy nightshade (*Solanum sarachoides*) harbored both TRV and PVX in leaves, roots, fruits and seeds. This weed was present in six of seven potato fields sampled. It was infected with TRV in six of these fields and infected with PVX in five fields. Redroot pigweed (*Amaranthus retroflexus*) occurred in all seven fields and was infected with PVX in two fields and infected with TRV in one field. Lambsquarter (*Chenopodium album*) was found in six of seven fields and was infected with PVX in three of these fields. They concluded that infected potato plants most likely provided the initial inoculum that infected the weeds.

Locatelli *et al.* (1978) examined weeds as a reservoir of TRV and other viruses in eastern Oregon. In areas adjacent to potato fields they found two of 18 redroot pigweed and four of 18 *Amaranthus powellii* were infected with TRV and PVX, respectively. Seven of 14 hairy nightshade plants were infected with both TRV and PVX. Beemster (1977) reported that bromegrass (*Bromus commutatus*) could be infected with PVX. Infected bromegrass developed symptoms and PVX was recovered. However, only a low percentage of plants became infected and he concluded that this species had a low susceptibility to the virus. It is not known if bromegrass is naturally infected and serves as a source of PVX inoculum.

Smith *et al.* (1969) found that *Tobacco necrosis virus* genus *Necrovirus* (TNV), which is mechanically transmissible to potato, occurred in the soil in three of 35 fields in California. They found that symptomless roots of systemically infected hosts released *Cucumber necrosis, Petunia asteroid, Tomato bushy stunt, Tobacco mosaic,* and *Southern bean mosaic viruses*. Most of these viruses remained infectious for at least 25 days if the soils were kept moist, but less than one to three days if the soils were dried. The release of viruses from intact roots was probably small in comparison to release of viruses from decomposing root tissues. Regardless, with the exception of *Tobacco mosaic virus* (TMV) and TNV, the other viruses in this study are not reported pathogens of potatoes. Although TNV can be mechanically transmitted to potato, the principal means of transmission is by *Olpidium brassicae*.

Roberts (1950) reported root infection of tomato occurred with PVX when infected sap was poured over soil of potted plants and precautions were taken to prevent foliage contact with the soil. He also reported that potato was infected when roots were directly inoculated with PVX. He hypothesized that mechanical injury to roots may permit virus transmission from plant-to-plant in the soil environment. The possibility of a soil borne vector or vectors for PVX transmission among roots of potato was cited by Bercks (1970). The soil-borne fungus *Synchytrium endobioticum* was implicated as a potential vector of PVX (Nienhaus and Stille, 1965). However, Lange (1978) was unable to transmit PVX from infected potato plants to tomato using zoospores of *S. endobioticum.*

Banttari and Venette (1980) reported that mechanically transmissible viruses may spread via aerosols. Controlled studies using aerosol generators and mild air-flow velocities in a wind tunnel revealed that purified TMV and PVX formed aerosols that were infectious, after abrasion, to *Nicotiana glutinosa* (TMV) and *Gomphrena globosa* (PVX). Infectious TMV and PVX aerosols also were obtained directly from injured infected tomato and potato plants, respectively, that were placed in a closed chamber and then subjected to air-blast and water-blast treatments. Naturally formed TMV aerosols were detected in 14 of 30 sampling periods in a 0.25 ha field plot of *N. tabacum.* However, PVX aerosols were not detected during twenty 24-hr sampling periods in a 40 ha field of Russet Burbank potatoes completely infected with PVX. In contrast to PVX in potato, *Tobacco mosaic* virions accumulate in large numbers (60 to 70 X 10^6 virions per cell) in tobacco epidermal cells. Also, tobacco leaves are characterized by the presence of abundant trichomes, which contain large virus inclusions when plants are TMV-infected. Trichomes are easily broken under various weather conditions and virus exposure is likely. In field experiments, TMV was most frequently collected when the wind velocity was at least 25 km per hr with or without precipitation or overhead irrigation. Although PVX aerosols were not detected from infected potato plants in the field, more severe meteorological events such as hail, driving rain, and wind, similar to those conditions simulated in the closed chamber experiments, might result in PVX aerosol formation under natural conditions. If virus aerosols are disseminated in the field, infection could result from any type of foliar abrasion.

8.1.4 Disease Management

Control of mechanically transmissible viruses of potato is largely dependent on several methods also used for general control of all potato viruses. These methods include certification and limited generation production of healthy seed stocks produced by federal, state or private firms. Conventional breeding, though somewhat limited in use, has been employed for development of cultivars resistant to mechanically transmissible viruses (Banttari *et al.* 1993). When sensitive and practical serological assays became available to certification programs in the 1970's, and the use of thermotherapy and tissue culture-generated seed stocks became available to growers, PVX eventually was eliminated from most major North American potato cultivars. Other viruses mechanically transmissible to potato, but that also have vectors (e.g., PVS, PVY, and TRV), are generally more difficult to control.

Sanitation and certain other cultural practices helpful for control of PVX and PVS will also aid in the control of other mechanically transmissible viruses. As mentioned above, PVS and PVX persist for considerable lengths of time on various surfaces contacted during potato production, including those made of metal, wood, and plastic. When cutting tubers into seedpieces for early generation seed increase, disinfestation of cutting knives between cuts of successive tubers is recommended. Heat, whether by dipping cutting knives into alcohol and flaming or by the application of steam or by other means, as well as chemical disinfestation are options generally considered to be effective. However, flaming was not completely effective for inactivation potato spindle tuber viroid (PSTV) from tools (Roistacher *et al.* 1969). Chemical disinfestation with 3% trisodium orthophosphate (TSP) in water (Broadbent, 1963) or 10% TSP plus detergent (Mulholland, 1962) is generally effective for virus inactivation.

Maximum isolation of early generation seed stocks from other potato seed stocks was an important factor in reducing reinfection by PVX (Cockerman, 1958) and other mechanically transmissible viruses. Whenever possible, isolation of seed production areas from commercial fields is highly desirable. To further reduce the potential for harboring inoculum and reintroducing viruses, some seed certification programs specify that all potatoes grown by a single farming operation must be inspected. This approach encourages isolation and prevents contact of the certified seed crop with storage areas and farming implements previously exposed to commercial potatoes.

Tuber-unit planting of early generation seed stocks permits identification and removal of virus-infected, off-type or otherwise diseased units. Detection is by visual inspection or by serological tests, and diseased units are removed by roguing to eliminate inoculum. Between-plant spacings are increased to minimize vine contact between tuber-units to help reduce movement of mechanically transmissible viruses. However, greater plant spacings attract aphids which increases the potential for the reintroduction of vectored viruses such as PVY and PLRV. In areas in which over-wintering of tubers occurs, volunteer plants may serve as important sources of inoculum. Avoidance of planting of seed stocks in such fields or the elimination of volunteers by herbicides or other means, is imperative. Use of true seed has been tested and utilized in some South American countries (Salazar, 1996). With the exceptions of *Potato virus T* genus *Trichovirus*

(PVT) and PSTV, potato viruses are not carried in true potato seed. However, seedlings derived from true seed were found to be more susceptible to virus infection compared with plants propagated from tubers.

Table 8.1.1. Viruses reported to infect potato (*Solanum tuberosum*) via mechanical means.

Mechanically Transmissible Viruses Infecting Potato:

Under Natural Field Conditions

Alfamoviruses
Alfalfa mosaic virus Syn: Calico (Aphids, non-persistently) (1)

Carlaviruses
Potato virus M Syn: potato paracrinkle virus (Aphids, non-persistently) (2)
Potato virus S (Some strains aphids, non-persistently) (Wetter, 1971) (2)
Potato latent virus (Goth *et al.* 1999)

Comoviruses
Andean potato mottle virus (Cucumber beetle) (1)

Cucumoviruses
Cucumber mosaic virus (Aphids, non-persistently) (1)

Furoviruses
Potato mop-top virus (*Spongospora subterranea*) (1)

Geminiviruses
Potato yellow mosaic virus (Whiteflies) (1)

Ilarviruses
Tobacco streak virus (Thrips) (1,2)

Necroviruses
Tobacco necrosis virus (*Olpidium brassicae*) (1,2)

Nepoviruses
Potato black ringspot virus; strain of tobacco ringspot virus (Nematodes, Aphids, Seed-borne) (1)
Tomato black-ring virus (Nematodes, Seed-borne) (1)
Potato virus U (Nematodes) (1)
Arrachacha virus B (Seed-borne) (1)

During Host Range Studies

Dodder latent mosaic virus (3)

Carlaviruses
Carnation latent virus (3)
Cassava mild mottle mosaic virus (1)

Carmovirus
Pelargonium line pattern virus (1)

Geminiviruses
Tomato Australian leafcurl virus (1)

Nepoviruses
Arrachacha virus A (1)
Cassava green mottle virus (1)
Tomato ringspot virus (1)

Potexviruses
Asparagus 3 virus (1)
Pepino mosaic virus (1)

Potyviruses
Henbane mosaic virus (2)
Melilotus mosaic virus (1)
Pepper veinal mottle virus (1)

Tombusvirus
Tomato bushy stunt virus (1)

Tymoviruses
Eggplant mosaic virus (1)

Varicosavirus
Tobacco stunt virus 1

<u>Under Natural Field Conditions</u>

Potexviruses

Potato virus X (Grasshoppers and *Synchytrium endobioticum* have been implicated but not confirmed as vectors) (2)

Potato aucuba mosaic virus Syn: potato virus F, potato virus G (Aphids, with the aid of a helper virus: PVA or PVY) (1,2)

Potyviruses

Potato virus Y (Aphids, non-persistently) (2)

Potato virus A (Aphids, non-persistently) (2)

Potato virus V (Aphids, non-persistently) (1), Fribourg and Nakashima (1984)

Rhabdoviruses

Potato yellow dwarf virus (Agallian leafhoppers) (1)

Eggplant mottled dwarf virus (1)

Tobamoviruses

Tobacco mosaic virus (Seed-borne) (2)

Tomato mosaic virus (Seed-borne) (2)

Tobraviruses

Tobacco rattle virus Syn: Corky ringspot virus (Nematodes) (1,2)

Tospoviruses

Tomato spotted wilt virus (Thrips) (1)

Trichoviruses

Potato virus T (Seed-borne) (1)

Tymoviruses

Andean potato latent virus (Flea beetles, Seed-borne less than 1%) (1)

---

(1) Brunt, *et al.* 1996.

(2) Gibbs, *et al.* 1970-1989.

(3) Smith, 1972.

References

Allen, T.C. and Davis, J.R. 1982. Distribution of tobacco rattle virus and potato virus X in leaves, roots, and fruits and/or seeds of naturally-infected weeds. *Am. Potato J.* 59:149-153.

Angell, S.M., Davies, C. and Baulcombe, D.C. 1996. Cell-to-cell movement of potato virus X is associated with a change in the size exclusion limit of plasmodesmata in trichome cells of *Nicotiana clevelandii*. *Virology* 215: 197-201.

Banttari, E.E., Anderson, N.A., Jevning, J. and Sushak, R. 1978. Reinfection rates and performance of disease tested potato seedstocks. pp. 39-42. *Proc. Annu. Potato Res. Planning and Rptg. Conf., Red River Valley Potato Growers Assoc.* (133 p.)

Banttari, E.E., and Venette, J.R. 1980. Aerosol spread of plant viruses. Potential role in disease outbreaks. *Ann. NY Acad. Sci.* 353: 167-173.

Banttari, E.E., Ellis, P.J. and Khurana, S.M.P. 1993. Management of diseases caused by viruses and virus-like pathogens. Pages 127-133 In: *Potato Health Management.* (ed.) R.C. Rowe. APS Press, St. Paul, MN. (178 pp).

Beemster, A.B.R. 1958. Some aspects of mature plant resistance to viruses in the potato. *Proc. Third Conf. Potato Virus Diseases.* Lisse-Wageningen 1958: 212-217.

Beemster, A.B.R. 1961. Translocation of leaf roll and virus Y in the potato. *Proc. Fourth Conf. Potato Virus Diseases.* Braunschweig, 1960: 60-67.

Beemster, A.B.R. 1969. Virus translocation in two-stemmed potato plants. *Euro. Potato J.* 12: 20-25.

Beemster, A.B.R. 1976. Translocation of potato viruses Y^N and Y^O in some potato varieties. *Potato Res.* 19: 169-172.

Beemster, A.B.R. 1977. *Bromus commutatus*: a new host for potato virus X. *Gent.* 42: 1207-1210.

Bercks, R. 1970. Potato virus X. *CMI/AAB Descriptions of plant viruses.* No. 4. UK Wm Culross and Son, Ltd. Coupar Angus, Perthsire, Scotland.

Bode, O. and Weidemann, H.L. 1971. Untersuchungen zur Blattlausubertragbarkeit von Kartoffel-M-und-S-Virus. *Potato Res* 14: 119-129.

Brants, D.H. 1964. The susceptibility of tobacco and bean leaves to tobacco virus infection in relation to the condition of ectodesmata. *Virology* 23: 588-594.

Broadbent, L. 1963. The epidemiology of tomato mosaic. III. Cleaning virus from hands and tools. *Ann. appl. Biol.* 52: 225-232.

Brunt, A.A., Crabtree, K., Dallwitz, M.J., Gibbs, A.J. and Watson, L. 1996. *Viruses of Plants.* Descriptions and Lists from the VIDE Database. CAB International, Wallingford, Oxon OX10 8 DE, UK

Cockerman, G. 1958. Observations on the spread of virus X. *Proc. Third Conf. on Potato Virus Diseases.* Lisse-Wageningen, 1957:144-148.

De Bokx, J.A. 1968. The translocation of various isolates of potato virus S in potato plants with primary infection. *Ghent Rijksf. Landbouwwet.Meded.* 33: 1179-1185.

De Bokx, J.A. 1972. Spread of potato virus S. *Potato Res* 15: 67-70.

De Bokx, J.A. and Waterreus, H.A.J.I. 1967. The translocation of potato virus S, isolate "Eersteling", in potato plants. *Neth. J. Plant Pathol.* 73: 55-57.

Dedic, P. 1979. Detectability and translocation of potato virus A in potato plants after artificial inoculation. *Potato Res.* 22: 357-360.

Draper, M.A. and Gudmestad, N.C. 1992. Mechanical transmission of PVY by seed cutting. *Am. Potato J.* 69: 577-578.

Franc, G.D. and Banttari, E.E. 1984. The transmission of potato virus S by the cutting knife and retention time of infectious PVS on common surfaces. *Am. Potato J.* 61: 253-260.

Franc, G. D. and Banttari, E.E. 1996. Translocation and mechanical spread of a Minnesota isolate of potato virus S in potatoes. *Am. Potato J.* 73: 123-133.

Fribourg, C.E., and Nakashima, J. 1984. Characteristics of a new potyvirus from potato. Phytopatholgy 74:1363-1369.

Gibbs, A.J., Harrison, B.D. and Murant, A.F. 1970-1989. *Descriptions of plant viruses*. Vol 1-354. Commonwealth Mycological Institute, Ferry Lane Kew, Surrey, England.

Gilbertson, R.L. and Lucas, W.J. 1996. How do viruses traffic on the 'vascular highway'? *Trends Plant Sci.* 1: 260-267.

Goth, R.W., Ellis, P.J., deVilliers, G., Goins, E.W. and Wright, N.S. 1999. Characteristics and distribution of potato latent carlavirus (Red LaSoda virus) in North America. *Plant Dis.* 83: 751-753.

Hahm Y., Slack, S.A. and Slattery, R.J. 1981. Reinfection of potato seed stocks with potato virus S and potato virus X in Wisconsin. *Am. Potato J.* 58: 117-125.

Johnson, J. 1925. Transmission of viruses from apparently healthy potatoes. *Res. Bul. No.* 63. *Agric. Exp. Stat., U. of Wisc.* (12 p.)

Khalil, E.M. and Shalla, T.A. 1982. Detection and spread of potato virus S. *Plant Dis.* 66: 368-371.

Kostiw, M. 1975. Transmission of potato virus S by *Aphis nasturtii* Kalt. *Potato Res.* 18: 641-643.

Krantz, F.A. and Bisby, G.R. 1921. Relation of mosaic to running-out of potatoes in Minnesota. *Bull. 197. U. Minn. Agric. Exp. Stat., St. Paul, MN.*

Lange, L. 1978. *Synchytrium endobioticum* and potato virus X. *Phytopath. Z.* 92: 132-142.

Larson, R.H. 1950. The spread of ringspot virus X by cutting knife. *Am Potato J.* 27:53-54.

Locatelli, E.A., Allen, T.C., Koepsell, P.A. and Appleby, A.P. 1978. Diagnosis of tobacco rattle virus (TRV) and other viruses in weed and rotation crops in potato fields. *Am. Potato J.* 55: 249-257.

Malcuit, I, Marano, R.M., Kavanagh, T.A., De Jong, W., Forsyth, A. and Baulcombe, D.C. 1999. The 25-kDa movement protein of PVX elicits Nb-mediated hypersensitive cell death in potato. *Molec. Plant-Microbe Inter.* 12: 536-543.

Mandahar, C.L. and Garg, I.D. 1990. Virus Transport in Plants. In: *Plant Viruses*, Vol. II. *Pathology*. Ed. C.L. Mandahar. CRC Press, Boca Raton, FL. (371 p.)

Manzer, F.E., Merriam, D.C and Hepler, P.R. 1978. Effects of potato virus S and two strains of potato virus X on yields of Russet Burbank, Kennebec and Katahdin cultivars in Maine. *Am. Potato J.* 55: 601-609.

Matthews, R.E.F. 1970. *Plant Virology*. Academic Press, Inc., New York, NY. (778 p.)

McDonald, J.G. 1987. Comparative levels of potato viruses S and Y infection of microplants and tuber-propagated plants in the field. *Am. Potato* J 64: 517-521.

Mulholland, R.I. 1962. Control of the spread of mechanically transmitted plant viruses. *Commonwealth Phytopathol. News* 8:60-61.

Nienhaus, F. and Stille, B. 1965. Ubertragung des Kartoffel-X. Virus durch Zoosporen von *Synchytrium endobioticum. Phytopathol. Z.* 54: 335-337.

Roberts. F.M. 1950. The infection of plants by viruses through roots. *Ann. appl. Biol.* 37: 385-396.

Roistacher, C.N., Calavan, E.C., and Blue, R.L. 1969. Citrus exocortis virus-chemical inactivation on tools, tolerance to heat and separation of isolates. *Plant Dis Reptr.* 53: 333-336.

Salazar, L.F. 1996. *Potato Viruses and their Control*. Int. Potato Cent. Lima, Peru. (214 p).

Schulz, E.S. and Folsom, D. 1920. Transmission of the mosaic disease of Irish potatoes. *J. Agr. Res.* 19: 315-337

Schulz, E.S., Folsom, D., Hildebrandt, F.M. and Hawkins, L.A. 1919. Investigations on the mosaic disease of the Irish potato. *J. Agr. Res.* 17: 247-274.

Smith K.M. 1972. *A Textbook of Plant Virus Diseases*. 3rd ed. T & A. Constable Ltd. Edinburgh, Great Britain. 684 p.

Smith P.R., Campbell, R.N. and Fry, P.R. 1969. Root discharge and soil survival of viruses. *Phytopathology* 59: 1678-1687.

Teri, J.M., Thurston, H.D. and Plaistad, R.L. 1977. The effect of potato virus X on the yield of the potato variety Hudson. *Am. Potato J.* 54: 271-275.

Thomas, P.E. 1983. Sources and dissemination of potato viruses in the Columbia basin of the northwestern United States. *Plant Dis.* 67: 744-747.

Thomas, P.E. and Fulton, R.W. 1968. Correlation of ectodesmata number with nonspecific resistance to initial virus infection. *Virology* 34: 459-469.

Todd, J.M. 1958. Spread of potato virus X over a distance. *Proc. of the Third Conf. on Potato Virus Diseases.* Lisse-Wageningen, 1957:132-143.

Wardrop, E.A., Gray, A.B., Singh, R.P. and Peterson, J.F. 1989. Aphid transmission of potato virus S. *Am. Potato J.* 66: 449-459.

Wetter, C. 1971. Potato virus S. *CMI/AAB Descriptions of Plant Viruses.* No. 60. Wm. Culross and Son Ltd. Coupar Angus, Scotland.

Whipple, O.B. 1919.Degeneration in potatoes. *Bull. 130. Agr. Exp. Sta. Bozeman, MT. (29p.)*

Wright, N.S. 1970. Combined effects of potato viruses X and S on yield of Netted Gem and White Rose potatoes. *Am. Potato J.* 47: 475-478.

Wright, N.S. 1974. Retention of infectious potato virus X on common surfaces. *Am. Potato J.* 51: 251-253.

Wright, N.S. 1977. The effect of separate infections by potato viruses X and S on Netted Gem potato. *Am. Potato J.* 54: 147-149.

Wright, N.S. and Bishop, G.W. 1981. Volunteer potatoes as a source of potato leafroll virus and potato virus X. *Am. Potato J.* 58: 603-609.

8.2 Potato Viruses with Soil-borne Vectors

D. P. Weingartner
University of Florida, Institute of Food and Agricultural Sciences, Hastings Research and Education Center, Hastings, Florida, 32145-0728, USA

8.2.1 Introduction

Only general aspects on the biology of soil-borne vectors and their viruses relative to seed potato production will be addressed in this chapter. More detailed information on this topic can be found in several excellent review articles and texts published in recent years (Harrison, 1977; Trudgill *et al.* 1983; de Bokx and van der Want, 1987; Jones, 1988; Brown *et al.* 1989; Adams, 1991; Ploeg *et al.* 1992; Mayo *et al.* 1995; Brown *et al.* 1995; Decraemer, 1995; Taylor and Brown, 1997).

There are six viruses of potato reported to have soil inhabiting vectors. These viruses are *Tobacco rattle virus* (TRV) (Robinson and Harrison, 1989), *Potato mop-top virus* (PMTV) (Harrison, 1974), *Tomato black ring virus* (ToBRV) (Murant, 1970), *Tobacco necrosis virus* (TNV) (Kassanis, 1970), *Potato black ringspot virus* (PBRV) (Salazar and Harrison, 1979), and *Potato virus U* (PVU) (Jones *et al.* 1983). Based on world-wide distribution and economic damage in potato, TRV and PMTV are by far the

G. Loebenstein et al. (eds.),
Virus and Virus-like Diseases of Potatoes and Production of Seed-Potatoes, 177–194.

most important of these viruses in potato. Although TNV is widely distributed, this virus, ToBRV, PBRV, and PVU are of minor importance in potato and will be only briefly addressed here. Also, viruses such as *Potato virus Y and* tuber necrotic strain (PVY^{NTN}), which affect potato tubers (Le Romancer *et al.* 1994), but do not have soil-inhabiting vectors, are likewise beyond the scope of this chapter.

8.2.2 *Potato black ringspot virus, Potato virus U,* and *Tomato black ring virus*

These viruses are all nepoviruses (i.e. nematode transmitted polyhedral viruses), reported to be transmitted by nematodes. However, only the vectoring of ToBRV by *Longidorus attenuatus* (de Man, 1876), (Thorne and Swanzer, 1936) and *Longidorus elongatus* Hooper, 1961 have been adequately verified (Trudgill *et al.* 1983; Brown *et al.* 1995; Taylor and Brown, 1997). As is the case for most nepoviruses, PBRV, PVU, and ToBRV each have wide host ranges and infect many species of monocotyledonous and dicotyledonous plants (Murant, 1970; Salazar and Harrison, 1979; Jones *et al.* 1983). The infection of true seed of many different host plants has been demonstrated for PVU (Jones *et al.* 1983) and ToBRV (Murant, 1970) and of true potato seed by PBRV (Slack, 2001).

The three viruses have isometric particles of 25nm, 28nm, and 30nm diameter, respectively for PBRV (Salazar and Harrison, 1979), PVU (Jones *et al.* 1983), and ToBRV (Murant, 1970), and bipartite genomes consisting of two functional RNA species (Taylor and Brown, 1997). Generally in nepoviruses, the genomic determinants for host range, seed transmissibility, and symptom expression are carried in RNA-1 and those for nematode transmissibility and serological specificity reside in RNA-2 (Taylor and Brown, 1997).

The geographic distribution of PBRV (Salazar and Harrison, 1979) and PVU (Jones *et al.* 1983) is limited to Peru and the latter virus has been found in nature systemically infecting potato only once (Jones *et al.* 1983). *Tomato black ring virus* is widely distributed in Europe and parts of the former Soviet Union. Although ToBRV has also been reported in Canada, Brazil, India, Japan, Kenya, and the USA, these reports were likely of imported infected propagation material (Taylor and Brown, 1997).

The symptoms of the three nepoviruses in potato vary. Symptoms of PBRV consist of necrotic spotting of foliage and intense generalised yellowing caused by the calico strain in some cultivars (Salazar and Harrison, 1979). Similar bright yellow or calico leaf markings were associated with PVU in Peru (Jones *et al.* 1983). Primary systemic symptoms of ToBRV appear as necrotic spots and rings on leaves, however, few plants express symptoms in the first year of infection. Chronic symptoms in the second year result in 'bouquet' symptoms consisting of necrotic spotting, cupping of tip leaves, and stunting of plants. The pseudo-aucuba strain produces bright yellow leaf markings (Murant, 1970; Jones and Fribourg, 1981).

The two species of nematodes which vector ToBRV inhabit sandy soils and the greatest population densities correspond with the availability of young actively growing roots. Both nematodes are generally distributed in Europe and although wide distribution of *L. elongatus* has been reported in North America, many of the reports are invalid. This species has been verified in British Columbia and Ontario, Canada, and in Oregon, USA. There have also been reports of *L. elongatus* in India, Pakistan, South Africa, and New Zealand (Taylor and Brown, 1997).

The high degree of vector specificity with nepoviruses, especially with virus-nematode associations in Europe (Brown *et al.* 1995), has also been reported for ToBRV. The English and German strains of ToBRV are most efficiently transmitted by *L. attenuatus* whereas *L. elongatus* transmits Scottish strains of the virus (Taylor and Brown, 1997).

Due to the restricted geographic distribution of PBRV and PVU, exclusionary measures restricting the importation of potentially infected planting material from Peru are possible. Similar measures for ToBRV and its vector in North America may also help limit the spread of this virus.

8.2.3 *Tobacco necrosis virus*

Tobacco necrosis virus (TNV) is a *Necrovirus*, transmitted in a nonpersistent manner by zoospores of *Olpidium brassicae* (Woron.) Dang (class Chytridiomycetes) (Harrison, 1977; Adams, 1991). The vector is favored by low soil temperatures (10-20C) and high soil moisture (Harrison, 1977; de Bokx and van der Want, 1987). TNV consists of RNA-containing, isometric particles, approximately 26nm in diameter. The virus and vector are worldwide in distribution and both have wide host ranges (Kassanis, 1970; Harrison, 1977). There is significant vector specificity between strains of TNV and those of *O. brassicae* (Harrison, 1977).

TNV rarely infects potato, however, occasionally it has been reported in the cultivars Duke of York (Eersteling) in the Netherlands and Sieglinde in Italy causing ABC disease in tubers (Peters, 1981). Similar diseases have been reported in Tunisia and the USA (de Bokx and van der Want, 1987). Systemic symptoms occur rarely and have not been reported in potato. The tuber symptoms include dark brown lesions with radial or reticular cracks which can resemble scab lesions caused by *Streptomyces scabies* (Thaxter)

Waksman and Henri (Syn. *Actinomyces scabies* (Thaxter) Gussow). Circular or band-like lesions and light brown patches develop on the tuber surface. Sometimes blisters, which later become sunken and cover the entire tuber surface, are seen at harvest or may develop in storage (Peters, 1981; de Bokx and van der Want, 1987).

8.2.4 *Potato mop-top virus*

Potato mop-top virus (PMTV) is a pomovirus (Pringle, 1998) and is vectored in a persistent manner (Adams, 1991) by the powdery scab fungus, *Spongospora subterranea* f. sp. *subterranea* Tomlinson (Jones, 1988; Arif, *et al.* 1995. The virus has been reported in potato growing regions of northern and central Europe, Andean regions of South America, China, Israel, Japan, and Canada (Harrison, 1974; Jones, 1988; Arif, *et al.* 1995). The experimental host range of PMTV is restricted to three families of dicotyledonous plants, *Aizoaceae, Chenopodiaceae* and *Solanaceae* (Jones, 1988).

8.2.4.1 *Spongospora subterranea* f. sp. *Subterranea,* the Vector of PMTV

The fungal vector of PMTV, *S. subterranea* is an obligate zoosporic parasite of plant roots. It is a member of the Plasmodiophoromycetes and has a biflagellite zoospore in contrast to the monoflagellite zoospore of *O. brassicae.* The organism persists in soil as spore balls consisting of thick-walled resting spores called sporangiosori or cystosori which are released into the soil from the powdery lesions on the surface of tubers or when infected roots decay (Hims and Preece, 1975; Jones, 1988; Adams, 1991).

The life cycle of *S. subterranea* can be divided into two phases. The primary phase of sporangial plasmodia (or zoosporangia) that infect root hairs and epidemal cells and produce more zoospores; and a cystogenous phase that produces resting spores. The primary stage of the life cycle is initiated with the germination of resting spores as zoosporangia which swim to root hairs or epidermal tissue of roots, stolons or tubers where they encyst. Following adhesion to surface tissue of the host and secretion of a thin wall, the protoplast of the encysted zoospore is injected into host tissue. A multinucleate plasmodium develops in the host cell and forms a sac-like zoosporangium which releases secondary zoospores. The secondary zoospores which are morphologically identical to primary zoospores are released into the soil water where they can swim to new infection sites or they can penetrate into deeper host tissue and continue the infection cycle. It is not known what triggers the cystogenous stage and development of resting spores. The stages leading to resting spore development include zoospore encystment, penetration and plasmodial development. Following a nuclear division interpreted as meiosis, a cystogenous plasmodium is formed which produces resting spores in cystosori. (Jones, 1988; Adams, 1991).

Cool, wet conditions are essential for development of *S. subterranea.* Soil temperatures of 12-20°C are favourable and little infection occurs at temperatures above 20°C (Jones, 1988). Infection can occur, however, when normal soil temperatures exceeding 20°C are reduced to the 12-20°C range by irrigation (Jones, 1988).

The spread of *S. subterranea* zoospores was inhibited by low soil pH. Reducing soil pH to 5.0 by applying sulfur decreased development of powdery scab. Conversely, applying lime before planting increased the incidence of powdery scab on tubers (Cooper *et al.* 1976). Soil acidity, however, did not influence incidence of spraing caused by PMTV in Denmark (Nielsen and Molgaard, 1997).

S. subterranea is world-wide in distribution and infects a wide range of plant hosts. Although the pathogen can produce zoosporangia in dicotyledons, monocotyledons, and gymnosperms, formation of spore balls and completion of the life cycle is apparently limited to tuber bearing Solanum spp., and nontuber bearing hosts such as *Solanum nigrum, Nicotiana rustica , Lycopersicum esculentum , Datura, stramonium,* and *Capsicum annuum* (Harrison *et al.* 1997).

8.2.4.2 Potato Mop-top -- the Disease

Symptoms of PMTV are influenced by cultivar and environment. Primary infection from soil to tubers rarely becomes systemic. Primary infected tubers can be assymptomatic (Sokmen *et al.* 1998) or in sensitive cultivars raised rings on the tuber surface or necrotic dark brown arcs (spraing) in the tuber flesh resembling some symptoms of corky ringspot. The necrotic arcs form at the boundary of viral invasion in the tuber flesh and are induced by a sudden change in temperature. The necrotic arcs do not prevent further extension of the virus into the tuber (Jones, 1981; Jones, 1988).

Cool temperatures favour development of secondary foliar symptoms in daughter plants grown from primary infected tubers. Foliar symptoms are variable including aucuba symptoms consisting of bright yellow blotches, rings, or diagnostic chevrons or v-shaped yellow markings. Aucuba symptoms are most common in lower leaves. Pale v-shaped markings appear on upper leaves. The name mop-top comes from the stunting of stems and shortening of internodes involving some or all of the stems of the plant. Yield losses up to 20% can occur in sensitive cultivars, mostly due to a reduction in the number of tubers produced, and severe effects on tuber quality occur (Jones, 1981; Jones, 1988). Total yield and dry matter, however, were unaffected by incidence of spraing in Denmark (Nielsen and

Molgaard, 1997).

The spraing symptoms in tubers are especially severe in cool weather potato production regions such as Scandinavia. Incidence of spraing in the sensitive cultivar Saturna which is widely planted for potato chip (i.e. crisp) production in Scandinavia, often exceeds 25% in Sweden and 30-50% in Denmark (Reavy *et al.* 1997).

Movement of PMTV to new geographic sites can result from the planting of virus infected tubers with powdery scab lesions (Jones and Harrison, 1969). Once established in a field, PMTV, even in absence of potato, has persisted for up to 18 years (Calvert, 1968).

8.2.4.3 Management and Control of PMTV

Successful management of PMTV encompasses managing both the virus and its vector. The virus is best managed through exclusion and sanitation practices. Since PMTV has a relatively limited geographic distribution (Jones, 1988; Arif *et al.* 1995) and depends upon *S. subterranea* for field transmission, preventing movement of the virus and its vector from affected to unaffected regions through quarantine and certification of seed tubers is a viable option. This is particularly true in the USA where PMTV has not been reported and powdery scab is widely distributed.

Sanitary practices aimed at reducing interregional and field to field movement of spore balls on infected tubers, in soil adhering to seed or field equipment, or in irrigation water are important in reducing the spread of sponospora and PMTV. Fortunately, PMTV is naturally self eliminating in seed lots. The virus has a slow pattern of systemic invasion and plants grown from infected tubers have a variable number of shoots and daughter tubers free of virus infection. When planted in soil free of *S. subterranea*, the virus was eliminated from a partially infected seed lot in Scotland in three years. Also in cool climates distinct foliar systems facilitate successful rouging of infected plants further reducing incidence of PMTV (Jones, 1988).

Jones (Jones, 1988) reviews chemical controls, however, most of the products mentioned such as formaldehyde and mercurials are prohibited from use or are regulated in modern agriculture. Chemical control has not been a viable commercial option in countries in which PMTV is a problem (Reavy *et al.* 1997).

There is presently no good source of resistance or tolerance in commercially acceptable potato cultivars. Recent resistance trials with a transgene encoding for the coat protein gene of a Scottish isolate of PMTV in *Nicotiana benthemiana* (Reavy *et al.* 1995; Reavy *et al.* 1997) and in potato cultivars Saturna and Pentland Marble (Barker *et al.* 1998) show promise for resistance to a wide range of PMTV isolates.

8.2.5 *Tobacco rattle virus*

Tobacco rattle virus (TRV) *Tobravirus* is vectored in potato by "stubby root"nematodes in the genera *Paratrichodorus* and *Trichodorus* (Fig.8.2.1) (Sol *et al.* 1960; Walkinshaw *et al.* 1961; Taylor and Brown, 1997). The virus is related to but considered taxonomically distinct from pea early browning virus (PEBV) (Robinson *et al.* 1987; MacFarlane *et al.* 1999). The virus is made up of straight tubular particles with a diameter of 23 nm and two predominant lengths (Fig.8.2.2). Depending upon the TRV isolate, the longer (L) rods are 185 to196 nm and the shorter (S) vary from 50 to 115 nm in length (Robinson and Harrison, 1989). TRV has a single stranded, positive sense, bipartite RNA genome. The L and S particles contain the different species of genomic RNA designated, respectively, as RNA-1 and RNA-2. RNA-1 encodes for infectivity, systemic movement within plants, replicase and a small non-structural protein with a hypothesised function in seed transmission. The nucleotide sequence of RNA-1 is highly conserved among isolates from the same TRV subgroup (Robinson and Harrison, 1989; Visser and Bol, 1999). Conversely, RNA-2 is highly variable in length and in the sequence of nucleotides between isolates. The RNA-2 alone is noninfective, however, encodes for coat protein, serological specificity, and may code for one to three non-structural proteins, which, in conjunction with coat protein, are believed to play a role in the process of vector transmission. There is a wide range of serological variants of TRV due to the high variability of RNA-2 (Robinson and Harrison, 1989; Visser and Bol, 1999).

TRV has the widest host range of any plant virus and is known to infect naturally or experimentally more than 400 species of 50 different plant families. The virus is often confined to the root system, however is systemic and seed borne in many plant species. The virus has been found in Europe, former Soviet Union, Japan, New Zealand, and North America (Robinson and Harrison, 1989). Isolates of TRV are made up of two types. The M or normal multiplying type which consists of both L and S particles and RNA-1 and RNA-2; and the NM or nonmultiplying type, which consists of RNA-1. The M types are easily transmitted by artificial inoculations with sap, are serologically detectable and are vectored by nematodes in family *Trichodoridae*. Conversely, the NM types are difficult to transmit with sap inoculations, cannot be detected serologically, and are not transmitted by nematodes (Robinson and Harrison, 1989; Taylor and Brown, 1997).

Fig. 8.2.1. Anterior region of a trichodorid nematode illustrating the typical curved onchiostyle. Bar= ca 12 Φm. Courtesy Society of Nematologists.

Fig. 8.2.2. *Tobacco rattle virus.* Bar= ca 115 nm. Courtesy Clemson University.

Fig. 8.3.3 Severe CRS surface lesions following early season TRV infection of cv. La Chipper tubers.

Fig. 8.4.4. "Textbook" necrotic arcs or spraing symptoms associated with TRV in many potato cultivars. Variety shown is La Chipper.

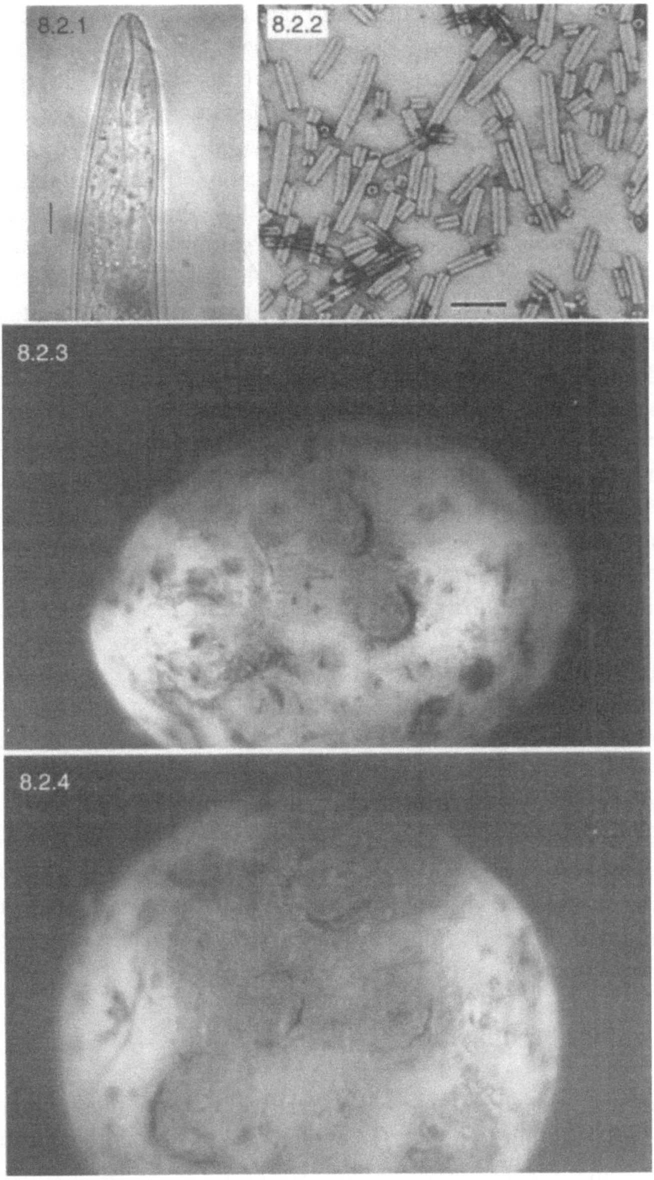
8.2.1
8.2.2
8.2.3
8.2.4

8.2.5.1 Trichodorid vectors of TRV

The first trichodorid nematodes were described in 1880 as *Dorylaimus primitivus* deMans (Decraemer, 1995), however, the group did not receive wide attention until Christie and Perry (Christie and Perry, 1951) demonstrated the ectoparisitism of *Paratrichodorus minor* (known as *Trichodorus christiei* at that time) on several vegetable crops. The abnormally stunted roots resulting from cessation of root elongation and overall root development caused by feeding of the nematode resulted in their being called stubby root nematodes. (Note: stubby root symptoms do not typically occur in potato. Weingartner, unpublished). Additional interest in trichodorids resulted from the discovery that they transmitted TRV (Walkinshaw *et al.* 1961; Sol *et al.* 1960).

Trichodorids are generally polyphagous ectoparisites, found in association with roots of many different plant species including potato. Various species of trichodorids have been recorded from many countries with most reports from Europe and North America. , In vegetable crops and potato trichodorids generally are most common in soil with high sand content (Harrison, 1977; Decraemer, 1995; Taylor and Brown, 1997).

The species of trichodorids which have been verified as vectors of TRV in potato include: *P. allius* (Jensen, 1963) Siddiqi, 1974, *P.* anemones (Loof, 1965) Siddiqi, 1974, *P. hispanus* (Roca and Arias, 1986), *P. minor* (Colbran, 1956) Siddiqi, 1974 (formerly *T. christiei*, Allen), *P. nanus* (Allen, 1957) Siddiqi, 1974, *P. pachydermus* (Seinhorst, 1954) Siddiqi, 1974, *P. teres* (Hooper, 1962) Siddiqi, 1974, *T. cylindricus* (Hooper, 1962)*T. primitivus* (de Man, 1880) Micoletzky, 1922, *T. viruliferous* Hooper, 1963, and *T. similis* Seinhorst, 1963 (Brown *et al.* 1989; Ploeg *et al.* 1992; Brown *et al*, 1995; Decraemer, 1995; Brown, personal correspondence). *Paratrichodorus allius* and *P. minor* are vectors in North America whereas the remaining species are vectors in Europe. Populations of *P. teres* reported in association with TRV in Oregon, USA have been recently identified as *P. allius* (Taylor and Brown, 1997).

8.2.5.2 Nematode Feeding Behaviour

Nematodes in the *Trichodorae* are relatively short, plump nematodes usually 350 to 1800 Φm in length (Decraemer, 1995) and although there are four genera, only *Paratrichodorus* and *Trichodorus* transmit tobraviruses (Taylor and Brown, 1997). Based on *in vitro* studies using other hosts, the nematodes frequently aggregate at or just behind the root elongation zone. Their unique feeding is accomplished by rapid thrusts of a solid onchiostyle or tooth and formation of a feeding tube, most likely derived from a hardening of salivary products. This substitutes for the lack of a hollow stylet. Feeding is mostly into epidermal cells and root hairs. Some individuals attack meristematic tissue only after root growth has slowed. Feeding consists of five phases: cell wall exploration and probing, wall perforation, salivation, ingestion, and departure (Wyss, 1982; Decraemer, 1995; Taylor and Brown, 1997). The entire feeding process during which the entire cell contents is ingested, often including the granulated, or optically empty nucleus, is completed within three minutes. The affected cells rapidly die, most likely due to destruction of the cell tonoplast at the onset of feeding. Cytoplasmic streaming is slowed within seconds after perforation of the cell wall and only resumes when nematodes occasionally abandon the feeding site without ingesting the cell contents (Wyss, 1982; Taylor and Brown, 1997).

Due to the rapid destruction of the cell and its contents it is hypothesised that the nematode transmits the virus either during rare occurrences of interrupted feeding (Wyss, 1982) or when epidermal cells are killed and the nematode thrusts deeper into the cortical tissue of the root and fails to ingest cellular contents because of inadequate length of the onchiostyle (Taylor and Brown, 1997). None the less, transmission is efficient and infection of multiple plants by a single nematode has been demonstrated (van Hoof, 1964b).

Particles of TRV in the trichodorid nematode vectors are retained along the cuticle lining of the oesophagus in the anterior end of the nematode and tend to lie parallel with the long axis of the food canal (Taylor and Robertson, 1970; Brown *et al.* 1996; MacFarlane *et al.* 1999). Recently, differences in the location of virus retention in the oesophagus by different virus serotypes have been hypothesised (Karanastasi *et al.* 2000).

There is a high degree of specificity between some TRV serotypes and individual species of trichodorids. The PRN serotype of TRV, for example, is apparently transmitted exclusively by *T. pachydermus*. There are exceptions, however, as TRV from a Belgium potato crop was transmitted by five different trichodorid species. Specificity of transmission must be determined experimentally using known serotypes of TRV and individual nematodes (Ploeg, *et al.* 1992; Taylor and Brown, 1997).

Once infested, field populations of trichodorids can remain infective for many years (Harrison, 1977; Taylor and Brown, 1997). Some potato fields in Florida, USA have remained infested since the mid 1940's (Weingartner, unpublished).

8.2.5.3 TRV Symptoms in Potato

Tuber symptoms vary depending upon the cultivar and strain of the virus. Tuber symptoms result from viruliferous nematodes feeding directly on tubers rather than from systemic movement of the virus from infected roots (van Hoof, 1964a). The disease has been reported in Europe and North America. The disease is called corky ringspot (CRS) or spraing (Figs. 3.2.3 and 3.2.4). The most conspicuous symptoms of CRS in red and white smooth-skinned cultivars are corky arcs or rings which develop around the feeding sites of viruliferous nematodes and which often extend into the tuber flesh. Circular lesions often have lenticels at the centre. The extent of tuber necrosis is dependent on cultivar and strain of TRV as well as on the time of infection relative to tuber development. Infection early in tuber development results in intensive internal necrosis, often extending throughout the tuber. Early infection can also cause tuber malformation. Infection late in tuber development results in superficial lesions confined to the periderm with no internal symptoms (Walkinshaw *et al.* 1961; Weingartner *et al.* 1983; Robinson and Harrison, 1989; Taylor and Brown, 1997). Tuber symptoms in many netted and russet-skinned cultivars vary from the typical spraing like symptoms. External lesions are often limited to nondescript cracks or are absent, especially in russet-skinned clones. Also, arcs and rings are often absent from tuber flesh and necrosis is more diffuse and generalised. These symptoms are easily confused with physiological disorders such as internal brown spot or heat necrosis (Mumford *et al.* 2000; Sterrett and Henninger, 1997). Virus isolates from CRS infected tubers are usually of the NM-type. Recently, using reverse transcriptase polymerase chain reaction (Rt-PCR) assymptomatic tubers of several European cultivars (Arran Consul, Home Guard, King Edward, Romano, Sante, and Wilja) and symptomatic tubers of Pentland Dell and Maris Bard were shown to be infected with M-type TRV. (Xenophontos *et al.* 1998; Dale *et al.* 2000).

When planted, a small number of CRS infected tubers produce plants with stem mottle symptoms which consist of stunted plants, malformed and yellowed leaves. Stem mottle plants in the UK usually contain NM-type TRV, however, in other regions M-type TRV prevails. Tubers from stem mottle plants may be deformed with necrotic spots in the flesh or more normal in shape coupled with CRS-like symptoms (Taylor and Brown, 1997). In recent trials, assymptomatic tubers of the cv. Wilja infected with M-type TRV developed stunted plants which produced an increase in the number of tubers, coupled with a decrease in tuber size (Dale *et al.* 2000). Several other varieties grown from symptomatic tubers were also systemically infected with M-type TRV, but symptoms were not described (Xenophontos *et al.* 1998).

The discovery of M-type TRV in assymptomatic tubers and subsequent systemic infection of daughter plants is of profound importance in seed production (Xenophontos *et al.* 1998). It was previously believed that TRV in potato was self-eliminating because of the small number of stem mottle plants growing from CRS infected seed (Dale and Solomon, 1988a). The disease was therefore considered to be more important in consumer potatoes than in seed production. (Dale and Solomon, 1988a). Presence of M-type TRV in assymptomatic seed tubers is a potential means of disseminating the virus to previously unaffected sites (Xenophontos *et al.* 1998). For this to occur, however, the strain of the virus and the nematode species must be compatible (Taylor and Brown, 1997).

8.2.5.4 Management of Corky Ringspot

Successful management of CRS in potato production requires controlling the disease in fields already infested and preventing the movement of TRV to new sites. The virus persists in viruliferous nematodes and in a wide range of plant species which are often weeds or alternative crops (Harrison, 1977; Allen and Davis, 1982). A promising means of reducing incidence of CRS in potato crops grown in affected fields is the use of rotational or cover crops. There is evidence that planting a crop, which is a good host for the trichodorid vector and is at the same time a poor or non-host for TRV, substantially reduces CRS in subsequent potato crops. In the UK incidence of CRS (spraing) was substantially reduced in potato when the crop was planted following three years of barley (French and Wilson, 1976). Similarly in the USA, TRV could no longer be detected in a heavily infested field three years after it was seeded to alfalfa (Thomas *et al.* 1998).

The discovery that assymptomatic tubers can carry and transmit M-type TRV to trichodorids creates a major challenge for seed potato programs, especially in North America where CRS has recently emerged as a serious problem (Crosslin and Thomas, 1995; Mojtahedi and Santo, 1999, Crosslin *et al.* 1999). The high sampling level required for establishing detection and certification thresholds (Perez, 1997), even for symptomatic tubers harvested from fields with a high incidence of CRS, may be excessive for certification; and therefore may require establishing zero tolerances for seed producing fields found to be infested with TRV.

There are relatively few potato cultivars resistant to TRV with marketable horticultural traits (Eddins, 1959; Harrison, 1968; Richardson, 1970; Seppänen, 1972; Taylor and Brown, 1997). Resistance to TRV is relatively simply inherited (Dale and Solomon, 1988b; Weingartner *and Haynes,* 1999) and breeding lines with high levels of resistance are available (Weingartner *et al.* 1994; Brown *et al.* 1998). The challenge will be to incorporate resistance in acceptable commercial cultivars without developing assymptomatic or latent carriers of TRV (Xenophontos *et al.* 1998). There is no known resistance in potato to trichodorid nematodes (Taylor and Brown, 1997).

The effectiveness of chemicals and laws regulating their use vary with location. Soil fumigants such as 1,3-D, DD, metham sodium, and dazomet have proven effective in many production systems (Harrison, 1977; Weingartner *et al.* 1983). Soil fumigation is ineffective for CRS control in Florida, USA due to upward migration of viruliferous *P. minor* from deeper soil after dissipation of the fumigant (Weingartner *et al.* 1983). Deep (40-50 cm) injection of soil fumigants partially alleviates this problem (Livingston *et al.* 1976; Weingartner, unpublished). Nonvolatile nematicides including aldicarb, carbofuran, oxamyl, phenamiphos, and phorate also control CRS, aldicarb and oxamyl generally beeing the most effective (French and Wilson, 1976; Taylor and Brown, 1997; Weingartner *et al.* 1993). Substantial reductions in incidence of CRS were observed in Florida experiments following appropriately timed foliar applications of oxamyl (Weingartner *et al.* 1983). Corky ringspot has generally been controlled in the USA using chemicals.

8.2.6 Diagnosis of PMTV and TRV

It is apparent that diagnosis and detection of PMTV and TRV in potato tubers is a critical component in effective management of these viruses in seed potato production. The spraing symptoms in potato tubers caused by PMTV and TRV are easily confused, particularly in northern Europe where the viruses are known to coexist (Harrison, 1977; Sokmen *et al.* 1998; Mumford *et al.* 2000). Surface lesions of PMTV and TRV on tubers in absence of internal symptoms can also be confused with potato tuber necrotic ring disease caused by PVY-^{NTN} (Le Romancer *et al.* 1994; Mumford *et al.* 2000). Atypical necrosis associated with many potato clones (Weingartner, unpublished) can easily be mistaken for physiological disorders such as rust spot (Mumford *et al.*, 2000) or internal heat necrosis (Sterrett and Henninger, 1997). More over, tubers of some cultivars are symptomless for PMTV (e.g. King Edward, Maris Peer, Pentland Crown, Record, and Saturna) (Sokmen *et al.* 1998) and/or TRV (Xenophontos *et al.* 1998) but contain detectable virus when tested using Rt-PCR. Furthermore, trichodorid nematodes acquired TRV from plants grown from infected daughter tubers and thereafter successfully transmitted virus to healthy plants (Taylor and Brown, 1997; Xenophontos, *et al.* 1998; Sokmen *et al.* 1998). TRV was also detected using Rt-PCR in corky ringspot (Crosslin *et al.* 1999; Kawchuk *et al.* 1997) and assymptomatic (Crosslin *et al.* 1999) tubers in the USA and Canada. New technology including multiplex real time fluorescent Rt-PCR or Tac Man (Mumford *et al.* 2000) or Rt-PCR coupled with tissue blotting (Perez *et al.* 2000) show promise for detecting PMTV and TRV in potato tubers. These methods, however are still labour intensive and expensive which limits their use for detecting TRV in large samples of tubers.

8.2.7 Conclusions

Among the viruses of potato, which have soil-borne vectors, PMTV and TRV are of the greatest concern to seed potato production. The remaining viruses (PBRV, PVU, ToBRV, and TNV) presently are of minor concern because of the limited distribution of the viruses, vectors, or the virus-vector association. Nonetheless, appropriate measures should be maintained to continue the restricted distribution of these pathogens.

Cool production regions of North America should remain alert to the potential for PMTV because the vector is wide spread in North America and has been shown to over season even in subtropical Florida (Weingartner, unpublished). Introduction of the virus could therefore create problems in some production regions since the vector already exists.

The greatest concern in potato seed production, however, is TRV. It will be essential in breeding programs, especially in North America, to test all resistant clones for their potential to be assymptomatic carriers of NM-type TRV since trichodorids are present in many USA potato production regions.

References

Adams, M.J. 1991. Transmission of plant viruses by fungi. *Ann. appl. Biol.* 118: 479-492.

Allen, T.C. and Davis, J.R. 1982. Distribution of tobacco rattle virus and potato virus X in leaves, roots, and fruits and/or seeds of naturally-infected weeds. *Amer. Potato J.* 59: 149-153.

Arif, M., Torrance, L. and Reavy, B. 1995. Acquisition and transmission of potato mop-top furovirus by a culture of *Spongospora subterranea* f. sp. *subterranea* derived from a single cystosorus. *Ann. appl. Biol.* 126: 493-503.

Barker, H., Reavy, B. and McGeachy, K.D. 1998. High level of resistance in potato to potato mop-top virus induced by transformation with the coat protein gene. *Eur. J. Plant Pathol.* 104: 737-740.

Bokx, J.A. de and Want, J.P.H. van der 1987. *Viruses of potatoes and seed-potato production.* Pudoc, Wageningen. 254 pp.

Brown, C.R., Mojtahedi, H., Santo, G.S., Thomas, P.E. and Pavek, J.J. 1998. Breeding for resistance to corky ringspot disease of potato. *Amer. J. Potato Res.* 75: 271-272. Abstract.

Brown, D.J.F., Ploeg, A.T. and Robinson, D.J. 1989. A review of reported associations between *Trichodorus* and *Paratrichodorus* species (Nematoda: Trichodoridae) and tobraviruses with a description of laboratory methods for examining virus transmission by trichodorids. *Rev. Nematol.* 12: 235-241.

Brown, D.J.F., Robertson, W.M. and Trudgill, D.L. 1995. Transmission of viruses by plant nematodes. *Annu. Rev. Phytopathol.* 33: 223-249.

Brown, D.J.F., Robertson, W.M., Neilson, R., Bem, F. and Robinson, D.J. 1996. Characterization and vector relation of a serologically distinct isolate of tobacco rattle tobravirus (TRV) transmitted by *Trichodorus similis* in northern Greece. *Eur. J. Plant Pathol.* 102: 61-68.

Calvert, E.L. 1968. The reaction of potato varieties to potato mop-top virus. Res. Exp. Rec. Minist. Agric. North. Irel. 17: 31-40.

Christie, J.R. and Perry, V.G. 1951. A root disease of plants caused by a nematode of the genus *Trichodorus*. *Science.* 113: 491-493.

Cooper, J.I., Jones, R.A.C. and Harrison, B.D. 1976. Field and glasshouse experiments on the control of potato mop-top virus. *Ann. appl. Biol.* 83: 215-230.

Crosslin, J.M. and Thomas, P.E. 1995. Detection of tobacco rattle virus in tubers exhibiting symptoms of corky ringspot by polymerase chain reaction. *Amer. Potato J.* 72: 605-609.

Crosslin, J.M., Thomas, P.E. and Brown, C.R. 1999. Distribution of tobacco rattle virus in tubers of resistant and susceptible potatoes and systemic movement of virus into daughter plants. *Amer. J. Potato Res.* 76: 191-197.

Dale, M.F.B. and Solomon, R.M. 1988a. A glasshouse test to assess the sensitivity of potato cultivars to tobacco rattle virus. *Ann. appl. Biol.* 112: 225-229.

Dale, M.F.B. and Solomon, R.M. 1988b. Inheritance of insensitivity to tobacco rattle virus in potatoes. *Ann. appl. Biol.* 113: 313-318.

Dale, M.F.B., Robinson, D.J., Griffiths, D.W., Todd, D. and Bain, H. 2000. Effects of tuber-borne M-type strain of tobacco rattle virus on yield and quality attributes of potato tubers of the cultivar Wilja. *Eur. J. Plant Pathol.* 106: 275-282.

Decraemer, W. 1995. *The Family Trichodoridae: Stubby Root and Virus Vector Nematodes.* Kluwer Academic Publishers, Netherlands. 360 pp.

Eddins, A.H. 1959. Susceptibility of potato varieties and seedling selections to corky ringspot. *Amer. Potato J.* 36: 187-190.

French, N. and Wilson, W.R. 1976. Influence of crop rotation, weed control and nematicides on spraing in potatoes. *Plant Pathol.* 25: 167-172.

Harrison, B.D. 1968. Reactions of some old and new British potato cultivars to tobacco rattle virus. *Eur. Potato J.* 11: 165-176.

Harrison, B.D. 1974. Potato mop-top virus. *AAB Descriptions of Plant Viruses* No. 138, 4 pp.

Harrison, B.D. 1977. Ecology and control of viruses with soil-inhabiting vectors. *Annu. Rev. Phytopathol.* 15: 331-360.

Harrison, J.G., Searle, R.J. and Williams, N.A. 1997. Powdery scab of potato - a review. *Plant Pathol.* 46: 1-25.

Hims, M.J. and Preece, T.F. 1975. *Spongospora subterranea* f. sp. *subterranea*. In: *Descriptions of Pathogenic Fungi and Bacteria* No. 477, 2 pp.

Hoof, H.A. van 1964a. Het tijdstip van infectie en veranderingen in de concentratie van ratelvirus (kringerigheid) in de aardappelknol. *Meded. Landbouwhogesch. Gent* 29: 944-955.

Hoof, H.A. van 1964b. Serial transmission of rattle virus by a single male of *Trichodorus pachydermus* Seinhorst. *Nematologica* 10: 141-144.

Jones, R.A.C. 1981. Potato mop-top virus. In: *Compendium of Potato Diseases.* (ed) W.J. Hooker, APS Press, St. Paul. pp. 79-80.

Jones, R.A.C. 1988. Epidemiology and control of potato mop-top virus. In: *Viruses with Fungal Vectors.* (eds) J.I. Cooper and M.J.C. Asher, Wellerbourne, UK: Assoc. Appl. Biologists. pp. 255-270.

Jones, R.A.C. and Fribourg, C.E. 1981. Tomato black ring virus. In: *Compendium of Potato Diseases.* (ed) W.J. Hooker, APS Press, St. Paul. pp. 85-86.

Jones, R.A.C. and Harrison, B.D. 1969. The behavior of potato mop-top virus in soil, and evidence for its transmission by *Spongospora subterranea. Ann. appl. Biol.* 63: 1-17.

Jones, R.A.C., Fribourg, C.E. and Koenig, R. 1983. A previously undescribed nepovirus isolated from potato in Peru. *Phytopathology* 73: 195-198.

Karanastasi, E., Vassilakos, N., Roberts, I.M., MacFarlane, S.A. and Brown, D.J.F. 2000. Immunogold localization of tobacco rattle virus particles within *Paratrichodorus anemones*. *J. Nematology* 32: 5-12.

Kassanis, B. 1970. Tobacco necrosis virus. *AAB Descriptions Plant Viruses* No. 14, 5 pp.

Kawchuk, L.M., Lynch, D.R., Leggett, F.L., Howard, R.J. and McDonald, J.G. 1997. Detection and characterization of a Canadian tobacco rattle virus isolate using a PCR-based assay. *Can. J. Plant Pathology* 19:101-105.

Le Romancer, M., Kerlan, C. and Nedellec, M. 1994. Biological characterization of various geographical isolates of potato virus Y inducing superficial necrosis on potato tubers. *Plant Pathol.* 43: 138-144.

Livingston, C.H., Lambert, R., Kaufmann, M. and Knutson, K. 1976. Effects of soil fumigation with Telone-C on the field inoculum potential of tobacco rattle virus and potato yields. *Amer. Potato J.* 53: 81-86.

MacFarlane, S.A., Vassilakos, N. and Brown, D.J.F. 1999. Similarities in the genome organization of tobacco rattle virus and pea early-browning virus isolates that are transmitted by the same vector nematode. *J. gen. Virol.* 80: 273-276.

Mayo, M., Robertson, W.M., Legorburu, F.J. and Brierley, K.M. 1995. Molecular approaches to an understanding of the transmission of plant viruses by nematodes. In: *Adv. Molec. Plant Nematology.* (eds) F. Lamberti, C. de Giorgi and D.M. Bird, Plenum Press, London and New York. pp. 277-293.

Mojtahedi, H. and Santo, G.S. 1999. Ecology of *Paratrichodorus allius* and its relationship to the corky ring-spot disease of potato in the Pacific Northwest. *Amer. J. Potato Res.* 76: 273-280.

Mumford, R.A., Walsh, K., Barker, I. and Boonham, N. 2000. Detection of potato mop top virus and tobacco rattle virus using a multiplex real-time fluorescent reverse-transcription polymerase chain reaction assay. *Phytopathology* 90: 448-453.

Murant, A.F. 1970. Tomato black ring virus. *AAB Descriptions of Plant Viruses* No. 38, 4 pp.

Nielsen, S.L. and Molgaard, J.P. 1997. Incidence, appearance , and development of potato mop-top furovirus-induced spraing in potato cultivars and the influence on yield, distribution in Denmark and detection of the virus in tuber by ELISA. *Potato Res.* 40: 101-110.

Perez, E.E. 1997. *Sampling and Detection of Trichodorid Nematodes and Tobacco Rattle Virus in Corky-Ringspot-Affected Potato Tubers*. PhD Thesis, Univ. Florida, Gainesville, 167 pp.

Perez, E.E., Weingartner, D.P., Hiebert, E. and McSorley, R. 2000. Tobacco rattle virus detection in potato tubers from northeast Florida by PCR and tissue blotting. *Amer. J. Potato Res.* 77: In press.

Peters, D. 1981. Tobacco necrosis virus. In: *Compendium of Potato Diseases.* (ed) W.J. Hooker, APS Press, St. Paul. pp. 86-87.

Ploeg, A.T., Brown, D.J.F. and Robinson, D.J. 1992. The association between species of *Trichodorus* and *Paratrichodorus* vector nematodes and serotypes of tobacco rattle virus. *Ann. appl. Biol.* 121: 619-630.

Pringle, C.R. 1998. The universal system of virus taxonomy of the International Committee on Virus Taxonomy (ICTV), including new proposals ratified since publication of the Sixth ICTV Report in 1995. *Archives of Virology* 143: 203-210.

Reavy, B., Arif, M., Kashiwazaki, S., Webster, K.D. and Barker, H. 1995. Coat protein-mediated immunity to potato mop-top virus in nicotiana benthaniana is effective against fungal inoculation. *Mol. Plant-microbe Interact.* 8: 286-291.

Reavy, B., Sandgren, M., Barker, H., Heino, P. and Oxefelt, P. 1997. A coat protein transgene from a Scottish isolate of potato mop-top virus mediates strong resistance against Scandinavian isolates which have similar coat protein genes. *Eur. J. Plant Pathol.* 103: 829- 834.

Richardson, D.E. 1970. The assessment of varietal reactions to spraing caused by tobacco rattle virus. *J. Nat. Inst. Agric. Bot.* 12: 112-118.

Robinson, D.J. and Harrison, B.D. 1989. Tobacco rattle virus. *AAB Descriptions of Plant Viruses* No. 346 (No. 12 revised), 6 pp.

Robinson, D.J., Hamilton, W.D.O., Harrison, B.D. and Baulcombe, D.C. 1987. Two anomalous tobravirus isolates: evidence for RNA recombination in nature. *J. gen. Virol.* 68: 2551-2561.

Salazar, L.F. and Harrison, B.D. 1979. Potato black ringspot virus. *AAB Descriptions of Plant Viruses* No. 206, 4 pp.

Seppänen, E. 1972. The resistance of some potato varieties to spraing caused by tobacco rattle virus. *J. Scientific Soc. Finland* 44: 76-82.

Slack, S.A. 2001. Viruses and viroids. In: *Compendium of Potato Diseases. Second Edition.* (eds) W.J. Stephenson, R. Loria, G. Franc and D.P. Weingartner, APS Press, St. Paul. In press.

Sokmen, M.A., Barker, H. and Torrance, L. 1998. Factors affecting the detection of potato mop- top virus in potato tubers and improvement of test procedures for more reliable assays. *Ann. appl. Biol.* 133: 55-63.

Sol, H.H., Heuven, J.C. van and Seinhorst, J.W. 1960. Transmission of rattle virus and *Atropa belladonna* mosaic virus by nematodes. *Tijdschr. Plantenziekten* 67: 307-311.

Sterrett, S.B. and Henninger, M.R. 1997. Internal heat necrosis in the Midatlantic region -- influence of environment and cultural management. *Amer. Potato J.* 74:233-243.

Taylor, C.E. and Brown, D.J.F. 1997. *Nematode vectors of plant viruses.* CAB International, Wallingford, UK. 286 pp.

Taylor, C.E. and Robertson, W.M. 1970. Location of tobacco rattle virus in the nematode vector *Trichodorus pachydermus. Seinhorst J. gen. Virol.* 6: 179-182.

Thomas, P.E., Mojtahedi, H., Crosslin, J.M. and Santo, G.S. 1998. Eradication of tobacco rattle virus from soils by growing weed-free alfalfa. Abstr. In: *Proc. of VII International Plant Virus Epidemiology Symposium.* Aguadulce (Almeria), Spain. pp. 164-165.

Trudgill, D.L., Brown, D.J.F. and McNamara, D.G. 1983. Methods and criteria for assessing the transmission of plant viruses by longidorid nematodes. *Rev. Nematol.* 6: 133-141.

Visser, P.B. and Bol J.F. 1999. Nonstructural proteins of tobacco rattle virus which have a role in nematode-transmission: expression pattern and interaction with viral coat protein. *J. gen. Virol.* 80: 3273-3280.

Walkinshaw, C.H., Griffin, G.D. and Larson, R.H. 1961. *Trichodorus christiei* as a vector of potato corky ringspot (tobacco rattle) virus. *Phytopathology* 51: 806-808.

Weingartner, D.P. and Haynes, K.G. 1999. Inheritance of resistance to corky ringspot. *Amer. J. Potato Res. 76: 387.* Abstract.

Weingartner, D.P., Haynes, K.G., Goth, R.W. and Meldrum, J. 1994. High levels of field resistance to corky ringspot disease exist in USDA segregating families B1136 and B1176. *Amer. Potato J.* 71: 709. Abstract.

Weingartner, D.P., McSorley, R. and Goth R.W. 1993. Management strategies in potato for nematodes and soil-borne diseases in subtropical Florida. *Nematropica* 23: 233-245.

Weingartner, D.P., Shumaker, J.R. and Smart, G.C., Jr. 1983. Why soil fumigation fails to control potato corky ringspot disease in Florida. *Plant Disease* 67: 130-134.

Wyss, U. 1982. Virus-transmitting nematodes: feeding behavior and effect on root cells. *Plant Disease* 66: 639-644.

Xenophontos, S., Robinson, D.J., Dale, M.F.B. and Brown, D.J.F. 1998. Evidence for persistent, symptomless infection of some potato cultivars with tobacco rattle virus. *Potato Res.* 41: 255-265.

8. Transmission of Viruses

8.3 Aphid Transmission of Potato Viruses

Yvon Robert and Danièle Bourdin
Institut National de la Recherche Agronomique, UMR 'Biologie des Organismes et des Populations appliquée à la Protection des Plantes' - Domaine de la Motte au Vicomte, B.P. 29; 35653 Le Rheu Cédex, France

8.3.1 Introduction

At least eight viruses can be spread by aphids in potato crops and have a critical role in the production of seed potatoes, *Potato leafroll virus* genus *Polerovirus* (PLRV), *Potato virus Y* genus *Potyvirus* (PVY), *Potato virus A* genus *Potyvirus* (PVA), *Potato virus M* genus *Carlavirus* (PVM), *Potato virus S* genus *Carlavirus* (PVS), *Alfalfa mosaic virus* genus *Alfamovirus* (AlMV), *Potato aucuba mosaic virus* genus *Potexvirus* (PAMV) and *Cucumber mosaic virus* genus *Cucumovirus* (CMV) (Beemster and de Bokx, 1987).

For some 50 years after the discovery that potato viruses could be transmitted by aphids (Oortwijn Botjes, 1920), mainly biological aspects of their transmission have been described. Nevertheless this considerably helped to improve our knowledge of virus disease epidemiology, to provide seed potato growers with better virus control measures based mainly on cropping practices and chemical control, and to implement National certification schemes for seed potatoes. However, environmental problems were soon posed by the widespread use of pesticides, one of the main causes of concern being the development of insecticide resistance in *Myzus persicae* (Sulz.). Therefore there was a need for new strategies such as breeding for targetted virus-resistant cultivars. Among other requirements for this, is

G. Loebenstein et al. (eds.),
Virus and Virus-like Diseases of Potatoes and Production of Seed-Potatoes, 195–225.

obviously the need to know most precisely the nature of transmission mechanisms and plant-virus-aphid interactions involved, as well as the potential for virus populations to overcome resistance to virus infection.

In the late 1970s, it has become possible to monitor electrically the different phases of aphid feeding behaviour and to observe *in situ* virus particles in their vectors. Higher refinement in the accuracy and reliability of virus detection and the implementation of novel molecular technologies, allowed to go deeper in our understanding of the structure and expression of virus genomes, and in the identification of aphid vectors.

This chapter is intended to make an updated review about our current knowledge of the transmission processes of potato viruses, taking advantage of fruitful comparisons with other viruses when deemed necessary.

8.3.2 The Aphid Vectors

Aphids belong to a group of insects very well suited to transmit viruses. Their mouthparts have evolved into efficient piercing and sucking organs, the stylets, sheathed in a rostrum that holds them in a groove. This system allows them to penetrate plant tissues inter or intracellularly and eventually reach phloem tissues for feeding.

8.3.2.1 Resident and Transient Aphids

At least twelve aphid species that can settle and reproduce on potato as a plant host ("resident" aphids) have long been deemed to be the main if not the sole vectors of both persistently (PLRV) and non-persistently (namely PVY) transmitted viruses, with a wide range of vector efficiency (Kennedy *et al.* 1962; Robert, 1999). The most common species and species groups include members of the green peach aphid complex, *Myzus persicae* (Sulz.), the foxglove aphid, *Aulacorthum solani* (Kltb.), the green and pink potato aphid, *Macrosiphum euphorbiae* (Thos.), the buckthorn aphid, *Aphis nasturtii* Kltb. (=*Aphis rhamni*), the *Aphis frangulae* Kltb. group, including *Aphis gossypii* Glov., and the black bean aphid, *Aphis fabae* Scop.. Another species, the bulb and potato aphid, *Rhopalosiphoninus latysiphon* (Davids.) is found on sprouting potatoes.

However, in a search of the causes of PVY (PVY^O and PVY^N) epidemics which occurred in Europe in the last 30 years, more than 30 additional aphid species were shown to transmit that virus (van Hoof, 1977; Robert, 1978; Kostiw, 1979; Rydén, 1979; van Hoof, 1980; Rydén *et al.* 1983; Sigvald, 1984; Katis and Gibson, 1985; Harrington *et al.* 1986; Piron, 1986; Harrington and Gibson, 1989; de Bokx and Piron, 1990). These aphids visit potato plants when they are looking for new host plants but they can neither settle for long nor reproduce on them ("transient" or "itinerant" aphids). Overall, they are poor vectors yet they are able to spread PVY in natural conditions when large populations of winged individuals infest potato crops at an early stage.

8.3.2.2 Biological and Morphological Characteristics of the Main Species

8.3.2.2.1 Aphid Life Cycle and its Role on Virus Transmission and Spread

For most resident aphids, potato is usually just one among many other plant hosts on which they can thrive. Alate (winged) females colonize potato crops whereon they give birth to apterous (wingless) progeny larvae which develop into adults, able to reproduce on the plants. After a number of multiplication cycles of apterae, new alatae are produced and soon leave the plants in search for other hosts. The whole process is achieved through ovo-viviparous and parthenogenetic reproduction. The origin and abundance of alatae at infestation time depend upon the kind of life cycle they complete in a particular area.

"Potato" aphids show two types of life cycles that may be separated or coexist in the same region, depending on aphid species and environmental constraints. Holocyclic aphids alternate between primary hosts, usually woody plants, on which they overwinter as diapausing fertilized eggs, and secondary hosts, on which they reproduce entirely parthenogenetically. Conversely, anholocyclic aphids show continuous asexual reproduction throughout the year. Primary hosts are usually not virus sources whereas secondary ones often encompass weeds and crops that are susceptible to virus diseases. Therefore, holocyclic emigrant aphids flying directly from primary hosts to spring crops are unlikely to be viruliferous, contrary to anholocyclic ones that may be already infective when they alight on potatoes. Another important consequence of host alternation on the subsequent transmission success of non-persistent viruses is that prior to their landing on a potato crop, aphids have been kept fasting, depending upon how long they were aloft between two plants (see below). Finally, return migrants of holocyclic aphid vectors are much less likely to colonize and infect autumn-sown potato crops than anholocyclic alate aphids in areas where more than one crop is possible each year.

8.3.2.2.2 Aphid Morphology and Biology

During its development, each morph passes through four larval instars (=nymphs) before it becomes adult. Morphologically, alate nymphs cannot be distinguished from apterous ones until wing-pads develop on the thorax, generally in the third instar. Morphological separation of apterous nymphs from adult morphs is mainly based on the presence of a well-developed cauda in the latter. There are other less conspicuous features. Precise identification of the various species, their morphs and instars, is based on taxonomic characters requiring microscopic examination. However, lower magnification can be used to identify some useful morphological characteristics (Fig. 8.3.1) needed for determining apterous and alate adults reproducing on potato (Fig. 8.3.2).

Apterous viviparous females of the *M. persicae* group are usually from pale to dark green or yellow, or even pink (see below). The front of head shows prominent antennal tubercles with convergent inner margins, antennae are shorter than the body. Cornicles are pale, slightly swollen on apical half, with dusky apices ending with a distinct flange preceded by a slight constriction. Alate females look dark, due to black head and thorax, dusky antennae and legs and the presence of a large black dorsal patch with dark lateral patches on the abdomen (Fig. 8.3.1). Dusky cornicles are swollen on one side over apical half in individuals born on potato but cylindrical in those that migrated from peach trees and other primary hosts.

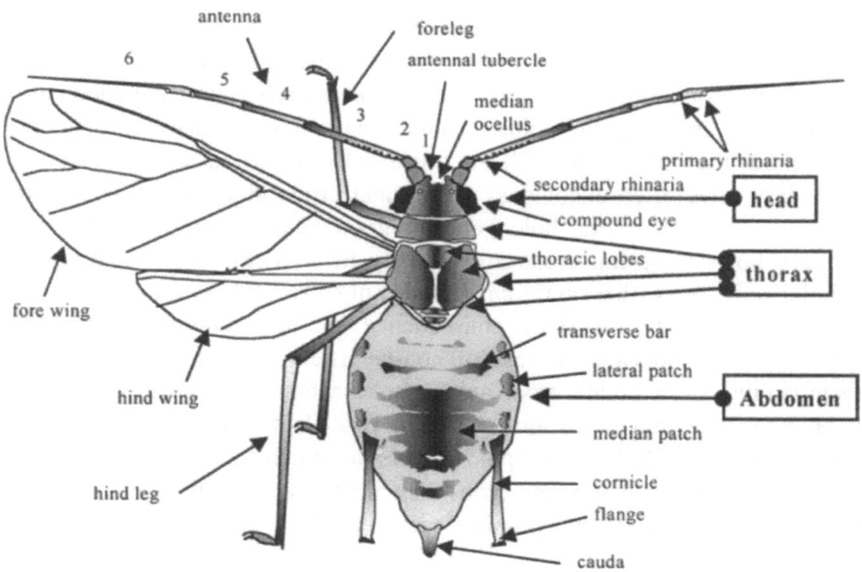

Fig. 8.3.1. Morphological characteristics of an alate viviparous female of *Myzus persicae*

The life cycle of *M. persicae* varies between areas. Usually both holocyclic and anholocyclic migrants colonize potato in a given region, depending for the former on the availability of primary hosts in cold areas (*Prunus persica* (L.) Batsch, *Prunus serotina* Ehrh. and other *Prunus* spp.; *Lycium halimifolium* Mill.) (Gabriel, 1969; Hille Ris Lambers, 1955) and for the latter on local possibilities to overwinter parthenogenetically (Turl, 1983; Bale *et al.* 1988).

A. solani apterae are bright yellowish to dark green globular individuals. Prominent antennal tubercles with parallel inner sides are present on the head. Between and before the base of cornicles, an area of green oily globules, clustered under the cuticle, is more or less visible. Cornicles are slightly tapering from base to their black apex showing a pronounced flange. Head of alate females is dusky brownish to almost black and antennae are rather dark. Transverse bands and bars are present on the abdomen. Holocyclic biotypes can overwinter on various herbaceous primary hosts (*Digitalis purpurea* L.; *Hieracium* sp.) but in many regions, potatoes are colonized by alatae produced by populations that have survived the winter anholocyclically on *D. purpurea* (foxglove).

M. euphorbiae apterous and alate viviparous females are either green or pink with long body and appendices. No antennal tubercles are present on the head. Abdomen is not pigmented, only some small lateral patches may appear in alate forms. Cornicles are long, cylindrical over most of their length, with a dark, slightly narrowed reticulated apical area. The cauda is long and acute. In North America, *M. euphorbiae* seems to be essentially holocyclic with *Rosa* spp. as primary hosts. In Europe, potatoes are most often colonized by anholocyclic migrants.

Both apterous and alate females of *R. latysiphon* are shiny black. Cornicles are cylindrical over basal third and apical part, being strongly swollen in between, which makes their identification easy. This species is strictly anholocyclic.

A. frangulae and *A. nasturtii* are small aphids with short appendices, rather difficult to identify and separate from each other. *A. frangulae* apterous adults are from yellowish brown to very dark green or blue-green whereas *A. nasturtii* ones are mostly yellowish. The former bears uniformly dark cornicles and the latter paler ones becoming darker towards the apical part. There is no dorsal pigmentation on the abdomen of either of them. Very small and weakly pigmented spots or bands may appear on the abdomen of *A. frangulae* alate females but not in *A. nasturtii*. Both species are holocyclic: primary hosts are *Frangula alnus* Mill. for *A. frangulae* and *Rhamnus cathartica* L. for *A. nasturtii*.

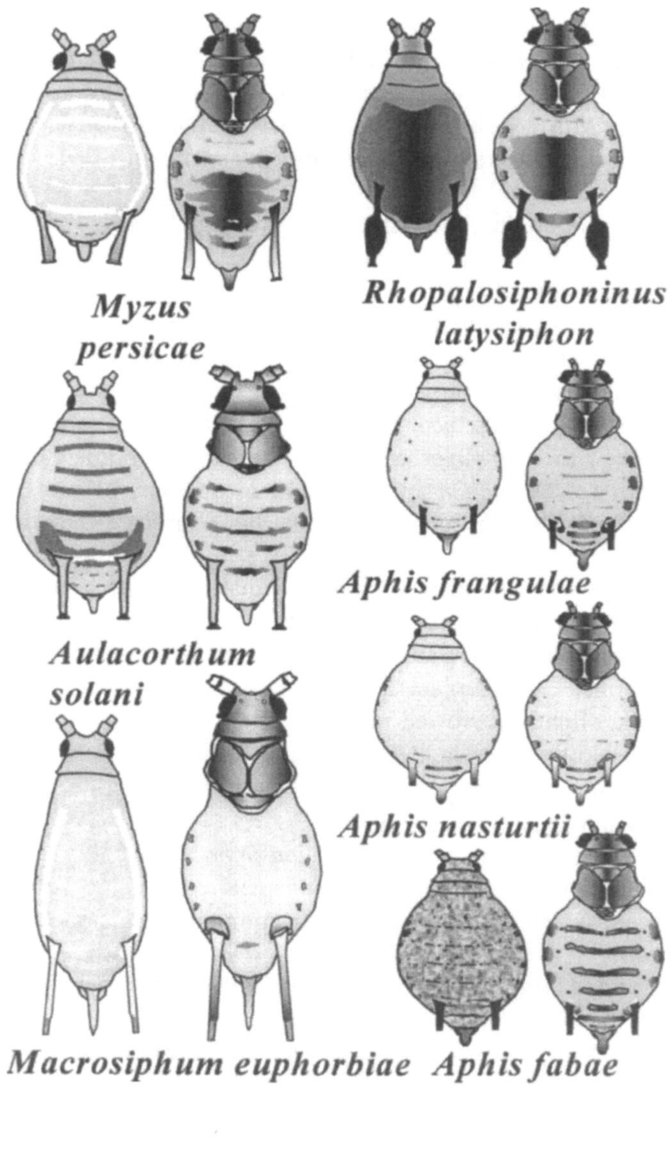

Fig. 8.3.2 Dorsal view of the body of apterous (left) and alate (right) females of various potato aphids (the antennae and legs of both apterae and alatae, and the wings of alatae have been omitted).

A. fabae apterous females are uniformly black to dark brownish green. Some white pulverulent areas can often be seen on the abdomen, corresponding to wax secretion. Black transverse bands are present on the abdomen of alate adults. Potato appears to be colonized by specifically adapted biotypes of the subspecies *A. fabae solanella* Theob. whose life cycle is not fully elucidated.

8.3.2.2.3 Inter - and Intraspecific Diversity

Myzus persicae has long been recognized to belong to a group of morphologically similar-looking species (van Emden *et al.* 1969). Particular forms differing by their colour, biology and physiological adaptation to feed on specific hosts were reported. A similar distinction, based on biological differences, was initially made between *A. frangulae* and *A. gossypii*. Such a difficulty to determine reliably individuals to species level has resulted in incorrect assessment of their vector efficiency and in contradictory results in the literature.

Three taxa previously ascribed to *M. persicae* have been described, *M. persicae*, *M. nicotianae* Blackman and *M. antirrhinii* (Macchiati) according to their colour, biology, karyotypes and morphometric components (Blackman and Paterson, 1986; Blackman, 1987). Green or pink forms of *M. nicotianae* were first reported from four continents on tobacco (Blackman, 1987) and then proved to be able to reproduce also on potato and transmit potato viruses (Boiteau and Lowery, 1989; Bourdin *et al.* 1998). However, the use of genetic markers subsequently showed that this "species" should be just considered as a tobacco-adapted form of *M. persicae* and not as a distinct species, whereas *M. antirrhinii* proved to be a genuinely different species (Blackman and Spence, 1992; Margaritopoulos *et al.* 1998; Terradot *et al.* 1999).

As for other aphids, very low genetic diversity was revealed by enzyme electrophoresis on the green peach aphid. Only esterases showed some variation in relation to resistance to insecticides. However, novel molecular techniques can be used to identify clones distinctively (Martinez-Torres *et al.* 1997; Fenton *et al.* 1998; Terradot *et al.* 1999). Differences in the ability of some of these clones to transmit particular virus strains (Bourdin *et al.* 1998) and the occurrence of highly efficient ones in seed producing regions can be of great concern if they are predominant (Fenton *et al.* 1998).

8.3.3 Modes and Mechanisms of Transmission

8.3.3.1 Aphid Feeding Behavioural Sequences

During their dispersal activity in search for new hosts, both apterous and alate aphids visit any plant species indiscriminately as a result of complex behavioural procedures, which appear to be predominantly based on responses to visual stimuli (Robert, 1987). However, by that time, they are still unable to discriminate between suitable and unsuitable hosts.

Once they have come into contact with a new plant substrate, they display a further series of behavioural sequences in response to a variety of physical and physiological stimuli received from the plant: only then will the insects be able to decide whether it is a suitable host (Pollard, 1973). During these steps, particular behavioural patterns may have important implications in both virus acquisition and inoculation processes.

First, the aphid shows an intense wandering activity, interrupted by short pauses of less than one minute during which it inserts its stylets into the superficial cell layers of plant tissues (epidermis, parenchyma) before resuming walking or flying. This probing activity is a reflex elicited by any solid surface encountered on the plant and is antagonistic to locomotion. Short probes serve as test-feeding punctures for a first-step recognition of the host by the aphid. Starved individuals tend to initiate penetration earlier and make shorter probes in larger number than non-fasted ones. Transient aphids, which are not prone to accept potato as a host, make a number of successive probes each becoming shorter and shorter while walking time increases, until they eventually leave the plant. Conversely, resident ones are more likely to settle soon for feeding and reproduction, depending on the level of resistance of the plant.

Most probes are initiated in intercellular grooves that are located by the apex of the rostrum tapping the surface of the substrate. Prior to a probe, a drop of gelling saliva is secreted and deposited on the plant surface, forming a kind of plug with an external flange and collar. During stylet penetration into the tissues, secretion of gelling saliva continues, resulting in a continuous sheath around the stylet bundle. Stylet insertion has been initially described as almost always intercellular, reaching the epidermis in less than one minute on average. In fact, epidermal cells may also be penetrated by the stylets. Evidence of cell membrane puncture can be obtained when stylet penetration activities are monitored electrically, using the Electrical Penetration Graph (EPG) technique (Tjallingii, 1978).

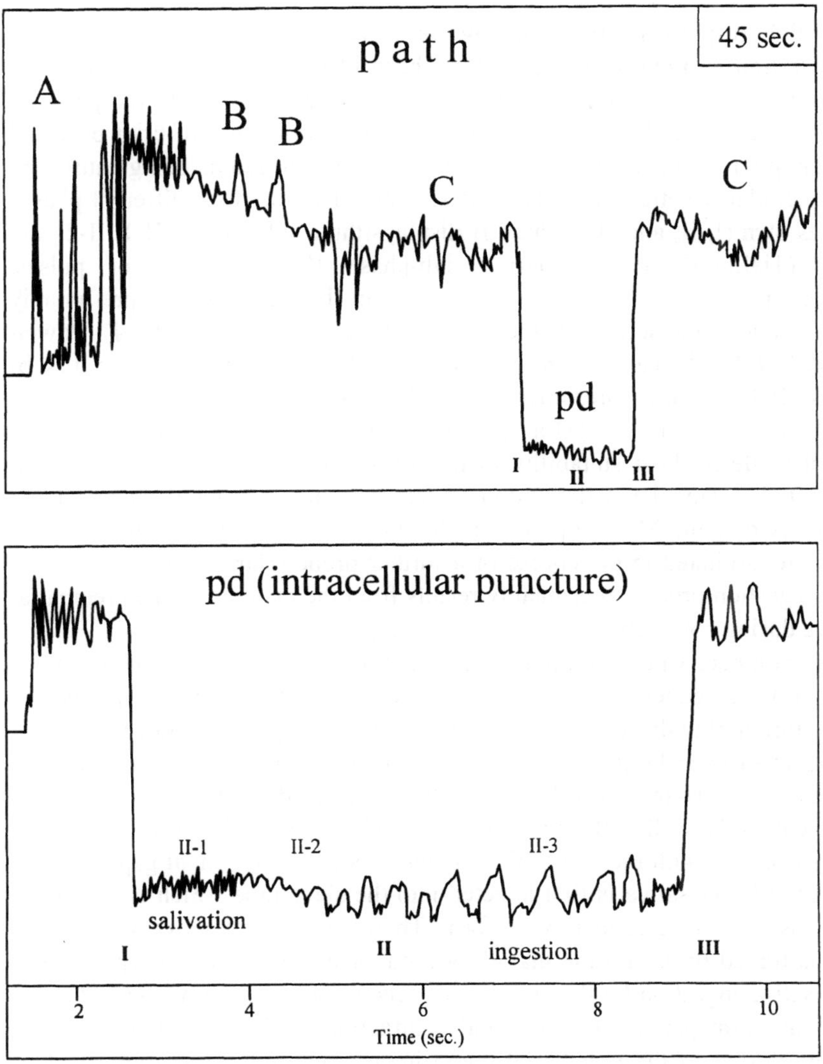

Fig. 8.3.3 Electrical Penetration Graphs (EPG) elicited by aphid probing behaviour showing the corresponding A, B, C waveform patterns of the pathway phase (path diagram) and the typical sub-phases (II-1, II-2, II-3) of a potential drop (pd) displayed during an intracellular puncture (pd diagram) (after Prado and Tjallingii, 1993; Martin *et al.*, 1997) (see text for details).

EPG responses corresponding to seven different steps of aphid penetration into the plant tissues have been identified as specific waveform patterns (A to G) (Tjallingii, 1988), the first three referring to the pathway phase in a probe (Fig. 8.3.3). During pattern C, a number of brief potential drops (pds) frequently occur: they have been recognized as being caused by intracellular penetration by the stylets. Each pd shows three different phases on its own (I, II, and III). Phase II displays three sub-phases, II-1, II-2, and II-3 (Tjallingii, 1985). During sub-phase II-3, a series of pulses, superimposed on the base waveform, can be seen which presumably correspond to the activity of the sucking (cibarial) pump (Fig. 8.3.4) (Powell *et al.* 1995). When a cell is pierced, watery saliva is first injected into it (sub-phase II-1) and then presumably actively sucked back with cell substances (pulses in sub-phase II-3) (Fig. 8.3.3) into the epipharyngeal area of the food canal to detect host suitability (sampling probe). If this sampled material cannot elicit any phagostimulation, stylets are withdrawn while watery saliva is secreted again. Most important is the possibility for this ingested material to be regurgitated in the course of a further probe (Harris and Bath, 1973). This egestion process presumably results from the opening of the precibarial valve during contraction of the cibarial pump.

In contrast, when preliminary information received by the aphid during the probing sequence shows that the plant may be suitable for feeding, penetration of underlying tissues towards phloem is pursued. Sieve tubes can be reached in at least 15 minutes but usually this is achieved only after a series of "trials-and-errors" that result in branched tracks. The mandibles pierce the tube wall of the phloem bundle, which allows the maxillae to enter the sieve tube with the end of the common food/salivary canal gaping in the lumen. Watery saliva is first injected into the sieve tube which is shown on EPG as a E1 sieve element waveform. Then a mixture of saliva and phloem sap is forced back into the maxillary food canal, likely due to sap pressure and active ingestion by the aphid. This second phase is shown on EPG as a E2 waveform pattern. Dilation and contraction of the cibarial pump with likely closure of the precibarial pump in-between, allow ingestion of plant sap and its expulsion into the foregut respectively (Fig. 8.3.4) (Forbes, 1977).

8.3.3.2 The Modes of Transmission

Overall, virus transmission process is made of a succession of events that encompass four critical phases, acquisition-, latent-, inoculation- and retention periods.

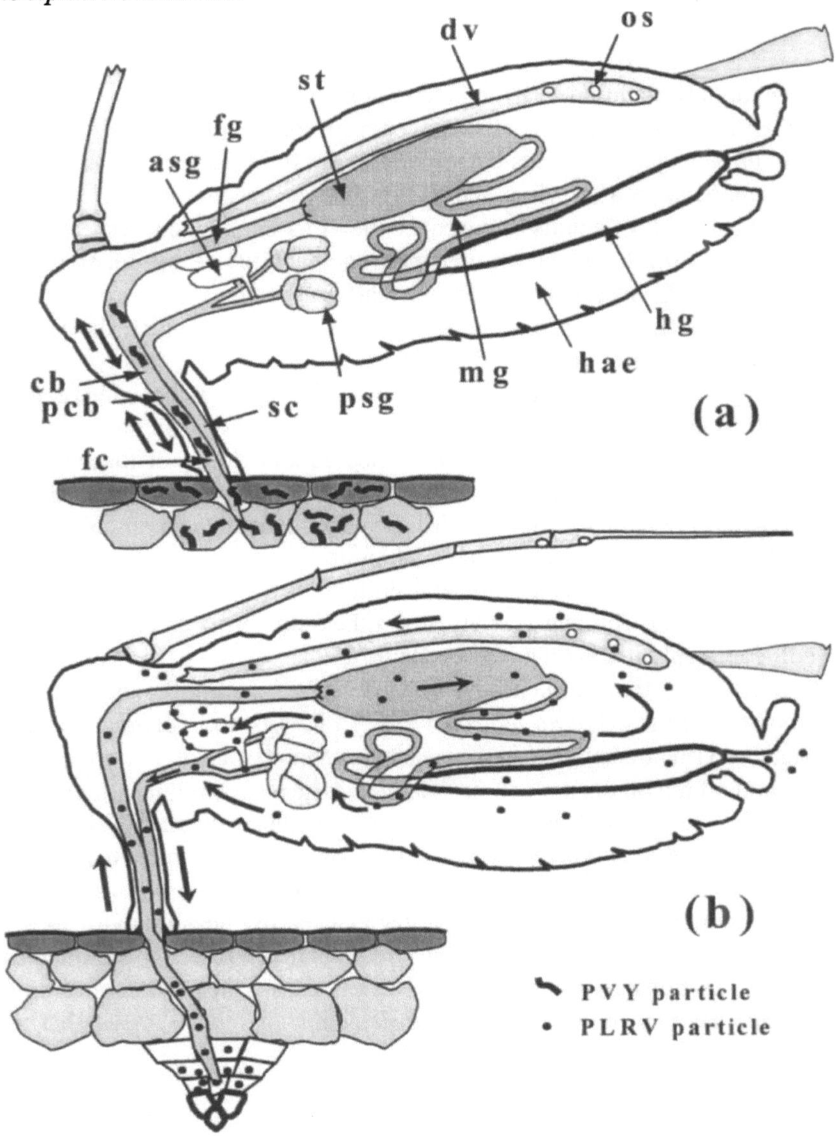

Fig. 8.3.4. Diagrams showing virus-vector relationships and putative routes (thick arrows) of (a) non-circulative PVY and (b) circulative PLRV in their vector (see text for explanation). fc, food canal; sc, salivary canal; pcb, precibarium; cb, cibarium; fg, foregut; st, stomach; mg, midgut; hg, hindgut; asg, accessory salivary gland; psg, principal salivary gland; hae, haemocoel; dv, dorsal vessel; o, ostium.

When aphids alight on virus infected plants, virus acquisition is achieved either while probing superficially or feeding from the phloem (Fig. 8.3.4). Latent and inoculation periods correspond to the time required for aphids to become inoculative after virus acquisition (i.e. to become able to infect a healthy plant) and for aphids to introduce the virus into a plant, respectively. The length of the latent period depends on the virus-vector relationship, being from not detectable for non-persistent viruses up to hours or days for persistent ones. Retention time refers to the time during which the vector retains infectivity after virus acquisition.

The length of these various phases has been first used to characterize non-persistent, semi-persistent and persistent transmission (Watson and Roberts, 1939). This classification was later modified as knowledge of mechanisms involved in the transmission process increased. Depending on the site of their retention within the aphid, viruses were defined as non-circulative (non-persistent and semi-persistent transmission) and circulative (persistent transmission). Among persistently-transmitted viruses, some do multiply within their vectors (propagative viruses), others do not. Both classifications are still commonly used (reviewed by Nault, 1997).

Two modes of transmission have been described among the eight aphid-borne potato viruses. *Potato leafroll virus* (PLRV) is a circulative and non propagative virus transmitted in a persistent manner whereas *Potato virus Y* (PVY) and all the others are non-circulative viruses that are transmitted non-persistently. Both PLRV and PVY have been extensively studied for their relationships with their vectors, because they are, by far, the most harmful to potato crops and also because they are representative of *Polerovirus* and *Potyvirus* genera. The transmission strategies used by these two viruses are the most common among aphid-borne viruses. Therefore we chose them to describe virus transmission processes. Nevertheless, when data concerning PLRV and PVY were lacking, results gained with other viruses sharing similar strategies were used.

8.3.3.2.1 Non-persistent Transmission of *Potato virus Y* (PVY)

8.3.3.2.1.1 The Transmission Characteristics

Non-persistent transmission is a very rapid and efficient process. Aphids are able to acquire PVY within a few seconds or minutes and require only a few seconds to inoculate it back to a plant immediately afterwards, as there is no discernible latent period. They rarely remain able to transmit it for long since retention time lasts less than 1 hour for feeding aphids and a little longer for non-feeding ones (Proeseler and Weidling, 1975; de Bokx and

Huttinga, 1981). However, the concept of short retention time has been questioned after much longer retention times from 19 to 70 hours have been found. Such a result could explain long-distance transport of *Maize dwarf mosaic virus* genus *Potyvirus* (MDMV) by aphids migrating over wide areas, depending on aphid species and environmental conditions (Berger *et al.*, 1987; Zeyen and Berger, 1990). This concept may also apply to PVY.

Non-persistent acquisition and inoculation have long been associated with aphid stylet penetration into epidermal tissues (Fig. 8.3.4a). The location of these brief probes, either between or into epidermal cells has remained more controversial, particularly because initially, most stylet pathways of *M. persicae*, one of the main vectors of PVY, were reported to be intercellular. However, using electron microscopy, Lopez-Abella *et al.* (1988) showed that only aphids making brief intracellular punctures to sample cell sap were able to transmit PVY. This was confirmed by Powell (1991) who found a positive correlation between the detection of epidermal cell membrane punctures and PVY transmission by *M. persicae*, using the EPG technique (Tjallingii, 1978). Moreover, a comparison of the probing behaviour between *Brachycaudus helichrysi* (Kltb.) and *Drepanosiphum platanoidis* (Schrk.) led to the conclusion that the latter species was presumably a poorer vector of PVY than the former one because individual aphids punctured the epidermal cell membrane only rarely (Powell *et al.* 1992). It still remains a possibility for the virus to be acquired and inoculated via broken plasmodesmata when stylets penetrate between cells, but this is presumably a much less efficient mechanism (Powell, 1991).

An interesting characteristic of non-persistent transmission is that pre-acquisition starvation of aphids has a positive effect on transmission rates (Watson, 1938). A number of hypotheses suggested that aphid saliva, secreted during feeding activity, contains inhibitors. Their presence in stylets of non-starved aphids would inhibit the virus or its retention (reviewed by Harris, 1977). This was supported by recent studies that failed to detect any difference in the probing behaviour of starved and non-starved *M. persicae* using EPG (Powell, 1993). Starvation did not affect the occurrence of electrically recorded membrane punctures during acquisition of PVY from infected plants. Therefore it was concluded that the conditions of the aphid mouthparts/foregut immediately following feeding might account for lower PVY transmission by non-starved *M. persicae*. This contrasted with electron microscopical examination that revealed that starved *M. persicae* inserted their stylets into epidermal cells of PVY infected plants more frequently than did non-starved ones (Lopez-Abella *et al.* 1988). Similarly, analyses of EPG results have shown that potential drops typical of sampling activity during probing were more frequent with starved *M. persicae* and *A. gossypi* than non-starved ones (Collar *et al.* 1998). Therefore, in spite of its paramount epidemiological consequences, the pre-acquisition starvation effect is still not fully explained. However, behavioural differences still remain more plausible to explain other transmission characteristics such as transmission efficiency of various vector species.

Other factors can affect non-persistent virus transmission efficiency, and hence can have an effect on virus spread during a cropping season. However, other than differences in aphid probing behaviour and plant cell penetration as mentioned above, the mechanisms that are involved in the broad range of transmission rates are poorly understood. Overall, acquisition from and inoculation to young plants usually result in higher transmission rates, irrespective of aphid species, clones, morphs, or virus strains. Large differences exist in the rate by which various virus strains are transmitted by aphids. Thus, PVY^N isolates are better transmitted than any other PVY strain (Proeseler and Weidling, 1975). This was also observed with PVM and PVS strains (Bode and Weidemann, 1971). However, it should be stressed that it may often be difficult to compare published results. The choice of different environmental regimes during any transmission phase can affect transmission efficiency (Simons, 1966; Simons and Eastop, 1970; Proeseler and Weidling, 1975) and, most important, the virus source and test plant. For example, contradictory conclusions were reached regarding the usefulness of tobacco plants as source and/or test plants when assessing PVY^O and PVY^N transmission efficiency by *Rhopalosiphum padi* (Rydén, 1979; van Hoof, 1980; Katis and Gibson, 1985)

8.3.3.2.1.2 Current Hypotheses on Non-persistent Transmission Mechanisms

Kennedy *et al.* (1962), argued whether non-persistent transmission could be a passive mechanism in which virus particles would remain attached to the inner or outer surface of the stylet tips. This was supported by early observations of virus particles on the surface of aphid stylets and, although lacking conclusive data, this hypothesis was not ruled out by more recent ones (Martín *et al.* 1997). However, most characteristics of the non-persistent transmission such as starvation effect and specificity cannot be explained entirely by such a passive mechanism.

The ingestion-egestion hypothesis (Harris, 1977) is more widely accepted to explain transmission efficiency. It relies on specific patterns of natural aphid behaviour corresponding to the ability of aphids first to penetrate epidermal cells as evidenced by typical potential drops (pds) on EPGs (Tjallingii, 1985), and second to regurgitate ingested cell material from the food canal during further probing (Fig. 8.3.4a). Dramatic increase in the transmission efficiency has been associated with the occurrence, during acquisition, of sub-phases II-3 when aphids were feeding on plants (Fig. 8.3.3) and of sub-phase II-3-like pulses when they were fed on membrane. The duration of this phase did not influence transmission (Powell *et al.* 1995; Martín *et al.* 1997). The inoculation part of the process has also been correlated to pds (Powell, 1991; 1993) but the exact role of egestion lacked convincing experimental evidence. Recently, using EPG, Martín *et al.* (1997) established that PVY inoculation was associated with the salivation sub-phase II-1 of the typical salivation – ingestion - egestion sequence, occurring during short probes (Fig. 8.3.3). Therefore, they suggested that ingestion - salivation is also involved in non-persistent transmission. However, since salivary and food canals become confluent about 8μm from the stylet tips, inoculation via salivation could be possible only for those virus particles that are retained in this common part but not for those present further up in the foregut (Hébrard *et al.*, 1999).

Thus, under these assumptions, non-persistent transmission appears to be associated with aphid probing behaviour during which acquisition would occur by ingestion whereas inoculation would be associated with egestion and/or salivation on a healthy plant (Fig. 8.3.4a).

8.3.3.2.1.3 The Molecular Interactions

In fact, the success of non-persistent transmission is not only determined by aphid behaviour but also by the various interactions that allow the retention of virus particles at and their subsequent release from specific sites in their vectors.

Potyvirus properties responsible for aphid transmissibility have been extensively studied using, in most cases, non-transmissible variants. As expected, the coat protein is involved in the process leading to efficient transmission. Evidence for this came from comparisons of coat protein amino acid sequences between transmissible and non-transmissible potyvirus isolates. In most cases, loss of transmissibility was correlated with a mutation in a highly conserved amino acid triplet aspartic acid - alanine - glycine (DAG in single-letter code) located near the coat protein N-terminus (reviewed by Pirone and Blanc, 1996). However, several aphid transmissible potyviruses have been described that contain a motif other than DAG at this location. This observation led to more precise studies suggesting that, in addition to the motif itself, the context in which it lies plays an important role in the transmission process (López-Moya *et al.*, 1999). Beside the coat protein, another virus protein is crucial for potyvirus transmissibility. Early studies had shown that transmission of the aphid non-transmissible *Potato virus C* genus *Potyvirus* (PVC, now identified as a strain of PVY, PVY^C)

could occur when aphids were previously fed on PVY-infected plants (Kassanis and Govier, 1971a). Similarly, aphids were unable to transmit PVY from purified preparations unless they had previously been fed on PVY-infected plants irradiated with UV to inactivate the virus. These authors later showed that a helper component (HC), produced in PVY-infected plants allowed PVY^{C} transmission and also assisted PVY transmission from purified preparations (Kassanis and Govier, 1971b; Govier and Kassanis, 1974). Subsequently, the HC was shown to be a virus-encoded protein of 58kDa, and the biologically active form to be dimeric (Govier *et al.* 1977; Thornbury and Pirone, 1983; Thornbury *et al.* 1985). Comparison of amino acid sequences of the HC between non transmissible potyviruses and transmissible ones revealed that various mutations could be associated with the loss of transmissibility. The HC of PVY^{C} differed from that of PVY in the highly conserved KITC sequence, by a mutation that could result in preventing dimer formation (Thornbury *et al.*, 1990). Other non-transmissible potyviruses also display mutations in the same KITC box or in another highly conserved motif with a PTK sequence (Pirone and Blanc, 1996). Recently, two changes in conserved amino acid residues located outside these highly conserved motifs, have been associated with the loss of transmissibility of a PVY strain (Llave *et al.*, 1999).

Govier and Kassanis (1974) hypothesized that HC acts as a bridge between the aphid and the virus. This was first supported by experiments in which ^{125}I-labelled purified *Tobacco etch virus* genus *Potyvirus* (TEV) and PVY were used to feed aphids in the absence or in the presence of HC. Only aphids fed with an HC-virus mixture were able to retain label in the stylets and cibarium and to transmit the viruses (Berger and Pirone, 1986). Subsequent electron microscopic observations revealed that mutations in the coat protein DAG motif of *Tobacco vein mottling virus* genus *Potyvirus* (TVMV) and in the KITC motif of PVY^{C}-HC led to non-transmissibility because they affected retention of virions in the food canal of the stylets (Wang *et al.* 1996). An alternative hypothesis, supported by competition assays, is that interaction between coat protein and HC of potyviruses modifies folding of the coat protein, allowing a direct interaction between a surface domain, perhaps including the DAG motif, and the aphid mouthparts (Salomon and Bernardi, 1995).

Both hypotheses are based on a possible interaction between virus and HC. Such an interaction was recently shown *in vitro*. Binding assays with various potyviruses revealed successively that (i) the DAG motif was directly involved in a specific interaction with HC (Blanc *et al.* 1997), (ii) the PTK motif of HC is involved in binding of HC to virions (Peng *et al.* 1998), and (iii) modifications of the KITC motif of HC did not alter the *in vitro* interaction between HC and the coat protein (Blanc *et al.* 1998). Moreover, immunogold labelling, followed by transmission electron microscopic observations have shown that only functional PVY-HC with an intact KITC motif was retained in the stylets of *M. persicae* (Blanc *et al.* 1998). When mutations affect other conserved amino acid residues, the helper component was shown to be still able to bind to virions, suggesting that these mutations rendered the HC incapable of interacting with aphid stylets or affected its stability in plant tissues (Llave *et al.*, 1999).

Although relying on *in vitro* experiments, these results strongly support the bridge hypothesis. On the one hand, HC would interact with the DAG motif of the coat protein through its central domain containing the PTK motif. On the other hand, HC would be involved in interactions with the aphid mouthparts through its N-terminal domain containing the KITC motif. Both these interactions would allow the retention of virus particles in the stylets. The process leading to the release of virus particles into epidermal cells during short probes is not known.

8.3.3.2.2 Persistent Transmission of *Potato Leafroll Virus* (PLRV)

8.3.3.2.2.1 Transmission Characteristics

Potato leafroll virus, like other members of the *Polerovirus* and *Luteovirus* genera is phloem limited. These viruses have evolved unique relationships with their vectors on which they depend almost exclusively for their transmission. Persistent transmission is characterized by the circulation of virus particles throughout the aphid's body. Once ingested, virus particles pass from the gut lumen into the haemocoel and diffuse in the haemolymph. They are pulsed towards the anterior part of the aphid through dilation-contraction of the dorsal vessel and eventually penetrate into the accessory salivary glands (ASGs) before being inoculated to a new plant (Fig. 8.3.4b). This complex process implies an active involvement of the aphid and should be considered as virus transport rather as a passive movement of the virus in the aphid (Gildow, 1999).

Virus acquisition can be split into (i) virus ingestion into the alimentary canal and (ii) actual virus acquisition which corresponds to the transport from gut lumen into haemocoel (Gildow, 1999). Electro-monitoring systems (EMS) and EPG analysis showed that virus ingestion was correlated with phloem feeding for various polero- and luteoviruses, including PLRV (EPG pattern E2) (Leonard and Holbrook, 1978; Prado and Tjallingii, 1994). Acquisition times varying from a few minutes to several hours have been reported for PLRV. Once salivation-ingestion is established in phloem cells, 1.6 min seems sufficient for a healthy aphid to ingest enough virus to become infective. However, there is a direct relationship between the length of the acquisition access period and transmission efficiency (Leonard and Holbrook, 1978; Kostiw, 1991). The virus content of *M. persicae* has been shown to increase linearly with an increasing acquisition access period up to 5 days (Tamada and Harrison, 1981).

Barley yellow dwarf virus genus *Luteovirus* (BYDV) inoculation has been correlated by EPG analysis to salivation in the phloem (EPG pattern E1) (Prado and Tjallingii, 1994). A similar phenomenon is probably involved in PLRV transmission. Inoculation periods can vary from a few minutes to several hours. Leonard and Holbrook (1978) reported that an inoculation access period as short as 2.5 min allowed aphids to transmit PLRV but that 30 min were needed for transmission to be completed. Maximum transmission efficiency is obtained after several days, which suggests that aphids spend most of this time in selecting penetration sites and in locating phloem bundles.

A further period of time is required between acquisition and inoculation, before the first transmission can be successfully achieved. This latent period corresponds to the time needed for the virus to be transported from the gut lumen to the salivary canal of ASGs, via the haemolymph. For PLRV, two days were deemed necessary for the latent period to be completed in *M. persicae* (Tamada and Harrison, 1981).

Aphids usually remain infective throughout their life. Overall, virus content of aphids increases with longer acquisition access periods and decreases with increasing time after they have left the virus source, but neither quickly nor completely enough for them to become inefficient at transmitting (Tamada and Harrison, 1981).

Acquisition, latent period and inoculation are dependent on a number of factors. Aphids acquire and transmit the virus more efficiently at higher temperatures than at lower ones (Webb, 1956; Robert and Rouzé-Jouan, 1971), and high temperatures reduce the latent period (Tamada and Harrison, 1981; Syller, 1987). Such effects have been related to modifications of either aphid behaviour or virus availability from infected plants (Syller, 1994). PLRV latent period is also shorter in younger nymphs and when virus particles are acquired from young leaves or from virus purified from young leaves. The latter effect suggests that intrinsic properties of virus particles can affect the transmission process (van den Heuvel *et al.* 1991). Transmission has been found to be closely related to virus content in membrane feeding tests. In source plants, the relation between virus kinetics and titre depends indirectly on plant status (*e.g.* susceptible or resistant genotypes), and on physiological changes in ageing leaves which contribute to increasing mature plant resistance (van den Heuvel *et al.* 1993b).

Acquisition depends on aphid species, their distribution on the plant and the amount of sap they can take up. Rather than virus concentration *per se*, it is the ability of feeding aphids to gain easy access to biologically active virus particles in the plant during the acquisition period, which seems to be the major component to successful transmission (van den Heuvel *et al.* 1993b).

8.3.3.2.2.2 Virus - Vector Specificity

The virus-vector relationship in persistent transmission is highly specific. This means that a given virus is efficiently transmitted by only a few aphid species and that the necessary transport of the virus throughout the aphid necessitates its recognition by aphid structures.

As already mentioned, PLRV is transmitted only when aphids feed in the phloem. As a consequence, aphid species that can transmit it are only those that settle and reproduce on potato plants, yet not every species of such resident species that colonize susceptible crops is efficient. Therefore, PLRV is spread by about 12 aphid species of which *M. persicae* is generally considered the most efficient.

Transmission efficiency depends first on intrinsic properties of the vector species (MacCarthy, 1954; Björling and Ossiannilsson, 1958; Kennedy *et al.* 1962). Within the same species it varies within aphid clones (Björling and Ossiannilsson, 1958; Hinz, 1966a, b; Robert, 1971; Upreti and Nagaich, 1971), morphs (either winged or apterous viviparous forms) and instars (either nymphs or adults) (Robert and Maury, 1970). Virus strain is also a deciding component for transmission efficiency and some strains have been characterized by the ease with which they can be transmitted by *M. persicae* (Tamada *et al.* 1984).

However, Bourdin *et al.* (1998) showed that a poorly aphid-transmissible (PAT) PLRV isolate was very poorly transmitted by some clones of *M. persicae* and at up to 80% by others. At the same time, all the clones were efficient vectors of a highly aphid-transmissible (HAT) PLRV isolate. The molecular characterization of the aphid clones used in this study revealed that most poor vectors belonged to the *M. antirrhinii* taxon within the *M. persicae* group (Terradot *et al.* 1999). Overall, these data show that PLRV transmission by aphids presumably depends more on the interaction between aphid clones and virus isolates than on the sole properties of either virus or vector.

8.3.3.2.2.3 The Virus Route within the Aphid

It was established very early that once an aphid had become inoculative, its ability to transmit the virus was not influenced by moulting and that the aphid retained its infectivity for the rest of its life (Elze, 1927). Therefore,

Elze (1931) suggested that PLRV was ingested with the food, passed from the gut into the haemolymph and, from there, into the salivary glands whence it would be injected into a plant with saliva (Fig. 8.3.4b). Various studies further supported these assumptions (Day, 1955; Stegwee and Ponsen, 1958). The cellular mechanisms involved in this circulation were studied only recently, using immune electron microscopy (IEM) observations on aphids having acquired the virus.

Gildow (1985) was the first to obtain evidence for an association of *Cereal yellow dwarf virus Polerovirus-RPV* (CYDV-RPV) particles with apical plasmalemma of the gut lumen of *Rhopalosiphum padi* L. Virus particles were visualized in coated pit invaginations and in coated vesicles in the cytoplasm of apical plasmalemma cells. Tubular vesicles containing virions were also observed fused to the basal lamina. These observations strongly suggested that virus transport through the gut membrane occurred via a receptor-mediated endocytosis-exocytosis mechanism. Such a mechanism was later confirmed and further observations revealed that whereas BYDV and CYDV were acquired through the hindgut membrane, PLRV was acquired through the midgut membrane (Gildow, 1993; Garret *et al.* 1993). Chronological IEM observations of *M. persicae* from feeding onset on purified PLRV have shown that there was a rapid translocation of the virus through the midgut cytoplasm and that the virus could accumulate and persist in the gut epithelial cells for many days (Garret *et al.* 1996)

Virus particles have never been observed free in the haemolymph (Gildow, 1993; Garret *et al.* 1993). Using a virus-overlay assay of aphid protein blots, van den Heuvel *et al.* (1994) first showed that the virus could interact *in vitro* with five aphid proteins. One of these proteins was further identified as symbionin, a groEL homologue produced by the primary endosymbionts of the aphid and abundantly present in the haemolymph. The observed interaction was shown to be a common phenomenon among polero- and luteoviruses (Filichkin *et al.* 1997; van den Heuvel *et al.* 1997). Symbionin belongs to a family of molecular chaperones, which are able to both bind and stabilize newly synthesized peptides and to allow their functional folding (van den Heuvel, 1999). Therefore, the virus presumably associates with symbionin during its diffusion in the haemolymph and this interaction would be essential for virion stability by retarding proteolytic breakdown. It has also recently been proposed that this interaction could help the virus to escape a putative immune system in the aphid (van den Heuvel, 1999). However, only two lines of evidence have suggested that this interaction can also occur *in vivo* and is involved in the transmission process. First, van den Heuvel *et al.* (1994) showed that antibiotic treatment, which reduced levels of aphid symbionts and symbionin, reduced the ability of *M.*

persicae nymphs to transmit PLRV and was associated with loss of capsid integrity. Secondly, mutant *Beet western yellows virus* genus *Polerovirus* (BWYV) particles, unable to interact with symbionin were also shown to persist for shorter periods of time in *M. persicae* than wild-type particles when they were injected directly into the haemolymph (van den Heuvel *et al.* 1997).

The role of the ASGs in the transmission process was first shown with BYDV, based on indirect evidence (Rochow *et al.* 1975). Direct IEM observations later confirmed that BYDV attaches specifically to ASG membrane (Gildow and Rochow, 1980a). Only a few studies have been conducted with PLRV, but such an attachment was also observed for PLRV and BWYV, suggesting that it is a general phenomenon among polero- and luteoviruses (Gildow, 1982). The whole process at the ASG level has been studied only with CYDV and BYDV. Virus particles have to (i) specifically attach to and penetrate into the basal lamina surrounding the ASG, (ii) come into contact with basal plasmalemma and accumulate within membrane invaginations and (iii) penetrate basal plasmalemma in order to be released into the salivary canal. A receptor-mediated endocytosis-exocytosis mechanism, similar to that observed at gut level allows virus particle transport through the ASG membrane (Gildow and Gray, 1993).

Along this route, three barriers of specificity have been identified, namely the gut membrane, the ASG basal lamina, and ASG basal plasmalemma (reviewed by Gildow, 1999). *M. persicae* for example, is able to acquire *Soybean dwarf virus* (SbDV) through the hindgut membrane and PLRV through the midgut one. In these cases, there is a high degree of tissue specificity which is probably determined by the viral capsid structure. In contrast, based on results obtained with BYDV and SbDV, virus attachment to ASG basal lamina would be virus-, tissue- and aphid- specific. Viruses can interact only with ASG basal lamina and not with that of other adjacent tissues. This interaction can be inhibited by pre-incubating the virus with monoclonal antibodies (Mabs) directed against the virus capsid and the type of interaction is dependent on the aphid species. The basal plasmalemma has been clearly identified as the last barrier of specificity, virus particles being able to traverse it or not once they have penetrated the basal lamina and accumulated at the plasmemma.

8.3.3.2.2.4 The Molecular Basis for Virus Circulation within the Aphid

The successful circulation of polero- and luteoviruses in their vectors depends on specific interactions between virus particles and aphid membranes. The importance of the viral capsid in determining transmissibility has been convincingly demonstrated for BYDV and CYDV

(Rochow, 1970; Gildow and Rochow, 1980b). For PLRV, HAT strains were shown to differ from PAT ones in their antigenic properties, thus confirming the role of the viral capsid in the transmission efficiency (Massalski and Harrison, 1987; Jolly and Mayo, 1994). The transmission by *M. persicae* of *Potato spindle tuber viroid* (PSTVd) through transcapsidation of its RNA in PLRV capsid has provided more circumstantial evidence for the importance of this structure in aphid transmission efficiency and specificity (Querci *et al.* 1997).

Besides a major capsid protein (CP), encoded by Open Reading Frame 3 (ORF3), PLRV particles, like those of other polero- and luteoviruses, also contain a minor component, the readthrough protein (RTP). The RTP is expressed as a result of translational readthrough of ORF 3 termination codon into ORF 5 (Bahner *et al.* 1990). Because of its presence on the surface of the virus particle, these authors hypothesized that the RTP could play a role in virus transmission. Using membrane feeding experiments, van den Heuvel *et al.* (1993a) were able to inhibit PLRV transmission after incubation of purified virus with a Mab that reacted with the CP but not with the RTP. They concluded that the CP itself was governing transmission. In contrast, Jolly and Mayo (1994), comparing the ORF 3 and 5 sequences between PLRV-HAT and -PAT isolates suggested two amino acid changes in the 3' region of the RTP gene as possible transmission determinants. The importance of this region in transmission efficiency is, however questionable because it is cleaved off in purified virus preparations that are still transmissible after membrane acquisition. Brault *et al.* (1995) found, by mutational analysis, that the C-terminal part of the RTP was dispensable for BWYV transmission. Moreover, it has been shown that CYDV particles containing a C-terminal truncated form of the RTP were efficiently transmitted by aphids (Filichkin *et al.* 1994; Wang *et al.* 1995) and that those devoid of RTP were acquired into the haemocoel but not transmitted (Chay *et al.* 1996). Therefore this shows that conserved domains within the coat protein should be sufficient for acquisition through the hindgut (Chay *et al.* 1996), whereas, depending on virus species, either the C- or the N-terminal part of the RTP would be involved in the transport from the haemolymph into the salivary canal. More recently, PLRV-VLPs (virus-like particles) lacking the RTP have been obtained using a recombinant baculovirus system (Gildow, 1999). Ultrastructural visualisations have revealed that these VLPs were able to complete their route in the aphid *M. persicae*. Although these conflicting results may reflect virus-vector specificity, it is quite difficult to obtain a clear idea on the respective involvement of each of the structural proteins in the transmission process.

Little is known about the potential receptors allowing virus transport through the gut and ASG membranes. Among the five proteins of *M. persicae* showing an *in vitro* affinity for PLRV, only symbionin, which interacts with the N-terminal region of the RTP of BWYV through its so-called equatorial domain, has been identified (van den Heuvel *et al.* 1997; Hogenhout *et al.* 1998). The four other proteins, with molecular weights ranging from 49 to 85 kDa might well be membrane receptors but are still to be identified. Although powerful, virus-overlay experiments such as those used to identify symbionin, have their own limitations. Most probably some of the interactions revealed on membrane do not occur *in vivo* and *vice versa*.

8.3.4 Conclusion

Significant advances have been made over the last 20 years, improving our understanding of the mechanisms of transmission and plant-virus-vector interactions. The EPG technique has allowed to go much deeper into the interpretation of successive sequences implied in the feeding behaviour of aphid vectors and to relate specific waveform patterns to various phases of the transmission process. The demonstration of the relation between epidermal cell penetration by aphid stylets and non-persistent transmission and of the role of egestion and/or salivation during the inoculation of both non-circulative and circulative viruses is an important step forward towards the understanding of the functional aspects of feeding behavioural sequences.

Some, if not all, virus proteins and their respective domains involved in the transmission process have been identified. However a possibility remains that, in addition to these proteins, others may be involved in the transmission process. *Cauliflower mosaic virus* genus *Caulimovirus* (CaMV) shares with PVY an HC-assisted strategy of transmission. Recently, it has been shown that a third virus protein (PIII) was required for transmission. This PIII was suggested to act as a bridge between the HC and the CP or induce structural changes in the HC, allowing it to bind to the CP (Leh *et al.*, 1999).

The critical role of specific interactions between viral proteins and some particular structures present in the aphid vectors has been emphasized. However, most of our knowledge relies on *in vitro* experiments assessing protein/protein interactions. Therefore the results may be strongly biased since the conditions met by virus particles in the aphid stylets, gut, haemolymph or ASGs are not necessarily comparable to experimental conditions. Obviously, there is a need for complementary techniques to validate *in vivo* the interactions detected *in vitro*. More information on the aphid saliva and haemolymph composition would be useful.

Some important steps of the non-persistent and persistent transmission processes remain to be clarified. For instance there is a need to understand more accurately how potyviruses are released during egestion and/or salivation from their site of retention in aphid stylets. For PLRV, the barriers the virus needs to traverse within its vector have been clearly identified, yet the exact nature and functioning of aphid membrane receptors at each barrier remain to be defined precisely.

Better knowledge of virus-vector-plant interactions could undoubtedly result in improving strategies for plant resistance to viruses through host gene- and/or transgene-mediated resistance. For instance, non-persistent virus spread could be limited by using cultivars that decrease the chance for a vector to penetrate into the epidermal cells during probing. Also, the expression in potato plants of aphid's genes corresponding to gut or ASG membrane receptors could interfere with the necessary circulation of PLRV in its vector, thus reducing its spread under field conditions. Such alternative approaches might be among the next challenges to take up for better control strategies in the future.

References

Bahner, I., Lamb, J., Mayo, M.A. and Hay, R.T. 1990. Expression of the genome of potato leafroll virus: readthrough of the coat protein termination codon *in vivo*. *J. gen. Virol.* 71: 2251-2256.

Bale, J.S., Harrington, R. and Clough, M.S. 1988. Low temperature mortality of the peach-potato aphid *Myzus persicae*. *Ecol. Entomol.* 13: 121-129.

Beemster, A.B.R. and Bokx, J. A. de 1987. Survey of properties and symptoms. In: *Viruses of potatoes and seed-potato production.* (eds) J.A. de Bokx and J.P.H. van der Want, Pudoc, Wageningen. pp. 84-113.

Berger, P.H. and Pirone, T.P. 1986. The effect of helper component on the uptake and localization of potyviruses in *Myzus persicae*. *Virology* 153: 256-260.

Berger, P.H., Zeyen, R.J. and Groth, J.V. 1987. Aphid retention of maize dwarf mosaic virus (potyvirus): epidemiological implications. *Ann. appl. Biol.* 111: 337-344.

Björling, K. and Ossiannilsson, F. 1958. Investigations on individual variations in the virus-transmitting ability of different aphid species. *Socker* 14: 1-13.

Blackman, R.L. 1987. Morphological discrimination of a tobacco-feeding form from *Myzus persicae* (Sulzer) (Hemiptera: Aphididae), and a key to New World *Myzus* (*Nectarosiphon*) species. *Bull. Entomol. Res.* 77: 713-730.

Blackman, R.L. and Paterson, A.J.C. 1986. Separation of *Myzus* (*Nectarosiphon*) *antirrhinii* (Macchiati) from *Myzus* (*N.*) *persicae* (Sulzer) and related species in Europe (Homoptera: Aphididae). *Syst. Entomol.* 11: 267-276.

Blackman, R.L. and Spence, J.M. 1992. Electrophoretic distinction between the peach-potato aphid, *Myzus persicae*, and the tobacco aphid, *M. nicotianae* (Homoptera: Aphididae). *Bull. Entomol. Res.* 82: 161-165.

Blanc, S., Ammar, D., Dolja, V.V., Llave, C., Baker, J. and Pirone, T.P. 1998. Mutations in potyvirus helper component protein: effects on interactions with virions and aphid stylets. *J. gen. Virol.* 79: 3119-3122.

Blanc, S., Lopez-Moya, J.J., Wang, R.Y., Garcia-Lampasona, S., Thornbury, D.W. and Pirone, T.P. 1997. A specific interaction between coat protein and helper component correlates with aphid transmission of a potyvirus. *Virology* 231: 141-147.

Bode, O. and Weidemann, H.L. 1971. Untersuchungen zur Blattlausübertragbarkeit von Kartoffel-M-und -S- Virus. *Potato Res.* 14: 119-129.

Boiteau, G. and Lowery, D.T. 1989. Comparison of a yellow form of the green peach aphid, *Myzus persicae* (Sulzer), and a green form of the tobacco aphid, *Myzus nicotianae* Blackman, coexisting on greenhouse potato in New Brunswick. *Can. Entomol.* 121: 1029-1035.

Bokx, J.A. de and Huttinga, H. 1981. Potato Virus Y. *CMI/AAB Descriptions of Plant Viruses* n° 242, 6 pp.

Bokx, J.A. de and Piron, P.G.M. 1990. Relative efficiency of a number of aphid species in the transmission of potato virus Y^N in the Netherlands. *Neth. J. Plant Pathol.* 96: 237-246.

Bourdin, D., Rouzé, J., Tanguy, S. and Robert, Y. 1998. Variation among clones of *Myzus persicae* and *Myzus nicotianae* in the transmission of a poorly- and a highly- aphid-transmissible isolate of potato leafroll luteovirus (PLRV). *Plant Pathol.* 47: 794-800.

Brault, V., Heuvel, J.F.J.M. van den, Verbeek, M., Ziegler-Graff, V., Reutenauer, A., Herrbach, E., Garaud, J.-C., Guilley, H., Richards, K. and Jonard, G. 1995. Aphid transmission of beet western yellows luteovirus requires the minor capsid read-through protein P74. *EMBO J.* 14: 650-659.

Chay, C.A., Gunasinge, U.B., Dinesh-Kumar, S.P., Miller, W.A. and Gray, S.M. 1996. Aphid transmission and systemic plant infection determinants of barley yellow dwarf luteovirus-PAV are contained in the coat protein readthrough domain and 17-kDa protein respectively. *Virology* 219: 57-65.

Collar, J.L., Martín, B. and Fereres, A. 1998. Nonpersistent transmission of plant viruses determined by aphid probing behaviour. In: *Aphids in natural and managed ecosystems*. (eds) J.M. Nieto Nafría and A.F.G. Dixon, Universidad de León, León. pp. 161-165.

Day, M.F. 1955. The mechanism of the transmission of potato leaf roll virus by aphids. *Aust. J. biol. Sci.* 8: 498-513.

Elze, D.L. 1927. De verspreiding van virusziekten van de aardappel (*Solanum tuberosum* L.) door insekten. *Meded. Landbouwhogesch. Wageningen* 31: 1-90.

Elze, D.L. 1931. The relation between insect and virus as shown in potato leafroll, and a classification of viruses based on this relation. *Phytopathology* 21: 675-686.

Emden, H.F. van, Eastop, V.F., Hughes, R.D. and Way, M.J. 1969. The ecology of *Myzus persicae*. *Annu. Rev. Entomol.* 14: 197-270.

Fenton, B., Woodford, J.A.T. and Malloch, G. 1998. Analysis of clonal diversity of the peach-potato aphid, *Myzus persicae* (Sulzer), in Scotland, UK and evidence for the existence of a predominant clone. *Mol. Ecol.* 7: 1475-1487.

Filichkin, S.A., Lister, R.M., McGrath, P.F. and Young, M.J. 1994. *In vivo* expression and mutational analysis of the barley yellow dwarf virus readthrough gene. *Virology* 205: 290-299.

Filichkin, S.A., Brumfield, S., Filichkin, T.P. and Young, M.J. 1997. In vitro interactions of the aphid endosymbiotic SymL chaperonin with barley yellow dwarf virus. *J. Virol.*, 71: 569-577.

Forbes, A.R. 1977. The mouthparts and feeding mechanism of aphids. In: *Aphids as Virus Vectors*. (eds) K.F Harris and K. Maramorosch, Academic Press, New York. pp. 83-103.

Gabriel, W. 1969. L'influence de la distribution des hôtes primaires des pucerons vecteurs du virus Y et de l'Enroulement sur la propagation de ces virus dans les cultures de la pomme de terre dans une région sub-carpathique. *Ziemniak* 1-17.

Garret, A., Kerlan, C. and Thomas, D. 1993. The intestine is a site of passage for potato leafroll virus from the gut lumen into the haemocoel in the aphid vector, *Myzus persicae* Sulz. *Arch. Virol.* 131: 377-392.

Garret, A., Kerlan, C. and Thomas, D. 1996. Ultrastructural study of acquisition and retention of potato leafroll luteovirus in the alimentary canal of its aphid vector, *Myzus persicae* Sulz. *Arch. Virol.* 141: 1279-1292.

Gildow, F.E. 1982. Coated vesicle transport of luteovirus through the salivary gland of *Myzus persicae*. *Phytopathology* 72: 1289-1296.

Gildow, F.E. 1985. Transcellular transport of barley yellow dwarf virus into the hemocoel of the aphid vector *Rhopalosiphum padi*. *Phytopathology* 75: 292-297.

Gildow, F.E. 1993. Evidence for receptor-mediated endocytosis regulating luteovirus acquisition by aphids. *Phytopathology* 83: 270-277.

Gildow, F.E. 1999. Luteovirus transmission and mechanisms regulating vector specificity. In: *The Luteoviridae*. (eds) H.G. Smith and H. Barker, CABI Publishing, CAB International, Wallingford, pp. 88-111.

Gildow, F.E. and Gray, S.M. 1993. The aphid salivary gland basal lamina as a selective barrier associated with vector-specific transmission of barley yellow dwarf luteovirus. *Phytopathology* 83: 1293-1302.

Gildow, F.E. and Rochow, W.F. 1980a. Role of accessory salivary glands in aphid transmission of barley yellow dwarf virus. *Virology* 104: 97-108.

Gildow, F.E. and Rochow, W.F. 1980b. Importance of capsid integrity for interference between two isolates of barley yellow dwarf virus in an aphid. *Phytopathology* 70: 1013-1015.

Govier, D.A. and Kassanis, B. 1974. A virus-induced component of plant sap needed when aphids acquire potato virus Y from purified preparations. *Virology* 61: 420-426.

Govier, D.A., Kassanis, B. and Pirone, T.P. 1977. Partial purification and characterisation of the potato virus Y helper component. *Virology* 78: 306-314.

Harrington, R. and Gibson, R.W. 1989. Transmission of potato virus Y by aphids trapped in potato crops in southern England. *Potato Res.* 32: 167-174.

Harrington, R., Katis, N. and Gibson, R.W. 1986. Field assessment of the relative importance of different aphid species in the transmission of potato virus Y. *Potato Res.* 29: 67-76.

Harris, K.F. 1977. An ingestion-egestion hypothesis of non-circulative virus transmission. In: *Aphids as Virus Vectors*. (eds) K.F Harris and K. Maramorosch, Academic Press, New York. pp. 165-220.

Harris, K.F. and Bath, J.E. 1973. Regurgitation by *Myzus persicae* during membrane feeding: its likely function in transmission of nonpersistent plant viruses. *Ann. entomol. Soc. Am.* 66: 793-796.

Hébrard, E., Froissart, R., Louis, C. and Blanc, S. 1999. Les modes de transmission des virus phytopathogènes par vecteurs. *Virologie* 3: 35-48.

Heuvel, J. F. J. M. van den 1999. Fate of a luteovirus in the haemolymph of an aphid. In: *The Luteoviridae*. (eds) H.G. Smith and H. Barker, CABI Publishing, CAB International, Wallingford, pp. 112-119.

Heuvel, J. F. J. M. van den, Boerma, T. M. and Peters, D. 1991. Transmission of potato leafroll virus from plants and artificial diets by *Myzus persicae*. *Phytopathology* 81: 150-154.

Heuvel, J. F. J. M. van den, Verbeek, M. and Peters, D. 1993a. The relationship between aphid-transmissibility of potato leafroll virus and surface epitopes of the viral capsid. *Phytopathology* 83: 1125-1129.

Heuvel, J. F. J. M. van den, Verbeek, M. and Wilk, F. van der 1994. Endosymbiotic bacteria associated with circulative transmission of potato leafroll virus by *Myzus persicae*. *J. gen. Virol.* 75: 2559-2565.

Heuvel, J. F. J. M. van den, Dirven, J.A.A.M., Os, G.J. van and Peters, D. 1993b. Acquisition of potato leafroll virus by *Myzus persicae* from secondarily-infected potato plants of different genotypes. *Potato Res.* 36: 89-96.

Heuvel, J. F. J. M. van den, Bruyère, A., Hogenhout, S.A., Ziegler-Graff, V., Brault, V., Verbeek, M., Wilk, F. van der and Richards, K. 1997. The N-terminal region of the luteovirus readthrough domain determines virus binding to *Buchnera* GroEL and is essential for virus persistence in the aphid. *J. Virol.* 71: 7258-7265.

Hille Ris Lambers, D. 1955. Potato aphids and virus diseases in the Netherlands. *Ann. appl. Biol.* 42: 355-360.

Hinz, B. 1966a. Beiträge zur Analyse der Vektoreignung einiger wirtschaftlich wichtiger Blattlausarten und -rassen. I. Versuche zur Ermittlung der Vektoreigenschaften für das Blattrollvirus der Kartoffel bei Rassen von *Myzus persicae* (Sulz.). *Phytopathol. Z.* 56: 54-77.

Hinz, B. 1966b. Beiträge zur Analyse der Vektoreignung einiger wirtschaftlich wichtiger Blattlausarten und -rassen. II. Versuche zur Ermittlung der Vektoreigenschaften für das Enationenvirus der Erbse bei Rassen von *Myzus persicae* (Sulz.), *Acyrthosiphon pisum* (Harris) und *Macrosiphum euphorbiae* (Thomas). *Phytopathol. Z.* 56: 123-140.

Hogenhout, S.A., Wilk, F. van der, Verbeek, M., Goldbach, R.W. and Heuvel, J. F. J. M. van den 1998. Potato leafroll virus binds to the equatorial domain of the aphid endosymbiotic GroEL homolog. *J. Virol.* 72: 358-365.

Hoof, H.A van 1977. Determination of infection pressure of potato virus Y^N. *Neth. J. Plant Pathol.* 83: 123-127.

Hoof, H.A van 1980. Aphid vectors of potato virus Y^N. *Neth. J. Plant Pathol.* 86: 159-162.

Jolly, C.A. and Mayo, M.A. 1994. Changes in the amino acid sequence of the coat protein readthrough domain of potato leafroll luteovirus affect the formation of an epitope and aphid transmission. *Virology* 201: 182-185.

Kassanis, B. and Govier, D.A. 1971a. New evidence on the mechanism of aphid transmission of potato C and potato aucuba mosaic viruses. *J. gen. Virol.* 10: 99-101.

Kassanis, B. and Govier, D.A. 1971b. The role of the helper virus in aphid transmission of potato aucuba mosaic virus and potato virus C. *J. gen. Virol.* 13: 221-228.

Katis, N. and Gibson, R.W. 1985. Transmission of potato virus Y by cereal aphids. *Potato Res.* 28: 65-70.

Kennedy, J.S., Day, M.F. and Eastop, V.F. 1962. *A conspectus of aphids as vectors of plant viruses*. Commonwealth Institute of Entomology, London, 114 pp.

Kostiw, M. 1979. Transmission of potato virus Y by *Rhopalosiphum padi* L. *Potato Res.* 22: 237-238.

Kostiw, M. 1991. Influence of the duration of acquisition and inoculation feeding on the effectiveness of potato leafroll virus transmission by *Myzus persicae* Sulz. *Potato Res.* 34: 41-45.

Leh, V., Jacquot, E., Geldreich, A., Hermann, T., Leclerc, D., Cerutti, M., Yot, P. and Blanc, S. 1999. Aphid transmission of cauliflower mosaic virus requires the viral P III protein. *EMBO J.* 18: 7077-7085.

Leonard, S.H. and Holbrook, F.R. 1978. Minimum acquisition and transmission times for potato leaf roll virus by the green peach aphid. *Ann. entomol. Soc. Am.* 71: 493-495.

Llave, C., Martínez, B., Díaz-Ruíz, J.R. and López-Abella, D. 1999. Helper component mutations in nonconserved residues associated with aphid transmission efficiency of a pepper isolate of potato virus Y. *Phytopathology* 89: 1176-1181.

Lopez-Abella, D., Bradley, R.H.E. and Harris, K.F. 1988. Correlation between stylet paths made during superficial probing and the ability of aphids to transmit nonpersistent viruses. *Adv. Dis. Vect. Res.* 5: 251-285.

López-Moya, J.J., Wang, R.Y. and Pirone, T.P. 1999. Context of the coat protein DAG motif affects potyvirus transmissibility by aphids. *J. Gen. Virol.* 80: 3281-3288.

MacCarthy, H.R. 1954. Aphid transmission of potato leafroll virus. *Phytopathology* 44: 167-174.

Margaritopoulos, J.T., Mamuris, Z. and Tsitsipis, J.A. 1998. Attempted discrimination of *Myzus persicae* and *Myzus nicotianae* (Homoptera: Aphididae) by Random Amplified Polymorphic DNA Polymerase Chain Reaction technique. *Ann. entomol. Soc. Am.* 91: 602-607.

Martín, B., Collar, J.L., Tjallingii, W.F. and Fereres, A. 1997. Intracellular ingestion and salivation by aphids may cause the acquisition and inoculation of non-persistently transmitted plant viruses. *J. gen. Virol.* 78: 2701-2705.

Martinez-Torres, D., Carrió, R., Latorre, A., Simon, J.C., Hermoso, A. and Moya, A. 1997. Assessing the nucleotide diversity of three aphid species by RAPD. *J. evol. Biol.* 10: 459-477.

Massalski, P.R. and Harrison, B.D. 1987. Properties of monoclonal antibodies to potato leafroll luteovirus and their use to distinguish virus isolates differing in aphid transmissibility. *J. gen. Virol.* 68: 1813-1821.

Nault, L.R. 1997. Arthropod transmission of plant viruses: a new synthesis. *Ann. entomol. Soc. Am.* 90: 521-541.

Oortwijn Botjes, J.G. 1920. De Bladrolziekte van de aardappel plant. *Landbouwhogesch.* Thesis, Wageningen.

Peng, Y., Kadoury, D., Gal-On, A., Huet, H., Wang, R.Y. and Raccah, B. 1998. Mutations in the HC-Pro gene of zucchini yellow mosaic potyvirus: effects on aphid transmission and binding to purified virions. *J. gen. Virol.* 79: 897-904.

Piron, P.G.M. 1986. New aphid vectors of potato virus Y^N. *Neth. J. Plant Pathol.* 92: 223-229.

Pirone, T. and Blanc, S. 1996. Helper-dependent vector transmission of plant viruses. *Annu. Rev. Phytopathol.* 34: 227-247.

Pollard, D.G. 1973. Plant penetration by feeding aphids (Hemiptera: Aphidoidea): a review. *Bull. entomol. Res.*, 62: 631-714.

Powell, G. 1991. Cell membrane punctures during epidermal penetrations by aphids: consequences for the transmission of two potyviruses. *Ann. appl. Biol.* 119: 313-321.

Powell, G. 1993. The effect of pre-acquisition starvation on aphid transmission of potyviruses during observed and electrically recorded stylet penetrations. *Entomol. exp. appl.* 66: 255-260.

Powell, G., Harrington, R. and Spiller, N.J. 1992. Stylet activities and potato virus Y vector efficiencies by the aphids *Brachycaudus helichrysi* and *Drepanosiphum platanoidis*. *Entomol. exp. appl.* 62: 293-300.

Powell, G., Pirone, T. and Hardie, J. 1995. Aphid stylet activities during potyvirus acquisition from plants and an *in vitro* system that correlate with subsequent transmission. *Eur. J. Plant Pathol.* 101: 411-420.

Prado, E. and Tjallingii, W.F. 1994. Aphid activities during sieve element punctures. *Entomol. exp. appl.* 72: 157-165.

Proeseler, G. and Weidling, H. 1975. Die Retentionszeit von Stämmen des Kartoffel-Y-Virus in verschiedenen Aphidenarten und Einfluß der Temperatur. *Archiv. Phytopathol. PflSchutz* 11: 335-345.

Querci, M., Owens, R.A., Bartolini, I., Lazarte, V. and Salazar, L.F. 1997. Evidence for heterologous encapsidation of potato spindle tuber viroid in particles of potato leafroll virus. *J. gen. Virol.* 78: 1207-1211.

Robert, Y. 1971. Epidémiologie de l'Enroulement de la pomme de terre: capacité vectrice de stades et de formes des pucerons *Aulacorthum solani* Kltb, *Macrosiphum euphorbiae* Thomas et *Myzus persicae* Sulz. *Potato Res.* 14: 130-139.

Robert, Y. 1978. Rôle épidémiologique probable d'espèces de pucerons autres que celles de la pomme de terre dans la dissémination intempestive du virus Y depuis 4 ans dans l'Ouest de la France. *Abstr. 7th trien. Conf. E.A.P.R.*, Warsaw, pp. 242-243.

Robert, Y. 1987. Dispersion and migration. In *"World Crops Pests". Vol. 2A*: *"Aphids, their Biology, Natural Enemies and Control"*. (eds) A.K. Minks and P. Harrewijn, Elsevier, Amsterdam, pp. 299-313.

Robert, Y. 1999. Epidemiology and control strategies: Epidemiology of potato leafroll disease. In: *The Luteoviridae*. (eds) H.G. Smith and H. Barker, CABI Publishing, CAB International, Wallingford, pp. 221-228.

Robert, Y. and Maury, Y. 1970. Capacités vectrices comparées de plusieurs souches de *Myzus persicae* Sulz., *Aulacorthum solani* Kltb et *Macrosiphum euphorbiae* Thomas dans l'étude de la transmission de l'Enroulement de la pomme de terre. *Potato Res.* 13: 199-209.

Robert, Y. and Rouzé-Jouan, J. 1971. Premières observations sur le rôle de la température au moment de la transmission de l'Enroulement par *Aulacorthum solani* Kltb, *Macrosiphum euphorbiae* Thomas et *Myzus persicae* Sulz.. *Potato Res.* 14: 154-157.

Rochow, W.F., 1970. Barley yellow dwarf virus: phenotypic mixing and vector specificity. *Science*, 167: 875-878.

Rochow, W.F., Foxe, M.J. and Muller, I. 1975. A mechanism of vector specificity for circulative aphid-transmitted plant viruses. *Ann. N. Y. Acad. Sci.* 266: 293-301

Rydén, K. 1979. Havrebladlusen, *Rhopalosiphum padi*, kan sprida potatisvirus Y. *Växtskyddsnotiser* 3: 51-53.

Rydén, K., Brishammar, S. and Sigvald, R. 1983. The infection pressure of potato virus Y^O (PVY^O) and the occurrence of winged aphids in potato fields in Sweden. *Potato Res.* 26: 229-235.

Salomon, R. and Bernardi, F.1995. Inhibition of viral aphid transmission by the N-Terminus of the maize dwarf mosaic virus coat protein. *Virology* 213: 676-679.

Sigvald, R. 1984. The relative efficiency of some aphid species as vectors of potato virus Y^O (PVY^O). *Potato Res.* 27: 285-290.

Simons, J.N. 1966. Effects of temperature and length of acquisition feeding time on transmission of nonpersistent viruses by aphids. *J. econ. Entomol.* 59: 1056-1062.

Simons, J.N. and Eastop, V.F. 1970. Temperature effects on aphid transmission of nonpersistent viruses with notes on morphological variation within clones of aphids with differing vector efficiencies. *J. econ. Entomol.* 63: 484-490.

Stegwee, D. and Ponsen, M.B. 1958. Multiplication of potato leafroll virus in the aphid *Myzus persicae* (Sulz.). *Entomol. exp. appl.* 1: 291-300.

Syller, J. 1987. The influence of temperature on transmission of potato leafroll virus by *Myzus persicae* Sulz.. *Potato Res.* 30: 47- 58.

Syller, J. 1994. The effects of temperature on the availability and acquisition of potato leafroll luteovirus by *Myzus persicae*. *Ann. appl. Biol.* 125: 141-145.

Tamada, T. and Harrison, B.D. 1981. Quantitative studies on the uptake and retention of potato leafroll virus by aphids in laboratory and field conditions. *Ann. appl. Biol.* 98: 261-276.

Tamada, T., Harrison, B. D. and Roberts, I. M. 1984. Variation among British isolates of potato leafroll virus. *Ann. appl. Biol.* 104: 107-116

Terradot, L., Simon, J.C., Leterme, N., Bourdin, D., Wilson, A.C.C., Gauthier, J.P. and Robert, Y. 1999. Molecular characterization of clones of the *Myzus persicae* complex (Hemiptera: Aphididae) differing in their ability to transmit the potato leafroll luteovirus (PLRV). *Bull. entomol. Res.* 89: 355-363.

Thornbury, D.W. and Pirone, T.P. 1983. Helper components of two potyviruses are serologically distinct. *Virology* 125: 487-490.

Thornbury, D.W., Hellmann, G.M., Rhoads, R.E. and Pirone, T.P. 1985. Purification and characterization of potyvirus helper component. *Virology* 144: 260-267.

Thornbury, D.W., Patterson, C.A., Dessens, J.T. and Pirone, T.P. 1990. Comparative sequence of the helper component (HC) region of potato virus Y and a HC-defective strain, potato virus C. *Virology* 178: 573-578.

Tjallingii, W.F. 1978. Electronic recording of plant penetration behaviour by aphids. *Entomol. exp. appl.* 24: 721-730.

Tjallingii, W.F. 1985. Membrane potential as an indication for plant cell penetration by aphid stylets. *Entomol. exp. appl.* 38: 187-195.

Tjallingii, W.F. 1988. Electrical recording of stylet penetration activities. In *"World Crops Pests". Vol. 2B: "Aphids, their Biology, Natural Enemies and Control"*. (eds) A.K. Minks and P. Harrewijn, Elsevier, Amsterdam, pp. 95-108.

Turl, L.A.D. 1983 The effect of winter weather on the survival of aphid populations on weeds in Scotland. *EPPO Bull.* 13: 139-143.

Upreti, G. C. and Nagaich, B. B. 1971. Variations in the ability of *Myzus persicae* Sulz. to transmit potato viruses. I. Leaf roll. *Phytopathol. Z.* 71: 163-168.

Wang, J.Y., Chay, C., Gildow, F.E. and Gray, S.M. 1995. Readthrough protein associated with virions of barley yellow dwarf luteovirus and its potential role in regulating the efficiency of aphid transmission. *Virology*, 206: 954-962.

Wang, R.Y., Ammar, E.D., Thornbury, D.W., Lopez-Moya, J.J. and Pirone, T.P. 1996. Loss of potyvirus transmissibility and helper-component activity correlate with non-retention of virions in aphid stylets. *J. gen. Virol.* 77: 861-867.

Watson, M.A. 1938. Further studies on the relationship between *Hyoscyamus* virus 3 and the aphis *Myzus persicae* (Sulz) with special reference to the effects of fasting. *Proc. r. Soc. Lond. B* 125: 144-170.

Watson, M.A. and Roberts, F.M. 1939. A comparative study of the transmission of *Hyoscyamus* virus 3, potato virus Y and cucumber virus 1 by the vectors *Myzus persicae* (Sulz), *M. circumflexus* (Buckton) and *Macrosiphum gei* (Koch). *Proc. r. Soc. Lond. B* 127: 543-577.

Webb, R. E. 1956. Relation of temperature to transmission of the potato leafroll virus. *Phytopathology* 46: 470.

Zeyen, R.J. and Berger, P.H. 1990. Is the concept of short retention times for aphid-borne nonpersistent plant viruses sound ? *Phytopathology* 80: 769-771.

8. Transmission

8.4 Transmission of Viruses by Leafhoppers and Thrips

M. Klein
Department of Entomology, Agricultural Research Organization, Bet Dagan, 50-250, Israel

8.4.1 Leafhopper Transmission

Most leafhopper vectors transmit viruses in a circulative manner and many of these viruses are propagative in their vector. For some of the viruses (*Wound tumor* and *Potato yellow dwarf viruses*, for example) there is proof of transovarial passage (Sinha and Shelley, 1965; Black, 1953). Viruses of this group are characterized by having a defined latent period (LP) during which the leafhopper cannot transmit virus after an acquisition feed (AF). AF with these viruses is of at least 30 minutes. The LP may be relatively short (hours) or long (at least a week). Most of the leafhopper species which transmit the circulative/propagative viruses remain viruliferous for most of their lives after a single AF on a virus source (Harris, 1979). With a few exceptions (e.g., *Sugar beet curly top* and *Potato yellow dwarf viruses*) none of these viruses is mechanically transmitted. Viruses such as *Rice tungro* and *Maize chlorotic dwarf viruses* are transmitted by their leafhopper vectors in a semipersistent manner (Ling, 1966; Nault *et al.* 1973) without a detectable LP or evidence for their circulation within the insect bodies; they enter the hemocoele and are

G. Loebenstein et al. (eds.),
Virus and Virus-like Diseases of Potatoes and Production of Seed-Potatoes, 227–236.

transmitted via the salivary glands. The vectors of these two viruses become non-viruliferous within a very short time. Inoculation access (IA) of more than 30 minutes was needed for a reasonable rate of transmission by the vector.

Potato yellow dwarf virus (genus *Nucleorhabdovirus*; family *Rhabdoviridae*) (PYDV) was first reported in potatoes from North-eastern United States and Canada by Barrus and Chupp (1922).The main symptoms of PYDV in potatoes are vein clearing, necrosis of apical pith, gradual yellowing of leaves with little or much vein necrosis. The tops of the plants eventually die, axillary shoots develop chronic or acute stages of the disease or the plants soon die. Progress of the disease depends on the virus strain. Development of viable and normal-shaped tubers also depend on the virus strain. Infection in potatoes occurs sporadically, but other hosts are naturally infected every year in the North-central USA (Lockhart, 1989) and may serve as sources for infection of potatoes. Some potato cultivars show differences in disease incidence.

In addition to potato, PYDV affects several other hosts (e.g., *Nicotiana* species, *Trifolium, Chrysanthemum, Tagetes, Zinnia, etc.*). The disease occurred naturally in potatoes in the US and Canada in the 1930s and 1940s, and is disseminated there mainly by infected tubers. True seeds do not transmit PYDV.

PYDV can be mechanically transmitted to *Nicotiana rustica* leaves, and to *Trifolium incarnatum* plants by pricking inoculum into crowns. In nature transmission is by leafhopper vectors. Two serotypes of PYDV, SYDV and CYDV, are designated according to their specific transmission by agallian leafhoppers. SYDV is transmitted efficiently by the leafhopper *Agallia sanguinolenta* and several other *Aceratagallia* species, poorly by *Agallia quadripunctata,* and not at all by *Agallia constricta*. CYDV is transmitted by *A. constricta* and by *A. quadripunctata* but not by *A. sanguinolenta*. The virus requires a relatively long incubation period in the vector and has been recovered infective from hemolymph and internal organs of viruliferous leafhoppers (Sinha, 1965), which suggests that PYDV multiplies in the vectors. This enables the vector to remain viruliferous throughout its life, so that the virus can overwinter in the adult leafhopper vector during harsh periods without the need to be re-acquired by the vector.

PYDV was successfully cultured in cells of the vectors. The ability to grow leafhopper cells in culture medium opened a new area in the research on virus multiplication (Black, 1969).

Seven distinct PYDV strains were isolated from *N. glutinosa* inoculated by the leafhopper *A. sanguinolenta* and from *T. incarnatum* plants inoculated by a single leafhopper of the same species (Black, 1940).

The virus has rhabdo- or bullet-shaped (380 x 75 nm) particles and its genome consists of single-stranded RNA. The virion contains 2% RNA, 78% protein and 20% lipids. The total genome size is 12.6 kb. There are five different proteins in the virion: the largest 180 kDa is the polymerase, and the others are an 80-kDa spike protein, a 55-kDa nucleoprotein, a 31-kDa matrix protein and 26 kDa protein (Brunt *et al.* 1997a). The virus has been shown to be associated with the nuclei of the hosts (plants and vectors) (Black, 1970). The replication of the virus does not depend on a helper virus.

Beet curly top virus (genus *Curtovirus*; family *Geminiviridae*) (BCTV). First reported in *Beta vulgaris* by Ball (1909) in the Western USA.

Symptoms in infected potatoes appear as vein clearing, swelling and malformation in young leaves. Older leaves become rigid, dwarfed, yellowed and malformed. Axillary buds are proliferated and phloem necrosed. Sometimes fluid exudes from phloem and plants wilt. The damage caused by this virus is usually of minor importance, and published data on BCTV in potatoes are scarce. Damage might sometimes be great especially for certified seeds.

In 1928 Severin and Henderson published a list of field and garden plants of three families, highly susceptible to BCTV in nature (*Chenopodiaceae, Leguminoseae* and *Cucurbitacae*). In 1936 this list was enlarged, and up to 73 genera belonging to 33 plant families were proven to be disease hosts (Freitag and Severin, 1936). Non-cultivated plants also became naturally infected by BCTV, and the virus overwinters in some of them (Severin, 1934).

BCTV occurs in arid and semi-arid regions of western North America (from Mexico to Canada), regions of South and Central America (Argentina to Costa Rica), in North Africa, the Mediterranean area and Australia.

BCTV is not transmitted mechanically, unless special procedures are used, nor is it transmitted by seeds. It is graft transmissible, and by dodder. BCTV is transmitted in North America by the beet leafhopper, *Circulifer tenellus* (Shaw, 1910). In the Old World, the virus is transmitted by leafhoppers, members of the *C. tenellus* complex (Freitag *et al.* 1955) and by members of the *Circulifer haematoceps* complex (Bennett and Tanrisever, 1957). The leafhoppers *Agalliana ensigera* (Fawcett, 1927) and *Agallia albidula* (Sauer, 1946) transmit BCTV in Central and South America and *Orosius argentatus* in Australia (Bowyer and Atherton, 1971). BCTV is transmitted by these leafhoppers in a persistent manner. It circulates, but does not multiply within the insect body. The virus has spread eastward in N. America to the migration limit of *C. tenellus*.

Thomas and Mink (1979) reviewed many genetically stable strains of BCTV which differed in virulence, symptoms and host range. One of these strains caused damage to certified seed potatoes (Milbrath, 1946; Giddings, 1954).

The virions of BCTV are geminate, not enveloped, isometric, round in profile with a diameter of 18–22 nm. After purification particles appear as monomers or dimers and only the latter form is infectious (Larsen and Duffus, 1984). The genome consists of single-stranded DNA. The total genome size is 2.8 kb (Boswell, 1985). The particles are restricted to the nuclei of phloem parenchyma cells (Esau and Hoefert, 1973).

8.4.2 Thrips Transmission

Tospoviruses are the most important group of viruses transmitted by thrips. The virus can be acquired only by the first larval stage shortly after the larvae cleave from the eggs and take their first feeds on infected plants. Because of the LP of the virus (3–10 days), transmission is mainly by adult thrips. During the LP the virus circulates and replicates in the vector (Ullman *et al.* 1993; Wijkamp *et al.* 1993). Inoculations take place by gentle probes during which plant cells continue to function. Once acquired, the virus is retained throughout life of the viruliferous thrips, but it does not pass transovarially to the progeny (Wijkamp *et al.* 1996).

Tomato spotted wilt virus (genus *Tospovirus*: family *Bunyaviridae*) (TSWV) was first recorded in tomatoes in Australia (Brittlebank, 1919), and in 1931 in England (Smith, 1931). The virus is known at present to occur worldwide. TSWV belongs to the genus Tospovirus, a group of plant-infecting ambisense RNA viruses, which form a separate genus within the arthropod-borne family of the *Bunyaviridae* (Phleboviruses).

Symptoms of TSWV in infected plants vary according to the species, age, and growing conditions of the plant. Initial symptoms are pale spots on leaves, which become ring-like and later convert into irregular necrotic lesions (Fig. 8.4.1). Such lesions may also appear on stems. Lesions sometimes coalesce causing wilting and death of the whole plant (Ie, 1970).

Cho *et al.* (1987) noted approximately 200 plants, di- and monocotyledonous species, belonging to more than 30 botanical families, which are susceptible to TSWV. Since 1987 additional natural hosts of TSWV have been reported (Peters *et al.* 1991; Latham and Jones, 1997). Sixteen species commonly serve as experimental hosts, with significant variations in virus titer (Wijkamp, 1995). In *Petunia* x *hybrida* "Blue Magic" several TSWV strains cause conspicuous local lesions, thus providing an indicator for thrips inoculation (Wijkamp and Peters, 1993).

Other useful diagnostic hosts are:

Catharanthus roseus – local black spots, yellow leaves, systemic mosaic and leaf deformation;

Cucumis sativus – chlorotic spots with necrotic centres on inoculated cotyledon , not systemic;

Pentunia x hybrida – necrotic local lesions, not systemic;

Nicotiana clevelandii, N. glutinosa, N. tabacum – necrotic local lesions, systemic necrotic patterns and leaf deformation;

Tropaeolum majus – systemic necrotic spotting and streaking (Brunt *et al.* 1997b).

Maintenance and propagation hosts are *Gomphrena globosa, N. glutinosa, N. rustica, N. benthamiana and T. majus.*

All tospoviruses are transmitted by thrips. In Australia TSWV is transmitted by *Thrips tabaci* (Pitman, 1927) and *Frankliniella schultzei*, in England and in Hawaii by *T. tabaci* (Best, 1968). The western flower thrips, *Frankliniella occidentalis* was recognized as a very efficient vector of TSWV in 1962 (Sakimura, 1962) and this thrips is now considered to be the most efficient vector of TSWV (van de Wetering, 1999). The endemic region of *F. occidentalis* was originally the western part of the North American continent. Since the mid-1980s, *F. occidentalis* has spread to most countries in the world (Anonymous, 1993) and tospoviruses with it. Since 1962 eight more species of thrips belonging to three separate genera (*Frankliniella, Thrips* and *Scirtothrips*) have been shown to be vectors of tospoviruses. Some of these species are very specific vectors and transmit only one tospovirus but not TSWV. Three more species of *Frankliniella, F. bispinosa, F. fusca, F. intonsa* have been found to be vectors of TSWV, as well as *T. setosus, T. palmi* and *T. flavus*, and, *S. dorsalis*. The high resistance of thrips to pesticides and their distribution with plant material from country to country are probably the causes of the emergence of tospoviruses and isolates (Goldbach and Peters, 1994) that may severely affect potato. TSWV is transmitted mechanically, but not by simple contact between plants, and by grafting, but not by seeds. In nature, transmission is probably by thrips only (Brunt *et al.* 1997b).

TSWV is the type species of the genus *Tospovirus*, which currently contains 13 different viruses, divided into 10 serogroups. Of all tospoviruses identified so far, TSWV is the most frequently encountered, and it is spread by most species of thrips (van de Wetering, 1999).

Purification – according to Tas *et al.* 1977.

TSWV has spherical membrane-bound particles, 80–100 nm in diameter, covered with surface projections consisting of two glycoproteins (Tas *et al.* 1977), but without a conspicuous capsomer arrangement. Purified preparations contain one sedimenting component of 550 S.

Density - 1.21 g cm^{-3} in sucrose. Virions contain 5% nucleic acid; 70% protein; 20% lipid and 5% carbohydrate.

The genome consists of three single-stranded linear RNA segments of 17.2 kb. These segments consists of a small RNA (S) of 2.916 kb, a medium-size RNA (M) of 5.4 kb and a large RNA (L) of 8.897 kb. Base composition: 16.2% G; 31.6% A; 19.3% C and 32.9% U. The nucleotide sequence of the genome was determined by De Haan *et al.* (1990, 1991) and Kormelink *et al.* (1992). Purified virus contains at least four structural proteins: the nucleocapsid protein, N (28.8 kDa), two glycoproteins, G1 (78 kDa) and G2 (58 kDa) and a large protein, L (331.5 kDa) which is the viral RNA polymerase. The N protein is tightly associated with three RNA segments. The S and M RNA display ambisense gene arrangements and they code for two non-structural proteins – NS_M and NS_S . One of these non-structural proteins is the viral movement protein (Kormelink *et al.* 1994; Storms, 1998).

Detection methods are based on ELISA (van de Wetering, 1999).

Control of virus spread demands crop management practices that will be to the disadvantage of the pest. This is difficult to achieve in open fields. Control of the vector requires frequent application of pesticides, but chemicals have proved to be inadequate, because thrips rapidly develop resistance and are difficult to reach, because of their cryptic habits. Also, thrips can successfully transmit virus after a short inoculation access period before being killed by the pesticide.

Tobacco streak virus (genus *Ilarvirus*; family: *Bromoviridae*) (TSV). First reported in tobacco by Johnson (1936) from Wisconsin, USA.

TSV has rarely been found in potato in nature, and only in South America (Costa *et al.* 1964; Salazar *et al.* 1982). TSV has never been seen to cause severe reduction of yield in South America.

Symptoms in potato plants mechanically infected with the potato strain of TSV showed only mild mosaic. Tubers planted from these infected plants developed severe leaf deformation and vein-thickening (Costa *et al.* 1964; Salazar *et al.* 1982). The symptoms produced on potato by two other strains of TSV in Brazil comprised necrotic spots and necrotic rings, which were never caused by potato strain SB10 (Salazar *et al.* 1982).

Several strains of TSV are known: the type strain from tobacco (Fulton, 1948); bean red node strain (Virgin, 1943); black raspberry latent strain (Converse and Lister, 1969); and the potato strain SB10 from Peru (Salazar *et al.* 1982). The first three strains have so far not found in potatoes in nature, but could be experimentally induced (Brunt *et al.* 1997c).

The potato strain was maintained by successive transfers to *Chenopodium quinoa* and from this plant it was mechanically transferred to 20 of 27 plant species in four botanical families (Salazar *et al.* 1982).

TSV could be transmitted by grafting, but not by contact between plants. It is transmitted, within certain plant species, by seeds (Johnson *et al.* 1984). It is also transmitted by pollen to the pollinated plants. TSV is transmitted by several species of thrips (Kaiser *et al.* 1982). Stoodee and Teakle (1987) suggested a unique mechanism of TSV transmission by thrips, which necessitates the involvement of pollen-borne virus at the site of infection, and wounding by the thrips. No such vector has yet been found for the potato strain of TSV in the mountains of Peru, nor has the source of pollen-borne TSV which would have to take part in the transmission to potatoes there (Salazar *et al.* 1982).

TSV virions are isometric rounded in profile, 27, 30 and 35 nm in diameter. The virions have been found in all parts of the host cells. Inclusions of different structures, lacking the virions, could be detected in certain cells of the infected plants.

TSV is a single-stranded RNA virus. The genome consists of four RNA's of total size 7.915kb (Clark and Lister, 1971). Base Composition of these RNA's: 25%G; 25%A; 20%C and 30%U.

References

Anonymous, 1993. Distribution maps of pests, No. 538. *Frankliniella occidentalis. CAB International Institute of Entomology, London.*

Ball, E.D. 1909. The leafhoppers of sugar beet and their relation to the "curly leaf" condition. *Bull. USDA Bur. Entomol.* 66: 33-52.

Barrus, M.F. and Chupp, C.C. 1922. Yellow dwarf of potatoes. *Phytopathology* 12: 123.

Bennett, C.W. and Tanrisever, A. 1957. Sugar beet curly top disease in Turkey. *Pl.Dis. Reptr* 42: 721-725.

Best, R.S. 1968. Tomato spotted wilt virus. *Adv. Virus Res.* 13: 65-145.

Black, L.M. 1940. Strains of potato yellow dwarf virus. *Am. J. Bot.* 27: 386-392.

Black, L.M. 1953. Occasional transmission of some plant viruses through eggs of their insect vectors. *Phytopathology* 43: 9-10.

Black, L.M. 1969. Insect tissue culture as tools in plant virus research. *Ann. Rev. Phytopathol.* 7: 73-100.

Black, L.M. 1970. Potato yellow dwarf virus. *C.M.I./A.A.B. Description of Plant Viruses No.35,* 4 pp.

Boswell, K. 1985. Beet curly top hybrigemini virus in: *Plant Virus Online: Descriptions and Lists from the Vide Database* (eds.) A.A. Brunt, K. Crabtree, M.J. Dallwitz, A.J. Gibbs, L. Watson and E.J. Zurcher (1996 onwards). Version January 1997.

Bowyer, J.W. and Atherton, J.G. 1971. Summer death of French bean: new hosts of the pathogen, vector relationship and evidence against mycoplasmal etiology. *Phytopathology* 61:1451-1455.

Brittlebank, C.C. 1919. Tomato diseases. *J. Agric. Victoria* 27: 213-235.

Brunt, A.A., Crabtree, K., Dallwitz, M.J., Gibbs, A.J., Watson, L. and Zurcher, E.J. 1997a. Potato yellow dwarf nucleorhabdovirus in: *Plant Virus Online: Descriptions and Lists from the Vide Database*. Version January 1997.

Brunt, A.A., Crabtree, K., Dallwittz, M.J,, Gibbs, A.J., Warson, L. and Zurcher, E.J. 1977b.Tomato spotted wilt Tospovirus. In : *Plant Virus Online: Descriptions and Lists from the Vide Database*. Version January 1997.

Brunt, A.A., Crabtree, K., Dallwitz, M.J., Gibbs, A.J., Watson, L., and Zurcher, E.J. 1997c. Tobacco streak virus in: *Plant Virus Online: Descriptions and List from the Vide Database*. Version January 1997.

Clark, M.F. and Lister, R.M. 1971. Preparation and some properties of the nucleic acid of tobacco streak virus. *Virology* 45: 61-74.

Cho, J.J., Mau, R.F.L., Mitchell, W.C., Gonsalves, D. and Yudin, L.S. 1987. Host list of plants susceptible to tomato spotted wilt virus. *HITAHR, College of Tropical Agriculture and Human Resources, University of Hawaii. Res. Ext. Series 078, 10 pp.*

Converse, R.H. and Lister, R.M. 1969. The occurrence and some properties of black raspberry latent virus. *Phytopathology* 59: 325-333.

Costa, A.S. and Carvalho, A.M.B. 1961. Studies on Brazilian tobacco streak. *Phytopath. Z.* 42: 113-138.

Costa, A.S., Carvalho, A.B. and Deslandes, J. 1964. Occorencia do virus de necrosa branca du fumo em plantacóes de batatinha. *Bragantia* 23:1-8.

De Haan, P., Wagemakers, L., Peters, D. and Goldbach, R. 1990. The S RNA segment of tomato spotted wilt virus has an ambisense character. *J. gen. Virol.* 71: 1001-1007.

De Haan, P., Kormelink, R., De O. Resende, R., Van Poelwijk, F., Peters, D. and Goldbach, R. 1991. Tomato spotted wilt virus L RNA encodes a putative RNA polymerase. *J. gen. Virol.* 71: 2207-2216.

Esau, K. and Hoefert, L.L. 1973. Particles and associated inclusions in sugar beet infected with the curly-top virus. *Virology* 5: 454-464.

Fawcett, G.L. 1927. The curly top of sugar beet in Argentina. *Phytopathology* 17: 407-408.

Freitag, J.H. and Severin, H.H.P. 1936. Ornamental flowering plants experimentally infected with curly top. *Hilgardia* 10: 263-302.

Freitag, J.H., Frazier, N.W. and Huffaker, C.B. 1955. Cross breeding beet leafhoppers from California and French Morocco. *J. Econ. Entomol.* 48: 341-342.

Fulton, R.W. 1948. Hosts of tobacco streak virus. *Phytopathology* 38: 421-428.

Giddings, N.J. 1954. Some studies of curly top on potatoes. *Phytopathology* 44: 125-128.

Goldbach, R.W. and Peters, D. 1994. Possible causes of the emergence of tospovirus diseases. *Seminars in Virology* 5: 113-120.

Harris, K.F. 1979. Leafhoppers and aphids as biological vectors: vector-virus relationships. In: *Leafhopper Vectors and Plant Disease Agents*. Maramorosch, K. and Harris, K.F. (eds.) pp. 217-308. Academic Press, New York.London.

Ie, T.S. 1970. Tomato spotted wilt virus. *CMI/AAB Plant Virus Description No. 39,* 4 pp.

Johnson, H.A. Jr, Converse, R.H., Amorao, A., Esperjo, J.I. and Frazier, N.W. 1984. Seed transmission of tobacco streak virus in strawberry. *Plant Dis.* 68: 390-392.

Johnson, J. 1936. Tobacco streak, a virus disease. *Phytopathology* 26: 285-292.

Kaiser, W.J., Wyatt, S.D. and Pesho, G.R. 1982. Natural hosts and vectors of tobacco streak virus in eastern Washington. *Phytopatholoty* 72: 1508-1512.

Kormelink, R., De Haan, P., Meurs, C., Peters, D. and Goldbach, R. 1992. The nucleotide sequence of the M RNA segment of tomato spotted wilt virus, a bunyavirus with two ambisense RNA segments. *J. gen. Virol.* 73: 2795-2804.

Kormelink, R., Storms, M., Van Lent, J., Peters, D. and Goldbach, R. 1994. Expression and subcellular location of the NS_M protein of tomato spotted wilt virus (TSWV), a putative viral movement protein. *Virology* 200: 56-65.

Larsen, R.C. and Duffus, J.E. 1984. A simplified procedure for the purification of curly top virus and isolation of its monomer and dimer particles. *Phytopathology* 74: 114-118.

Latham, L.J. and Jones, R.A.C. 1997. Occurrence of tomato spotted wilt tospoviruses in native flora, weeds and horticultural crops. *Aust. J. Agric. Res.* 48: 359-369.

Ling, K.C. 1966. Nonpersistence of the tungro virus of rice in its leafhopper vector, *Nephotettix impicticeps. Phytopathology* 65: 1252-1256.

Lockhart, B.E.L. 1989. Recurrence of natively occurring potato yellow dwarf virus in Minnesota. *Plant Dis.* 73: 321-323.

Milbrath, J.A. 1946. Green dwarf, a virus disease of potatoes. *Phytopathology* 36: 671-674.

Nault, L.R., Styer, W.E., Knoke, J.K. and Pitre, H.N. 1973. Semipersistent transmission of leafhopper-borne maize chlorotic dwarf virus. *J. Econ. Entomol.* 66:1271-1273.

Peters, D., De Avila, A.C., Kitajima, E.W., De O. Resende, R., De Haan, P. and Goldbach, R. 1991. An overview of tomato spotted wilt virus. In: *Virus-Thrips-Plant Interaction oj TSWV, Proc. USDA Workshop, Beltsville*, MD, USA, H.-T. Hsu and R.H. Lawson (eds.), pp. 1-14. Springfield, National Technology Information Service.

Pitman, H.A. 1927. Spotted wilt of tomatoes. Preliminary note concerning the transmission of the "spotted wilt" of tomatoes by an insect vector (*Thrips tabaci* Lind). *Bull. Commonw. Australia,* CSIR 1: 74-77.

Sakimura, K. 1962. *Frankliniella occidentalis* (Thysanoptera: Thripidae), a vector of tomato spotted wilt virus with special reference to the color forms. *Ann. Ent. Soc. Am.* 55: 387-389.

Salazar, L.F., Abad, J.A. and Hooker, W.J. 1982. Host range and properties of a strain of tobacco streak virus from potatoes. *Phytopathology* 72: 1550-1554.

Sauer, H.F.G. 1946. A cigarrinha *Agallia albidula* Uhl. (Hom. Cicadel.) vetora de una doencade virus do tomaterio. *Biologico* 12: 176-178.

Severin, H.H.P. 1934. Weed host range and overwintering of curly-top virus. *Hilgardia* 8: 263-277.

Severin, H.H.P. and Henderson, C.F. 1928. Some host plants of curly top. *Hilgardia* 3: 339-393.

Shaw, H.B. 1910. The curly-top of sugar beets. *Bull. USDA Bur. Plant Industry* 181: 1-46.

Sinha, R.C. 1965. Recovery of potato yellow dwarf virus from hemolymph and internal organs of an insect vector. *Virology* 27: 118-119.

Sinha, R.C. and Shelley, S. 1965. The transovarial transmission of wound-tumor virus. *Phytopathology* 55: 324-327.

Smith, K.M. 1931. Studies on potato virus diseases. VIII. On a ringspot virus affecting solanaceous plants. *Ann. appl. Biol.* 18: 1-15.

Stoodee, R. and Teakle, D.S. 1987. Transmission of tobacco streak virus by *Thrips tabaci*: a new method of plant virus transmission. *Plant Pathol.* 36: 377-380.

Storms, M. 1998. The role of NS_M during tomato spotted wilt virus infection. *Ph.D. Dissertation, Agricultural University Wageningen, The Netherlands,* 128 pp.

Tas, P.W.L., Boergan, M.L. and Peters, D. 1977. The structural proteins of tomato spotted wilt virus. *J. gen. Virol.* 36: 267-279.

Thomas, P.E. and Mink, G.I. 1979. Beet curly top virus. *CMI/AAB Description of Plant Viruses. No. 210, 6 pp.*

Ullman, D.E., German, T.L., Sherwood, D.M., Westcot, D.M. and Cantone, F.A. 1993. Tospovirus replication in insect vector cells: Immuno-cytochemical evidence that the non-structural protein encoded by the S RNA of tomato spotted wilt tospovirus is present in thrips vector cells. *Phytopathology* 83: 456-463.

Van de Wetering, F. 1999. Effects of thrips feeding on tospovirus transmission in Chrysanthemum. *Ph.D. Dissertation, Agricultural University, Wageningen, The*

Netherlands, 120 pp.
Virgin, W.J. 1943. An unusual bean disease. *Phytopathology* 33: 743-745.
Wijkamp, I. 1995. Virus vector relationships in the transmission of tospoviruses. *Ph.D. Dissertation, Agricultural University, Wageningen, The Netherlands*, 143 pp.
Wijkamp, I. and Peters, D. 1993. Determination of the median latent period of two tospoviruses in *Frankliniella occidentalis*, using a novel leaf disk assay. *Phytopathology* 83: 986-991.
Wijkamp, I., Goldbach, R. and Peters, D. 1996. Propagation of tomato spotted wilt virus in *Frankliniella occidentalis* does neither result in pathological effects nor in transovarial passage of the virus. *Entomol. Exp. Applic.* 81: 285-292.
Wijkamp, I., Van Lent, J., Kormelink, R., Goldbach, R. and Peters, D. 1993. Multiplication of tomato spotted wilt virus in its insect vector, *Frankliniella occidentalis*. *J. gen. Virol.* 74: 341-349.

Fig. 8.4.1 Lesions induced by *Tomato spotted wilt virus* on potato leaves (Courtesy Dr. C. Wilson, Hobart, Australia).

9. Epidemiology and Field Control of PVY and PLRV

D.W. Ragsdale[1], E.B. Radcliffe[1] and C.D. DiFonzo[2]

[1]University of Minnesota, Department of Entomology St. Paul, Minnesota USA 55108 and
[2]Michigan State University, Department of Entomology, East Lansing, Michigan USA 48824

9.1 Introduction

Viral diseases of potatoes can cause significant yield and quality losses for the commercial producer but are of particular concern to the seed potato producer. *Potato virus Y* (PVY) and *Potato leafroll virus* (PLRV) are the most common of the 28 viral diseases known to infect potato (Salazar, 1996). Current season (primary) infection with PVY causes little yield loss (Hane and Hamm, 1999; Salazar, 1996) especially if initial infection occurs after flowering. Quality loss caused by PVY results from an increase in the number of undersized tubers (Hane and Hamm, 1999). With PLRV, primary infection rarely reduces yield (Beemster and Rozendaal, 1972) but it can affect tuber quality by causing phloem net necrosis. Net necrosis is expressed in tubers as a darkening of the vascular bundle that becomes more pronounced during storage. Net necrosis is particularly severe in the cultivars Russet Burbank, Norgold Russet, and Green Mountain (Douglas and Pavek, 1972), the former being the most widely grown cultivar in the U.S.

Because potatoes are grown from vegetative seed pieces, daughter tubers are only as healthy as the mother plant. Planting seed already infected with PLRV or PVY (secondary or chronic infection) can cause substantial yield loss and compensation by adjacent plants is less than 100% (Reestman, 1972). For some cultivars that tolerate infection and show few or mild foliar

G. Loebenstein et al. (eds.),
Virus and Virus-like Diseases of Potatoes and Production of Seed-Potatoes, 237–270.

symptoms, compensation by neighbouring plants is reduced even further (Hane and Hamm, 1999).

Even before the etiology of potato virus diseases was understood, potatoes were recognised to degenerate after several years of successive cultivation. The most likely cause of this "running out" or degeneration was an increase in the proportion of virus infected tubers. If measures are not employed to reduce the incidence of tuber-borne infection, seed potato stocks deteriorate and planting heavily infected seed results in substantially depressed yield and tuber quality in the commercial crop (Salazar, 1996; van der Want, 1972).

To maintain healthy seed stocks, seed potato certification programs have been developed. Most certification programs use a limited generation production system that keeps virus incidence low by restricting continuous cultivation of a particular seed lot for a specified number of years, usually from 5 to 6 years (Slack, 1993). Seed used to plant the commercial crop generally has the highest tolerance for virus incidence, typically from 1 to 4% (Marco, 1980; Woodford, 1988), whereas, seed potato lots to be increased another year generally have virus tolerances ranging from 0 to 1%. A zero tolerance is usually applied to seed lots derived from tissue culture where there is essentially no chance of becoming infected and permitted to increase for successive generations in a limited generation certification program (Hiddema, 1972; Woodford, 1988). If virus levels increase beyond tolerances, seed lots either advance to the next lower grade or are rejected outright and may not be replanted as seed or perhaps even sold as certified seed depending upon the level of virus infection.

Our focus in this chapter is to discuss different management strategies used to reduce spread of aphid-transmitted viruses in the field, specifically PVY and PLRV. Commercial producers can adopt many of the control strategies discussed here if there is a concern about tuber quality such as net necrosis. Typically, it is only seed producers who must meet the strict phytosanitary standards set by certification agencies; thus they are more likely to use a variety of these control strategies in an integrated approach to disease management.

9.2 Aphids as Virus Vectors

For aphid-transmitted viruses, the classic disease triad of susceptible host, available inoculum and permissive environmental conditions requires a fourth dimension, the insect vector. Given that we have no crop protection chemicals that can lessen the impact of virus infection once a plant is inoculated with a virus, the only means of control are: decreasing susceptibility of the host, intervening with vectors before transmission occurs, reducing inoculum, or by isolation from sources of virus or vectors.

For PVY and PLRV, transmission is primarily through aphid feeding. PVY can be spread mechanically but in the field this mode of transmission is a minor component of viral epidemiology (Banttari, 1994). Monitoring aphid density in the crop and invasion by winged aphids can give producers an early warning permitting appropriate management strategies to be used to minimise virus spread. Which strategy is most appropriate depends on a variety of factors some of which include physiological age of the crop, the estimated value, the ultimate use of the crop, and management practices at the producer's disposal.

Aphid life cycles are complex and often involve multiple hosts. There are excellent reviews of aphid biology to which the reader is referred (Dixon, 1985; Minks and Harrewijn, 1987). Aphids have biological attributes that make them ideal virus vectors. Indeed, the majority of plant viruses which have an arthropod vector are transmitted by aphids (Nault, 1997). These attributes include a short life cycle with the ability to produce multiple generations during the summer facilitated by parthenogenic reproduction. If conditions are favourable, immense aphid populations can develop. Aphids cause little cellular damage with their piercing sucking mouthparts, and they use a unique host finding behaviour of sampling epidermal cell contents in brief feeding probes. Once aphids settle on a host they feed primarily on phloem where viruses like PLRV are located.

In mid-summer the most common form found on annual plants are wingless parthenogenic aphids (apterae). Apterae are stimulated to produce winged progeny (alatae) in response to environmental cues of shortened daylength and lower temperature. Other stimuli that result in alate production include senescence of the host plant and crowding. Alatae leave their host in search of a suitable plant to begin new colonies. Alatae can fly short distances (<5 km) or become transported long distances (from a few to several hundred kilometres) with the aid of low level winds (Taylor, 1986a). Little is known about what determines the proportion of an alate aphid population that undergo long distance migration or move only short distances in the landscape (Taylor, 1986a). Local movement is thought to be important in viral epidemics when there is abundant inoculum present (Garrett and McLean, 1983). This is not to minimise the fact that long distance movement has triggered viral epidemics in potatoes and other crops (Berger *et al.* 1987; Hille Ris Lambers, 1972; Taylor, 1986b). Monitoring aphid movement and dispersal in the crop is paramount in providing producers with information on potential for virus spread with sufficient lead time to implement management strategies to reduce tuber infection.

9.3 Integrated Control of Aphid-Transmitted Viruses

Methods to control virus spread in potatoes fall into two broad categories: those that are preventative and are implemented before the onset of virus spread during a given growing season, and those that are therapeutic and are applied in response to presence of aphids or infected plants. A primary focus of seed certification programs is to control virus spread by reducing inoculum. Certification programs accomplish this by inspecting seed potato fields multiple times during the growing season, removing (roguing) diseased plants, and testing daughter tubers for viruses after harvest to insure the quality of the seed for next year's crop. These programs generally work well (Garrett, 1986; Slack, 1993) and seed health is maintained, but on occasion an unexpected increase in disease incidence occurs and other control methods must be used in conjunction with a seed certification program to reduce disease incidence. Once seed is heavily infected it often takes several growing seasons to effectively reduce viral inoculum (Harrison, 1971).

Potato virus Y and PLRV differ in their transmission characteristics and therefore control options vary depending upon which virus predominates. *Potato virus Y* is a nonpersistently transmitted, stylet-borne virus (Nault, 1997) that is acquired by a naive vector aphid in short feeding probes lasting only seconds to minutes. Once virus is acquired, the now viruliferous aphid can immediately infect the next susceptible plant it feeds upon. When a viruliferous aphid feeds on an uninfected plant it is likely to lose its charge of virus and does not become infective until it again feeds on a virus infected plant. Over 50 aphid species or unique genera have been shown to acquire and transmit PVY (Table 9.1). A majority of these potential vectors do not colonise potato. The green peach aphid, *Myzus persicae*, is the most efficient PVY vector although many inefficient vectors are important because of their propensity to develop alatae and their abundance commonly exceeds that of *M. persicae*. As winged aphids are searching for a suitable host, they will discriminate between host and non-host by feeding intercellularly on epidermal cells in a process called sap sampling. Acquisition and transmission of virus can occur within the brief sap sampling probes that aphids use in host finding, which explains why the most important PVY vectors are aphids that do not colonise potato. Apterae are less important than alatae in PVY spread (Ragsdale *et al.* 1994). Insecticides are regarded as ineffective in controlling spread of nonpersistent, stylet-borne viruses although there are some notable exceptions.

Potato leafroll virus is a persistently transmitted, circulative virus (Nault, 1997) that does not replicate in the vector (nonpropagative). A naive vector must feed for several minutes to a few hours on phloem to acquire the phloem-restricted PLRV. Once a vector acquires PLRV it is incapable of transmitting for 24-72 hours until virus has passed through the hindgut of the aphid, into the hemocoel and is eventually taken up by the accessory salivary glands. Once the aphid is able to transmit PLRV, it usually remains viruliferous for life even following molts. Aphids that are important in PLRV spread are those species that colonise and reproduce on potato (Table 9.2) although some species listed are experimental vectors because they are rarely found colonising potato. Apterae and alatae are important in PLRV spread. Insecticides can be effective in reducing in-field spread of PLRV, but most insecticides do not kill quickly enough to prevent transmission caused by an incoming viruliferous alate aphid. A possible exception is the insecticide pymetrozine. This material is highly selective and once an aphid imbibes a toxic dose, it ceases to feed (Harrewijn and Kayser 1997). In replicated field trials pymetrozine treated plants had a lower incidence of PLRV infected daughter tubers (Ragsdale and Radcliffe, unpublished data).

Green peach aphids prefer the lower one-third of the potato canopy and thus foliar insecticides must reach the lower canopy to provide adequate control. Aphids also respond to increased fertilisation levels and other factors affecting host quality. Castle and Berger (1993) showed that plants infected with either PLRV or PVY but not *Potato virus X* (non-aphid vectored) were superior hosts when measuring aphid performance. Virus infection can also affect host selection possibly due to volatiles emitted from an infected plant because green peach aphids preferentially landed on PLRV infected potato in a dark arena (Castle *et al.* 1998). Clearly, the association of aphids with their hosts is complex and this interaction becomes even less intuitive when plants are infected with aphid-transmitted viruses.

9.3.1 Inoculum Sources

9.3.1.1 Within the Crop

Although both PVY and PLRV are capable of infecting other hosts, the most common source of inoculum is the potato crop itself. Roguing is often practised in seed fields, and is especially important in small, early-generation seed fields where each row can be surveyed several times during the growing season. Roguing should begin as soon as symptoms of secondary (tuber-borne) infection can be seen, typically when plants are 15-20 cm tall. Roguing is easier to accomplish before the canopy closes (Woodford and Gordon, 1990). The goal must be to remove all infected plants before winged aphids arrive (Salazar, 1996; Woodford and Gordon, 1990). When roguing PLRV infected plants occurs after aphid colonisation, plants surrounding the symptomatic plant need to be removed because these neighbouring plants within the row and in adjacent rows may be infected but not yet symptomatic (Mowry, 1994). It is important to remove these plants because spread can occur from these plants later in the season. If plants are rogued after aphids colonise the crop, one risks dislodging viruliferous aphids from these plants in the process of removal. However, treating a field with insecticide one week prior to roguing did not reduce virus spread, so it is unclear if treating a field with an effective aphidicide prior to roguing is necessary (Woodford and Gordon, 1990). If the seed field is heavily infected (>1% virus) it is generally impractical to rogue effectively because infected plants will be missed. Moreover, removing many infected plants leaves large gaps in the field that increase the apparency of the plants on the perimeter because of the contrast between green plants and fallow soil is attractive to winged aphids (Bell, 1989). To facilitate finding all virus-infected plants, a tuber-unit planting scheme can be used where seed pieces cut from a single tuber are planted in succession rather than having them randomly dispersed throughout the field. A drawback of a tuber-unit planting scheme is that it is labour intensive and slows planting. Delaying planting and thereby emergence can result in plants that are younger and more susceptible to virus infection when aphids begin to colonise the field. In practice, tuber-unit planting is often restricted to the earliest generation seed potato fields or small acreage of later generation fields.

Unfortunately, not all cultivars express strong symptoms of virus infection. In North America, the cultivars Shepody and Russet Norkota are well known as symptomless carriers of PVY. A recent epidemic of PVY in the Red River Valley of Minnesota and North Dakota was linked to the increased popularity of these cultivars before it was widely recognised that absence of symptoms did not assure they were virus free (DiFonzo *et al.* 1996). Similarly, there are other cultivars, such as Goldrush and Anoka, that poorly express PLRV symptoms. Secondarily infected plants of these cultivars may appear healthy the entire season, or only briefly express symptoms early in their vegetative growth phase. Most potato breeding programs now recognise the importance of understanding the range of symptom expression of PVY and PLRV and routinely use serological tests (ELISA) to screen all advanced selections that appear to be virus resistant and are likely to be released as a named cultivar (Bagnall and Tai, 1986). Serological tests are frequently used by seed certification agencies on known symptomless cultivars in lieu of observing symptoms in winter trials.

Transgenic cultivars expressing genes that confer pathogen-derived resistance to PVY and PLRV are now available (Chapter 13). Transgenic cultivars offer producers a means of eliminating effects of PVY and PLRV, thus reducing the risk associated with these two pathogens. To date, transgenic plants developed are highly resistant to infection, but not immune (Thomas *et al.* 1997b). In a practical sense immunity may not be necessary. Plant to plant spread in the field is less with virus resistant cultivars partly because of lower virus titre (Barker and Woodford, 1992; Sigvald, 1984; Souza Dias *et al.* 1987; Thomas *et al.* 1997b). DiFonzo *et al* (1995) showed that with conventional cultivars known to be PLRV resistant, mature plant resistance was more durable and more viruliferous aphids were needed to infect plants. Similarly, Wilson and Jones (1993) observed that resistance could be overcome if inoculum dose was sufficient. Certification programs must be aware that transgenic plants can be infected, and because virus titre is low symptoms may not express well. However, transgenic plants pose no special threat to the certification system because, as with any new cultivar, its range of symptoms must be assessed (Slack and German, 1998). Should changes in resistance status occur in transgenic cultivars as a result of selection for a new virus strain, such events would be detected in the current seed certification protocols.

9.3.1.2 Outside the Crop:

In temperate climates, most of the known weed hosts of PVY or PLRV are annuals and transmission via true seed does not occur (Salazar, 1996), effectively eliminating these weeds as virus sources. In Canada, no weed is considered a virus reservoir (Singh, 1987), although a systematic survey has not been done. Other winter annuals like shepherd's purse (*Capsella bursa-pastoris*) and Jim Hill mustard (*Sisymbrium altissimum*) are known PLRV hosts, but appear not to be important in spread of the virus to potatoes (Thomas *et al.* 1997a; Woodford, 1988). In Morocco, where winters are mild, PLRV-infected jimsonweed (*Datura stramonium*) was shown to support aphid populations before the winter potato crop emerged and was implicated along with volunteer potatoes as being a principal source of viruliferous aphids and inoculum (Hanafi *et al.* 1995). Volunteer potatoes that emerge as weeds in rotation crops in the Pacific Northwest are considered important PLRV reservoirs (Thomas, 1983), and cull piles have been implicated as a source of virus in Canada (Frazer, 1987). In regions where winter frost extends below the root zone of potatoes, volunteer potatoes are rare and appear to be inconsequential sources of potato viruses (DiFonzo *et al.* 1997). Potatoes are planted at a rate of approximately 15,000 plants per acre. Since it is not uncommon for commercial (ware) producers to plant seed with 1-5% virus (150-750 infected plants per acre), the potato crop itself overshadows any perennial weeds as a source of potato viruses, even when virus levels in seed potatoes are what growers commonly consider acceptable.

9.3.2 Methods to Minimise Aphid Transmission

Aphid abundance and species diversity vary greatly from one growing region to another, as does the importance of principal vector species (Harrington *et al.* 1986). *Myzus persicae* is the most important PVY and PLRV vector world-wide (Radcliffe, 1982). Also important, however, are a long list of vector aphids, many of which do not colonise potato (Tables 9.1 and 9.2). Much of the early work on potato virus epidemiology focused on green peach aphid or other potato colonising species. As our understanding of the importance of non-colonising aphids for PVY spread increased, aphid-trapping networks began routinely identifying these aphids (DiFonzo *et al.* 1997; Harten, 1983; Kurppa and Rajala, 1986; Sigvald, 1984; Sigvald, 1986). These data have been used to alert growers of high aphid numbers and initiation of an action such as mineral oil sprays or early harvest of the seed crop (Boiteau and Parry, 1985; Hille Ris Lambers, 1972; Ragsdale and Radcliffe, 1998; Woiwood *et al.* 1984).

9.3.2.1 Monitoring Aphid Flight

To determine aphid arrival dates, alate aphids can be monitored using a variety of passive or active traps or, alternatively, apterae can be counted on potato foliage (Byrne and Bishop, 1979; Flanders *et al.* 1991; Hille Ris Lambers, 1972; Woodford *et al.* 1977). The most common type of alatae monitoring tool is a pan trap. Pan traps are usually yellow to be attractive to aphids or green to mimic foliage. Green traps (Fig. 9.1) reportedly provide a less biased estimate of landing rates than do yellow traps in some crops (Irwin, 1980). In potato, green tile traps provided a biased estimate of landing rates when compared to similarly designed traps using potato foliage (Boiteau, 1990). For most aphid species, green tile traps captured fewer aphids than did yellow traps (Eastop and Raccah, 1988), but only rarely are coloured traps species specific (DiFonzo *et al.* 1997). Trap placement tends to have more to do with performance than does colour. Green tile traps placed in a fallow field border caught significantly more aphids than traps placed in the middle of the field, while traps placed in the first potato row captured an intermediate number of aphids (DiFonzo *et al.* 1996).

Suction traps (Fig. 9.2) have been used extensively in England and Europe for several decades (Woiwood *et al.* 1984) and for over 15 years in the Pacific Northwest (U.S.) (Halbert *et al.* 1998) to monitor a variety of aphid species. These 12.2-m high traps were designed to capture migrating aphids with the expectation that they are monitoring long distance migrants. Data available from these traps suggest they were capturing aphids produced on vegetation no more than 30-60 km from the trap rather than long distant migrants and that surrounding topography substantially impacts trap catch (Halbert *et al.* 1998). Traps are useful tools in determining approximate arrival date of alatae, but data show they provide no earlier warning of aphid arrival than do pan traps or beating foliage (Halbert *et al.* 1990; Hille Ris Lambers, 1972) and often apterae are detectable in foliage before alatae are found in traps.

Regardless of the monitoring methodology used (suction traps, pan traps, or sampling plants at regular intervals) throughout a given region, there is always a delay in processing the samples, summarizing data, and alerting growers of the findings. Woiwood *et al.* (1984) noted that it takes nearly 1 week for suction trap samples to be processed and reports sent out for aphids that were trapped the preceding 5 to 12 days. Our experience in the Upper Midwest (U.S.) with an aphid trapping network is similar. We operated a network of 75 suction and pan traps in Minnesota and North Dakota for the past two years. A weekly report was posted each Friday on the internet for aphids captured the preceding 7 days (http://ipmworld.umn.edu/alert.htm) (Ragsdale and Radcliffe, 1998). Printed copies of the report were mailed and received by growers in an additional 3 days. Clearly, by the time growers and pest management advisors receive reports, they are making management decisions on aphid flight activity that occurred 1-3 weeks earlier. Any additional post-trapping processing of the samples, for example, detection of PLRV in the aphid itself, may make these data more valuable but it is important that such processing not further delay reporting. PLRV and PVY can be reliably detecting in single aphids using immunocapture RT-PCR (Singh *et al.* 1995, 1996, 1997). However, it is unclear if detection of virus within a single aphid actually measures those aphids that are likely to transmit virus or whether such monitoring results in overestimation of risk (Leclercq-LeQuillec *et al.* 1995; Pirone and Thornbury, 1988).

9.3.2.2 Cultural Management Options

9.3.2.2.1 Isolation

Long before viruses were recognised as pathogens and before anyone realised aphids were involved in virus transmission, isolating seed fields was practised as a means of maintaining healthy seed stocks (van der Want, 1972). Isolation of seed potato fields is still an important concept in the overall management of PVY and PLRV, but what is considered an effective isolation distance varies from one area to another. Factors that influence the

Fig. 9.1
Green-tile pan trap in propylene glycol

Fig. 9.2
2m Suction Traps

Solar powered Battery powered (12v)

effective isolation distance include the time of year when aphids first colonise the crop, availability of host plants to arrest aphid movement when non-colonising vectors are flying, age of the crop when vectors begin to appear, the cultivar being grown, and environmental conditions during inoculation. If seed potatoes are produced in close proximity to commercial potato fields, seed production is more challenging. If any one component of the vectored disease tetrahedron (susceptible hosts, aphid vectors, abundant inoculum and permissive environment) is greatly diminished, seed can be successfully increased along side commercial production, e.g., when prevalence of virus is low in all potato fields. When an epidemic occurs, producing seed in the same area as commercial production makes it much more difficult to restore control.

Isolation as a preventative tactic can either be isolation in space or temporal isolation through modifying planting or harvesting dates. In 1998, for example, some Minnesota and North Dakota seed growers were able to produce virus-free seed lots even when all nearby seed farms had their seed lots rejected for exceeding virus tolerance. The single common factor for the successful seed farms was an early top kill date. To be economically feasible for grower to harvest early, they must plant early to assure sufficient yield if early top kill is necessary.

Early planting can be a useful strategy if principal vector species begin colonisation late in the growing season. The more advanced the crop growth stage at the time of inoculation, the less likely daughter tubers will become infected. Mature plant resistance is cultivar specific and even varies within a cultivar when inoculated with different PVY strains (Gibson, 1991). For example, in two susceptible cultivars mature plant resistance to PVY^{O} was evident with few infected daughter tubers produced when inoculation occurred 13 weeks after plant emergence, but mature plant resistance was even more pronounced when inoculated with PVY^{N} (Gibson, 1991). Mature plant resistance is pronounced in the PLRV-resistant cultivars Cascade and Abnaki (DiFonzo *et al.* 1995), but less evident in the PLRV-susceptible cultivar Russet Burbank (Flanders *et al.* 1990). Unfortunately, planting early to take advantage of mature plant resistance is rarely possible in most north temperate production areas because of a limited number of frost-free days (Singh and Boiteau, 1987), but growers need to be aware of an increased risk of infection associated with delayed planting.

A different temporal isolation strategy was used in the Netherlands in the early 1950's, with planting delayed so that plant emergence occurred after the green peach aphid colonisation flight was over. By delaying planting, the crop was not colonised by spring migrants and mid-summer migrants were either uncommon or virus inoculum was so low that the crop remained virus-free. This practice was eventually stopped because a new strain of late blight made this cultural practice impractical (Hille Ris Lambers, 1972). In the U.S., Shands *et al.* (1972b) found 50 percent less virus spread in late plantings compared to early plantings. They attributed this difference to late plantings missing early influxes of colonising aphids, resulting in a lower vector population developing in the field. Delayed planting was suggested as a PLRV control strategy for the spring-planted crop in one seed potato production area in Morocco (Hanafi *et al.* 1995) and in Cyprus (Ioannou, 1989), where a colonising aphid flight persists for only a few weeks beginning at crop emergence. In New Brunswick, late-planted potato plots had lower densities of *Aphis nasturtii* (buckthorn aphid) and *Macrosiphum euphorbiae* (potato aphid) suggesting that delayed planting could improve seed health. However, yield and tuber size were compromised with a late planting because haulms had to be killed before the late August migration of green peach aphids (Boiteau, 1984).

Spatial isolation is a common cultural control method. The recommended distance required for effective isolation varies. In the Netherlands, viral epidemics are known to occur when green peach aphids arrive from other parts of Europe and England, some 100 – 150 km distant (Hille Ris Lambers, 1972). In other areas it appears that effective isolation can be achieved with much less isolation. In eastern Idaho, the recommended isolation distance for seed potatoes is from 0.25 to 3 miles (400 m to 5 km) from a known virus source of PVY while as much as 20 miles (32 km) is needed for isolation from PLRV sources (Halbert *et al.* 1999). These distances are based in part on how far aphids fly as determined by an aphid trapping network (Halbert *et al.* 1990) and not by observing actual infection in fields located at specific distances from a known virus source. In England, the recommended separation from a PVY source is 800 m (Harrington *et al.* 1986) while in Denmark a distance of 40 m was shown to reduce spread of PVY (Hiddema, 1972). In British Columbia, Canada, elite seed is grown in the Pemberton Valley, some 150 km from commercial potato production in the Fraser Valley. Because land available for seed potato production was limited in the Pemberton Valley, seed production in the Fraser Valley along side commercial production occurs to a limited degree, but the effective isolation distance is unknown (Frazer, 1987).

Distances needed for effective isolation appear to depend upon whether the crop aphids are emigrating from is a host of PVY or PLRV and which virus is the primary concern. The degree of isolation required from known sources of PVY appears to be considerably less than for PLRV. Greater isolation distance is needed for PLRV because of the persistent, circulative mode of transmission that allows the vector to remain viruliferous for life. A change in cropping patterns may change the effective isolation distance. For example, when canola production increased in Scotland, virginoparous green peach aphid overwintered successfully on this crop resulting in earlier colonisation of potato than had occurred previously (Woodford, 1988). In Minnesota and North Dakota, there has been a recent, dramatic increase in the canola acreage and concomitant increase in the rejection rate of seed potato lots for excess virus. In the Upper Midwestern U.S., canola is an early spring-planted crop that matures in late July. Green peach aphids do not overwinter on canola or as eggs on primary hosts (*Prunus* spp.) in Minnesota or North Dakota. It appears that canola serves as a highly susceptible early season "bridging" host for migrant green peach aphids that arrive before potatoes have emerged. Canola is also an excellent host for the turnip aphid, *Lipaphis erysimi*. *Myzus persicae* and *L. erysimi* are known virus vectors (Tables 9.1 and 9.2) and have been implicated in a PVY epidemic in the Red River Valley (DiFonzo *et al.* 1997). In regions where normal winter temperatures fall below green peach aphid survival limits, aphids typically colonise potatoes later in the growing season (Table 9.3). Historically, such areas are among the better seed production regions in the world.

9.3.2.2.2 Early Harvest

Combining early planting with early haulm destruction has been recognized as an effective method of maintaining the health of elite seed stock in the Netherlands since 1810 (Hille Ris Lambers, 1972). Because yields are greatly compromised with an early harvest, aphid monitoring is used to advise certification agencies when haulms must be destroyed to prevent virus spread. These thresholds are based on the number of alate aphids captured in pan or suction traps. In New Brunswick, a cumulative total of 5 green peach aphids in any one pan trap in the seed growing region triggers a mandatory top-kill date for the entire region (Boiteau and Parry, 1985). A slightly different threshold is used in the Netherlands. Initially the threshold was when a trap caught 2 green peach aphids in a single day vines were to be killed within 10 days (Hille Ris Lambers, 1972). This threshold was later modified to include non-colonising aphids so that PVY spread was accounted for in determination of top kill dates. These thresholds are region specific because they take into account the average maturity of the crop, historical patterns of aphid movement and other factors such as weather and surrounding crops and vegetation. Application of thresholds for haulm destruction to regions outside where they were developed is unlikely to be useful. For example, if either of these thresholds were used in the Upper Midwest or the Pacific Northwest, top kill dates would likely occur before tubers were formed, making seed potato production impossible. In New

Brunswick, if aphid migration occurs before a harvestable yield is present, other management strategies are employed (Parry, 1987).

9.3.2.2.3 Mechanical Barriers and Border Crop

A number of aphid-transmitted viruses have been controlled experimentally using barriers and reflective mulches. Barriers such as nonwoven or extruded polyester or polypropylene sheets (Agronet, Reemay and Agryl) substantially reduced PVY and PLRV spread in potatoes (Harrewijn *et al.* 1991; Hemphill *et al.* 1988) and other crops (Avilla *et al.* 1997; Cohen, 1981). Floating row covers increase humidity which favour foliar pathogens; however, fungicides apparently penetrate the row covers sufficiently to provide adequate protection for late blight (Harrewijn *et al.* 1991). Temperature under row covers can increase substantially (2 – 9 °C) and may adversely impact plant and tuber development in growing areas with high daytime temperatures, but this may be an advantage in regions where temperatures are cool (Hemphill *et al.* 1988). Insects may gain entry through tears or emerge from underneath the row cover and because plants cannot be routinely sampled, infestations go undetected which may necessitate use of soil applied systemic insecticides at planting.

Aphids, particularly green peach aphids, are attracted to yellow and repelled by highly reflective surfaces (Kring, 1970). By exploiting this behaviour Cohen and Marco (1973) reduced PVY spread in peppers 51-80% by ringing plots with sticky yellow polyethylene sheets. PLRV was reduced by 79 to 88% by this technology (Zimmerman-Gries and Gries, 1979). Mulches, particularly aluminium foil (George and Kring, 1971), white plastic (Wyman *et al.* 1979), and oat straw (DiFonzo, 1995; Setiawan and Ragsdale, 1987) have been known to reduce insect transmitted pathogens in potato and other crops. As the season progresses and crop plants cover much of the mulch, the effectiveness of mulches is compromised. Altogether, the usefulness of floating row covers, barriers, and mulches in seed potato production is likely to be restricted to high value seed fields of small size such as the first field generation following propagation in the greenhouse (minitubers) or in the laboratory (plantlets or microtubers).

Barrier crops are more widely adaptable since they are easier to install and keep in place than mulches or floating row covers, and they don't loose their effectiveness as the canopy closes. Initially, it was thought that barrier crops should be taller than the crop being protected, but (DiFonzo *et al.* 1996) showed this was not a requirement. For seed potato production, barrier crops of forage sorghum, winter wheat, or soybean, all significantly reduced PVY spread (DiFonzo *et al.* 1996). Moreover, borders of winter wheat and forage sorghum that were routinely treated with insecticides to control aphids were no different in effectiveness in reducing PVY than barrier crops that were left untreated or with soybean which had no aphids. From this study it was learned that a key feature in application of barrier crops to control PVY spread is to have a fallow area adjacent to the outside edge of the barrier crop with no gap between the border crop and potatoes. Barrier crops planted in such a manner take advantage of aphid host finding behaviour of landing at the interface of fallow ground and green crop. With PVY if immigrating viruliferous alatae feed first on the border crop, they are likely to lose their charge of virus before continuing trivial flight and landing on and sap sampling potatoes. Barrier crops need not be more than a few meters wide to be effective (DiFonzo *et al.* 1996).

Seed certification agencies could take advantage of the general knowledge concerning aphid flight and feeding behaviour. For example, fallow ground or skip rows are frequently used to separate cultivars to maintain varietal purity. This planting scheme places seed lots at greater risk of infection because of their increased apparency to immigrating aphids. If a seed field does not have a non-potato border and there is inoculum nearby, treating outer rows as a separate seed lot from inner rows is advisable. In the Netherlands if a seed field is bordered by a virus source, the first 10-25 meters of the seed field adjacent to the source field is either downgraded or rejected (Hiddema, 1972). Given what is known about virus spread from inoculum sources outside a seed potato area, seed that is to be replanted on a seed farm should come from the centre of the field to reduce the chance it is infected.

9.3.2.3 Chemical Management Options

9.3.2.3.1 Insecticides

Insecticides can be useful for reducing in-field spread of PLRV especially if colonising aphids arrive virus-free. A nonviruliferous aphid needs several minutes to hours of continuous feeding to acquire virus and an 8 hour or longer latent period before it can transmit (Leonard and Holbrook, 1978). Recommendations as to the timing of insecticide applications for PLRV control are regionally specific due to differences in aphid biology, time of arrival with respect to average crop maturity, and the ultimate use of the crop. For PVY, insecticides do not act quickly enough to prevent virus transmission although some pyrethroids and botanical insecticides are repellent as well as toxic and these compounds may reduce spread of non-persistent viruses (Gibson 1983, Gibson *et al.* 1982).

Potato leafroll virus

If colonising alatae are rarely viruliferous, the control of apterae by systemic insecticides applied at planting can significantly reduce within-field spread of PLRV (DiFonzo *et al.* 1995; Flanders *et al.* 1991; Hanafi *et al.* 1989; Marco, 1980; Powell and Mondor, 1973; Webley and Stone, 1972; Woodford *et al.* 1983). However, properly timed foliar applied insecticides are equally effective in controlling PLRV spread (DiFonzo *et al.* 1995; Flanders *et al.* 1991; Hanafi *et al.* 1989). In those production areas where apterae density is correlated to final incidence of PLRV, action thresholds based on apterae counts can be developed to time foliar applications. In Scotland, a warning was issued to growers when two to five adult green peach aphids were found per 100 potato hills using a plant beating or leaf count sampling method (Woodford *et al.* 1977). In Europe, it was demonstrated that PLRV spread is best correlated with arrival of alatae (Broadbent, 1950; Harrison, 1971; Hille Ris Lambers, 1955; Hollings, 1955). However, Ribbands (1965) offered contrary evidence and showed that PLRV spread was related to apterae density. The most likely explanation for these contrary views is that importance of alatae varies from year to year and from one region to another.

In Minnesota, a threshold of 10 green peach aphid apterae per 100 leaves for seed potato production was derived by balancing the effect of reducing aphid numbers against the negative consequences of increased insecticide applications, such as decimating natural enemy populations and increasing the likelihood of developing insecticide resistant aphids (Cancelado and Radcliffe, 1979). Later, action thresholds developed in Minnesota were based on the amount of PLRV spread occurring in plots where aphid density was maintained below predetermined levels (DiFonzo *et al.* 1995; Flanders *et al.* 1991; Hanafi *et al.* 1989). Development of action thresholds based on apterae density was realistic in the Upper Midwest because under typical growing conditions there was a low level of inoculum (<1%) and immigrating alate green peach aphids that colonise potato in early July were rarely, if ever, carrying PLRV. In these studies, spread of PLRV was significantly correlated with green peach aphid apterae density ($r^2 = 0.39 - 0.81$) and not correlated with captures of alatae in pan traps ($r^2 = 0.01$-0.11) (Flanders *et al.* 1991; Hanafi *et al.* 1989). The action threshold proposed for seed potatoes of 3-10 green peach aphid apterae per 100 leaves was developed using the susceptible cultivar, Russet Burbank. Cultivar resistance status and initiation of mature plant resistance are important factors in producing infected daughter tubers. DiFonzo *et al.* (1995) showed that action thresholds could be effectively increased to 30 – 100 apterae per 100 leaves if a PLRV resistant cultivar was planted. No one has proposed modifying action thresholds based on crop maturity, but it seems reasonable to incorporate mature plant resistance that many cultivars exhibit later in the growing season (Bagnall and Tai, 1986).

The primary concern in commercial production with PLRV infection is expression of net necrosis in tubers. Byrne and Bishop (1979) used regression analysis to determine the number of green peach aphids apterae which resulted in unacceptable levels of net necrosis (>5%) and developed thresholds of 10 green peach aphids per 50 leaves for south-central Idaho and 40 green peach aphids per 50 leaves for south-western Idaho obtained in two consecutive samples (weeks) before August 1. In northern California, the recommended spray threshold to prevent net necrosis is 5% of leaves in a sample infested with green peach aphids (Bacon *et al.* 1978).

Once the action threshold is reached, insecticidal sprays should be applied to reduce the green peach aphid population. Unfortunately, the green peach aphid is often resistant to most insecticides leaving growers with few chemical options available to provide reliable aphid control. Foliar sprays and systemic insecticide applied to the soil either at planting or at plant emergence as a side-dress application, have been widely used to control aphids in potatoes. Interestingly, not all insecticides reduce PLRV spread even when they give adequate aphid control (Bacon *et al.* 1976; Powell and Mondor, 1973). The carbamate aldicarb (Temik®) does not prevent transmission by viruliferous alatae (Bacon *et al.* 1976), but it does give consistent control of in-field spread of PLRV by apterae (Hanafi *et al.* 1989; Powell and Mondor, 1973; Woodford *et al.* 1988). For much of the U.S., aldicarb is no longer registered for use on potatoes because of its potential for groundwater contamination.

In feeding studies using electronic monitoring of apterae, aphids reached the phloem of aldicarb treated plants before any toxic effects were seen, supporting the observation that aldicarb cannot prevent transmission by viruliferous alatae (Holbrook, 1977). Control of virus spread by aldicarb is partly due to reducing aphid populations but there are substantial sublethal effects of aldicarb (Boiteau *et al.* 1985). For example, over half the alatae exposed to aldicarb treated plants were unable to fly following an 18 h exposure and fewer than 10% of those that were unable to fly regained their ability to fly after a 24 h recovery period on untreated plants (Boiteau *et al.* 1985). As one would expect, naive aphids given access to PLRV infected plants treated with systemic insecticides were largely unable to acquire virus and transmission was significantly less than in the untreated control (Villacarlos, 1987). Other plant systemic insecticides are available, but insecticides such as phorate and disulfoton generally do not provide satisfactory aphid control nor do they prevent spread of PLRV (Powell and Mondor, 1973; Woodford *et al.* 1983). Research with the chloronicotinyl class of insecticides (imidacloprid) show that this insecticide reduces PLRV spread but have no effect on PVY (Boiteau and Singh, 1999). Imidacloprid has gained wide acceptance for Colorado potato beetle control and when used at the maximum labelled rate provides aphid control for much of the growing season. Imidacloprid has a relative slow mode of action on aphids that may limit its effectiveness in controlling PLRV by alatae arriving viruliferous (Boiteau *et al.* 1997), but the same was true with aldicarb. A new insecticide just entering the potato market is pymetrozine (Fulfill), and it has a dramatic effect on the feeding mechanism of aphids and interferes with stylet penetration (Harrewijn and Kayser, 1997) which could interfere with virus transmission. With all insecticides the risk of resistance is present, especially with the green peach aphid. Frequent use and reliance on a single compound to provide aphid control will result in failure of the product to provide reliable control.

Potato virus Y

Potyviruses such as PVY are transmitted in short feeding probes lasting only seconds, and insecticides cannot prevent transmission of PVY because they do not kill quickly enough to prevent aphids from sap sampling. Important PVY vectors, unlike those for PLRV, are typically non-colonising species and are only transient visitors, so the likelihood a non-colonising aphid will remain on the treated crop long enough to be killed is low. Some insecticides, particularly pyrethroids, cause hyperactivity in the first stage of poisoning followed by rapid incapacitation, thereby reducing potyvirus spread (Gibson, 1983; Gibson *et al.* 1982; Sawicki *et al.* 1983). The use of pyrethroids has been suggested as a means of reducing spread of PVY and other potyviruses, but in potato frequent use of any pyrethroid usually flares green peach aphids which in turn could increase virus spread (Cancelado and Radcliffe, 1979; Foster, 1986; Harrington *et al.* 1989). Some insecticides are irritating to aphids and can increase spread of PVY when applied to potatoes (Gabriel *et al.* 1981).

Control of PVY and other potyviruses is possible using non-toxic foliar sprays. Bradley *et al.* (1962) demonstrated in laboratory and greenhouse studies that plants treated with mineral oils reduced PVY transmission by 90%. Field application of this discovery provided some level of control (43 – 75% reduction of PVY) in potatoes but did not approach the level of control obtained in laboratory studies (Bell, 1989; Boiteau and Singh, 1982; Gibson *et al.* 1988; Shands, 1977). Reasons for this lower efficacy in the field are unclear, but likely include weathering of oil deposits (Boiteau and Wood, 1982; Gibson *et al.* 1988), plant growth occurring in the interval between applications, and incomplete coverage. If high concentrations of oil are used (>3%) they can be phytotoxic, especially when mixed with fungicides or when applied when temperatures are high (Boiteau and Singh, 1982). Oils appear to offer one of the better methods of controlling PVY spread, but because of some of the negative interactions with fungicides, the need for repeated application, and the relatively low level of control, adoption has been slow. Combining oils with pyrethroids (Collar *et al.* 1997; Gibson and Cayley, 1984), plant derived antifeedants (Powell *et al.* 1998), and whitewash (Marco, 1986; Marco, 1993) all improved PVY control over mineral oils alone. The mechanism involved in control of PVY by mineral oils is not fully understood. Both acquisition and inoculation are inhibited by mineral oil, but this is not related to differences in feeding behaviour that can be detected when monitoring stylet penetrations electronically (Powell, 1992). Loebenstein *et al.* (1964) demonstrated that when oil and purified virus are mixed mechanical inoculation is inhibited. It may be that oils interfere with a physiological process in the early stages of the infection process. Regardless, if PVY is of primary concern and there is an abundance of immigrating aphids with inoculum present in the field, oils may assist in preventing further spread, but they should not be used as the sole means of control.

9.3.2.4 Negative Interactions with Common Cultural and Chemical Controls.

Chemical control of a variety of pests is an essential practice in modern potato production. Because the crop takes a few weeks to emerge following planting, herbicides and cultivation are considered necessary components of crop management. A common practice in many growing regions is to mechanically control weeds just as plants are beginning to emerge. This operation is commonly referred to as "drag-off" cultivation. If drag-off is done after plants have emerged it can result in mechanical spread of PVY (Banttari, 1994). It is unclear if drag-off cultivation or other mechanical means of spreading PVY plays a significant role in viral epidemiology.

Insects other than aphids are key pests in most potato production regions (Radcliffe *et al.* 1991). One insect that stands out as a major pest in North America and much of Europe is the Colorado potato beetle, *Leptinotarsa decimlineata*. The Colorado potato beetle is a pest that, like the green peach aphid, has become resistant to hundreds of insecticides. Unfortunately, insecticides used to control Colorado potato beetles are usually poor aphidicides because of resistance. These same insecticides devastate predators and parasitoids of the green peach aphid causing aphid populations to flare. When this occurs virus spread can reach epidemic proportions. Insecticides targeting other mid-season pests common to the midwestern and eastern U.S. such as the potato leafhopper, *Empoasca fabae*, can also trigger aphid outbreaks. One strategy that can be employed to reduce the risk of flaring aphids is to use below-label rates of insecticides to control potato leafhoppers (Suranyi *et al.* 1999). By using below-label rates (1/10 to 1/4 the lowest label rate) potato leafhoppers were controlled and green peach aphid populations were not flared. More recently, aphid outbreaks have been linked to increased use of fungicides in potato production in the Upper Midwest (Lagnaoui and Radcliffe, 1998).

Potatoes are routinely sprayed with various fungicides for control of foliar pathogens, especially early blight, *Alternaria solani,* and late blight, *Phytophthora infestans*. Since the metalaxyl (Ridomil®) resistant late blight strain US8 (A2 mating type) has largely replaced the metalaxyl susceptible strain US1 (A1 mating type) in North America, fungicide use on potato has dramatically increased. Instead of relying on one or two applications of metalaxyl (a curative treatment) when late blight symptoms are first noticed, now growers must have a protective cover of fungicide to prevent late blight spores from germinating and establishing an infection. It is now not uncommon to have as many as 16 applications of fungicide applied during a growing season for irrigated potatoes, depending upon weather conditions.

In Minnesota, Nanne and Radcliffe (1971) showed that three fungicides, then commonly used on potato, captafol, mancozeb, and Bordeaux mixture, all favoured green peach aphid by suppressing entomophthoran fungi. In that experiment, fungicide-treated potatoes had late-season green peach aphid populations 1.7-2.6 times greater than did nonfungicidal controls. Aphids in nonfungicidal controls were 22.4% infected by entomophthoran fungi, whereas those in treated plots were only 4.0-5.2% infected. More recent work done in our laboratory (Lagnaoui and Radcliffe, 1998) showed that in a nonirrigated production system, fungicides commonly used on potatoes negatively impacted three entomopathogenic fungi: *Pandora neoaphidis* (syn. *Ernyia neoaphidis*), *Entomophthora planchoniana*, and *Conidiobolus obscurus*. In commercial potato production, combining frequent insecticide application to control beetles, leafhoppers or other insect pests with routine application of fungicide has created unprecedented increases in green peach aphid populations. Aphid populations growing unchecked can double their numbers every 2 – 3 days, eventually producing enormous numbers of alatae that can immigrate to seed fields contaminating even the cleanest seed lot in short order. Seed growers must remain ever vigilant and be ready to act if large influxes of green peach aphids occur, and monitoring aphid flight activity in the Upper Midwest has become of paramount importance.

9.4 Forecasting and Simulation Models

Many crop loss models have been developed to aid in our understanding of disease losses and a few models have attempted to evaluate losses caused by insect vectored pathogens (Gamez and Saavedra, 1986; Madden and Campbell, 1986; Ruesink and Irwin, 1986). The added complexity of vectored pathosystems makes model development a challenge (Madden and Campbell, 1986). The most useful application of forecasting in potato viruses has been the use of relative transmission efficiency factors for vectors of PVY. Sigvald (1986, 1987) predicted risk of PVY spread in Sweden using aphid trap catch data and applying different transmission efficiencies to the key vector species. Relative transmission efficiency of PVY by green peach aphid was 1.0 and all other species were given a lower transmission efficiency. Unfortunately, trapping schemes are rarely predictive for individual farms and they make the assumption that what was caught in relatively few traps can be extrapolated over a large area. As with many biological systems, simple predictive models are useful in demonstrating where additional information is needed, but rarely are they useful in management decisions because data needed for real-time input are difficult to collect.

9.5 Conclusions

Control of aphid-transmitted viruses in potatoes is complex and a multitude of preventative and therapeutic tactics need to be employed to keep virus spread to a minimum. Not all control methods discussed here are applicable to every production area. At a minimum, the degree of isolation, the source of inoculum, the timing and intensity of aphid flights along with identifying the predominant aphid species is vital information needed to develop an integrated control program. In the absence of such knowledge the focus of a seed production program should be on minimising the amount of inoculum in seed potatoes.

Crop protection chemicals can provide some level of control of PLRV, because aphid vectors colonise potatoes and chemical control of colonising aphids reduces plant-to-plant spread. However, if inoculum is arriving via viruliferous alatae, these insecticides do not kill quickly enough to prevent transmission of PLRV. Insecticides are of little value in preventing spread of PVY, whether the inoculum source is outside the crop or the crop itself. Although some pyrethroids have reduced spread of PVY experimentally, these same insecticides when applied at frequent intervals can flare colonising aphid populations. Cultural practices represent a collection of valuable tools to reduce PVY spread, but because practices such as early vine kill and harvest can dramatically reduce yield, information on risk of virus spread is needed to convince growers to harvest early. Other cultural practices, such as using crop borders and not using fallow ground to separate seed lots are inexpensive alternatives that can be applied with little cost. Altogether, seed potato production will continue to be a challenge for growers, certification agencies, and research scientists because of the complex nature of aphid-transmitted viruses.

Table 9.1. Aphid species known to transmit *Potato virus Y*, using live trapping in the field or laboratory transmission tests.

Aphid Species	% Transmission Efficiency	Reference(s)
Acyrthosiphon pisum (Harris)	3.8 - 14	Fereres *et al.* 1993; Harrington and Gibson, 1989; Harrington *et al.* 1986; Piron, 1986; van Hoof, 1980
Acyrthosiphon primulae (Theobald) (=*Metopolophium*)	15.0	Heinze, 1960
Aphis citricola (Van der Goot)	6.2	Raccah *et al*, 1985
Aphis craccivora (Koch)	4.0	Fereres *et al.* 1993; Perez *et al.* 1995
Aphis fabae Scolopi	7.6 - 24	Fereres *et al.* 1993; Harrington and Gibson, 1989; Harrington *et al.* 1986; Piron, 1986; van Hoof, 1980
Aphis gossypii (Glover)	12 - 31	Fereres *et al.* 1993; Raccah *et al.* 1985
Aphis helianthi Monell	0.3	DiFonzo *et al.* 1996
Aphis nasturtii Kaltenbach	19.0 - 50	Harrington and Gibson, 1989; Harrington *et al.* 1986; Piron, 1986
Aphis pomi De Geer	1.7 - 9	Harrington and Gibson, 1989; Harrington *et al.* 1986; van Hoof, 1980
Aphis spp.	5.9 - 6.7	Harrington and Gibson, 1989; Harrington *et al.* 1986; Perez *et al.* 1995; Piron,1986; Raccah *etal.*1985
Aulacorthum solani (Kaltenbach)	5.0	van Hoof, 1980
Brachycaudus (Thuleaphis) amygdalinus (Schouteden)	not reported	Perez *et al.* 1995

Aphid Species	% Transmission Efficiency	Reference(s)
Brachycaudus (Thuleaphis) rumexicolens (Patch)	not reported	Perez *et al.* 1995
Brachycaudus cardui (L.)	not reported	Perez *et al.* 1995
Brachycaudus helichrysi (Kaltenbach)	4.8 - 12.5	Harrington and Gibson, 1989; Piron, 1986; Powell *et al.* 1992
Brachycaudus spp.	14.7	Piron, 1986
Capitophorus hippophaes (Walker)	3.0 - 3.1	Piron, 1986; van Hoof, 1980
Capitophorus elaeagni (del Guercio)	not reported	Halbert *et al.* 1999
Capitophorus spp.	not reported	Perez *et al.* 1995
Cavariella aegopodii (Scopoli)	0.2 - 0.4	Harrington and Gibson, 1989; Piron, 1986
Cryptomyzus ballotae	100	Harrington and Gibson, 1989; Harrington *et al.* 1986
Cryptomyzus galeopsidis (Kaltenbach)	17.4	Piron, 1986
Cryptomyzus ribis (L.)	14.0 - 15.4	Harrington and Gibson, 1989; Piron, 1986
Diuraphis noxia (Mordvilko)	4.0 - 7.0	Perez *et al.* 1995
Drepanosiphum platanoides (Schrank)	0.6	Powell *et al.* 1992
Dysaphis spp.	1.8	Harrington and Gibson, 1989
Hayhurstia atriplicis (L.)	not reported	Perez *et al.* 1995
Hyadaphis foeniculi (Passerini)	14.7	Piron, 1986
Hyalopterus pruni (Geoffroy)	13.9	Perez *et al.* 1995
Hyperomyzus lactucae (L.)	0.4 - 17.4	Harrington and Gibson, 1989; Harrington *et al.* 1986; Piron, 1986
Lipaphis erysimi (Kaltenbach) (=*Hyadaphis*) (=*Rhopalosiphum pseudobrassicae*)	10.0	Heinze, 1960; Sigvald, 1989
Macrosiphoniella sanborni (Gillette)	not reported	Kennedy *et al.* 1962
Macrosiphum euphorbiae (Thomas)	4.0 - 29.0	Harrington and Gibson, 1989; Harrington *et al.* 1986; Piron, 1986;

Aphid Species	% Transmission Efficiency	Reference(s)
		Raccah *et al.* 1985; Singh and Boiteau, 1986; van Hoof, 1980
Metopolophium albidum	11.0	van Hoof, 1980
Metopolophium dirhodum (Walker)	0.5 - 10	Harrington and Gibson, 1989; Perez *et al.* 1995; Piron, 1986; van Hoof, 1980
Metopolophium festucae (Theobald)	0.4 - 0.5	Harrington and Gibson, 1989; Harrington *et al.* 1986
Myzaphis rosarum (Kaltenbach)	10.0 - 14.3	Harrington and Gibson, 1989; Harrington *et al.* 1986
Myzus cerasi (Fabricius)	3.2	Harrington and Gibson, 1989
Myzus certus (Walker)	34.3 - 71	Piron, 1986; van Hoof, 1980
Myzus ligustri	50 -100	Harrington and Gibson, 1989; Harrington *et al.* 1986
Myzus nicotianae Blackman	not reported	Boiteau and Lowery, 1989
Myzus myosotidis	100	Harrington and Gibson, 1989; Harrington *et al.* 1986
Myzus persicae (Sulzer) (=*Myzodes*)	4.7 - 71.1	Harrington and Gibson, 1989; Harrington *et al.* 1986; Piron, 1986; Powell *et al.* 1992; Raccah *et al.*1985; Singh and Boiteau, 1986; van Harten, 1983; van Hoof, 1980
Phorodon humuli (Schrank)	4.9 - 35	Harrington and Gibson, 1989; Harrington *et al.* 1986; Piron, 1986; van Hoof, 1980
Rhopalosiphoninus staphyleae tulipaellus (Theobald) (=*Hyperomyzus*, =*Myzotoxoptera*)	100	Heinze, 1960
Rhopalosiphum insertum (Walker)	0.8 - 50.0	Harrington and Gibson, 1989; Harrington *et al.* 1986; Piron, 1986; van Hoof, 1980
Rhopalosiphum padi (L.)	0.5 - 11.5	Harrington and Gibson, 1989; Perez *et al.* 1995; Piron, 1986; van Hoof, 1980; Harrington *et al.* 1986
Rhopalosiphum maidis (Fitch)	not reported	Halbert *et al.* 1999
Schizaphis graminum	not reported	Halbert *et al.* 1999; Perez *et al.* 1995

Aphid Species	% Transmission Efficiency	Reference(s)
(Rondani)		
Sitobion avenae (F.)	0.06 - 1.8	Harrington and Gibson, 1989; *Perez et al.* 1995; Piron, 1986; Harrington *et al.* 1986
Sitobion fragariae (*Walker*)	0.09 - 10	Harrington and Gibson, 1989; Piron, 1986; Harrington *et al.* 1986
Therioaphis spp.	not reported	Perez *et al.* 1995
Uroleucon spp.	0.5 - 8.3	Harrington and Gibson, 1989; Harrington *et al.* 1986; Piron, 1986

Table 9.2. Aphid species shown to transmit *Potato leafroll virus* (PLRV) in laboratory transmission studies.

Aphid Species	% Transmission Efficiency	Reference(s)
Aphis fabae Scopoli	not reported	Elze 1927 (in Sylvester, 1980)
Aphis gossypii Glover	4 - 74	Singh *et al.* 1988; Woodford *et al.* 1995
Aphis nasturtii Kaltenbach (= *rhamni*) and often referred to as the *nasturtii-frangulae* complex	20	Loughnane, 1943
Aulacorthum (Neomyzus) circumflexum (Buckton)	75	Heinze, 1960
Aulacorthum solani (Kaltenbach)	not reported	Robert and Rouze-Jouan, 1971 in Singh *et al.* 1988
Macrosiphum euphorbiae (Thomas)	0 - 25	Tamada *et al.* 1984
Myzus ascalonicus Doncaster	not reported	Hille Ris Lambers, 1972
Myzus nicotiannae Blackman	30.1	Halbert *et al.* 1995
Myzus ornatus Laing	not reported	Loughnane, 1939 in Sylvester, 1980
Myzus persicae (Sulzer)	2.4 - 83.8	Halbert *et al.* 1995
Phorodon humuli (Schrank)	10	Heinze, 1960
Rhopalosiphoninus	not reported	Roland, 1952 in Kennedy *et al.* 1962

latysiphon (Davidson)		
Rhopalosiphoninus staphyleae tulipaellus (Theobald)(=*Hyperomyzus*) (=*Myzotoxoptera*)	10	Heinze, 1960

Table 9.3. Colonization of green peach aphids in different potato growing regions in relation to crop growth stage

Country, Region, or State	**Average First Arrival of *Myzus persicae***	**Stage of. Potato Crop**	**Reference**
The Netherlands	late June	mid vegetative growth to tuber bulking	Hille Ris Lambers, 1972
Scotland	late June	mid vegetative growth to tuber bulking	Turl, 1978; Woodford *et al.* 1977
Sweden	late June to early July	mid vegetative growth to tuber bulking	Sigvald, 1986
USA Washington	lateApril,	plant emergence	Thomas *et al.* 1997a
Idaho (western)	mid June	early vegetative growth	Halbert *et al.* 1990
Idaho (eastern)	mid July	mid vegetative growth to tuber bulking	Halbert *et al.* 1999
Minnesota	late June to early July	mid vegetative growth to tuber bulking	DiFonzo *et al.* 1996; Flanders *et al.* 1991; Hanafi *et al.* 1989
Maine	early July	early to mid vegetative growth	Shands *et al.* 1972a
New Brunswick	late July to early August	tuber bulking	Parry, 1987
British Columbia	late May to early June	plant emergence	Frazer, 1987
Morocco	early March	plant emergence	Hanafi *et al.* 1995
Cyprus	late March – early April	plant emergence	Ioannou, 1989

References

Avilla, C., Collar, J. L., Duque, M., Perez, P. and Fereres, A. 1997. Impact of floating row covers on bell pepper yield and virus incidence. *J. Agric. Food Chem.* 45: 882-883.

Bacon, O. G., Burton, V. E., McLean, D. L., James, R. H., Riley, W. D., Baghott, K. G. and Kinsey, M. G. 1976. Control of the green peach aphid and its effect on the incidence of potato leaf roll virus. *J. Econ. Entomol.* 69: 410-414.

Bacon, O. G., Burton, V. E. and Wyman, J. A. 1978. Management of insect pests on potatoes. *California Ag.* 32: 26-27.

Bagnall, R. H. and Tai, G. C. C. 1986. Potato leafroll virus: evaluation of resistance in potato cultivars. *Plant Dis.* 70: 621-623.

Banttari, E. E. 1994. PVY, verticillium wilt, purple top research. *Valley Potato Grower.* 60: 36-37.

Barker, H. and Woodford, J. A. T. 1992. Spread of potato leafroll virus is decreased from plants of potato clones in which virus accumulation is restricted. *Ann. appl. Biol.* 121: 345-354.

Beemster, A. B. R. and Rozendaal, A. 1972. Potato viruses: properties and symptoms. In: *Viruses of Potatoes and Seed-potato Production.* (ed) J. A. de Bokx. pp. 144-151. Centre for Agricultural Publishing and Documentation, Wageningen.

Bell, A. C. 1989. Use of oil and pyrethroid sprays to inhibit the spread of potato virus Y^N in the field. *Crop Prot.* 8: 37-39.

Berger, P. H., Zeyen, R. J. and Groth, J. V. 1987. Aphid retention of maize dwarf mosaic virus (potyvirus): epidemiological implications. *Ann. appl. Biol.* 111: 337-344.

Boiteau, B. and Lowery, D. T. 1989. Comparison of a yellow form of the green peach aphid, *Myzus persicae* (Sulzer), and a green form of the tobacco aphid, *Myzus nicotianae* Blackman, coexisting on greenhouse potato in New Brunswick. *Can. Entomol.* 121: 1029-1035.

Boiteau, G. 1984. Effect of planting date, plant spacing, and weed cover on populations of insects, arachnids, and entomophthoran fungi in potato fields. *Environ. Entomol.* 13: 751-756.

Boiteau, G. 1990. Effect of trap color and size on relative efficiency of water-pan traps for sampling alate aphids (Homoptera: Aphididae) on potato. *J. Econ. Entomol.* 83: 937-942.

Boiteau, G., King, R. R. and Levesque, D. 1985. Lethal and sublethal effects of aldicarb on two potato aphids (Homoptera: Aphidae): *Myzus persicae* (Sulzer) and *Macrosiphum euphorbiae* (Thomas). *J. Econ. Entomol.* 78: 41-44.

Boiteau, G., Osborn, W. P. L. and Drew, M. E. 1997. Residual activity of imidacloprid controlling Colorado potato beetle (Coleoptera: Chrysomelidae) and three species of potato colonizing aphids (Homoptera: Aphidae). *J. Econ. Entomol.* 90: 309-319.

Boiteau, G. and Parry, R. H. 1985. Monitoring of inflights of green peach aphids, *Myzus persicae* (Sulzer), in New Brunswick potato fields by yellow pan from 1974 to 1983: Results and degree-day simulation. *Amer. Pot. J.* 62: 489-496.

Boiteau, G. and Singh, R. P. 1982. Evaluation of mineral oil sprays for reduction of virus Y spread in potatoes. *Amer. Pot. J.* 59: 253-262.

Boiteau, G. and Singh, R. P. 1999. Field assessment of imidacloprid to reduce the spread of PVY^O and PLRV in potato. *Amer. J. Potato Res.* 76: 31-36.

Boiteau, G. and Wood, F. A. 1982. Persistence of mineral oil spray deposits on potato leaves. *Amer. Pot. J.* 59: 55-63.

Bradley, R. H. E., Wade, C. V. and Wood, F. A. 1962. Aphid transmission of potato virus Y inhibited by oils. *Virology.* 18: 327-329.

Broadbent, L. 1950. The correlation of aphid numbers with the spread of leafroll and rugose mosiac in potato crops. *Ann. appl. Biol.* 37: 58-65.

Byrne, D. N. and Bishop, G. W. 1979. Relationship of green peach aphid numbers to spread of potato leaf roll virus in southern Idaho. *J. Econ. Entomol.* 72: 809-811.

Cancelado, R. E. and Radcliffe, E. B. 1979. Action thresholds for green peach aphid on potatoes in Minnesota. *J. Econ. Entomol.* 72: 606-609.

Castle, S. J. and Berger, P. H. 1993. Rates of growth and increase of *Myzus persicae* on virus-infected potatoes according to type of virus-vector relationship. *Entomol. Exp. Appl.* 69: 51-60.

Castle, S. J., Mowry, T. M. and Berger, P. H. 1998. Differential settling by *Myzus persicae* (Homoptera: Aphididae) on various virus infected host plants. *Ann. Entomol. Soc. Amer.* 91: 661-667.

Cohen, S. 1981. Reducing the spread of aphid-transmitted viruses in peppers by coarse-net cover. *Phytoparasitica*. 9: 69-76.

Cohen, S. and Marco, S. 1973. Reducing the spread of aphid-transmitted viruses in peppers by trapping the aphids on sticky yellow polyethylene sheets. *Phytopathology* 63: 1207-1209.

Collar, J. L., Avilla, C., Duque, M. and Fereres, A. 1997. Behavioral response and virus vector ability of *Myzus persicae* (Homoptera: Aphididae) probing on pepper plants treated with aphicides. *J. Econ. Entomol.* 90: 1628-1634.

DiFonzo, C. D. 1995. Epidemiology and control of potato virus Y (PVY) in the Red River Valley of Minnesota and North Dakota. *Ph.D. dissertation, University of Minnesota, St. Paul.*

DiFonzo, C. D., Ragsdale, D. W. and Radcliffe, E. B. 1995. Potato leafroll virus spread in differentially resistant potato cultivars under varying aphid densities. J *Amer. Potato* J. 72: 119-132.

DiFonzo, C. D., Ragsdale, D. W., Radcliffe, E. B., Gudmestad, N. C. and Secor, G. A. 1996. Crop borders reduce potato virus Y incidence in seed potato. *Ann. appl. Biol.* 129: 289-302.

DiFonzo, C. D., Ragsdale, D. W., Radcliffe, E. B., Gudmestad, N. C. and Secor, G. A. 1997. Seasonal abundance of aphid vectors of potato virus Y in the Red River Valley of Minnesota and North Dakota. *J. Econ. Entomol.* 90: 824-831.

Dixon, A. F. G. 1985. *Aphid Ecology* Blackie, Glasgow.

Douglas, D. R. and Pavek, J. J. 1972. Net necrosis of potato tubers associated with primary, secondary and tertiary infection of leafroll. *Amer. Potato J.* 49: 330-333.

Eastop, V. F. and Raccah, B. 1988. Aphid and host plant species in the Arava Valley of Israel: epidemiological aspects. *Phytoparasitica.* 16: 23-32.

Fereres, A., Perez, P., Gemeno, C. and Ponz, F. 1993. Transmission of Spanish pepper- and potato-PVY isolates by aphid (Homoptera: Aphididae) vectors: epidemiological implications. *Environ. Entomol.* 22: 1260-1265.

Flanders, K. L., Radcliffe, E. B. and Ragsdale, D. W. 1991. Potato leafroll virus spread in relation to densities of green peach aphid (Homoptera: Aphididae): implications for management thresholds for Minnesota seed potatoes. *J. Econ. Entomol.* 84: 1028-1036.

Flanders, K. L., Ragsdale, D. W. and Radcliffe, E. B. 1990. Use of enzyme linked immunosorbent assay to detect potato leafroll virus in field grown potatoes, cv. 'Russet Burbank'. *Amer. Pot. J.* 67: 589-602.

Foster, G. N. 1986. Effects of two insecticidal spray programmes on aphids and the spread of potato leafroll virus. *Ann. appl. Biol.* 108: 24-25.

Frazer, B. D. 1987. Aphid and virus management in British Columbia. In: *Potato Pest Management in Canada.* (ed) G. Boiteau, R. P. Singh and R. H. Parry. pp. 23-29. Agriculture Canada, Ottawa.

Gabriel, W., Szulc, J. and Wislocka, M. 1981. The influence of distance from source of infection on the effect of treatments with systemic insecticides on the spread of potato viruses Y and M. *Potato Res.* 25: 1-11.

Gamez, R. and Saavedra, F. 1986. Maize rayado fino: a model of a leafhopper-borne *outbreaks.* virus disease in the neotropics. In: *Plant Virus Epidemics Monitoring, modelling and Predicting* (ed) G. D. McLean, R. G. Garrett and W. G. Ruesink. pp. 550. Academic Press, Sidney.

Garrett, R. G. 1986. Prologue: a basis for control. In: *Plant Virus Epidemics Monitoring, Modelling and Predicting Outbreaks.* (ed) G. D. McLean, R. G. Garrett and W. G. Ruesink. pp. 1-10. Academic Press, New York.

Garrett, R. G. and McLean, G. D. 1983. The epidemiology of some aphid-borne viruses in Australia. In: *Plant Virus Epidemilogy.* (ed) R. T. Plumb and J. M. Thresh. pp. 199. Blackwell Scientific Publications, Oxford.

George, W. L. and Kring, J. B. 1971. Virus protection of late-season summer squash with aluminum mulch. Conn. Agr. Exp. Sta. Circular 239. pp. 1-8, New Haven.

Gibson, R. W. 1983. The ability of different pyrethroids to control spread of potato virus Y by aphids. *10th Inter. Cong. Plant Protection* 1983. England, pp. 1192.

Gibson, R. W. 1991. The development of mature plant resistance in four potato cultivars against aphid-inoculated potato virus Y^{O} and Y^{N} in four potato cultivars. *Potato Res.* 34: 205-210.

Gibson, R. W. and Cayley, G. R. 1984. Improved control of potato virus Y by mineral oil plus the pyrethroid cypermethrin applied electrostatically. *Crop Prot.* 3: 469-478.

Gibson, R. W., Payne, R. W. and Katis, N. 1988. The transmission of potato virus Y by aphid of different vectoring abilities. *Ann. appl. Biol.* 113: 35-43.

Gibson, R. W., Rice, A. D. and Sawicki, R. M. 1982. Effects of the pyrethroid deltamethrin on the acquisition and inoculation of viruses by *Myzus persicae*. *Ann. appl. Biol.* 100: 49-54.

Halbert, S., Corsini, D., Sandvol, L. and Nolte, P. 1999. Epidemiology and control of aphid transmitted potato viruses. *Idaho Cooperative Extension.* BUL 809, 13pp.

Halbert, S. E., Castle, S. J. and Mowry, T. M. 1995. Do *Myzus* (*Nectarosiphon*) species other than *M. persicae* pose a threat to the Idaho potato crop? *Amer. Potato J.* 72: 85-97.

Halbert, S. E., Connelly, J. and Sandvol, L. 1990. Suction trapping of aphids in western North America (emphasis on Idaho). *Acta Phyto. et Ent. Hung.* 25: 4111-4422.

Halbert, S. E., Elberson, L. R., Feng, M., Poprawski, T. J., Wraight, S., Johnson, J. B. and Quisenberry, S. S. 1998. Suction trap data: implications for crop protection forecasting. In: *Response Model for an Introduced Pest - the Russian Wheat Aphid.* (ed) S. S. Quisenberry and F. R. Peairs. pp. 392-408. *Entomol. Soc. Amer.*, Lanham.

Hanafi, A., Radcliffe, E. B. and Ragsdale, D. W. 1989. Spread and control of potato leafroll virus in Minnesota. *J. Econ. Entomol.* 82: 1201-1206.

Hanafi, A., Radcliffe, E. B. and Ragsdale, D. W. 1995. Spread and control of potato leafroll virus in the Souss Valley of Morocco. *Crop Prot.* 14: 145-153.

Hane, D. C. and Hamm, P. B. 1999. Effects of seedborne potato virus Y infection in two potato cultivars expressing mild symptoms. *Plant Dis.* 83: 43-45.

Harrewijn, P., Ouden, H. d., Piron, P. G. M. and Den Ouden, H. 1991. Polymer webs to prevent virus transmission by aphids in seed potatoes. *Ent. Exp. Appl.* 58: 101-107.

Harrewijn, P. and Kayser, H. 1997. Pymetrozine, a fast-acting and slective inhibitor of aphid feeding. In-situ studies with electronic monitoring of feeding behaviour. *Pestic. Sci.* 49: 130-140.

Harrington, R., Bartlet, E., Riley, D. K., Ffrench Constant, R. H. and Clark, S. J. 1989. Resurgence of insecticide-resistant *Myzus persicae* on potatoes treated repeatedly with cypermethrin and mineral oil. *Crop Prot.* 8: 340-348.

Harrington, R. and Gibson, R. W. 1989. Transmission of potato virus Y by aphids trapped in potato crops in southern England. *Plant Cell, Tissue and Organ Culture*. 32: 167-174.

Harrington, R., Katis, N. and Gibson, R. W. 1986. Field assessment of the relative importance of different aphid species in the transmission of potato virus Y. *Potato Res.* 29: 67-76.

Harrison, B. D. 1971. Potato viruses in Britain. In: *Diseases of Crop Plants.* (ed) J. H. Western. pp. 123-159. MacMillan, London.

Harten, A. v. 1983. The relation between aphid flights and the spread of potato virus Y^N (PVY^N) in the Netherlands. *Potato Res.* 26: 1-15.

Heinze, K. 1960. Versuche zur Ubertragung nichtpersistenter und persistenter Viren durch Blattlause. *NachBl. dtsch. PflSchDienst,* Stuttgart. 12: 119-121.

Hemphill, D. D., Reed, G. L., Wilson, R. C., Gutbrod, O. and Allen, T. C. 1988. Prevention of potato virus Y transmission in potato seed stock with direct covers. *Plasticulture.* 79: 31-36.

Hiddema, J. 1972. Inspection and quality grading of seed potatoes. In: *Viruses of Potatoes and Seed-potato Production.* (ed) J. A. de Bokx. pp. 206-215. Centre for Agricultural Publishing and Documentation, Wageningen.

Hille Ris Lambers, D. 1955. Potato aphids and virus diseases in the Netherlands. *Ann. appl. Biol.* 42: 355-360.

Hille Ris Lambers, D. 1972. Aphids: their life cycles and their role as virus vectors. In: *Viruses of Potatoes and Seed-potato Production.* (ed) J. A. d. Bokx. pp. 36-56. Centre for Agricultural Publishing and Documentation, Wageningen.

Holbrook, F. R. 1977. Aldicarb and thiofanox: effect on the feeding activity of green peach aphids. *J. Econ. Entomol.* 70: 742-744.

Hollings, M. 1955. Aphid movement and virus spread in seed potato areas of England and Wales, 1950-53. *Plant Pathol.* 4: 73-82.

Ioannou, N. 1989. The infection pressure of potato leafroll virus and potato virus Y in relation to aphid populations in Cyrpus. *Potato Res.* 32: 33-47.

Irwin, M. E. 1980. Sampling aphids in soybean fields. In: *Sampling Methods in Soybean Entomology.* (ed) M. Kogan and D. C. Herzog. pp. 239-259. Springer-Verlaug, New York.

Kennedy, J. S., Day, M. F. and Eastop, V. F. 1962. *A Conspectus of Aphids as Vectors of Plant Viruses.* Commonwealth Agricultural Bureaux, London.

Kring, J. B. 1970. Determining the number of aphids over reflective surfaces. *J. Econ. Entomol.* 63: 1350-1353.

Kurppa, S. and Rajala, P. 1986. Occurrence of winged aphids on potato plants and pressure for potato virus Y transmission in Finland. *Ann. Agr. Fenniae* (Finland). 25: 199-214.

Lagnaoui, A. and Radcliffe, E. B. 1998. Potato fungicides interfere with entomopathogenic fungi impacting population dynamics of green peach aphid. *Amer. J. Potato Res.* 75: 19-25.

Leclercq-LeQuillec, F., Tanguy, S. and Dedryver, C. A. 1995. Aerial flow of barley yellow dwarf viruses and of their vectors in western France. *Ann. appl. Biol.* 126: 75-90.

Leonard, S. H. and Holbrook, F. R. 1978. Minimum acquisition and transmission times for potato leaf roll virus by the green peach aphid. *Ann. Entomol. Soc.* Amer. 71: 493-495.

Loebenstein, G., Alper, M. and Deutsch, M. 1964. Preventing aphid-spread cucumber mosaic virus with oils. *Phytopathology* 54: 960-962.

Loughnane, J. B. 1939. *Myzus ornatus* a vector of potato viruses. *Nature*. 144: 785-786.

Loughnane, J. B. 1943. *Aphis rhamni* Boyer; its occurrence in Ireland and its efficiency as a vector of potato viruses. *J. Dep. Agric.* Eire. 40: 291-298.

Madden, L. V. and Campbell, C. L. 1986. Descriptions of virus disease epidemics in time and space. In: *Plant Virus Epidemics Monitoring, Modelling and Predicting Outbreaks*. (ed) G. D. McLean, R. G. Garrett and W. G. Ruesink. pp. 273-293. Academic Press, Sidney.

Marco, S. 1980. The use of insecticides to control potato leafroll virus in seed-potato crops on the Golan Heights. *Phytoparasitica*. 8: 61-71.

Marco, S. 1986. Incidence of aphid-transmitted virus infections reduced by whitewash sprays on plants. *Phytopathologt* 76: 1344-1348.

Marco, S. 1993. Incidence of nonpersistently transmitted viruses in pepper sprayed with whitewash, oil and insecticide, alone or combined. *Plant Dis*. 77: 1119-1122.

Minks, A. K. and Harrewijn, P. 1987. *Aphids: Their Biology, Natural Enemies and Control.* Elsevier, Amstermam.

Mowry, T. M. 1994. Potato leafroll virus management in the Pacific Northwest (USA). In: *Advances in Potato Pest Biology and Management.* (ed) G. W. Zehnder, M. L. Powelson, R. K. Jansson and K. V. Raman. pp. 111-123. APS Press, St. Paul.

Nanne, H. W. and Radcliffe, E. B. 1971. Green peach aphid populations on potatoes enhanced by fungicides. *J. Econ. Entomol.* 64: 1569-1570.

Nault, L. R. 1997. Arthropod transmission of plant viruses: a new synthesis. *Ann. Entomol. Soc. Amer*. 90: 521-541.

Parry, R. H. 1987. Aphid and virus management in potatoes in eastern Canada. In: *Potato Pest Management in Canada.* (ed) G. Boiteau, R. P. Singh and R. H. Parry. pp. 9-22. Agriculture Canada, Ottawa.

Perez, P., Collar, L., Avilla, C., Duque, M. and Fereres, A. 1995. Estimation of vector propensity of potato virus Y in open-field pepper crops of central Spain. *J. Econ. Entomol.* 88: 986-991.

Piron, P. G. M. 1986. New aphid vectors of potato virus Y(N). *Neth. J. Plant Pathol.* 92: 223-229.

Pirone, T. P. and Thornbury, D. W. 1988. Quantity of virus required for aphid transmission of a potyvirus. *Phytopathology* 78: 104-107.

Powell, D. M. and Mondor, T. W. 1973. Control of the green peach aphid and suppression of leaf roll on potatoes by systemic soil insecticides and multiple foliar sprays. *J. Econ. Entomol.* 66: 170-177.

Powell, G. 1992. The effect of mineral oil on stylet activities and potat virus Y transmission by aphids. *Ent. Exp. Appl.* 63: 237-242.

Powell, G., Hardie, J. and Pickett, J. A. 1998. The effects of antifeedant compounds and mineral oil on stylet penetration and transmission of potato virus Y by *Myzus persicae* (Sulz.) (Hom., Aphididae). *J. Appl. Entomol.* 122: 331-333.

Powell, G., Harrington, R. and Spiller, N. J. 1992. Stylet activities and potato virus Y vector efficiencies by the aphids *Brachycaudus helichrysi* and *Drepanosiphum platanoidis*. *Ent. Exp. Appl.* 62: 293-300.

Raccah, B., Gal-on, A. and Eastop, V. F. 1985. The role of flying aphid vectors in the transmission of cucumber mosaic virus and potato virus Y to peppers in Israel. *Ann. appl. Biol.* 106: 451-460.

Radcliffe, E. B. 1982. Insect pests of potato. *Ann. Rev. Entomol.* 27: 173-204.

Radcliffe, E. B., Flanders, K. F., Ragsdale, D. W. and Noetzel, D. M. 1991. Pest management systems for potato insects. In: *CRC Handbook of Pest Management in Agriculture,* 2nd edition. (ed) D. Pimentel. pp. 587-621. Vol. III. CRC Press, Boca Raton.

Ragsdale, D. W. and Radcliffe, E. B. 1998. Aphids cause big problems for industry. *Valley Potato Grower 63(110): 4, 6, 31.*

Ragsdale, D. W., Radcliffe, E. B., DiFonzo, C. D. and Connelly, M. S. 1994. Action thresholds for an aphid vector of potato leafroll virus. In: *Advances in Potato Pest Biology and Management.* (ed) G. W. Zehnder, M. L. Powelson, R. K. Jansson and K. V. Raman. pp. 99-110. APS Press, St. Paul.

Reestman, A. J. 1972. Incidence of infection in commercial crops and consequent losses. In: *Viruses of Potatoes and Seed-potato Production.* (ed) J. A. de Bokx. pp. 152-157. Centre for Agricultural Publishing and Documentation, Wageningen.

Ribbands, C. R. 1965. The significance of apterous aphids in the spread of viruses within agricultural crops. *Proc. XIIth Inter. Cong. Entomol.* London, pp. 525.

Robert, Y. and Rouze-Jouan, J. 1971. Premiere observation sur la role de la temperature au moment de la transmission de l'enroulement par Aulacorthum solani Kltb., Macrosiphum euphorbiae Thomas et *Myzus persicae* Sulz. *Potato Res.* 14: 154-157.

Roland, G. 1952. Quelques recherches sur l'enroulement de la pomme de terre (Solanum viurs 14, Appel & Quanjer). *Parasitica.* 8: 150-158.

Ruesink, W. G. and Irwin, M. E. 1986. Soybean mosaic virus epidemiology: a model and some implications. In: *Plant Virus Epidemics Monitoring, Modelling and Predicting Outbreaks.* (ed) G. D. McLean, R. G. Garrett and W. G. Ruesink. pp. 550. Academic Press, Sidney.

Salazar, L. F. 1996. *Potato Viruses and their Control.* International Potato Center, Lima.

Sawicki, R. M., Rice, A. D. and Gibson, R. W. 1983. Insecticide-resistance in the peach-potato aphid *Myzus persicae* and the prevention of virus spread. *Aspect Appl. Biol.* 2: 29-33.

Setiawan, D. P. and Ragsdale, D. W. 1987. Use of aluminum-foil and oat-straw mulches for controlling aster leafhopper, *Macrosteles fascifrons* (Homoptera: Cicadellidae), and aster yellows in carrot. *Great Lakes Entomol.* 20: 103-109.

Shands, W. A. 1977. Control of aphid-borne potato virus Y in potatoes with oil emulsions. *Amer. Potato J.* 54: 179-187.

Shands, W. A., Simpson, G. W., Murphy, H. J., Seaman, B. A., Roberts, F. S. and Flynn, C. M. 1972a. Effects of cultural methods for controlling aphids on potatoes in northeastern Maine. *Tech. Bull., Life Sci. Agri. Exp. Stat.* University of Maine at Orono. No. 57: 31 pp.

Shands, W. A., Simpson, G. W., Seaman, B. A., Roberts, F. S. and Flynn, C. M. 1972b. Effects of differing abundance levels of aphids and of certain virus diseases upon yield and virus disease spread in potatoes. *Tech. Bull., Life Sci. Agri. Exp. Stat.,* University of Maine at Orono. No. 56: 40 pp.

Sigvald, R. 1984. The relative efficiency of some aphid species as vectors of potato virus Y (PVY). *Potato Res.* 27: 285-290.

Sigvald, R. 1986. Forecasting the incidence of potato virus Y^O. In: *Plant Virus Epidemics: Monitoring, Modelling and Predicting Outbreaks.* (ed) *G. D. McLean,* R. G. Garrett and W. G. Ruesink. pp. 419-439. Academic Press, Sidney.

Sigvald, R. 1987. Forecasting potato virus Y. *Vaxtskyddsnotiser.* 51: 131-137.

Sigvald, R. 1989. Relationship between aphid occurrence and spread of potato virus Y^O (PVY^O) in field experiments in southern Sweden. *J. Appl. Entomol.* 108: 35-43.

Singh, M. N., Khurana, S. M. P., Nagaich, B. B. and Agrawal, H. O. 1988. Environmental factors influencing aphid transmission of potato virus Y and potato leafroll virus. *Potato Res.* 31: 501-509.

Singh, R. P. 1987. Role of weeds in potato virus spread. In: *Potato Pest Management in Canada.* (ed) G. Boiteau, R. P. Singh and R. H. Parry. pp. 355-362. Agriculture Canada, Ottawa.

Singh, R. P. and Boiteau, G. 1986. Reevaluation of the potato aphid, *Macrosiphum euphorbiae* (Thomas), as vector of potato virus Y. *Amer. Potato J.* 63: 335-340.

Singh, R. P. and Boiteau, G. 1987. Control of aphid borne diseases: nonpersistent viruses. In: *Potato Pest Management in Canada.* (ed) G. Boiteau, R. P. Singh and R. H. Parry. pp. 30-53. Agriculture Canada, Ottawa.

Singh, R. P., Kurz, J. and Boiteau, G. 1996. Detection of stylet-borne and circulative potato viruses in aphids by duplex reverse transscription polymerase chair reaction. *J. Virol. Meth.* 59: 189-196.

Singh, R. P., Kurz, J., Boiteau, G. and Bernard, G. 1995. Detection of potato leafroll virus in single aphids by the reverse transcription polymerase chain reaction and its potential epidemiological application. *J. Virol. Meth.* 55: 133-143.

Singh, R. P., Kurz, J., Boiteau, G. and Moore, L. M. 1997. Potato leafroll virus detection by RT-PCR in field-collected aphids. *Amer. Pot. J.* 74: 305-313.

Slack, S. A. 1993. Seed certification and seed improvement programs. In: *Potato Health Management.* (ed) R. C. Rowe. pp. 61-65. American Phytopatholoical Society, St. Paul.

Slack, S. A. and German, T. L. 1998. Impact of transgenic viral resistance on seed potato certification. *Amer. J. Potato Res.* 75: 265-268.

Souza Dias, J. A. C., Slack, S. A. and Dias, J. A. C. S. 1987. Relation of potato leafroll virus concentration in potatoes to virus concentration in aphids. *Amer. Potato J.* 64: 459.

Suranyi, R., Longtine, C., Ragsdale, D. and Radcliffe, E. 1999. Controlling leafhoppers with below-label rates. *Valley Potato Grower.* 64: 11-13, 16.

Sylvester, E. S. 1980. Circulative and propagative virus transmission by aphids. *Ann. Rev. Entomol.* 25: 257-286.

Tamada, T., Harrison, B. D. and Roberts, I. M. 1984. Variation among British isolates of potato leafroll virus. *Ann. appl. Biol.* 104: 107-116.

Taylor, L. R. 1986a. The distribution of virus disease and the migrant vector aphid. In: *Plant Virus Epidemics Monitoring, Modelling and Predicting Outbreaks.* (ed) G. D. McLean, R. G. Garret and W. G. Ruesink. pp. 35-57. Academic Press, New York.

Taylor, L. R. 1986b. The distribution of virus disease and the migrant vector aphid. In: *Plant Virus Epidemics: Monitoring, Modelling and Predicting Outbreaks.* (ed) G. D. McLean, R. G. Garrett and W. G. Ruesink. pp. 35-57. Academic Press, New York.

Thomas, P. E. 1983. Sources and dissemination of potato viruses in the Columbia Basin of the Northwestern United States. *Plant Dis.* 67: 744-747.

Thomas, P. E., Pike, K. S. and Reed, G. L. 1997a. Role of green peach aphid flights in the epidemiology of potato leaf roll disease in the Columbia Basin. *Plant Dis.* 81: 1311-1316.

Thomas, R. E., Kaniewski, W. K. and Lawson, E. C. 1997b. Reduced field spread of potato leafroll virus in potatoes transformed with the potato leafroll virus coat protein gene. *Plant Dis.* 81: 1447-1453.

Turl, L. A. D. 1978. Epidemiology of potato aphids in 1975-77 with regard to the incidence of potato leaf roll virus in Scotland. In: *Plant Disease Epidemiology.* (ed) P. R. Scott and A. Bainbridge. pp. 235-242 van der Want, J. P. H. 1972. Introduction to plant virology. In: *Viruses of Potatoes and Seed-potato Production.* (ed) J. A. de Bokx. pp. 19-25. Centre for Agricultural Publishing and Documentation, Wageningen.

van Harten, A. 1983. The relation between aphid flights and the spread of potato virus Y^N (PVY^N) in the Netherlands. *Potato Res.* 26: 1-15.

van Hoof, H. A. 1980. Aphid vectors of potato virus Y^N. Neth. J. Pl. Pathol. 86: 159-162.

Villacarlos, L. T. 1987. Effect of systemic insecticides on the feeding behavior and mortality of green peach aphid and its transmission of potato leaf roll virus. *Philippine Entomol.* 7: 121-128.

Webley, D. P. and Stone, L. E. W. 1972. Field experiments on potato aphids and virus spread in South Wales, 1966/9. *Ann. appl. Biol.* 72: 197-203.

Wilson, C. R. and Jones, R. A. C. 1993. Resistance to potato leafroll virus infection and accumulation in potato cultivars, and the effects of previous infection with other viruses on expression of resistance. *Australian J. Agric. Res.* 44: 1891-1904.

Woiwood, I. P., Tatchell, G. M. and Barrett, A. M. 1984. A system for the rapid collection, analysis and dissemination of aphid-monitoring data from suction traps. *Crop Prot.* 3: 273-288.

Woodford, J. A. T. 1988. The impact of cropping policy on methods to control potato leafroll virus. *Aspects appl. Biol.* 17: 163-171.

Woodford, J. A. T. and Gordon, S. C. 1990. New approaches for restricting spread of potato leafroll virus by different methods of eradicating infected plants from potato crops. *Ann. Biol.* 116: 477-487.

Woodford, J. A. T., Gordon, S. C. and Foster, G. N. 1988. Side-band application of systemic granular pesticides for the control of aphids and potato leafroll virus. *Crop Prot.* 7: 96-105.

Woodford, J. A. T., Harrison, B. D., Aveyard, C. S. and Gordon, S. C. 1983. Insecticidal control of aphids and the spread of potato leafroll virus in potato crops in eastern Scotland *Myzus persicae, Macrosiphum euphorbiae. Ann. appl. Biol.* 103: 117-130.

Woodford, J. A. T., Jolly, C. A. and Aveyard, C. S. 1995. Biological factors influencing the transmission of potato leafroll virus by different aphid species. *Potato Res.* 38: 133-141.

Woodford, J. A. T., Shaw, M. W., McKinlay, R. G. and N., F. G. 1977. The potato aphid spray warning scheme in Scotland, 1975 - 1977. *Proc. 1977 British Crop Protection Conference - Pests and Diseases* 247-254.

Wyman, J. A., Toscano, N. C., Kido, K., Johnson, H. and Mayberry, K. 1979. Effects of mulching on the spread of aphid-transmitting watermelon mosaic virus to summer squash. *J. Econ. Entomol.* 72: 139-143.

Zimmerman-Gries, S. and Gries, S. Z. 1979. Reducing the spread of potato leaf roll virus in seed potatoes by sticky yellow polyethylene sheets. *Phytoparasitica.* 7: 41.

10. Detection and Identification of Viruses in Potatoes

Abed Gera[1] and Shlomo Marco[2]
[1]Department of Virology, Agricultural Research Organization, the Volcani Center, Bet Dagan 50250, Israel. [2]Viran, Plant Disease Diagnosis, Neve Yarak 49945, Israel.

The early and accurate diagnosis of viral diseases is critical for effective management of most crop systems. Plant viruses are generally identified by symptoms, particle morphology, host range, serological properties and nucleic acid sequence. Symptoms are of major importance, because they are the main means by which a virus disease is diagnosed. Transmission of a virus from infected to healthy tissue is a fundamental procedure used to isolate viruses from diseased plants, to study viral symptoms and to assay for infectivity.

Electron microscopy is the only direct method used for determining the shape and size of the virus particle. In many cases it serves as a rapid method for placing a virus in its taxonomic group. Serological procedures form the most reliable method for identification and quantitative assay of viruses. Serological identification greatly improved with the application of the enzyme-linked immunosorbent assay (ELISA) for the detection and identification of plant viruses in the mid- 1970s. Cross-reactivity of antisera raised against viruses from different groups has frequently been used for classification of plant viruses. However, nucleic acid sequence data are accumulating rapidly and allow more accurate relationships to be established between the individual members of virus groups when compared to serological methods.

Identification of a virus, by sequencing parts of its genome, is often done if extensive serological analysis cannot provide conclusive data about the nature of the virus. This approach requires purification of the virus particles and the subsequent cloning of part of the virus genome. However, with novel molecular such as virus group-specific polymerase chain reaction (PCR)

G. Loebenstein et al. (eds.),
Virus and Virus-like Diseases of Potatoes and Production of Seed-Potatoes, 271–283.

techniques, sequence data from a new virus can be obtained, even without the need to purify the virus or to clone parts of its genome. Once identification of a virus has been accomplished by nucleic acid sequencing, specific PCR primer sets can be designed for sensitive detection of the virus. Alternatively, the application of nucleic acid probes and related techniques based on properties of viral nucleic acid is becoming common and is providing new possibilities for the detection of plant viruses. The cloning and expression of virus coat protein as recombinant fusion products in an expression vector can provide a source of pure antigens. Such antigens will provide an infinite source for the production of new generations of virus specific-antisera.

In this chapter the various techniques commonly used for the detection and identification of potato viruses are described.

10.1 Test Plants

In the field, symptoms give the first clue to the identity of a virus. In the laboratory and greenhouse, symptoms produced in a range of test plants may be of considerable diagnostic value but insufficient to allow positive identification. However, visual inspection for symptoms is often not adequate, since the symptom expression can be highly variable.

Experimental host plants, under standardized conditions, will exhibit consistent and characteristic disease symptoms when infected with a particular virus. Knowledge of the host range of a particular virus and the symptoms it elicits are usually essential for studying new viruses or virus strains and may provide useful clues to the identity of an unknown virus.

Several herbaceous plants are susceptible to a large number of viruses. Some plant species react to mechanical inoculation with potato viruses by exhibiting local lesions or systemic symptoms. The following test plants can be used for detecting, and in many instances, diagnosing potato virus infections. *Physalis floridana, Gomphrena globosa, Datura stramonium, D. metel, Chenopodium amaranticolor Lycopersicon esculentum, Solanum demissum x S. tuberosum* and *Nicotiana glutinosa.*

Biological techniques are usually accurate for disease diagnosis and still widely used for both detection and diagnosis of some viruses and viroids. These methods are, however, not suitable for processing a large number of samples in a short time and have the disadvantage that strains of the same virus may produce different symptoms on the same host.

10.2 Electron Microscopy

Plant viruses in infected tissues often exist in sufficient concentrations that they can be extracted and examined by standard electron microscopy (EM). Observation of the shape and size of a virus particle is a basic criterion for virus identification. In many cases EM enables information on virus morphology to be obtained within minutes after sampling a diseased plant. EM is used to examine viruses in crude extracts from infected plants (Hill, 1984). Crude sap preparations are obtained by grinding or chopping an infected leaf, stem or root tissue in buffer or directly in a negative stain. Alternatively, a fragment of epidermis can be stripped from the leaf and passed through a drop of water or stain on a specimen grid. The technique facilitates simple, rapid and direct examination of crude extracts of infected plants without the necessity of using a purified virus. The use of EM in combination with serology for detecting and identifying plant viruses will be discussed later.

10.3 Serology

Most plant viruses are strong immunogens: when injected into an experimental animal they stimulate the production of specific antibodies that can be used in various serological tests. The production of virus-specific antisera is the basis for a successful serological procedure. Antisera are produced by immunizing animals with a highly purified virus or viral coat protein. It is important to purify the virus to such a degree that only minimal amounts of host proteins will remain in the preparation to be injected, since the presence of host proteins would lead to non-specific reactions. Therefore, it is recommended to check the virus preparations by sodium dodecyl sulfate –polyacrylamide gel electrophoresis (SDS–PAGE), and to continue purification until no host protein bands are visible. There is considerable divergence of opinion in the literature concerning the most appropriate schedule of immunization. The production of high-titered antisera depends on the amount of virus injected and viral antigenicity. The number of injections will also affect the specificity of antisera. When few injections are given, highly specific antisera are produced. Multiple injections may produce broad-spectrum antisera without the specificity required for strain differentiation.

In our laboratory, antisera are routinely produced in rabbits by four to six biweekly intramuscular injections of 1 ml of virus preparation (0.5-2 mg), emulsified with an equal volume of Freund's complete or incomplete adjuvant. Bleedings are taken 10-20 days after the last injection, when serum titer reaches its maximum. Non-specific antibodies can be removed from the antiserum by cross-absorption with any contaminating antigens.

Polyclonal antisera specific to viral coat protein treated with SDS and separated on PAGE, have been successfully produced by immunizing rabbits with protein bands cut from gels (Caroll *et al.* 1978). Recently we have shown that booster immunization with a partially purified *Potato virus Y* (PVY, genus *Potyvirus*, family *Potyviridae*) preparation, after priming with SDS- PAGE coat protein, enhances the binding capacity of capture antibodies as measured by ELISA (Gera *et al.* 1999).

Antibodies produced by a hybrid cell (hybridoma) formed by fusion of a B lymphocyte cell with a mouse myeloma cell, are called monoclonal antibodies (Mabs). Since Mabs against *Tobacco mosaic virus* (TMV, family *Tobamoviridae*) were first described, they have been prepared against 50 different plant viruses, including important viruses infecting potato, such as *Potato leafroll virus* (PLRV, genus *Polerovirus*, family *Luteoviridae*), *Potato virus A* (PVA, genus *Potyvirus*, family *Potyviridae*), *Potato virus M* (PVM, genus *Carlavirus*), *Potato virus S* (PVS, genus *Carlavirus*), *Potato virus X* (PVX, genus *Potexvirus*) and PVY. These antibodies are valuable for detecting viruses that are difficult to purify. In practice, however, production of Mabs is often labor-intensive.

Among the early immunological test procedures were chloroplast agglutination, microprecipitation tests and gel immunodiffusion. These techniques will not be reviewed here; for information on the procedures see Hampton *et al.* (1990).

In order to increase the sensitivity of serological tests, the use of solid phase to adsorb an antibody or antigen with subsequent attachment of an antigen or antibody and the use of an antibody-enzymes conjugate to detect the antigen has become an important procedures in plant virus detection. In this chapter only methods which are commonly used for potato virus detection will be discussed.

10.3.1 Enzyme-linked Immunosorbent Assay (ELISA)

The ELISA technique, which was first applied to plant virus detection by Clark and Adams (1977), has significantly increased the ability to detect and study plant viruses, and is currently the most widely used method for the detection of potato viruses. Variations of ELISA have become the preferred serological test for potato viruses because of their simplicity, adaptability, rapidity, sensitivity and accuracy. ELISA has been reviewed elsewhere (Converse and Martin, 1990).

The double antibody sandwich (DAS-ELISA) test on a solid phase (usually plastic) is most commonly used. Virus is first selectively trapped by a specific antibody adsorbed on a solid surface, a specific enzyme-labeled antibody (conjugate) is added to the immobilized virus, and the reaction is measured visually or spectrophotometrically, after adding a suitable enzyme substrate.

1. Microtiter plates are coated by incubation of 100 μl of immunoglobulins (1-10 μg/ml) in carbonate buffer, pH 9.6 for 2-6 h at 37°C.
2. Three washing steps with phosphate-buffered saline, pH 7.4, containing 0.05% Tween-20 (PBST).
3. Incubation with 100 μl of crude plant extract diluted in PBST containing 1-2% polyvinylpyrrolidone (PVP), for 2 h at 37°C or overnight at 5°C.
4. Three washing steps with PBST.
5. Incubation with 100 μl of alkaline phosphatase-labeled antivirus immunoglobulin diluted 1/1000-1/5000 in PBST-PVP, for 3-5 h at 37°C.
6. Three washing steps with PBST.
7. Incubation at room temperature with 100 μl of (1 mg/ml) p-nitrophenylphosphate in diethanolamine buffer, pH 9.8. Absorbance at 405 nm is read after 30 min to 2 h of incubation of the substrate. It is customary to consider the result positive if the absorbance is twice that of the healthy control.

A variation of the above method is the indirect ELISA, in which plates are coated with antigen, and the primary antiviral antibody of one animal species (e.g., rabbit) is added. A secondary commercial antibody conjugate (e.g., goat anti-rabbit) which reacts with the first antibody is then added.

DAS-ELISA is especially useful for detecting antigens in complex mixtures. This is because the bound antibody specifically traps the antigens of interest, while non-specific antigens are removed in the wash step. Although the indirect ELISA is considered less strain-specific, the test is simple to perform and the enzyme-antibody conjugates can be used for detecting many different viruses.

ELISA is the most common method for diagnosing the main potato viruses. Commercial kits for PLRV, PVM, PVS, PVX, PVY, and other viruses are available, and give reliable results when potato leaves are tested. For routine testing it is preferable to use polyclonal antisera. Care has to be taken to include both positive and negative controls in the ELISA tests. The minimum level of virus detection by ELISA is about 2 ng/ml, which is not sensitive enough to detect PLRV and sometimes PVY in potato tubers.

10.3.2 Dot ELISA

Assays in which antibodies or antigens are bound to nitrocellulose or nylon membranes have been used to detect PVS, PVX and PVY (Banttari and Goodwin, 1985) and PLRV (Smith and Banttari, 1987). In principle, solutions containing purified antigen or crude sap from infected plants are spotted directly onto a membrane and air dried. The membrane surface is then saturated with bovine serum albumin (BSA) and a specific viral antibody is added. Finally, a second antibody conjugate and substrate are added. The enzyme reacts with the soluble substrate to form an insoluble colored product at the site of the reaction. The method is readily adaptable to field application.

10.3.3 Tissue Blotting and Tissue Squashes

Tissue blotting is a serological technique similar to ELISA (Lin *et al.* 1990). Blots are made by pressing the freshly cut tissue surface gently but firmly on a nitrocellulose membrane. Antigens in tissue blots are detected by enzyme-labeled probes. The technique has been used to detect both PVX and PVY from tubers in the field (Bravo-Almonacid, *et al.* 1992).

Nucleic acids of plant viruses were also detected specifically and sensitively by hybridization of infected plant tissues squashed onto a nylon membrane (squash blot) with a specific radioactive probe (Navot *et al.* 1989). The method provides a specific, rapid and simple tool for large-scale diagnosis of plant viruses.

10.3.4 Immunosorbent Electron Microscopy (ISEM)

Techniques involving the detection and identification of plant viruses by combining electron microscopy and serology are highly sensitive. ISEM was introduced by Derrick (1973). In this procedure, an electron microscope grid is coated with virus-specific antiserum. Grids are floated on purified virus or crude sap from an infected plant, and antibodies on the grid surface trap virus particles from solution. Improved detection was achieved by combining ISEM with a decoration step (Fig. 10.1), whereby the virus particles are coated with a second layer of antibodies following virus attachment to a grid (Milne, 1992). The application of ISEM for diagnosis is, however, limited to only a small number of samples.

10.4 Nucleic Acid Hybridization

Detection of plant viruses by nucleic acid hybridization is based on pairing between a specific nucleic acid probe and viral nucleic acid immobilized on filters. Depending on the initial reagents, the products of the reaction may be RNA:RNA, RNA:DNA, or DNA:DNA duplexes.

For plant virus diagnosis, mixed-phase hybridization is the most commonly used. Viral nucleic acid is first denatured and immobilized on filter papers (nitrocellulose or nylon), which are immersed in solutions of a labeled probe under conditions which favor hybridization. If the solution contains labeled nucleic acid sequences which are complementary to the viral nucleic acid immobilized on the filter annealing will occur. Autoradiography is then used to detect the sample which shows a positive reaction. The nucleic acid probe can be DNA or RNA. Although the use of cDNA probes is more common, RNA probes (riboprobes) have the advantage of forming stronger duplexes with the target, and thus being more sensitive. A cRNA probe is made by cloning the cDNA of interest into a transcription vector and to synthesizing cRNA *in vitro*. cRNA probes have been shown to be more sensitive than cDNA probes (Robinson and Romero, 1991).

The use of nucleic acid hybridization for the detection of plant viruses has been reviewed by Hull (1993). Because of the problems associated with the use of radioactivity, there is an increasing number of commercially available nonradioactive systems. Digoxigenin labeled dUTP appears to be the most widely used nonradioactive tag for labeling probes for detection of plant viruses. The method has been used successfully for routine diagnosis of PLRV in non-sprouting tubers (Loebenstein *et al.* 1997).

10.5 Polymerase Chain Reaction

Following the development of the polymerase chain reaction (PCR) in 1983, and the first reports in 1990 of adapting PCR for the detection of plant viruses (Vunsh *et al.* 1990), the application of PCR for diagnosing plant viruses and viroids has rapidly gained a foothold. The method allows amplification of low concentrations of specific target sequences of DNA. A prerequisite for the use of PCR for virus detection is the availability of sequence data on the viral genome. The genome of most plant viruses is RNA, and the detection of an RNA sequence by this method requires some modifications. RNA has to be reverse-transcribed (RT) into cDNA before the PCR amplification process (RT-PCR). The method has been applied for the detection of PLRV and PVY in potato leaves and tubers (Glias *et al.* 1996; Singh and Singh, 1996; Spiegel and Martin, 1993). A "single tube"

RT-PCR was developed (Wetzel *et al.* 1992) and successfully utilized for the detection of PSTV (Shamloul *et al.* 1997). A number of techniques deriving from the original PCR protocol have been presented (Nolasco *et al.* 1993; Barker, 1995). While few of these variants have been applied, some of them offer great potential for the detection of virus strains. For example, a procedure for differentiating PVY^{NTN} from PVY^{N} based on the unique cleavage of their respective PCR products with strain specific restriction endonucleases was recently described (Rosner and Maslenin, 1999).

The specificity of the method depends on the choice of the synthetic oligonucleotide primers derived from the sequence of the target DNA. The use of highly conserved sequences for primer design is recommended for detecting all individuals of a particular virus. Alternatively, degenerate primers derived from a conserved amino acid sequence of viral coat protein may be used for the detection of all members of a virus group.

Although highly sensitive, PCR is too laborious and allows processing of only a limited number of samples. It has been reported that a preliminary immunocapture (IC) step, directly performed in the tube used for PCR, gives a highly sensitive assay. This variant of PCR (IC-PCR) seems well-suited for the large-scale detection of plant viruses. Although it has the drawback of requiring an antiserum against the virus to be detected, IC-PCR combines the advantage of providing a simple way of preparing the plant extracts before amplification, with increased sensitivity (Schoen *et al.* 1996).

10.6 Detection of the Main Viruses

ELISA is the most common method for diagnosing viruses in potatoes intended for seed production. Commercial kits for diagnosing PVA, PVM, PLRV, PVS, PVX, PVY, and other viruses are available. Samples for testing may be taken from foliage, sprouting eyes or dormant tubers. In principle, foliage is more suitable. Care has to be taken to include both positive and negative controls in the ELISA tests. The viruses which are commonly tested according to the seed potato grade, are the worldwide-spread potato viruses: PVA, PVM, PVS, PVX, PVY and PLRV. Among these, by far the most important economically are PLRV and PVY.

10.6.1 *Potato leafroll virus* (PLRV)

Since PLRV is not mechanically transmissible, DAS-ELISA is the method of choice for routine testing. From our experience, a good antiserum will detect the virus in tubers, though some background may form. Using reducing agents, such as 1% DIECA, may decrease this background. The best indicator plant for grafting is *Physalis floridana*.

Recently, single-chain variable fragment antibodies (scFv) obtained from phage display libraries were used to detect PLRV in potato (Toth *et al.* 1999). A fully recombinant ELISA was devised that detects PLRV in extracts of infected potato, as readily as the standard ELISA.

Various PCR procedures have been applied to the detection of PLRV in potato leaves and tubers (Glias *et al.* 1996; Singh and Singh, 1996). More recently, a new procedure which increases the sensitivity of IC-PCR detection of PLRV has been described (Schoen *et al.* 1996). It consists of pre-PCR immunocapture of the virus from tuber extracts by paramagnetic beads carrying an antiserum against the virus, RT of a specific portion of the viral coat protein ORF and a fluorogenic 5' nuclease detection assay with the TaqMan^{TM} System (Applied Biosystems, Foster City, CA) to detect the amplified transcript.

A non-radioactive digoxigenin-labeled cRNA probe for PLRV in non-sprouted tubers has been described (Loebenstein *et al.* 1997). The detection limit of PLRV with this probe was 1 pg/ml. PLRV could be detected in a combined sample of one infected plus 9 healthy dormant tubers (Marco and Alon, unpublished data).

10.6.2 *Potato virus Y* (PVY)

PVY is divided into three groups, PVY^O, PVY^N, and PVY^C of which the first two are the most common. PVY^O can be readily detected in foliage and sprouts by ELISA, and by the indicator plants *Chenopodium amaranticolor* and *C. quinoa* which react with typical local lesions. Singh (1994) reported the use of *Solanum brachycarpum* or *N. tabacum* var. Samsun which react with the typical vein necrosis to PVY^O.

The virulent variant of PVY^N- PVY^{NTN} cannot be distinguished from other members of the N group by conventional serological methods (van den Heuvel *et al.* 1994). Differentiation of PVY^{NTN} from PVY^N by restriction fragment length polymorphism (RFLP) of PCR products amplified from the 5' end region of the virus genome was demonstrated by Glias *et al.* (1996). More recently, a procedure for differentiating PVY^{NTN} from PVY^N based on the unique cleavage of their respective PCR products with strain-specific restriction endonucleases was described (Rosner and Maslenin, 1999). The procedure provides a convenient and reliable method for the differentiation of PVY^{NTN}.

10.6.3 *Potato virus X* (PVX)

PVX is transmitted only mechanically and induces local lesions with a red halo when inoculated on *G. globosa*. Tuber sap can be used for detection. DAS-ELISA is the method of choice for routine testing. The virus has been detected in tubers immediately after harvest or after storage at 4°C for 30-40 weeks, but testing of tubers was not as effective as testing foliage from growing plants.

10.6.4 *Potato virus A* (PVA)

PVA is transmitted mechanically and induces local lesions when inoculated on *Physalis angulata* or *Solanum demissum*. Tuber sap or cut tuber pieces can be used for detection. The virus can be detected in foliage from growing plants by means of ELISA. Attempts to detect the virus in dormant tubers have not been satisfactory.

10.6.5 *Potato aucuba mosaic virus* (PAMV, Genus *Potexvirus*)

The virus could be detected in foliage from growing plants by means of ELISA, but sometimes cross-reactivity with PVX may occur. Mechanical inoculation on *C. annuum* cv. Long Red resulted in necrotic local lesion reactions, which killed young test plants.

10.6.6 *Potato virus S* (PVS) and *Potato virus M* (PVM)

DAS-ELISA is the method of choice for routine testing of PVS and PVM. However, it is recommended to validate a certain percentage of the tests by biological assays.

The two viruses can be distinguished by inoculation to tomato, as only PVM, but not PVS, infects tomato. *D. metel* and *G. globosa* are useful for detecting PVM and *C. quinoa* for PVS.

10.6.7 *Potato mop top virus* (PMTV, Genus *Pomovirus*)

This virus can be detected by ELISA with some difficulty depending on the specificity of antisera. Mechanical inoculation on *C. amaranticolor* elicited the development of concentric local lesions, and killed young test plants.

10.7 Conclusions

For diagnostic studies, sensitive assays in which a large number of samples can be tested quickly, are required. Serological or nucleic acid-based tests are employed for detection of well-characterized viruses which are known to occur in the crop, but tests that detect a wide range of viruses are also desirable. To date most such tests make use of ELISA and dot blot assays, since these are suited to handling large number of samples in a short time. A standardized international test protocol should be introduced to ensure consistent results between laboratories. Standardization parameters should include plates and antiserum type, part of plant to be sampled, buffers and reagents used, temperature and length of each incubation step, positive and negative controls.

Nucleic acid-based tests also work well in these situations, but many laboratories are not equipped to work with radiolabeled probes. However, as nonradioactive probes become more accepted, dot blot assays for nucleic acids will become more commonly used. A positive diagnosis should involve the application of two or more of the available techniques. Nevertheless, biological assays on sensitive indicator plants will remain vital for the diagnosis of plant virus diseases.

Acknowledgement

Contribution No. 532/1999 series from the Agricultural Research Organization, The Volcani Center, Bet Dagan, Israel.

References

Banttari, E.E. and Goodwin, P.H. 1985. Detection of potato viruses S, X, Y by enzyme-linked immunosorbent assay on nitrocellulose membranes (dELISA). *Plant Dis.* 69:202-205

Barker, I. 1995. Detection of PVY strains in potato by using a simplified method of PCR. *Potato Res.*38: 427

Bravo-Almonacid, F., Hain, L. and Mentaberry, A. 1992. Rapid immunological detection of potato viruses in plant tissue squashes. *Plant Dis.* 76:574-578.

Caroll, R.B., Goldfine, S.M. and Melero, J.A. 1978. Antiserum to polyacrylamide gel-purified simian virus 40 antigen. *Virology* 87: 194-198.

Clark, M.F. and Adams, A.N. 1977. Characteristics of the microplate method of enzyme-linked immunosorbent assay (ELISA) for the detection of plant viruses. *J. Gen. Virol.* 34:475-483.

Converse, R.H. and Martin, R.R. 1990. ELISA for plant viruses Pages 179-196. In: Serological Methods for Detection and Identification of Viral and Bacterial Plant Pathogens. R. Hampton, E. Ball and S. DeBoer, eds. APS Press, St. Paul. MN.

Derrick, K.S. 1973. Quantitative assay for plant viruses using serologically specific electron microscopy. *Virology* 56:652-658.

Gera, A., Cohen, J., Kritzman, A., Beckelman, E. and Bar-Joseph, M. (1999). Production of diagnostic antibodies to plant viruses utilizing denatured coat proteins and booster immunization with partially purified virus. *J. Plant Pathol.* 3 (In press).

Glias, L., Kerlan, C., Tribodet, M., Tordo, V. M. J., Robaglia, C. and Astier-Manifacier, S. 1996. Molecular characterization of potato virus Y^N isolated by PCR-RFLP. *Europ. J. Plant Pathol.* 102:655-662.

Hampton, R., Ball, E. and DeBoer, S. eds. 1990. Serological Methods for Detection and Identification of Viral and Bacterial Plant Pathogens. APS Press, St. Paul.

Henson, J.M. and French, R. 1993. The polymerase chain reaction and plant disease diagnosis. *Annu. Rev. Phytopathol.* 31:81-109.

Heuvel, van den J.F.J.M., R.A.A., Van der Vlugt, M., Verbeek, O.T. De Haan and H. Huttinga 1994. Characteristics of resistance-breaking isolate of potato virus Y causing potato tuber necrotic ringspot disease. *European J. Plant Pathol.*100: 347-356.

Hill, S. A. 1984. Electron microscopy. Pages 130-153. In: Methods in Plant Virology. T. F. Preece, ed. Blackwell Scientific Publication, Oxford.

Hull, R. 1993. Nucleic acid hybridization procedures. Pages: 253-272. In: Diagnosis of Plant Virus Diseases. ed. Mathews, R.E.F. CRC Press, Boca Raton, FL.

Lin, N.S., Hsu, Y.H. and Hsu, H.T. 1990. Immunological detection of viruses and mycoplasma-like organisms by direct tissue blotting on nitrocellulose membranes. *Phytopathology* 80:824-828.

Loebenstein, G., Akad, F., Filatov, V., Sadvakasova, G., Manadilova, A., Bakelman, H., Teverovsky, E., Lachman, O. and David, A. 1997. Improved detection of potato leafroll luteovirus in leaves and tubers with a digoxigenin-labeled cRNA probe. *Plant Dis.* 81: 489-491.

Mathews, R.E.F. 1993. Diagnosis of Plant Virus Diseases. CRC Press, Boca Raton, FL.

Miller, S.A. and Martin, R.R. 1988. Molecular diagnosis of plant diseases. *Annu. Rev. Phytopathol.* 26:409-432.

Milne, R.G. 1992. Immunoelectron microscopy of plant viruses and mycoplasmas. *Adv. Dis. Vector Res.* 9: 283-297.

Navot, N., Ber, R. and Czosnek, H. 1989. Rapid detection of tomato yellow leaf curl virus in squashes of plants and insect vectors. *Phytopathology* 79:562-568.

Nolasco, G., de Blas, C., Torres, V. and Ponz, F. 1993. A method combining immunocapture and PCR amplification in a microtiter plate for the detection of plant viruses and subviral pathogens. *J. Virol. Meth.* 45:201-218.

Podleckis, E. V., Hammond, R. W., Hurt, S. and Hadidi, A. 1993. Chemiluminescent detection of potato and pome fruit viroids by digoxigenin-labeled dot blot and tissue blot hybridization. *J. Virol. Meth.* 43:147-158.

Robinson, D.J. and Romero, J. 1991. Sensitivity and specificity of nucleic acid probes for potato leafroll luteovirus detection. *J. Virol. Meth.* 34:209-214.

Rosner, A. and Maslenin, L. 1999. Differentiating PVY^{NTN} by unique single-restriction cleavage PCR products. *Potato Res.* (In press)

Salazar, L.F., Balbo, I. and Owens, R.A.. 1988. Comparison of four radioactive probes for the diagnosis of potato spindle tuber viroid by nucleic acid hybridiazation. *Potato Res.* 31:431-442.

Sano, T., Smith, C.L. and Cantor, C.R. 1992. Immuno-PCR: Very sensitive antigen detection by means of specific antibody-DNA conjugates. *Science* 258:120-122.

Schoen, C. D., Knorr, D., and Leone, G. 1996. Detection of potato leafroll virus in dormant potato tubers by immunocapture and a fluorogenic 5' nuclease RT-PCR assay. *Phytopathology* 86:993-999.

Shamloul, A. M., Hadidi, A., Zhu, S-F., Singh, R. P. and Sagredo, B., 1997. Sensitive detection of potato spindle tuber viroid using RT-PCR and identification of a viroid variant naturally infecting pepino plants. *Can. J. Plant Pathol.* 19:89-98

Singh, M. 1994. A fast reacting bioassay for the tobacco veinal necrosis strain of potato virus Y (PVY^{N}) *Plant Dis.* 78: 775-778

Singh, M. and Singh, R. P. 1996. Factors affecting detection of PVY in dormant tubers by reverse-transcription polymerase chain reaction. *J. Virol. Methods* 60:47-57

Smith, T.D. and Bantarri, E.E. 1987. Dot-ELISA on nitrocellulose membranes for detection of potato leafroll virus. *Plant Dis.* 71:795-799.

Speigel, S., and Martin, R. R. 1993. Improved detection of potato leafroll virus in dormant potato tubers and microtubers by the polymerase chain reaction and ELISA. *Ann. Appl. Biol.* 122:493:500.

Toth, R.L., Harper, K., Mayo, M.A. and Torrance. 1999. Fusion proteins of single-chain variable fragments derived from phage display libraries are effective reagents for routine diagnosis of potato leafroll virus infection in potato. *Phytopathology* 89:1015-1021.

Vunsh, R., Rosner, A. and Stein, A. 1990. The use of polymerase chain reaction (PCR) for the detection of bean yellow mosaic virus in gladiolus. *Ann. Appl. Biol.* 117:561:569.

Wetzel, T., Candresse, T., Ravelonandro, M. and Dunez, J. 1992. A highly sensitive immunocapture polymerase chain reaction method for plum pox potyvirus detection. *J. Virol. Methods* 39:27-37.

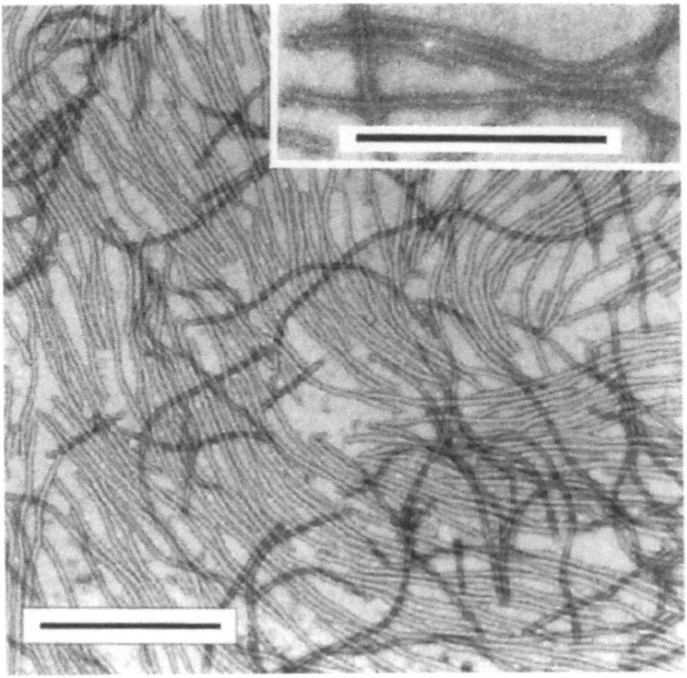

Fig. 10.1. Electron micrograph of purified particles of *Potato virus Y* (PVY) stained with uranyl acetate. Insert (right top) after decoration with polyclonal antibodies (diluted 1:50). Bars= 500 nm.

11. Isolation and Purification

P. E. Thomas[1] and W.K. Kaniewski[2]
[1]*Research Plant Pathologist, Vegetable and Forage Crop Production, Agricultural Research Service, U.S. Department of Agriculture, 24106 N. Bunn Road, Prosser, WA 99350, USA.*
[2]*Scientist, Monsanto Co.,700 Chesterfield Village Parkway, St. Louis, MO 63198, USA.*

11.1 Introduction

The isolation and purification of plant viruses are necessary for the initial identification and characterization of viruses, for the production of antisera, and for a multitude of other studies. The terms "isolation" and "purification" often are used interchangeably by virologists to describe separation of virus particles in a pure form, free from plant host constituents. In this chapter the term "purification" will be used to describe this activity. The term "isolation" will be used to describe the pure culture isolation of a single virus from a mixture of co-infecting viruses.

Potato (*Solanum tuberosum*) is a host of 54 viruses (Brunt *et al.*1996 onwards; Dallwitz, 1980; Dallwitz *et al.* 1993) representing 21 different genera or groups. There are 11 nepo-, 7 poty-, 6 gemini-, 4 luteo-, 4 carla-, 4 potex-, 3 tymo- and 2 tobamoviruses, and 13 individual viruses, each representing a different group. Of these viruses, 37 have been found infecting potato in the field (See Table 8.1.1 of Chapter 8.1). This work will be restricted to 23 of those viruses and one viroid (Table 11.1). Procedures for purification of all except two of these viruses, *Potato deforming mosaic* and *Solanum apical leaf curl,* have been published (Table 11.1). We will

G. Loebenstein et al. (eds.),
Virus and Virus-like Diseases of Potatoes and Production of Seed-Potatoes, 285–310.

present general methods for isolation and purification of plant viruses with some comments specific to potato viruses followed by our procedures for purifying the most common potato viruses.

11.2 Virus Isolation

Because viruses that infect potato are passed from one vegetative generation to the next in the tubers used for propagation, potato clones tend to accumulate viruses and often are infected by complexes of several viruses. Historically, all popular potato clones were infected with mild strains of at least two viruses, *Potato virus S* (PVS) and *Potato virus X* (PVX). Multiple infection is a problem because purification and characterization of a virus normally require its availability in a pure, homogeneous culture. Fortunately, most viruses have been eliminated from seed stocks of popular potato cultivars in recent years using a combination of therapeutic and meristem culture methods. This has generally simplified the problem of pure culture isolation, but the possibility of mixed infection continues to exist, particularly in areas of the world where advanced seed certification systems are not in use.

There are numerous approaches to the isolation of a particular virus from a mixture. All depend upon physical separation of the viruses based on differences in their key properties. Viruses that are very different from each other are easy to isolate, while isolation of closely related viruses or strains of a specific virus from a mixture may present a substantial challenge.

Differences in modes of transmission may be used to separate specific viruses from mixed infections. Modes of potato virus transmission are presented in Table 11.1. Most potato viruses are mechanically transmissible using buffered sap as inoculum, but three (BCTV, PLRV, and PYVV) are dependent on specific biological vectors. These viruses are readily separated from those that are transmitted mainly or entirely by contact and from others that require a different vector by employing the appropriate vector. Viruses that are transmitted by the same insect vector but in a different mode (persistent or non-persistent) are easily separated by manipulation of acquisition and transmission access periods so as to exclude either persistent or non-persistent transmission. Short access periods preclude both acquisition and transmission of persistent viruses, such as PLRV, but favour transmission of non-persistent viruses, such as PVY. The potential for transmission of non-persistent viruses by aphids that have also acquired a persistent virus may be eliminated by restricting aphids on a healthy plant for an hour or so after the acquisition feeding period. Subsequently, the aphids are transferred to hosts of the persistent virus for transmission.

Differences in host range and in systemic movement of viruses in the same host are often used to isolate viruses. The host ranges and systemic distribution patterns for mechanically transmissible viruses on key diagnostic hosts are summarised in Table 11.2. The best strategy for isolation of viruses based on host reactions is to find a host plant susceptible to only one virus of a mixture. Another is to find a host that permits systemic movement of one virus but restricts others to the site of inoculation. Thus, AMV may be isolated from many other potato viruses by its systemic infection of *Gomphrena globosa.* Another strategy is to isolate viruses in local lesions. Thus, *Gomphrena globosa* may individually isolate PVM, PVS, PVX, and TNV from a mixture of all four viruses in different local lesions on the same plant. The potato clones, Saco and Seedling 41956 are particularly useful for elimination of PVS and PVX from mixtures of potato viruses. Highly virus resistant transgenic potato clones may also be used to eliminate the specific viruses to which they are resistant. It is sometimes necessary to use a series of transmissions or to use mode of transmission in combination with host range and systemic distribution to isolate specific viruses. It is difficult to separate different strains of a virus from hosts that are systemically infected by all strains, but individual strains may be isolated in specific local lesions.

Differential stability of viruses in sap or in specific buffers may also be used to separate some viruses, though complete separation is only effective if differences in stability are substantial. Density gradient centrifugation, as a final step in purification, may separate virus mixtures, but complete separation may not be achieved even when two viruses sediment in distinctly separate zones.

Successful isolation of a virus should be confirmed by several passages through a host plant, preferably a specific indicator plant to test homogeneity of the culture.

11.3 Virus Purification

General strategies for purification of potato viruses do not differ from those used for other common plant viruses. Most potato viruses have been purified and citations to published purification procedures and useful virus propagation hosts are given in Table 11.1. A review of the literature shows a striking variability in the methods that are successful in different laboratories. However, the art and science of virus purification have advanced considerably since the earlier works cited in Table 11.1, and modern methods have become considerably more uniform. We will present general guidelines and some specific purification procedures for the major potato viruses.

Table 11.1. Summary of potato virus information pertinent for isolation and purification

Virus Name	Acronym	Rate[a]	Biological Trans Mode[b]	Sap Trans[c]	Purification methods[d]	Prop Host[e]	Extinc. Coef.[f]	Ratio A260/A280[g]
Alfalfa Mosaic	AMV	1	Aphid, ts	X	31	N.tabacum	4,9	1,7
Andean Potato Latent	APLV	3	Beetle, ts	X	6, 8, 9	N. bigelovii N. clevelandii		1,5
Andean Potato Mottle	APMV	3	Beetle	X	3, 32	N benthamiana N. clevelandi	m6.2 - b10	m1.6 - b1.74
Beet Curly Top	BCTV	1	Leafhopper	N	20, 16	Phaseolus vulgaris C.bursa-pastoris	6,3	1,46
Potato Acuba Mosaic	PAMV	2	Aphid+hv	X	13	N. tabacum	2,2	1,2
Potato Leafroll	PLRV	5	Aphid	0	27	D.stram. var.tatula, Physalis floridana	6,55	1,8
Potato Mop-Top	PMTV	4	Fungus	X	14, 15	N. benthamiana N. debnei	3,1	1,2
Potato A	PVA	3	Aphid	X	5	N. tabacum	2,35	1,21
Potato M	PVM	3	Aphid	X	22, 23	S. tuber. cv.Saco L. esculentum	1,77	1,17
Potato S	PVS	2	Aphid	X	22, 34	N. clevelandi	1,77	1,17
Potato T	PVT		Pollen, ts	X	25	C. quinoa	1,8	1,18
Potato V	PVV	3	Aphid	X	7	N. clevelandii N. glutinosa	2,35	1,21
Potato X	PVX	2	None	X	4	N. tabacum	2,2	1,2
Potato Y	PVY	5	Aphid	X	1, 2, 12, 17, 21	N. tabacum	2,35	1,21
Potato Spindle Tuber	PSTV		ts	X	23	L.esculentum		
Potato Yellow Dwarf	PYDV	3	Leafhopper	X	11	N. rustica		
Solanum Apical Leaf Curling	SALCV	?	?	0		L.esculentum	6,3	1,46
Tobacco Necrosis	TNV	1	Fungus	X	24	N. tabacum	5,2	1,7
Tobacco Rattle	TRV	4	Nematode, ts	X	10, 18	N. clevelandii	3	1,15
Tobacco Ringspot	TRSV	2	Nematode, ts	X	26, 29	Cucumis sativus N. tabacum	b10	b1.84
Tomato Black Ring	TBRV	1	Nematode, ts	X	19	Petunia hybrida N. clevelandii	b10	b1.84

Tomato Spotted Wilt	TSWV	2	Thrip	X	28, 30	N. glutinosa		

[a]Rating of importance as a potato disease in areas where the virus is present from 1 = not important to 5 = very important.
[b]Abreviations: hv = helper virus required, ts = transmitted in true seed
[c]Symbols indicate the following: X = sap transmissible, **X** mainly or entirely sap transmitted, 0 = no sap transmission,
N = transmitted by needle puncture, ? = unknown
[d]Citations to purification methods
[e]Genus abbreviations: C. = Chenopodium, Cap. = Capsella, L. = Lycopersicon, N. = Nicotiana, S. = Solanum, ? = unknown
[f]Extinction coeffiecient = Absorbance at 260 of 1mg of virus/ml.
[g]Characteristic ratio of light absorbance at A260/A280. Middle (m) and bottom (b) components are given for multipartite viruses

Purification of a plant virus typically involves a number of overlapping, but fairly clearly defined steps. These include production of the virus in a selected propagation host, trituration of the virus-infected tissue in an extraction medium, removal of most plant pigments and other plant constituents in a process loosely referred to as clarification, concentration of the virus with further removal of plant constituents, and final steps in purification which usually involves density gradient centrifugation.

11.3.1 Propagation Host

The capacity of a host to generate high concentrations of the virus is a critical factor in selection of a propagation host for virus purification. Virus concentration in a host may be markedly affected by the specific isolate of the virus chosen for purification (Hassan and Thomas, 1989). It may also be affected by the time after inoculation when tissue is harvested and the conditions under which the host is grown. Optimum concentration in the propagation host is especially important for viruses, such as PLRV and BCTV, that are tissue-specific and tend to be very low in concentration under the best of circumstances.

A second important but less easily measured factor in selection of a propagation host is the degree to which the host produces substances that interfere with purification or promote inactivation, disintegration, or precipitation of the virus *in vitro*. These substances include phenolic compounds, certain proteins, ribonucleases and other enzymes. A few hosts have been identified that contain a minimum of such compounds and are distinctly superior to others for general virus purification (Matthews, 1991). In our experience, *Datura stramonium* var. *tatula* is a superior host for purification of PLRV and possibly other potato viruses. Virus preparations purified from *D. tatula* are unusually free of host constituents as compared with those from other hosts, and antisera prepared with such preparations produce a much lower non-specific background reaction. An antiserum prepared with virus purified from *D. tatula* was the only one among a collection of others purified from other hosts that did not show specificity for stress proteins in potatoes (Gunn and Pares, 1988).

11.3.2 Virus Stability in the Extraction Medium

Many factors affect virus stability at various stages in the purification process. It may be necessary to monitor viral stability during purification to determine the stage at which a virus loses stability if it is one that is not readily purified by standard procedures. The use of a local lesion host (Table 11.2) is a good way to monitor degree of infectivity. When an antiserum is already available, serological procedures are often most convenient.

The major problem in virus purification is to create and maintain an environment that protects the virus particles from destabilizing agents in the plant sap, maintains their physical integrity and infectivity, and prevents their aggregation during their separation from host constituents. This always involves the trituration of virus-source tissue in a buffered extraction medium that provides the appropriate environment. The volume of extraction medium used is typically one or two times the volume of tissue processed. Increasing the volume may reduce the ionic strength of buffering salts required and usually increases the efficiency of virus extraction. Constraints imposed by centrifuge rotors and processing equipment often limit the volume used. A variety of methods have been used to triturate the source tissue in the extraction medium. Thorough grinding and immediate contact between cell constituents and the protective influence of the extraction medium is desirable. We prefer to use a high speed, high capacity blender with sharp blades, but different methods may be chosen for convenience and effectiveness.

It is usually important to minimize exposure of virus particle to plant sap by processing the tissue as rapidly as possible. Freezing the tissue prior to extraction and processing the tissue at low temperature (0-4°C) reduces the activity of inactivating agents in the plant sap and facilitates extraction of most viruses, but low temperature reduces the purified yields of some other viruses. Freezing or storing PLRV-infected tissues at 4°C overnight or cooling tissue homogenates below 15°C during the early stages of extraction markedly reduces virus yield.

Stability of virus particles in the extraction medium may be affected by pH, the specific buffering salt employed, the ionic strength, and the presence of specific ions. Common buffering salts include potassium or sodium phosphates, borates, acetates or citrates. The effects of these variables on virus stability may be assessed prior to purification. Aggregation of virions is a particular problem of potyviruses affected by these factors. *Turnip mosaic virus* aggregated in low molarity buffers of phosphate and other common buffer salts except borate (Shepherd and Pound, 1960), while PVA aggregated even in borate (Fribourg and de Zoeten, 1970). Some viruses require the presence of divalent metal ions, Ca^{2+} or Mg^{2+} (Brakke, 1963). Others are precipitated or degraded by divalent ions. EDTA may be included in the extraction medium to minimize aggregation by divalent ions. At 0.01M in pH 7.4 buffer, EDTA causes disruption of ribosomes that may co-purify with the virus, but it can be used for this purpose only for viruses that do not require divalent metal ions for stability.

Stability of viruses is often affected when their coat proteins undergo chemical alteration by interaction with quinones as leaf phenols are oxidized in the extraction medium (Pierpoint *et al.* 1977). Many viruses are stabilized

Table 11.2. Some host range options useful for pure culture isolation of potato viruses

Plant Hosts	Host Reactions[a] to Plant Viruses[b]																	
	AMV	APLV	APMV	PAMV	PMTV	PVA	PVM	PVS	PVT	PVV	PVX	PVY	PYDV	TNV	TRV	TRSV	TBRV	TSWV
Chenopodium quinoa	Ll & Sc/nf	Ll & Sm	–	?	+	?	Ll	Ll	Sm & Tn	–	Ll	Ll	?	Ll	Ll	Ll	Ll & Sm/n	?
Datura tatula	–	+	+	A	+	–	+	?	Sm	–	Sm& Srs	–	?	Ll	–	+	+	+
Datura metal	?	?	?	?	?	?	Ll & Sn	Ll	+	+	?	?	?	?	–	+	?	?
Gomphrena globosa	Ll & Sn	+	–	Ll & Sn	–	–	Ll	Ll	+	–	Ll	–	+	Ll	+	+	+	S
Nicandra physaloides	–	+	+	Sp	+	Sm	–	–	+	?	Sm& Srs	Ll	?	?	+	+	+	+
Nicotiana tabacum	Ll & Sp	?	+	A	+	Sm	–	–	+	+	Sm& Srs	Sm	+	Ll	Ll & Snlp	Ll & Srs	Ll & Sm/n	Ll & Slp
Nicotiana debneyi	+	Sm	+	Sm	Ll & Sp	–	Ll	Svc	SM & N	Ll & Scl	Sm& Srs	Sm	?	?	+	+	?	–
Nicotiana glutinosa	Ll & Sp	+	Sm	Sm & vb	+	Sm	–	–	+	Sm	Sm& Srs	Sm	Ll & Svc	?	Ll & Snlp	Ll & Srs	–	Ll & Slp
Lycopersicon chilense	?	?	+	?	?	?	+	?	?	?	Sm	Sm	?	?	?	?	?	?
Lycopersicon esculentum	+	+	+	Ll & Sm	+	–	A	–	?	+	Sm	Sm	?	Ll	+	+	Ll & Sm/n	+
Solanum rostratum	?	?	?	Lyl	?	?	Ll & Sn	Ll & Sn	+	?	Sm	?	?	?	?	+	+	+
Solanum tuberosum var. Saco	Sym	+	+	+	+	Sm	+	–	+	–	–	Sm	+	?	Ll	+	+	+
Solanum tuberosum var. Seedling 41956	Sym	+	+	+	+	Sm	+	+	+	+	–	Sm	+	?	Ll	+	+	+
P. floridana	+	–	+	Sm	+	?	?	?	+	+	Sm	Sm	?	?	–	+	+	?

Tinantia erecta				–			–	–										
L. pimpinellifolium	?	+	+	Sm	?	+	+	?	?	?	Sm	?	?	?	+	?	–	+
N. clevelandii	+	Ll & Sm	Sm	?	+	Sm	?	+	+	+	Sm	?	?	?	Ll & Sm	Ll & Srs	Ll & Sm/n	Ll & Slp
N. megalosiphon	+	Sm	+	?	?	Sm	?	?	?	?	Sm	?	?	?	Ll & Sm	?	?	?
N. occidentalis	?	?	+	?	?	?	+	?	?	+	Sm	?	?	?	Ll & Snlp	Ll & Srs	?	?

[a]Host reactions: + = susceptible, - = Insusceptible, ? = host reaction unknown to authors. Capital letters indicate distribution, (L = local, S = systemic, T = tip); lower case letters indicate symptom type,(a = aysmptomatic, b = banding, c = clearing, f = flecks, l = lesion, lp = line pattern, m = mosaic or mottle, n = necrosis, p = pattern, r = ring, s = spot, v = vein, y = yellow).

[b]Refer to Table 11.2, to associate names with the virus acronyms given here.

by inhibiting phenol oxidation. Oxidation may be inhibited by adding reducing agents including sodium sulfite (Best and Samuels, 1936), sodium cyanide (Fulton, 1957; Hampton and Fulton, 1961), polythiourea (Hampton and Fulton, 1961), 2-mercaptoethanol (Fulton, 1966; Hougas, 1951), ascorbic acid (Best and Samuels, 1936), dithiothreitol (Paulsen and Fulton, 1968), and thioglycolic acid (Tomlinson, 1964) Copper chelating agents such as diethyldithiocarbamate (DIECA) and potassium ethyl xanthate may also inhibit phenol oxidation by inhibiting polyphenoloxidase (Hampton and Fulton, 1959; Hampton and Fulton, 1961; Harrison and Pierpoint, 1963). Caffeine (Diener and Weaver, 1959), alumina (Cadman, 1959) and polyvinyl pyrrolidone (PVP) minimize reactions of virus coat proteins with tannins. Bentonite and charcoal (Dunn and Hitchborn, 1965) may be added to the extraction buffer to remove ribonucleases, pigments, and other host constituents.

11.3.3 Release of Virus Particles from Plant Constituents

A number of enzymes may be used in extraction buffers to degrade various plant constituents and to release virus particles into the medium. Pectinase is used to degrade mucilage (Adomako *et al.* 1983), and the proteolytic enzyme trypsin was employed for purification of potyviruses (Thompson *et al.* 1988). Degradation of fibrous phloem tissues with pectinase-cellulase enzymes is essential for efficient purification of some phloem-restricted viruses such as PLRV (Singh et al. 1984; Hassan and Thomas, 1989;Takanami and Kubo, 1979).

Detergents, Triton X-100 and Tween 80, are sometimes added to the extraction medium to assist in release of virus particles from insoluble cell components, dissociate cellular membranes, assist in initial clarification of the tissue extract, and may assist in dissolution of high speed centrifuge pellets.

11.3.4 Clarification and Removal of Major Plant Constituents

After tissue trituration, most plant pigments and the major portion of host plant constituents are removed from the extraction medium in a step called clarification. A variety of procedures have been developed for this. The most common involves partitioning the components in a two-phase system. The aqueous tissue extract is emulsified with a number of different organic solvents, including include ether, carbon tetrachloride, chloroform (Schneider, 1953), n-butanol (Bachrach and Schwerdt, 1952), and a combinations of chloroform and n-butanol (Steer, 1956). The triturated tissue homogenate may be expressed through cheesecloth or muslin to remove large, insoluble host constituents prior to emulsification. We often eliminate that step and emulsify the tissue extract with the organic solvent immediately after trituration in the same blender. As the emulsion is broken by low-speed centrifugation, an aqueous phase containing the virus separates at the top, the solvent containing most plant pigments separates at the bottom, and a cake of coagulated plant constituents forms as an interphase. The virus-containing aqueous phase at the top is collected by decanting it from the sturdy interphase.

Other methods used to coagulate and precipitate plant constituents in the clarification procedure include heating the tissue extract to 50-60°C for a few minutes, freezing tissue and/or homogenate, and reducing pH to 5 or below. Some highly labile viruses have been purified by trapping plant constituents in hydrated calcium phosphate (Fulton, 1959). Earlier purification procedures were more dependent than modern methods on repeated cycles of high and low speed centrifugation for clarification of the extraction medium.

11.3.5 Concentration and Further Purification

Following clarification, the virus remains suspended in a large quantity of extraction medium. Some purification protocols depend on repeated cycles of high and low speed centrifugation to concentrate and further purify the virus. This approach eliminates low molecular weight materials that do not pellet with the virus at high speed. Plant constituents that pellet at high speed but remain insoluble are eliminated as pelleted material during the subsequent low-speed centrifugation, which leaves virus in suspension.

A more efficient approach is to precipitate the virus from the clarified extraction medium and follow this by high and low speed centrifugation to remove additional plant constituents that are not readily removed by precipitation.

Some viruses may precipitate at their isoelectric point under low speed centrifugation. The appropriate pH for sedimentation may be determined empirically by monitoring the virus content in the supernatant and pellet after low speed centrifugation over a range of pH values (Hassan and Thomas, 1989). Viruses may also be precipitated from a 50% saturated solution of ammonium sulfate (30.4 g of ammonium sulfate dissolved per 100 ml of clarified extract). Many plant proteins co-precipitate with the virus by this procedure. An improved method developed by Hebert (1963) precipitates virus particles in a single phase polyethylene glycol (PEG) system. Some host DNA co-precipitates with the virus, but most plant constituents remain suspended and are eliminated. Host DNA may be remove in the subsequent high speed sedimentation cycle. The exact concentration of PEG required varies depending on the virus, ionic strength of the medium, and other factors. Typically, PEG is dissolved in the clarified extract to 4% (w/v) to precipitate filamentous viruses and to 8% to precipitate spherical viruses. Sodium chloride is also added to at least 0.3 % (w/v) for either filamentous or spherical viruses. Precipitated virus pellets are collected by low speed centrifugation and dissolved in an appropriate buffer. The amount of buffer used is typically a fraction of the original extraction buffer, and its ionic strength may be reduced. PEG precipitation may be used with many viruses, including some that are quite fragile. It is now used in the purification of most potato viruses.

Virus may be lost in repeated cycles of high and low speed centrifugation either by partial dissolution of the pelleted virus before the supernatant can be removed or by aggregation of the virus particles in the high speed pellets. The addition of non-ionic detergent to the dissolution buffer may assist dissolution of aggregated pellets if binding with plant membranes is involved. The addition of a 0.5 ml cushion of 40% sucrose at the bottom of high speed tubes may decrease virus compaction and can result in better virus recovery from pellets.

11.3.6 Final Steps in Purification

Density gradient centrifugation (Brakke, 1960) is often applied as a final stage in purification. In the most common method, called rate zonal centrifugation, a small volume (1-2 ml) of concentrated, partially purified virus suspension is layered on the top of a preformed, buffered, sucrose gradient in a centrifuge tube. Sucrose concentration increases from 10 % (w/v) at the top to 40% at the bottom of the tube. The sucrose gradient remains relatively stable during centrifugation, and particles in the virus sample sediment through the gradient at different rates determined by their size, shape, and density. Groups of uniform particles sediment as zones and are separated from other groups of uniform particles based on rate of sedimentation. Centrifugation is stopped while the particles are still sedimenting. Rate zonal centrifugation eliminates particles from the virus preparation that do not sediment at the same rate as the virus. It may also give some indication of the purity of the preparation, allow correlation between particles and infectivity, reveal the presence of nonviral particles and multipartite viruses, and may be used to determine sedimentation coefficients. Sucrose is the most commonly used material to form rate zonal gradients, but glycerol or non-ionic media such as Ficol or Nycodenz can be also used.

Virus purification may involve isolation in isopycnic gradients (Brakke, 1960) either as an alternative to rate zonal centrifugation or as an additional step to achieve greater purity of the virus or to determine its buoyant density. The combination of both types of gradients always produces preparations of high purity. In isopycnic centrifugation, particulate components of the sample accumulate in areas of the gradient where density of the gradient is equal to that of the particle. The most common ingredients used to form isopycnic gradients for virus purification are cesium chloride or cesium sulfate, but other salts may also be used. Solutions of appropriate ionic strength may be prepared by dissolving the gradient salt directly in the virus sample. Alternatively, a preformed gradient may be prepared. In this case, the virus sample is layered on the top of the gradient. In contrast to sucrose gradients that remain stable during rate zonal centrifugation, the salt solutions used in isopycnic centrifugation form gradients while under centrifugal force. As force is increased, the density distribution of the gradient is increased. As the gradients form, particles of the sample migrate either up or down to areas of the gradient equal to their own densities. As the density distribution of the gradient increases with increasing rotor speed, the degree of separation between particles of different density decreases. Thus, for maximum separation of virus from contaminants, it is best to run at lower speeds. The buoyant density of the virus may be determined by measuring the refractive index of the cesium salt solution in the zone of the gradient that contains the virus with a refractometer. The refractive index reading is converted to density (g/cm^3) using a reference table. As gradients form during isopycnic centrifugation, salt concentrations at the bottom of the tube may reach saturation levels if rotor speeds and starting ionic concentrations are not appropriate. The changes in density distribution in the gradients as salts crystallize out of solution may cause rotor failure. Explicit directions from rotor manufacturers should always be followed when using isopycnic centrifugation.

Virus stability may be a problem in isopycnic centrifugation. A virus that is not stable in one isopycnic salt solution is often stable in another (Hull, 1976). The stability and buoyant density of viruses in strong cesium chloride solutions may depend on the pH and on other ions present (Matthews, 1974). In our experience, PLRV disintegrates when cesium chloride is dissolved directly in the virus sample. It remains stable when layered at the top of a partially preformed gradient prior to centrifugation.

The gradient ingredients present in virus zones collected from density gradients may be removed by dialysis or by high speed sedimentation from diluted samples. Virus buffer concentrations should decrease with increased purity of the virus since viruses are generally more stable at low ionic strength. Purified viruses may be stored as precipitates in ammonium sulfate, frozen in buffer, or stored at low temperature mixed with glycerol to prevent solidification depending on the characteristics of the virus.

Some additional purification procedures may be applied when simple methods are not successful. These may include gel filtration, electrophoretic, and various chromatographic methods. Immunoaffinity column chromatography may also be used in final steps of purification (Ronald *et al.* 1986).

11.4 Assessment of Virus Purity

The methods used in the final steps of purification will depend on the level of purity required for the intended application. Exceptional purity is required to characterize the chemical composition of a new virus. The presence of non-infectious virus particles might not constitute a major impurity for determination of chemical composition but would affect infectivity studies. For molecular applications, such as cloning and sequencing, homogeneity of the virus population is of highest importance. Mixtures of strains or presence of mutants could lead to cloning of a non prevalent virus isolate. For physicochemical studies, buffer composition and its pH might be of highest importance. Freedom from plant antigens is essential for antisera production to eliminate non-specific background reactions in serological assays.

Based on needs, various tests are applied to assess purity of purified virus preparations. The capacity of virus particles to crystallize is a measure of purity, but some viruses do not crystallize readily, and lack of crystallization for such viruses does not indicate impurity or lack of homogeneity. Conversely, successful crystallization may occur in the presence of impurities capable of co-crystallization.

The most common test for purity of a known virus preparation is based on UV spectral relationships. Nucleoproteins have specific A260/A280 ratio values depending on their content of nucleic acid and protein (Fig. 11.1) (Englander and Epstein, 1957; Warburg and Christian, 1941). Based on this relationship, either the A260/280 ratio or the nucleic acid content may be determined when the other value is accurately known (Fig.11.1). The actual A260/A280 ratio for a preparation of a known virus may be compared with empirical values published for the virus. A lower A260/A280 ratio than expected for the virus indicates that the preparation is contaminated with proteins or that it is a mixture of virus and empty capsids. A higher ratio than expected usually indicates contamination with nucleic acids, almost always RNA in contaminating ribosomes.

Many other techniques are used to detect specific types of impurities. Electron microscopy may identify impurities differing in size or shape from purified virus. Analytical ultracentrifugation can confirm the presence of a single component of appropriate sedimentation coefficient (Kaniewski, 1973). The proper value for the diffusion coefficient of a virus is an additional criterion for purity which can be measured using analytical ultracentrifugation (Kaniewski, 1974). Contaminants separated by density gradient centrifugation may be detected by absorbency zones in a density gradient as its contents is forced through a flow cell monitored spectrophotometrically. The sensitivity of detection in density gradients can be enhanced by using a gradient containing acrylamide. After centrifugation, the gradient is polymerized and stained for presence of non-viral proteins (Jolley *et al.* 1967). Gel electrophoresis separates many proteins that differ by size or charge (Kaniewski, 1973). Chemical analysis for nitrogen or phosphorus are rarely used, though more sophisticated analysis of amino acid composition or sequence and nucleotide composition or sequence may be useful.

11.5 Virus Quantitation and Nucleic Acid Content

The nucleic acid content of a new virus may be determined from its A260/A280 ratio (Fig. 11.1) or its molecular extinction coefficient (E = A260 of 1 mg/ml) (Fig. 11.2) (Englander and Epstein, 1957; Warburg and Christian, 1941). To determine the later, the A260 of a specific volume of virus preparation is measured. The concentration (C = mg/ml) of virus in a specific volume of the same virus sample is measured directly by weighing the lyophilized sample. Then E is calculated by the formula, E = A260/C. There are practical and theoretical limitations on the use of this formula (Englander and Epstein, 1957). The nucleic acid content of the unknown virus may be compared by both methods (Fig. 11.1 and Fig. 11.2).

Purified yields of different potato viruses very widely ranging from as much as 1000 mg/kg of leaf tissue for PVX to a maximum of 2 mg/kg for PLRV. Yields of known viruses are easily calculated from the extinction coefficient (Table 11.1) and an application of the formula, E = A260/C. The use of these methods to quantify viruses requires highly pure preparations since an accurate A260 determination is required. Serological methods, based on highly purified virus standards, are routinely used to determine virus content in partially purified samples and in crude plant extracts (Copeland, 1998). Standards are subject to degradation immediately after purification. We find that many potato viruses loose antigenicity very slowly when maintained in buffered, 50% glycerol at –20°C.

11.6 Purification Protocols

11.6.1 General Methods

Rate zonal sucrose gradients are prepared in the virus buffer specified in centrifuge tubes. A wide variety of rotors and tubes are available. We use Beckman rotor SW25.1 by layering 7, 7, 7, and 4 ml of 40, 30, 20, and 10 % (w/v) buffered sucrose solutions, respectively, from bottom to top of a centrifuge tube. The solutions are layered at the bottom of the tube with a 10 ml syringe. To allow diffusion between the zones of different sucrose concentration, the tubes are stored at 4°C at least 16 hr (overnight) and up to 40 hr prior to use. Alternatively, sucrose gradients may be prepared by thawing tubes that were filled with a uniform solution of 25% (w/v) sucrose prior to freezing. Uniform gradients form as the tubes thaw at room temperature.

Isopycnic centrifugation may be performed in any of a wide variety of rotors. Since the gradients are self generating, the gradient salts may be dissolved directly in virus suspensions or may be introduced as a concentrated solutions. We use Beckman rotor SW25.1 and prepare preformed gradients by layering 7, 7, 7, and 4 ml with 750, 650, 550, and 450 mg/ml of salt in virus buffer from the bottom to top of the gradient. About 4 ml of buffered virus suspension is layered on the top of the gradient. This creates a gradient with a mean cesium chloride concentration of 532.8 mg/ml and density of 1.402 g/cm^3. Buoyant densities of plant viruses generally fall in a range from 1.30 to 1.50 g/cm^3 depending on their nucleic acid content. The exact concentrations of salt used in the gradient may be adjusted so that the mean density matches the approximate buoyant density of the virus. The gradients are used immediately after preparation.

11.6.2 Purification of *Potato Leafroll Virus*

Grow *Datura stamonium* var. *tatula* plants under long day conditions. Inoculate when plants are about 10 cm tall using viruliferous *Myzus persicae*. Plants develop symptoms within 14 days and maintain high concentrations of virus for at least 6 wk. New leaves produced after the first harvest also contain high concentrations of virus.

To prevent virus aggregation, conduct all purification operations at or above 15°C until after rate zonal density gradient centrifugation.

Triturate freshly harvested leaves in two volumes (w/v) of 0.1 M citrate buffer, pH 6, containing EDTA (0.01M), DIECA (0.3% w/v), 2-mercaptoethanol (0.5% v/v) and either Driselase (0.2% w/w of tissue) or Rohament (1.5% w/w of tissue) in a high speed blender at room temperature.

Shake or stir gently at room temperature for at least 2.5 hr. Stirring may continue for 16 hr (overnight) with no reduction in virus yield. Add Triton X-100 to 1% (v/v), and continue stirring 30 min.

Add a 1:1 mixture of n-butanol-chloroform to 15% (v/v), emulsify in a high speed blender for 30 sec., and centrifuge at low speed (5000 r.p.m. in Beckman rotor JA-10) for 10 min at 15°C. Decant and process the upper aqueous phase.

Add PEG to 8% and NaCl to 1% (w/v) of the aqueous phase, stir gently at room temperature for 1 hr, and centrifuge at low speed for 20 min. Keep the pellets.

Shake or stir the PEG pellets in 25% volume (v/w of original tissue) of citrate buffer (0.1 M, pH 6.4) containing EDTA (0.01M) for 2 hr at room temperature. Stirring may continue overnight.

Centrifuge the dissolved pellets at low speed (10,000 **g**, 15 min, 15°C) and transfer the supernatant to a clean tube.

Sediment the virus by high speed centrifugation at 15°C for 2 h, 48 – 60,000 **g** (R_{min}). (We use Beckman rotor 30 at 29,000 r.p.m. or rotor 45 Ti at 38,000 r.p.m.). Layer 1 ml of 25% (w/v) sucrose in virus buffer at the bottom of each tube prior to centrifugation. Resuspend pellets in 5% volume (v/w of original tissue) of citrate buffer (0.01 M, pH 6.4) containing EDTA (0.001M). Use a small Nylon tube brush to disrupt pellets. Combine pellet extracts, stir for at least 1 hr, then centrifuge at low speed (7000 **g** R_{min}) to remove insoluble particles.

Subject the virus suspension to a second round of high- and low-speed centrifugation reducing the resuspension volume to 0.5-1% volume (v/w of original tissue).

Subject the virus suspension to rate zonal density gradient centrifugation in 10-40% (w/v) buffered (citrate buffer at 0.1 M, pH 6.4 containing 0.01 M EDTA). Layer *c* 2 ml of virus preparation on each gradient. Collect and combine virus fractions.

Subject the virus fractions from rate zonal density gradients to isopycnic density gradient centrifugation. Use preformed cesium chloride gradients (mean cesium chloride content = 532.8 mg/ml and density of 1.402 g/cm^3) prepared in citrate buffer at 0.1 M, pH 6.4 containing 0.01 M EDTA. Layer 4 ml of virus suspension collected from the rate zonal gradients on each cesium chloride tube. Centrifuge for 18 hr at 28,000 **g** at R_{min} (21,000 r.p.m. in Beckman rotor SW25.1) for 18 hr. Remove and combine the virus zone from each tube, dilute in buffer (citrate buffer at 0.1 M, pH 6.4 containing 0.01 M EDTA), and concentrate by high speed centrifugation as in Step No. 8.

11.6.3 Purification of *Potato A and Y Viruses*

Note. A number of different published purification methods were attempted including those cited in Table 11.1, and many variations on these procedures were tried. None was successful. The virus was always lost at some stage during the purification, presumably by aggregation. The following procedure was finally developed. A critical aspect of this procedure was the addition of 0.8 M urea in the buffering systems to prevent irreversible precipitation, presumably by aggregation, of the virus particles.

Use *Nicotiana tabacum* var. Xanthi-nc to propagate PVY. Use *N. megalosiphon* to propagation PVA. Grow plants in a heavily shaded greenhouse. Mechanically inoculate the young plants when *c.* 5 cm in height with buffered sap of an infected stock plant. Harvest symptomatic, systemically infected leaves at a stage when plants have produced about four to five new leaves about 14-20 days after inoculation. Cool the leaves to 4°C.

Triturate the leaves in a high speed blender with two volumes (v/w of leaf tissue) of 0.1 M, pH 7.4 potassium phosphate buffer containing 0.8 M urea, 0.01 M EDTA, 1% DIECA, and 1% 2-mercaptoethanol.

Add a cold 1:1 mixture of n-butanol:chloroform to 10% of the tissue plus buffer volume in the same blender, blend for an additional 30 sec, centrifuge the mixture at low speed (5000 r.p.m. in Beckman rotor JA-10) for 10 min at 4°C, decant and process the upper aqueous phase.

Add PEG to 4% and NaCl to 1% (w/v) of the aqueous phase, stir gently at room temperature for 1 hr, and centrifuge at low speed for 20 min. Keep the pellets.

Shake or stir the PEG pellets in 25% (v/w of original tissue) of 0.03 M, pH 7.4 potassium phosphate buffer containing 0.8 M urea, 0.01 M EDTA, 1% DIECA, and 1% 2-mercaptoethanol for 2 hr at 4°C.

Centrifuge the dissolved pellets at 29,000 r.p.m. in Beckman rotor 30 for 2 hr at 4°C. Layer 1 ml of 25% (w/v) sucrose in virus buffer at the bottom of each tube prior to centrifugation. Resuspend pellets in 1.5% (v/w of original tissue) of 0.03 M, pH 7.4 potassium phosphate buffer containing 0.8 M urea, 0.01 M EDTA, 1% DIECA, and 1% 2-mercaptoethanol. Use a small Nylon tube brush to disrupt pellets. Stir the pellet extraction for 1 hr. Centrifuge at low speed to remove insoluble particles.

Layer about 1 ml/gradient on 10-40% (w/v) sucrose gradients prepared in virus buffer in tubes for Beckman rotor SW25.1 and centrifuge 2 hr. Recover and combine the virus zones from the gradients.

Conduct isopycnic density gradient centrifugation as described above for purification of PLRV in gradients prepared in 0.03 M, pH 7.4 potassium phosphate buffer containing 0.8 M urea. Dilute the recovered virus zones in 0.03 M, pH 7.4 potassium phosphate buffer containing 0.8 M urea, and concentrate by high speed centrifugation.

Some isolates of PVA and PVY aggregate in cesium chloride gradients but not cesium sulfate. For isopycnic density gradient centrifugation in cesium sulfate, dissolve 7 g of cesium sulfate in the virus suspension and adjust the volume to 30 ml. Then centrifuge 18 hr at 28,000 g at R_{min} (21,000 r.p.m. in Beckman Rotor SW 25.1 or 30,000 r.p.m. in Beckman Rotor SW 55.

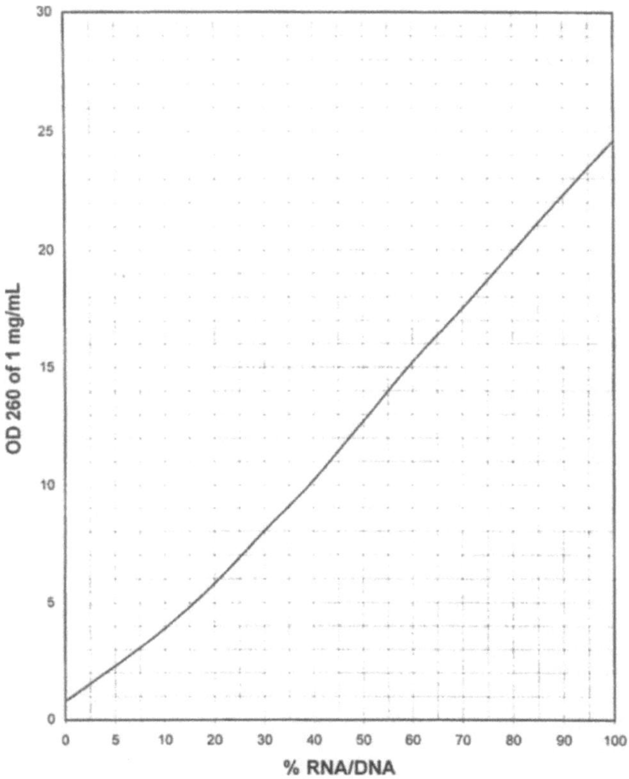

Fig. 11.1. Correlation between OD260/280 ratio of pure virus preparation and percent of nucleic acid in virions.

Citations in Fig 11.1:

(1) Damirdagh and Shepherd, 1970; (2) Delgado-Sanchez and Grogan, 1966; (3) Dunn and Hitchborn, 1965; (4) Francki and McLean, 1968; (5) Fribourg and de Zoeten 1970; (6) Fribourg *et al.*, 1977; (7) Fribourg and Nakashima, 1984; (8) Gibbs and Mackenzie, 1998; (9) Gibbs and Harrison, 1969; (10) Harrison and Nixon, 1959; (11) Hsu and Black, 1973; (12) Huttinga, 1973; (13) Kassanis and Govier, 1972; (14) Kassanis *et al.*, 1972; (15) Koenig *et al.*, 1984; (16) Larsen and Duffus, 1984; (17) Leiser and Richter, 1978; (18) Lister and Bracker, 1969; (19) Mayo *et al.*, 1979; (20) Mink and Thomas, 1974; (21) Moghal and Francki, 1976; (22) Proll and Richter, 1979; (23) Raymer and Diener, 1969; (24) Rezaian and Francki, 1973; (25) Salazar and Harrison, 1978; (26) Steer, 1956; (27) Takanami and Kubo, 1979; (28) Tas *et al.*, 1977; (29) Teakle, 1962; (30) Van den Hurk *et al.*, 1977; (31) Van Vloten-Doting and Jaspars, 1972; (32) Wellink, 1998; (33) Wetter, 1960; (34) Wetter, 1971.

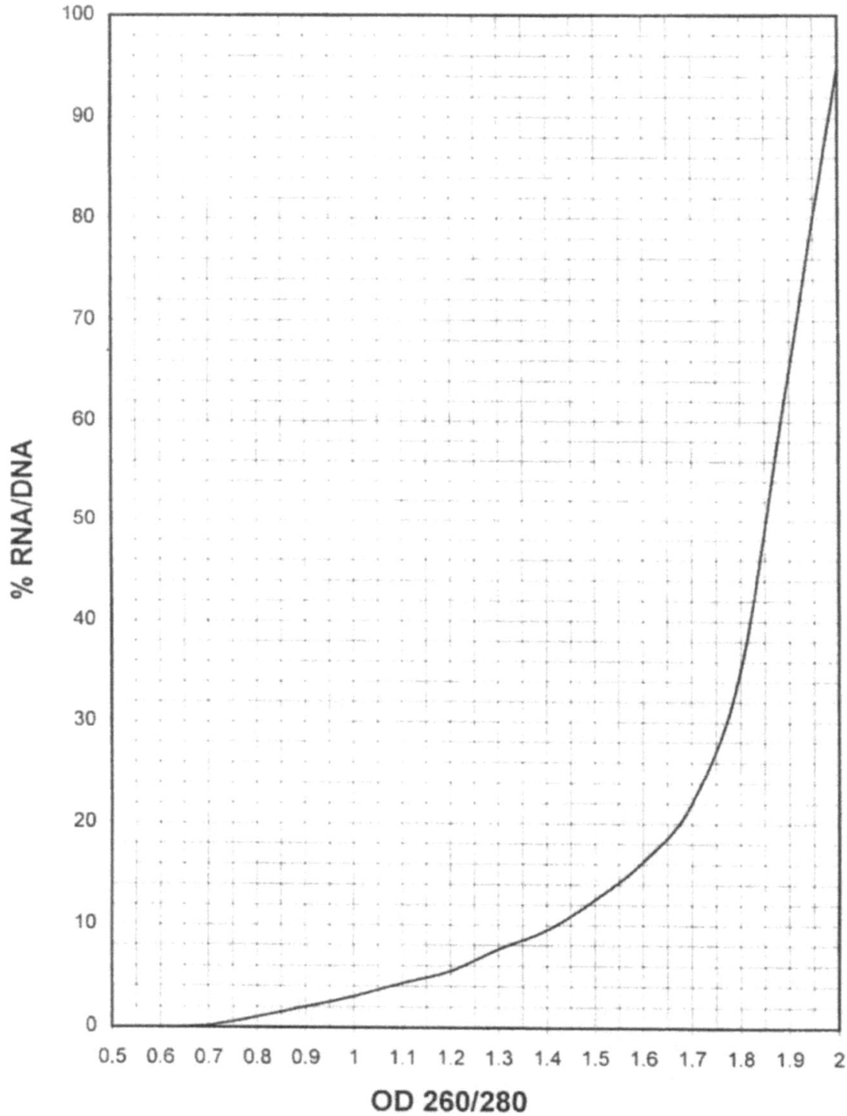

Fig. 11.2. OD260 of pure virus preparation at 1 mg/ml concentration as a function of percentage of nucleic acid in virions.

11.6.4 Purification of *Potato X, S, and M Viruses*

Use *Lycopersicon esculentum* to propagate PVM, *N. clevelandii* to propagate PVS, and *N. tabacum* var. Xanthi-nc or Samsun-NN to propagate PVX. Grow, inoculate, and harvest tissue as described for purification of PVY above. Process fresh leaves or freeze leaves individually, package, and hold frozen until used.

Triturate frozen or fresh, cooled leaves four volumes (w/v) of cold, 0.5 M borate buffer, pH 8.2, containing sodium sulfite (1% w/v) and Triton X-100 (1% v/v) in a high speed blender.

Add cold chloroform to 25% (v/v), emulsify an additional 30 sec, then centrifuge at low speed (5,000 r.p.m. in Beckman rotor JA-10) for 10 min at 15°C. Decant and process the upper, clear, yellow, aqueous phase.

Add PEG to 5% and NaCl to 4% (w/v) of the aqueous phase, stir gently at least 2.5 hr at 4°C, centrifuge at low speed (5,000 r.p.m. in Beckman rotor JA-10) at 4°C for 20 min. Keep the pellets.

Shake or stir the PEG pellets in 25% volume (v/w of original tissue) of borate buffer (0.05 M, pH 8.2) containing 1% Triton X-100. The dissolved pellets may be held at 4°C overnight.

Centrifuge the dissolved pellets at low speed (10,000g for 15 min) and transfer the supernatant to a clean tube.

Sediment the virus by high speed centrifugation for 2 h at 48 – 60,000 **g** (R_{min}) (We use Beckman rotor 30 at 29,000 r.p.m. or rotor 45 Ti at 38,000 r.p.m.). Layer 1 ml of 25% (w/v) sucrose in virus buffer at the bottom of each tube prior to centrifugation.

Resuspend pellets in 5% volume (v/w of original tissue) of borate buffer (0.05 M, pH 8.2) containing 1% Triton X-100. Use a small Nylon tube brush to disrupt pellets. Combine pellet extracts, stir for at least 1 hr, then centrifuge at low speed (7000 **g** R_{min}) to remove insoluble particles.

Subject the virus suspension to rate zonal density gradient centrifugation in 10-40% (w/v) buffered (borate, 0.05M, pH 8.2 containing 1% Triton X-100) gradients. Layer *c.* 2 ml of virus preparation on each gradient. Collect and combine virus fractions. Remove sucrose by sedimentation or dialysis.

Conduct isopycnic density gradient centrifugation as described for PLRV purification in gradients prepared in 0.05 M borate buffer at pH 8.2. Dilute recovered virus zones in 0.05 M borate buffer at pH 8.2, and concentrate by high speed centrifugation as in Step No. 7.

11.6.5 Purification of *Tobacco Rattle Virus*

Use *Nicotiana tabacum* cv. Samsun-NN to propagate the virus. Grow, inoculate, and harvest tissue as described above for purification of PVX.

Triturate frozen leaves in four volumes (w/v) of cold 0.03 M phosphate buffer, pH 8, containing sodium thioglycolate (0.01 M) and sodium diethyldithiocarbamate (0.01M) in a high speed blender.

Centrifuge at low speed (10,000 r.p.m. in Beckman rotor JA-10) for 10 min at 4°C. Discard pellets.

Add PEG 6,000 to 4% (w/v) and NaCl to 1.75% and stir for 4 hr at 4°C, and collect pellets after centrifugation at low speed (10,000 r.p.m. in Beckman rotor JA-10). Resuspend the pellets in 25% volume (w/v of original tissue) of trituration buffer by stirring overnight at 4°C.

Add an equal volume of cold chloroform, emulsify with an overhead stirrer, break the emulsion by centrifugation at low speed (5,000 r.p.m. in Beckman rotor JA-10) for 10 min at 4°C, and collect the upper aqueous phase.

Sediment the virus by high speed centrifugation for 2 h at 48 – 60,000 **g** (R_{min}) (We use Beckman rotor 30 at 29,000 r.p.m. or rotor 45 Ti at 38,000 r.p.m.). Layer 1 ml of 25% (w/v) sucrose in virus buffer at the bottom of each tube prior to centrifugation.

Resuspend pellets in 5% volume (v/w of original tissue) of phosphate buffer (0.02 M, pH 8). Use a small Nylon tube brush to disrupt pellets. Combine pellet extracts, stir for at least 1 hr, then centrifuge at low speed (7000 **g** R_{min}) to remove insoluble particles.

Subject the virus suspension to rate zonal density gradient centrifugation in 10-40% (w/v) buffered (phosphate, 0.03M, pH 8) gradients. Layer *c.* 2 ml of virus preparation on each gradient. Collect the upper and lower fractions and combine each fraction. Remove sucrose by sedimentation or dialysis.

We have not subjected this virus to isopycnic density gradient centrifugation. The methods described under General Methods may be useful.

11.6.6 Purification of *Tobacco Necrosis Virus*

Use *Nicotiana tabacum* cv. Samsun-NN to propagate the virus. The virus remains localized in discrete local lesions. Grow and inoculate plants as described above for PVX. Harvest inoculated leaves as local lesion begin to appear, freeze individually, package, and hold frozen until used.

Follow the purification schedule given for *Tobacco rattle virus* with the following exception: 1) Omit diethyldithiocarbamate from the trituration buffer. 2) Add PEG 6,000 to 6% (w/v).

References

Adomako, D., Lesemann, D. E., Paul, H. L. and Owusu, G. K. (1983). Improved methods for purification and detection of cacao swollen shoot virus. *Ann. appl. Biol.* 103: 109-116.

Bachrach, H. L. and Schwerdt, C. E. (1952). Purification studies on Lansing poliomyelitis virus: pH stability, CNS extraction, and butanol purification experiments. *J. Immunol.* 69: 551-561.

Best, R. J. and Samuels, G. (1936). The effect of various chemical treatments on the activity of the viruses of tomato spotted wilt and tobacco mosaic. *Ann. appl. Biol.* 23: 759-780.

Brakke, M. K. (1960): Density gradient centrifugation and its application to plant viruses. *Adv. Virus Res.* 7: 193-224.

Brakke, M. K. (1963). Stabilization of brome grass mosaic virus by magnesium and calcium. *Virology* 19: 367-374.

Brunt, A., Crabtree, K., Dallwitz, M., Gibbs, A.and Watson, L. (1996 onwards). Plant Viruses Online: Descriptions and Lists from the VIDE Database (http://biology.anu.edu.au/Groups/MES/vide/).

Cadman, C. H. (1959). Some properties of an inhibitor of virus infection from leaves of raspberry. *J. Gen. Microbiol.* 20: 113-128.

Copeland, R. (1998). Assaying levels of plant virus by ELISA. In: G. D. Foster and S. C. Taylor (Eds): *Plant Virology Protocols: From Virus Isolation to Transgenic Resistance*, Humana Press, Totowa, New Jersey, USA. pp. 571.

Dallwitz, M. J. (1980). A general system for coding taxonomic descriptions. *Taxon* 29: 41-46.

Dallwitz, M. J., Paine, T. A. and Zurcher, E. J. (1993). *User's Guide to the DELTA System: a general system for processing taxonomic descriptions.* CSIRO Division of Entomology. Canberra.

Damirdagh, I. S. and Shepherd, R. J. (1970). Purification of tobacco etch and other viruses of the potato Y group. *Phytopathology* 60: 132-142.

Delgado-Sanchez, S., and Grogan, R. G. (1966). Purification and properties of potato virus Y. *Phytopathology* 56: 1397-1404.

Diener, T. O. and Weaver, M. L. (1959). A caffeine additive to aid mechanical transmission of necrotic ringspot virus from fruit trees to cucumber. *Phytopathology* 49: 321-322.

Dunn, D. B. and Hitchborn, J. H. (1965). The use of bentonite in the purification of plant viruses. *Virology* 25: 171-192.

Englander, H. T. and Epstein, H. T. (1957). Optical methods for measuring nucleoprotein and nucleic acid concentrations. *Arch Biochem Biophys.* 68: 144-149.

Francki, R. I. B. and McLean, G. (1968). Purification of potato virus X and preparation of infectious nucleic acid by degradation with lithium chloride. *Aust. J. biol. Sci.* 21: 1311-1318.

Fribourg, C. E. and de Zoeten, G. A. (1970). Antiserum preparation and partial purification of potato virus A. *Phytopathology* 60: 1420-1421.

Fribourg, C. E., Jones, A. C. and Koenig, R. (1977). Host plant reactions, physical properties, and serology of three isolates of Andean potato latent virus from Peru. *Ann. appl. Biol.* 86: 373-280.

Fribourg, C. E. and Nakashima, J. (1984). Characterization of a new potyvirus from potato. *Phytopathology* 74: 1363-1369.

Fulton, R. W. (1957). Properties of certain mechanically transmitted viruses of Prunus. *Phytopathology* 47: 683-687.

Fulton, R. W. (1959): Purification of sour cherry necrotic ringspot and prune dwarf viruses. *Virology* 9: 522-535.

Fulton, R. W. (1966). Mechanical transmission of tatter leaf virus from cowpea to citrus. *Phytopathology* 56:, 575.

Gibbs, A. and Mackenzie, A. M. (1998). Tymovirus isolation and genomic RNA extraction. In: G. D. Foster, and S. C. Taylor (Eds): *Plant Protocols: From Virus Isolation To Transgenic Resistance*, Humana Press, Totowa, New Jersey, USA. pp. 571.

Gibbs, A. J. and Harrison, B. D. (1969). Eggplant mosaic virus, and its relationship to Andean potato latent virus. *Ann. appl. Biol.* 64: 225-231.

Gunn, L. V. and Pares, R. D. (1988). Effect of potato physiology on the interpretation of ELISA results for potato leafroll virus. *Plant Pathology* 37: 516-521.

Hampton, R. E. and Fulton, R. W. (1959). Factors responsible for the instability of some labile plant viruses. *Phytopathology* 49: 540.

Hampton, R. E. and Fulton, R. W. (1961). The relation of the polyphenol oxidase to instability in vitro of prune dwarf and sour cherrey necrotic ring spot viruses. *Virology* 13: 44-52.

Harrison, B. D. and Nixon, H. L. (1959). Separation and properties of particles of tobacco rattle virus with different lengths. *J. gen. Microbiol.* 21: 569-581.

Harrison, B. D. and Pierpoint, W. S. (1963). The relation of polyphenol oxidase in leaf extracts to the instability of cucumber mosaic and other plant viruses. *J. Gen. Microbiol.* 32: 417-427.

Hassan, S. and Thomas, P. (1989). Purification and some properties of tomato yellow top virus. *Sarhad J. Agric.* 5: 507-519.

Hebert, T. T. (1963). Precipitation of plant viruses by polyethylene glycol. *Phytopathology* 53: 362-363.

Hougas, R. W. (1951). Factors affecting sap transmission of potato yellow-dwarf virus. *Phytopathology* 41: 483-493.

Hsu, H. T. and Black, L. M. (1973). Polyethylene glycol for purification of potato yellow dwarf virus. *Phytopathology* 63: 692-696.

Hull, R. (1976). The structure of tubular viruses. *Adv. Virus Res.* 20: 1-32.

Huttinga, H. (1973). Properties of the viruses of the potyvirus group. I. A simple method to purify bean yellow mosaic virus, pea mosaic virus, lettuce mosaic virus, and potato virus YN. *Neth. J. Pl. Path.* 79: 125-129.

Jolley, W. B., Allen, H. W. and Griffith, O. M. (1967). Ultacentrifugation using acrylamide gel. *Anal. Biochem.* 21: 454-461.

Kaniewski, W. (1973). Wplyw parametrow ultrawirowania analitycznego na uzyskane ta metoda wyniki wspolczynnika sedymentacji. *Pr. Nauk. Inst. Ochr. Rosl.* 15: 127-145.

Kaniewski, W. (1974). Pomiar wspolczynnika dyfuzji wirusow przy pomocy ultrawirowki analitycznej. *Pr. Nauk. Inst. Ochr. Rosl.* 16: 143-149.

Kassanis, B. and Govier, D. A. (1972). Potato aucuba mosaic virus. *CMI/AAB Descr. Pl. Viruses* 98, 4 pp.

Kassanis, B., Woods, R. D. and White, R. F. (1972). Some properties of potato mop-top virus and its serological relationship to tobacco mosaic virus. *J. gen. Virol.* 14: 123-132.

Koenig, R., Lesemann, D. E. and Burgermeister, W. (1984). Beet necrotic yellow vein virus: purification, preparation of antisera, and detection by means of ELISA, immunosorbent electron microscopy, and electroblot immunoassay. *Phytopath. Z.* 111: 244-250.

Larsen, R. C. and Duffus, J. E. (1984). A simple procedure for purification of curly top virus and the isolation of its monomer and dimer particles. *Phytopathology* 74: 114-118.

Leiser, R. M. and Richter, J. (1978). Reinigung und einige Eigenschaften des Kartoffel-Y-Virus. *Arch. Phytopath. PflSchutz* 14: 337-350.

Lister, R. M. and Bracker, C. E. (1969). Defectiveness and dependence in three related strains of tobacco rattle virus. *Virology* 37: 262-275.

Matthews, R. E. F. (1974). Some properties of TYMV nucleoproteins isolated in cesium chloride density gradients. *Virology* 60: 54-64.

Matthews, R. E. F. (1991). *Plant Virology*. Academic Press, Harcourt Brace Jovanovich. New York, USA.

Mayo, M. A., Barker, H. and Harrison, B. D. (1979). Evidence for protein covalently linked to tobacco ringspot virus RNA. *J. gen. Virol* 43: 735-740.

Mink, G. I. and Thomas, P. E. (1974). Purification of curly top virus. *Phytopathology* 64: 140-142.

Moghal, S. M. and Francki, R. I. B. (1976). Towards a system for the identification and classification of potyviruses I. Serology and amino acid composition of six distinct viruses. *Virology* 73: 350-362.

Paulsen, A. Q. and Fulton, R. W. (1968). Hosts and properties of a plum line pattern virus. *Phytopathology* 58: 766-772.

Pierpoint, W. S., Ireland, R. J. and Carpenter, J. M. (1977). Modification of proteins during the oxidation of leaf phenols:Reaction of potato virus X with chlorogenoquinone. *Phytochemistry* 16: 29-34.

Proll, E. and Richter, J. (1979). Reinigung und einige Eigenschaften der Kartoffelviren M und S. *Arch. Phytopathol, Berlin* 15: 233-245.

Raymer, W. B. and Diener, T. O. (1969). Potato spindle tuber virus: A plant virus with properties of a free nucleic acid I. assay, extraction, and concentration. *Virology* 37: 343-360.

Rezaian, M. A. and Francki, R. I. B. (1973): Replication of tobacco ringspot virus I. detection of low molecular weight double-stranded RNA from infected plants. *Virology* 56: 238-249.

Ronald, W. P., Tremaine, J. H. and MacKenzie, D. J. (1986). Assessment of southern bean mosaic virus monoclonal antibodies for affinity chromatography. *Phytopathology* 76: 491-494.

Salazar, L. F. and Harrison, B. D. (1978). Host range, purification, and properties of potato virus T. *Ann. appl. Biol.* 89: 223-235.

Schneider, I. R. (1953). Solution of tobacco mosaic virus in the aqueus phase of a chloroform-water emulsion and application of this phenomenon in virus assay. *Science* 117: 30-31.

Shepherd, R. J. and Pound, G. S. (1960). Purification of turnip mosaic virus. *Phytopathology* 50: 797- 803.

Singh, S. K., Anjaneyulu, A. and Lapierre, H. (1984). Use of pectino-cellulitic enzymes for improving extraction of phloem-limited viruses as exemplified by the tungro virus complex. *Agronomie* 4: 479-484.

Steer, R. L. (1956). Purification and properties of tobacco ringspot virus. *Phytopathology* 46: 60-69.

Takanami, T. and Kubo, S. (1979). Enzyme-assisted purification of two phloem-limited viruses: tobacco necrotic dwarf and potato leafroll. *J. gen. Virol.* 44: 153-159.

Tas, P. W. L., Boerjan, M. L. and Peters, D. (1977). Purification and serological analysis of tomato spotted wilt virus. *Neth. J. Plant Path.* 83: 61-72.

Teakle, D. S. (1962). Necrotic symptoms of tobacco necrosis virus in roots. *Phytopathology* 52: 1037-1040.

Thompson, S., Fraser, R. S. S. and Barnden, K. L. (1988). A beneficial effect of trypsin on purification of turnip mosaic virus (TuMV) and other potyviruses. *J. Virol. Methods* 20: 57-64.

Tomlinson, J. A. (1964). Purification and properties of lettuce mosaic virus. *Ann appl. Biol.* 53: 95-102.

Van den Hurk, J., Tas, P. W. L. and Peters, D. (1977). The ribonucleic acid of tomato spotted wilt virus. *J. gen. Virol* 36: 81-91.

Van Vloten-Doting, L. and Jaspars, E. M. J. (1972). The uncoating of alfalfa mosaic virus by its own RNA. *Virology* 48: 699-708.

Warburg, A. and Christian, W. (1941). Wirkungsgruppe des Gärungsferments Zymohexase fum. *Biochem. Z.* 311: 209-210.

Wellink, J. (1998). Comovirus isolation and RNA extraction. In: G. D. Foster, and S. C. Taylor (Eds): *Plant Virology Protocols: From Virus Isolation to Transgenic Resistance*, Humana Press, Totowa, New Jersey, USA. pp. 571.

Wetter, C. (1960). Partielle Reinigung einiger gestreckter Pflanzenviren und ihre Verwendung als Antigene bei der Immunisierung Mittels Freundschen Adjuvans. *Arch. Microbiol.* 37: 278-290.

Wetter, C. (1971). Potato virus S. *CMI/AAB Descriptions Plant Viruses*, 3 pp.

12. Operation of a Commercial Enzyme-linked Immunosorbent Assay (ELISA) and Seed Testing Laboratory

Y. Alon[1] and I. Ben Zeev[2]

[1]*Phytodiagnostic Laboratory, Bactochem Ltd. Ness-Ziona 70400, Israel; and* [2]*Plant Protection and Inspection Services. Israel Ministry of Agriculture. Bet Dagan P.O.78 50250 Israel*

12.1 Introduction

Laboratory testing of potato plant material for the presence of specific pathogens is an essential component of any efficient seed potato certification program. A professional, commercial diagnostic laboratory must deal with several important issues including sampling, choice of assay method, sample preparation and evaluation of results, laboratory management and certification. A number of these issues will be discussed in this chapter.

12.2 Tolerances for Viruses and Sampling

Tolerance for pathogenic factors differs according to the disease potential of the pathogen and the value of the plant material. In testing for the presence of pathogenic factors, special attention should be given to

G. Loebenstein et al. (eds.),
Virus and Virus-like Diseases of Potatoes and Production of Seed-Potatoes, 311–322.

determining the sample size. In addition to statistical considerations, sample size is strongly influenced by practical issues, including the cost of testing.

The accepted tolerance for pathogens in mother plants, pre-nuclear material and nuclear stocks is zero. The tolerance for further generations rises as a function of time and genetic distance from the "zero tolerance" material. Each mother plant is tested separately and individually. Entry-level material, such as *in vitro* plantlets, *in vitro* microtubers, pre-nuclear plantlet nodal cuttings, and pre-nuclear sprouts, must undergo two complete testing cycles (Oregon, 1998). In progressive generations, it is sufficient to test samples from the plant population. Table 12.1 provides a summarization of the winter test tolerances for potato viruses according to the Oregon potato seed certification standards (Oregon, 1998).

Table 12.1. Winter test tolerances of viruses (% infection according to the Oregon potato seed certification standards)

Virus	Nuclear	Gen 1	Gen 2	Gen 3	Gen 4	Gen 5
Leafroll	0	0.25	0.30	0.75	1.00	*
Mosaic Other Varieties	0	0.25	0.50	1.00	2.00	
Mosaic Russet Norkotah	0	0.25	0.50	1.00	5.00	*
Spindle Tuber viroid	0	0	0	0	0	0
Other visible virus	0	0.25	0.75	2.00	2.00	*
Total visible virus-other	0	0.50	0.75	2.00	2.00	*
Total visible virus-R.Norkotah	0	0.50	0.75	2.00	2.00	*

*Acceptance of the seed lot will be based on the buyer/seller agreement.

Sample sizes according to the Oregon certification standards are determined by a combination of statistical analysis of tolerance ranges and field size. Results are provided in Table 12.2.

Table 12. 2. Size of sample as a function of the tested potato seed field according to the Oregon potato seed certification standards (Oregon, 1998).

Acres	Tubers	
0.5-1	At least	220
2-5	At least	320
6-10	At least	420
11-20	At least	620
21-30	At least	1,020
31-40	At least	1,220

Acres	Tubers	
41-50	At least	1,420
51-60	At least	1,620
61-70	At least	1,820
71-80	At least	2,020
81-90	At least	2,220
91-100*	At least	2,420

* For lots over 100 acres at least 2,420 tubers, plus 20 tubers for each additional acre have to be submitted for testing.

The seed certification program in the Netherlands is somewhat different (NAK, 1999a). Seed potatoes are divided into three categories, with subdivisions into classes:

- prebasic seed (minitubers and 1st and 2nd clonal generation)
- basic seed (BS): Classes S, SE, and E (3rd, 4th and 5th generation)
- certified seed (CS) Classes A, B, and C

After field inspection, post harvest tests are carried out for viruses by ELISA, 3-4 weeks after planting "eyes" from the rose end of the tuber in a greenhouse.

Classes S and SE are tested for PVA, PVS, PVX, PVY and PLRV (compulsory) as well as Class E for varieties susceptible to PVY.

Classes E, A, B, and C are tested for PVA, PVY and PLRV.

Sample size for basic seeds is 200 tubers per field (average field size is 1 hectare. Sample size for certified seeds is 100 seeds per field (average field size is 2.5 hectare).

Tolerances (total virus- no distinction between mild and severe viruses): Class S, SE: 0 in 200 tubers; Class E: 1 in 200 tubers; Class A: 5 in 100 tubers; Class B: 8 in 100 tubers and Class C: 10 in 100 tubers (NAK, 1999b).

12.3 Choosing the Proper Detection System

There are a number of detection systems available to the phytopathology laboratory. These detection systems may be divided into visual detection of disease symptoms on the potato plants or tuber seeds and detection of specific molecules as DNA sequences or coat proteins, which are associated with the pathogen.

Many certification authorities would prefer visual detection of pathogens on the potato plants or tuber seeds. These symptoms can be immediately detected and indicating, in a relatively quick fashion, the presence of a pathogen. Relying on detection of symptoms on plants is, however, not always conclusive. In many cases plants may carry a latent infection, which does not produce detectable symptoms, as with certain strains of *Potato virus X (*PVX) (Banttari *et al.* 1993). Therefore, plant viruses and some plant bacteria have to be diagnosed by laboratory methods. For virus diagnosis the common method is the enzyme-linked immunosorbent assay (ELISA). The sensitivity of ELISA is about 1 mg protein per ml. However, some viruses may be present in plant tissues below this level. Recent developments using detection of specific nucleic acid sequences by PCR or probes are beginning to be employed as an additional verification system. At present there is no international agreement concerning the reliability of PCR for routine virus tests for PVY strains (NAK, 1999b).

12.4 Determination of Positive and Negative Results in ELISA

A result may be considered as positive when the signal/backround ratio is above 6:1 and not less then 0.6 O.D., and readings of less then 0.1 O.D. in the negative controls. It is, therefore, necessary to use virus specific, high quality immunoglobulins, that are free from antibodies to host components. Monoclonal antibodies are considered to be high quality specific immunoglobulins, but polyclonal antibodies are generally preferred, because they will detect a wider range of virus strains.

Though it is theoretically desirable that the results indicate 0.6 O.D. or greater, in practice, results, which are lower than 0.6 O.D. for the positive controls, may occur. Therefore, clear cut-off values have to be formulated. Two formulas are used in certain ELISA systems: NC+0.1PC and 2NC where PC stands for Positive Control and NC stands for Negative Control. To assure a high level of confidence, any result that is higher than that determined by one of the above cut off formulas will be considered positive, and the potato seed sample will be considered virus contaminated.

High affinity between the specific antibody and the virus enables a higher dilution of both the antibody and the sample in the ELISA reaction mixture. As a result, nonspecific and possible masking materials in the sample are decreased, with an increase in confidence of the result.

Where only negative results are indicated in the test sample, the results should be considered with caution. Negative results indicate only that the specific assay used in the sample testing did not detect pathogens in the plant material. The sensitivity of any detection system is limited, particularly when biological material is not tested soon after it is sampled. At the level of sensitivity of ELISA, millions of virus particles may actually be present in the sample, but as discussed above, remain undetected. In addition, the test sample may be too small and not representative of the seed lot. Despite the fact that sample testing is imperfect at best, negative results indicated in sample testing greatly increases the confidence that there is no pathogen contamination.

12.5 Carrying out a Testing Program

12.5.1 Personnel

The management of a diagnostic laboratory must be placed in the hands of qualified personnel, experienced in biological laboratory work and in management. In addition, the laboratory director has to have a basic knowledge of virus transmittance, replication, virus-host interactions and conditions which favour high or low virus titers in the plant The director should also have a good knowledge of general plant protection procedures including the practical and the theoretical aspects of these procedures.

In addition to the professional director or manager, a diagnostic laboratory should employ at least one technician who is familiar with all relevant laboratory procedures. During the season(s), when demand for diagnosis is high, additional personnel, trained in sample preparation and simple routine procedures, will also be required.

12.5.2 Equipment and Facilities

In order to proceed with plant virus testing, the diagnostic laboratory must arrange for the following four categories of equipment and facilities:

1. Greenhouse or open field.
2. Virus extraction system/ plant homogenizer.
3. An ELISA reader and computer.
4. Basic reagents and equipment for preparation and maintenance of solutions.

Greenhouse or Open Field

Post harvest testing is done soon after harvesting, by breaking tuber dormancy, planting tubers in a greenhouse or open field, (temperatures permitting) and subsequently testing for pathogens (See Chapter 17).

Virus Extraction Methods

A plant homogenizer is needed to breakdown the plant material structure, and allow the virus to enter the sample buffer. Several types of homogenizers are common in diagnostics laboratories:

- Two cylinder homogenizer (about $ 4000-7000 US).
- Handheld model homogenizer (BIOREBA) (Fig.12.1) (for $ 75-120 US).
- Electrical homogenizer (for example BIOREBAs Homex 6) (for $ 6000-7000 US).
- A hammer for smashing hard plant tissue inside a plastic bag.

An additional method to release the virus into the detection medium is to cut the plant material into small pieces directly into the wells of the microtiter plate. This is followed by incubation overnight in the sample buffer. The advantage of this system is that most of the plant material remains intact and only soluble molecules and viruses are suspended in the sample buffer. Non-specific reactions are, therefore, minimized.

ELISA Reader and Computer

The use of an ELISA reader combined with a computer (Fig.12.2) enables the monitoring of up to 96 simultaneous enzymatic staining reactions. The ELISA reader should be connected to a PC, to enable data processing (prices vary between $4000 – 10,000 US).

The computer processing provides the following information:

- Averaging of the results from replicates of the sample.
- Data after subtraction of the blank results.
- Determination of results as positive or negative.
- Determination of standard deviation and coefficient of variance of replicate.
- Capability of validating the test.

A software package, which meets the above specifications, is available from Bio-Tek Instruments (Kc4 User Guide, 1997) (cost is several hundred US $).

Software for population analysis, as Microsoft Access is recommended.

Basic Reagents and Equipment for Preparation and Maintenance of Solutions.

Only high quality de-ionized water (calcium and magnesium free water) should be used. Key reagents should be of analytical grade, from reliable sources. It is advisable to use diagnostic kits from reliable sources. For example Agdia, BIOREBA or Loewe (price for a kit for 1000 tests is about $ 400 US).

The diagnostic laboratory must be equipped with a refrigerator, freezer, incubator, analytical and semi analytical balances, pH meter, magnetic stirrer, vortex mixer and sets of microtiter pipettes (about $ 5000 US).

12.5.3 Procedures in the ELISA Laboratory

The minimum time for running an assay is from 8 to 10 hours. It is usually more convenient to split the assay into two sequential working days.

There are several major steps in the ELISA assay:

12.5.3.1 Receiving and Registering Samples.

Each potato sample is registered, numbered and attached with a computerized barcode. Samples are then kept in cold storage (about 4°C).

12.5.3.2 Preparing Potato Tissue for Assay.

Leaf samples should be tested within a few days. Tubers should be tested either after dormancy, or dormancy should be broken artificially before testing. A small sample, including a budding "eye", is homogenised for ELISA, using one of the extraction methods described.

One technician can prepare between 100-150 samples per day and load them into two or three wells of the microtiter plates. Various kinds of flexible and rigid plates are used. Volume of wells varies from 0.2 ml to 0.35 ml. Attention should be paid to the width of the wells, which should be wider than the light beam source from the ELISA reader.

The extraction step is often the bottleneck of the ELISA procedure. Well-organized laboratory personnel can prepare more than one thousand samples per day.

12.5.3.3 Further Steps in the ELISA Procedure

The remaining steps in the ELISA procedure include coating the microtiter wells with antibody against the virus, loading the samples into the coated wells, adding an antibody against the virus conjugated to alkaline phosphatase, and finally, adding alkaline phosphatase substrate. Between these steps thorough washing with phosphate buffer is required. When color becomes visible, results are read and interpreted.

12.6 Quality Assurance

12.6.1 Procedures

In order to achieve and maintain high standards along with accreditation, a diagnostic laboratory should follow Good Laboratory Practices (GLP) as a measure of quality control (QC) and quality assurance (QA). The laboratory staff should use written obligatory guidelines and procedures, with established standards for all activities.

Critical activities necessary for the maintenance of high standards include the following:

- Registration of samples and matching them to proper tests (using barcodes).
- Use of appropriate storage facilities (usually 4^0C).
- Adequate procedures for handling samples including timely preparation and maintenance of fresh solutions and assays, according to validated internationally accepted procedures,

repetition of tests when indicated, and registration of results.

- The QC/QA procedures should include supervision of all laboratory activities. Ideally, supervision should be carried out at two levels: laboratory management and staff, as well as by an outside agency or authority for laboratory accreditation.

12.6.2 Positive (Pathogen) Controls

Activities of phytodiagnostic laboratories are based on quantitative or semi-quantitative tests like ELISA. One of the major problems that these laboratories face is the long-term maintenance of standard pathogenic plant material without loss of antigenic activity. Companies, which produce diagnostic kits, offer limited quantities of lyophilised, virus-free and virus-infected plant material as negative and positive controls, respectively. The source of this plant material is frequently unknown. Healthy lyophilised material is not a good alternative for fresh virus-free plant material from the respective plant.

It is quite difficult to quantify the amount of virus in the infected material. The distribution of the virus in the tissue is not uniform, and is difficult to measure. Therefore, a high signal is not always an indication for heavily infected plant tissue. ELISA results can be reliable only when the amount of virus is sufficient to cause a substantial specific signal. When the signal is low, it might fall in the area below the determined cut off level. Such a signal may be non-specific, or alternatively result from the presence of substances, which interfere with the detection of the virus, or may mask the presence of viral particles.

12.6.3 External Pathogens

To evaluate possible interference in measuring the quantity of virus in a sample, a certain small amount of infected plant material in suspension should be added. In parallel, the same amount of infected tissue should be added to and measured in the buffer alone. A signal, lower than the one caused by infected material in buffer, indicates that there is signal inhibition, meaning that the plant material interferes with the virus detection. A result where inhibition is high, (above 50%), indicates decreased reliability of the detection assay.

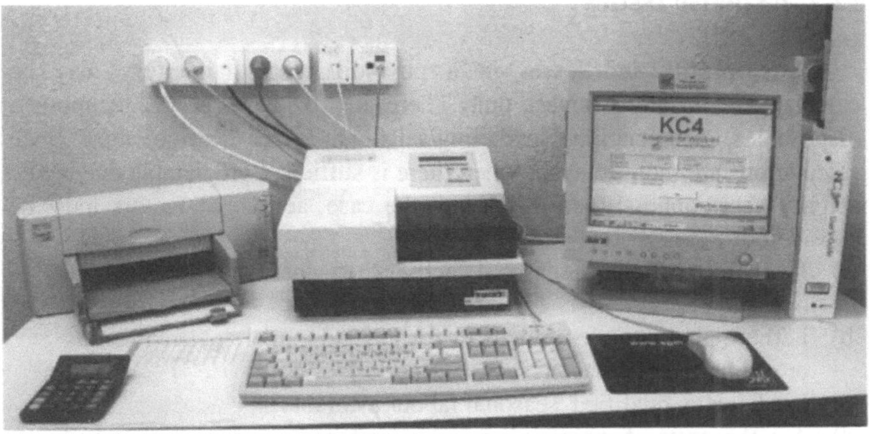

Fig. 12.1 An ELISA system including reader, computer (PC) and printer.

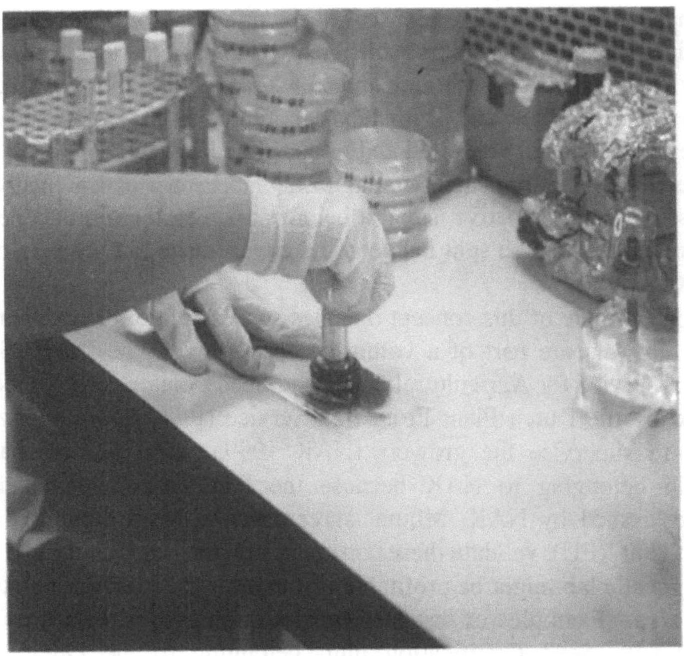

Fig. 12.2 Hand homogenizer (BIOREBA) for squeezing potato leaves placed in a plastic bag.

12.7 Expense Issues

Costs, profits and losses in a phytodiagnostic laboratory may be calculated as in any business unit. There are fixed overheads, manpower, disposables, depreciation, marketing/sales, customer service etc. Such calculations are more relevant when there is sufficient predictable diagnostic activity year round. When this is not the case, additional factors must be considered:

a) Is there sufficient seasonal activity to justify a commercial operation?
b) Is there alternative employment for key personnel between testing or during the off-season?
c) Is adequate, qualified, seasonal labour available?
d) Are the costs of maintaining and refreshing critical materials and reagents, solutions etc. during the off-season justified?

Given the conditions in many agricultural regions, a dedicated, private, commercial phytodiagnostic laboratory, which maintains high standards, is unlikely to be consistently profitable.

In many instances, in order to keep prices affordable, the actual diagnostic service does not meet a sufficiently professional level of testing or accuracy. As a result, the demand for diagnostic laboratory services diminishes even further, and ultimately, laboratory tests are performed only when they are absolutely required and/or controlled by law.

Rather than operating a phytodiagnostic laboratory as an independent profit/loss unit, an alternative is that it belongs to an agricultural cooperative. Ownership might also be split between the government and growers or their organizations.

A good example of this concept operates successfully in the Netherlands. Here the growers are part of a voluntary cooperative- The Dutch General Inspection Service for Agricultural Seeds and Seed Potatoes (NAK). NAK is controlled by the Dutch Plant Protection Service (P.D.) and authorized by the P.D. to supervise the growers (NAK 1999a). The farmers have an interest in belonging to NAK because the P.D. accepts pathogen free certificates issued by NAK. Mutual agreements between countries of the European Unit (EU), validate these certificates by the other countries.

A diagnostic lab might be profitable if it is located in a major potato seed growing area. Examples of test fees for detection of potato viruses in the Oregon Potato Seed Certification unit, belonging to the Oregon State University and the State Department of Agriculture are given in Table 12.3.

Table 12.3: Test fees for detection of Potato viruses (X,Y,S,LR,M,A)

	Minimum (US$)	Per Sample (US$)
ELISA	40	3
EM	15	10
PSTV	20	10

Fees by NAK in the Netherlands are 112.3 *f* ($ 47.2 US) for a sample of 100 tubers and 168.45 *f* ($ 70.9 US) for a sample of 200 tubers. These fees include sampling.

In Israel, higher fees are charged as the tests are not done regularly.

12.8 Laboratory Certification

Authorities in each country, grant laboratory accreditation and/or recognition according to their specific guidelines. In general, the guidelines include GLP requirements such as an experienced professional laboratory manager; printed protocols for each type of test; order and sanitation in the laboratory; environmental protection including decontamination equipment; and careful record keeping and documentation. The laboratory has to be open for inspection by the approving authorities and to succeed in audits for each recognized or certified test.

12.9 Conclusion

The operation of a reliable diagnostic laboratory requires the integration of several elements including updated expertise, manpower, equipment, internal and external quality control systems and proper management of the laboratory. When planning or improving diagnostic monitoring systems, such elements should be integrated properly by plant protection organizations. Proper integration and implementation of protocols of the diagnostic laboratory will facilitate the future production of healthy non-infected plant propagation material and the trading of pathogen- free plants.

References

Banttari, E.E., Ellis P.J. and Khurana S.M.P. (1993). Management of diseases caused by viruses and virus-like pathogens. In: *Potato Health Management*. Ed. R.C. Rowe American Phytopathological Soc.page 127- 133.

Gutbrod , O.A. and Mosley, A.R. (2000). Common seed potato certification schemes . See chapter 17.

.Kc4 User Guide (1997). Bio-Tek Instruments Inc. Highland Park, Winooski, Vermont USA.

NAK. (1999a). *Seed Certification in the Netherlands.* Dutch General Inspection Service.

NAK. (1999b). *Post Harvest Virus Tests.* Dutch General Inspection Servicee.

Oregon (1999). Potato Seed Certification. (1999).Oregon State University and Oregon Seed Services.
http://www.oscs.orst.edu/potato/standards.html.

13. Resistance

13.1 Genetics and Breeding of Virus Resistance: Traditional Methods

C. R. Brown and D. Corsini
USDA/ARS, Prosser, WA and Aberdeen, ID, USA

13.1.1 Introduction

The potato was introduced to Europe from South America around 1570. By the turn of that century botanical references were noted both in Continental Europe and the British Isles. Approximately a century later the potato was in the initial stages of being adopted as a foodstuff. The phenomenon of decline in performance of potato clonal cultivars has been noted thoughout the history of its utilization as a crop. Degenerated cultivars were replaced by new vigorous ones that, in time, also declined. Reference was also made to a condition called "Curl,'' which was undoubtedly indicative of the presence of potato leafroll virus (Salaman, 1985). It was not until the twentieth century, however, with the discovery of the existence of viruses in plants, that an understanding of the importance of virus in potato production began to emerge (Salaman, 1921). The instigation of hygienic seed production came from the recognition that seed grown in certain colder areas was much more vigorous than that from warmer areas. This was shortly followed by attempts to utilize wild species and primitive cultivars in breeding as sources of virus resistance. Expeditions to collect *Solanum* germplasm in a systematic fashion began in the 1930's and have continued to this day. Today, there is a large body of information available on virus resistance present in wild and cultivated germplasm, and the

G. Loebenstein et al. (eds.),
Virus and Virus-like Diseases of Potatoes and Production of Seed-Potatoes, 323–340.

breeding and genetics of virus resistance. The genetics and breeding of virus resistance in potato have recently been thoroughly reviewed by Swiezynski (1994) and Valkonen (1994), and this review will hopefully add further accessibility and interpretation to this growing body of knowledge. Table 13.1.1 presents a summary of the germplasm sources of virus resistance, type of resistance and gene symbols that have been applied.

13.1.2 *Potato leafroll virus*

Potato leafroll virus (PLRV) is among the most important viruses of potato worldwide. PLRV is a *Polerovirus* (formerly *Luteovirus*) transmitted by aphids (primarily *Myzus persicae* or green peach aphid) in a persistent manner (Hooker, 1981). There is no cultivar or advanced breeding clone that has extreme resistance to PLRV, but a range of resistance types are known from wild species and are in various stages of incorporation into the cultivated breeding pool.

Resistance to PLRV seems to have arisen from materials utilizing *S. demissum* as a source of late blight resistance. In Europe, the cultivars "Schwalbe" (aka "Lenino") and "Aquila" were products of late blight breeding and also showed partial resistance to PLRV (Davidson, 1980). Partial resistance to PLRV is defined as a relative resistance to infection by the virus. The absolute percentage of incidence will vary from year to year, but the relative ranking of genotypes remains mostly intact. Results from artificial inoculation are similar to field infection (Hamann *et al.* 1968; Baerecke, 1958, 1961; Bagnall and Tai, 1986; Corsini *et al.* 1994). Even cultivars with high levels of partial resistance may be completely infected in one season of high inoculum pressure (Wilson and Jones, 1993b). The American cultivars "Abnaki" and "Penobscot" were the results of direct selection for PLRV resistance and may derive this from parents (e.g., X927-3) originally selected for late blight resistance derived from *S. demissum* (Akeley *et al.* 1971).

Cultivars of *S. tuberosum* ssp *andigena* with resistance to PLRV have been noted (Brown, 1979; Rizvi, 1980). A diallel mating design, utilizing *S. tuberosum ssp. andigena* as parents, estimated a low narrow-sense heritability of resistance to artificial greenhouse aphid inoculation of the virus. However, significant differences in general combining ability of certain clones were found (Brown, 1979). The *ssp. andigena* PLRV resistant cultivar CUP199 emerging from these studies is available from the International Potato Center's *In Vitro* Pathogen-Free Collection. Seedling tests have been employed to evaluate parents for breeding value (Davidson, 1973; Brown, 1979, 1980, 1985; Brown *et. al.* 1982, 1984a; Chuquillanqui

Table 13.1.1. Inventory of virus resistances in different germplasm sources and gene symbols.

Virus	Germplasm Source	Type of resistance	Inheritance, gene symbols (if any)	Citations
PLRV	*S. tuberosum ssp tuberosum*	Relative resistance to infection, transport, virus accumulation	Polygenic	Davidson, 1980; Ross, 1986; Swiezynski,, 1994; Barker; 1987, Wilson and Jones, 1993a
	S. tuberosum ssp. andigena	Relative resistance	Polygenic	Brown, 1979; Rizvi, 1980; Brown *et al.* 1997,
	Diploid cultivated potato	Extreme resistance	Not determined	Swiezynski *et al.* 1989
	S. chacoense	Extreme resistance	Rl_{chc}	Brown and Thomas, 1994
	S. tuberosum ssp. tuberosum ("Apta," "Carla")	Intolerance or Hypersensitivity	Nl_{tbr}	Butkiewicz , 1978
	S. brevidens	Resistance to virus accumulation	Not determined	Jones, 1979; Austin *et al.* 1985; Gibson *et al.* 1988
	S. etuberosum	Resistance to virus accumulation	Not determined	Chavez *et al.* 1988a, 1988b
PVY	*S. stoloniferum*	Extreme resistance	Ry_{sto}	Ross, 1958
	S. tuberosum ssp. andigena (?)	Extreme resistance	Ry_{adg}	Muñoz *et al.* 1976
	S. phureja	Relative resistance	Polygenic	Vallejo *et al.* 1995
	Series Etuberosa	Variable	Not determined	Valkonen *et al.* 1992
	S. demissum,	Hypersensitivity to PVY and PVA	Ny_{dms}	Cockerham, 1970
	S. chacoense	Hypersensitivity	Ny_{chc}	Cockerham, 1970
	S. hougasii	Extreme resistance	Ry_{hou}	Cockerham, 1970
PVA	*S. stoloniferum*	Extreme resistance	Ra_{sto}	Barker, 1996
	S. tuberosum ssp andigena	Hypersensitivity	Na_{adg}	Hamalainen *et al.* 1997
PVS	*S. tuberosum ssp. andigena*	Hypersensitivity	Ns_{adg}	Baerecke, 1967
PVM	*S. gourlayi*	Relative resistance	*Gm*	Swiezynski, 1981
PVX	*S. tuberosum ssp. tuberosum*	Hypersensitivity to PVX groups	*Nx, Nb*	Cockerham, 1955; Santa Cruz *et al.* 1995
	Solanum tuberosum ssp andigena	Immunity to "X" class and some "B" class PVX strains	Rx_{adg}	Ross, 1954; Ritter *et al.* 1991
	Solanum acaule	Immunity to "X" class and some "B" class of PVX	Rx_{acl}	Ross, 1954; Ritter *et al.* 1991
	Solanum chacoense	Hypersensitivity to all groups of PVX	Nx_{acl}	Cockerham, 1970
	Solanum sucrense	Immunity to PVX_{HB}	Rx_{Hbsuc}	Brown *et al.* 1984b
	Solanum phureja	Hypersensitivity to PVX	Nx_{phu}	Tommiska *et al.* 1998
	Series Etuberosa	Extreme resistance	Not determined	Valkonen *et al.* 1992
TRV	*S. tuberosum ssp tuberosum*	Relative resistance	Not determined	Robinson and Dale, 1994; Brown *et al.* 2000

and Jones, 1980; Solomon-Blackburn *et al.* 1994). It has generally been concluded that such tests may accurately distinguish parents with greater combining ability for PLRV resistance through the mean performance of progeny, but do not accurately measure field resistance of individual progeny clones. It would appear that the complex processes involved in expression of partial resistance are best evaluated in plants growing through the full field cycle under aphid inoculation.

Partial resistance to PLRV has been further defined by a number of studies that have sought to partition the components of resistance. These components may be separated into three groups: 1) resistance to infection, 2) resistance to virus accumulation (i.e., titer build-up), and 3) resistance to virus translocation. Studies of restriction of virus accumulation have revealed that failure of virus to reach a high titer is one of the common means by which resistance is expressed (Barker and Harrison, 1985, 1986; Barker, 1987; Wilson and Jones, 1992; 1993a, 1993b). Genetic studies suggest that resistance to virus accumulation is under simple genetic control (Barker and Solomon, 1990). The combination of resistance to virus accumulation with resistance to infection by sexual breeding improves the effectiveness of the resistance (Solomon-Blackburn and Barker, 1993). Aphids acquire virus with low efficiency from clones with combined resistance, leading to an overall reduction of incidence in such a clone (Barker and Woodford, 1992). It has also been found that there is an additive interaction of this kind of host resistance with the coat-protein-mediated transgenic resistance (Barker *et al.* 1994a, 1994b). In a particularly insightful study it was found that transgenic resistance and resistance to titer buildup share the common attribute of restricting or preventing the spread of PLRV out of phloem tissue into surrounding companion cells and mesophyll (Derrick and Barker, 1997). This suggests that both natural host and transgenic coat protein mediated resistance may cause impairment of virus movement. In either case, reduction of disease may be manifested as reduced virus titers and movement of virus from the foliage to the tubers.

Field inoculation of progenies arising from pollination of a series of females with a composite pollen bulk showed distinct differences in combining ability. The Argentine variety "Serrana" was the best parent for genetic transference of PLRV resistance to progeny (Brandolini *et al.* 1992). A field inoculation study of a line x testor mating design consisting of parents crossed to two tester pollen parents was analyzed as parent-progeny regression. A high narrow-sense heritability (h^2 = 0.79) was found for resistance to cumulative natural field infection (Brown *et al.* 1997). The progenies of certain resistant clones (e.g., "Abnaki") showed such a high percentage of resistance that the genetics of resistance may have a relatively simple genetic control. Similar results were found with two parents derived from *S. tuberosum* ssp. *andigena.* Possible simple inheritance was confirmed by Flis and Wasilewics-Flis (1998) in progeny tests of diploid clones with extreme resistance derived from Polish breeding clone DW84-1457. Some clones segregated as if they were heterozygous for a dominant resistance monogene and one clone produced all resistant progeny as if it were homozygous for this resistance allele.

Intolerance to PLRV has been noted in certain varieties ("Apta" and "Carla"). Infected plants expressing intolerance show a severe growth retardation. Plants may fail to develop and produce tubers. Tubers arising from primarily infected intolerant mother plants may fail to sprout. Intolerance appears to be monogenically inherited (Nl_{tbr}). The combination of intolerance with at least moderate partial resistance has been the strategy of several workers (Baerecke, 1961; Butkiewicz, 1978; Zadina and Novak, 1983). The risk exists that a very high level of infection under extreme inoculum pressure would lead to crop failure unless the percentage of infection was low. Intolerance would, however, tend to lead to an automatic purging of PLRV infection from the seed for the following season, since little or no sprouting of the daughter tubers of infected plants occurs.

The search for more extreme forms of PLRV resistance has led to several results. Extreme resistance to PLRV was reported in *S. brevidens,* a Chilean non-tuber-bearing wild species (Jones, 1979; Austin *et al.* 1985; Gibson *et al.* 1988; Valkonen *et al.* 1994b). Subsequent studies found that *S. brevidens* and *S. etuberosum* became infected upon aphid inoculation, but maintained a titer undetectable by ELISA. Graft transmission of virus out of the aphid-inoculated *S. etuberosum* and hybrids with tuber bearing species was documented by Chavez *et al.* (1984a, 1984b). Swiezynski *et al.* (1988, 1989) reported extreme resistance, verified by graft inoculation and ELISA detection, in diploid cultivated potato populations. Ancestors of this material that may have contributed strong resistance were MPI 44.1016/10 (a much used source of resistance to PVX derived from *S. acaule*) and "Apta." Extreme resistance inherited in a dominant monogenic fashion was reported in *S. chacoense* (Brown and Thomas, 1994). Inoculation was performed by grafting infected *Datura tatula* onto test potato plants. The gene symbol Rlchc denotes to this source of resistance.

13.1.3 *Potato virus Y*

Potato virus Y genus *Potyvirus* (PVY) is transmitted in a nonpersistent manner by aphids or mechanically. Aphid transmission is the usual means of natural spread. Symptoms range from mild mosaic to severe necrosis leading to plant death. PVY is less easily controlled in seed programs than

PLRV since insecticides are not effective in preventing introduction or spread (see Chapter 9). Spread of PVY may occur particularly early in the growing cycle before roguing can remove infected plants that provide a source of inoculum.

Partial resistance to PVY has been noted in many varieties and has been characterized genetically in *Solanum phureja* and *S. stenotomum* populations (Vallejo *et al.* 1995). Extreme resistance to PVY has been found in *S. stoloniferum* (Ross 1957, 1960) and *S. tuberosum* ssp *andigena* (Muñoz *et al.* 1975; Galvez and Brown, 1980). Both sources of resistance provided a monogenic dominant source of resistance (Rysto and Ryadg, respectively). An unmapped gene, *Ny*, characterized by a necrotic response, segregates independently from Ry_{adg}. When Ry_{adg} and *Ny* are both present the phenotype is immunity. (Valkonen *et al.* 1994a). The *S. stoloniferum* source of resistance to PVY coincidently carries immunity to *Potato virus A* genus *Potyvirus* (PVA), with attributes similar to PVY (Ross, 1960). The *S. tuberosum* ssp *andigena* source displays a hypersensitive reaction to PVA (Fernandez-Northcote, 1990). Nearly all breeding clones and varieties possessing the *S. stoloniferum* source are male-sterile, presumably due to cytoplasmic male sterility. Both Ryadg and Rysto have been mapped to identical locations on chromosome 11 (Hamalainen *et al.* 1997; Brigneti *et al.* 1997). Although reactions to PVY and PVA were thought to pleiotropic effects, further mapping revealed that both sources of immunities to PVY and PVA are due to closely linked but distinct genes in both sources (Barker, 1996; Hamalainen *et al.* 1998). This region of chromosome 11 also has sequences of DNA, revealed by polymerase chain reaction (PCR) amplification, possessing sequence homology to the N gene from tobacco (Whitham *et al.* 1994) and kinase activities present in other cloned resistance genes (Sorri *et al.* 1999).

These "R-gene analogue primers" have been successfully used to detect Ryadg source of immunity in PCR amplification tests, and can be used as a highly accurate marker for this trait. The co-location of Ry_{sto} and Ry_{adg} appears to be a striking example of synteny, although there is some doubt that the *S. tuberosum* ssp *andigena* source is correctly ascribed. Despite a systematic search, no immunity to PVY has been found in the andigena germplasm of CIP. Additionally, the haploid source of immunity, 2x(v-2)7, used in the Ryadg mapping study (Hamalainen *et al.* 1997) was derived from clone XY13.15 (375335.1=V-2) developed at CIP. This clone was derived from Neo-Tuberosum (pedigree = N-558.14 x Neo-Tuberosum bulk) obtained from the Cornell University Potato Breeding Program. The Cornell population received introductions from Dr. Simmonds' Neo-Tuberosum population developed at the Scottish Plant Breeding Society in Pentlandfield, Scotland, which were among the only survivors after a PVY epidemic decimated the Cornell breeding program. Dr. Simmonds' breeding materials may have received migrant genes from Dr. Cockerham's virus resistance breeding program. The Ry_{adg} gene may be derived from an unknown wild source, possibly *S. stoloniferum*, although this puzzle may never be solved.

A number of other genes were detected in extensive studies of the wild species sources of PVY and PVA resistance by Cockerham (1970). Extreme resistance to PVY and PVA, controlled monogenically, was found in *S. hougasii* (Ry_{hou}). Hypersensitivity to both viruses controlled in a monogenic dominant fashion was found in *S. chacoense,* and *S. microdontum* (Ny_{chc}) and in *S. demissum* (Ny_{dms}). It is likely that further surveys of wild germplasm will find numerous monogenes controlling hypersensitive or extreme resistances to PVY, PVA, and possibly other potyviruses.

The immune reaction of potato breeding clones harboring the Ry_{sto} has been characterized histologically using *Tobacco etch virus* genus *Potyvirus* (TEV) that expresses a glucuronidase (GUS) gene when the genome is transcribed. After inoculation, systemic spread of the virus is not detectable although replication of virus occurs in initially infected cells. Necrotic streaks that become apparent in leaf veins starting 3 days after inoculation are the visible manifestation of a hypersensitive reaction that apparently stops spread of the virus. This reaction is elicited by several distinct potyviruses including TEV normally not found in potato (Hinrichs *et al.* 1998). Immunity to potato virus V (PVV) is also regulated by Ry_{sto} (Barker, 1997). Ry_{sto} is apparently effective against all known strains of PVY including the "common (PVY^{O})" "necrotic (PVY^{N})"and the causative agent of "tuber necrotic ringspot disease" (PVY^{NTN}) (Chrzanowska and Muchalski, 1999).

13.1.4 *Potato virus X*

Single genes for necrotic response to strain groups of *Potato virus X* genus *Potexvirus* (PVX) were discovered in potato varieties (Cockerham, 1954, 1970). Two dominant monogenes were named as controllers of these resistance, *Nx* and *Nb*. Extreme resistance was found in *S. acaule,* with the Max Planck Institute MPI 44.1016/10 breeding clone as the common ancestor for the introgression of this gene Rx_{acl} into advanced materials (Ross, 1954). A monogenic (Rx_{adg}) dominant source of immunity to PVX was identified in *S. tuberosum* ssp *andigena* derived from Commonwealth Potato Collection CPC 1693, and USDA 41956 have also been extensively used in breeding (Ross, 1986). Cockerham (1970) also described a monogene controlling hypersensitive resistance to all forms of PVX derived from *S. chacoense* (Nx_{chc}).

The mapping locations of Rx_{acl}, and Rx_{adg} were determined to be chromosomes 5 and 12, respectively (Ritter *et al.* 1991). Furthermore, the gene Nb has been mapped to the approximate location of Rx_{acl} on chromosome 5 (De Jong *et al.* 1997). A new dominant monogene extracted from *S. phureja* (a taxon encompassing some cultivated diploid cultivars native to the Andes) for hypersensitivity to PVX (Nx_{phu}) was described and mapped by Tommiska *et al.* (1998) to chromosome 9. Figure 13.1.1 presents a summary of the virus resistance genes that have been mapped to date.

Studies of the mechanism of action of Rx_{adg} have shown that resistance is activated by PVX coat protein even in the absence of other PVX encoded proteins (Bendahmane *et al.* 1995). The response is non-specific and can effect a prevention of infection by *Cucumber mosaic virus* genus *Cucumovirus* (unrelated to PVX) or the resistance breaking strain PVX_{HB} in mixed infections in protoplasts (Köhm *et al.* 1993). The synthesis of new RNA is not required for the mechanism to actuate, indicating perhaps a constitutive expression of Rx_{adg} (Gilbert *et al.* 1998). Resistance to Rx_{adg} and Rx_{acl} was broken by an aberrant strain PVX_{HB} discovered in Bolivia (Moreira and Jones, 1980; Jones, 1985). An examination of the amino acid sequence of twelve isolates of PVX lead to the designation of two groupings based on differences in 14 amino acid positions in their coat proteins. Those avirulent on *Nx* and Rx_{acl} genes were of class "X" while those capable of overcoming Nx were of type "B" (Santa Cruz and Baulcombe, 1995). Subsequent studies have shown that although PVX_{HB} differs from other PVX isolates at a number of amino acid residues, and one in particular plays a crucial role in the breaking of resistance. The coat protein of PVX_{HB} lacks the avirulence property that elicits the Rx_{adg}–mediated response. An accession of *S. sucrense* (11926.4) has shown monogenic dominant resistance to PVX_{HB} (Brown *et al.* 1984b), but was defeated by other "HB-like" coat protein mutant laboratory strains of PVX (Goulden *et al.* 1993; Querci *et al.* 1995).

Sequencing and subsequent determination of the Rx_{adg} gene product protein structure (Bendahmane *et al.* 1997; 1999) revealed similarities to other leucine-rich-repeat R-genes, such as the N gene from tobacco encoding resistance to *Tobacco mosaic* genus *Tobamovirus* (Whitham *et al.* 1994). A diagrammatic representation of this virus-host interaction is presented in model form in Figure 13.1.2. The key interaction is recognition between a viral avirulence factor and the host resistance factor. The PVX coat protein serves, apparently, as an avirulence factor for a constitutively expressed R_{adg} recognition gene product. Recognition invokes a distinct non-specific resistance response that functions against unrelated viruses and resistance breaking strains of the virus if present with an avirulence factor that accomplishes recognition. The ensuing response is hypersensitive in nature, involving death of cells immediately around the initial contact of host and virus. Susceptibility and systemic infection of the virus occurs when the host resistance factor does not recognize the viral avirulence factor.

13.1.5 *Potato virus S*

Potato virus S genus *Carlavirus* (PVS) is transmitted by contact and by aphids. It is the most ubiquitous of potato viruses, but usually presents mild symptoms or is latent. It is often the first virus to re-infect virus-free seedstocks after field multiplication has begun. A necrotic response to PVS was discovered in *S. tuberosum* ssp *andigena* native cultivar "*Huaca ñahui*" (PI 258907), and the inheritance was determined to be monogenic dominant (Ns_{adg}) by Baerecke (1967b) and Makarov (1975). The American variety "Saco" cannot be infected with PVS by mechanical inoculation. It was determined that resistance derived from "Saco" was not extreme and that the inheritance appears to be polygenic and recessive, and thus a poor genetic source of resistance in a breeding program (Baerecke, 1967a). Pietrak (1985) graft-tested breeding clones and varieties harboring the Ns_{adg} gene, finding variation in necrotic response severity. PVS was occasionally detected in daughter tubers that developed on graft inoculated plants. .

13.1.6 *Potato virus M*

Potato virus M genus *Carlavirus* (PVM) is transmitted by aphids and contact. It is economically important in Eastern Europe, particularly Poland. Resistance in the form of hypersensitivity was discovered in *S. megistacrolobum* and *S. stoloniferum (*Dziewonska and Ostrowska, *1977;* Dziewonska *et al.* 1978). Relative resistance to PVM, similar to the partial resistance to PLRV known in cultivated *S. tuberosum* ssp. *tuberosum* germplasm was discovered in *S. gourlayi* (Was *et al.* 1980). The inheritance was found to be monogenic dominant and the gene symbol *Gm* applied to this source (Swiezynski *et al.* 1981). Resistance was strong enough to be evaluated by graft inoculation.

13.1.7 *Tobacco rattle virus*

Tobacco rattle virus genus *Tobravirus* (TRV) is vectored exclusively by stubby root nematodes (*Trichodorus* and *Paratrichodorus* spp.). It is a locality-specific problem, threatening to become more important in sandy

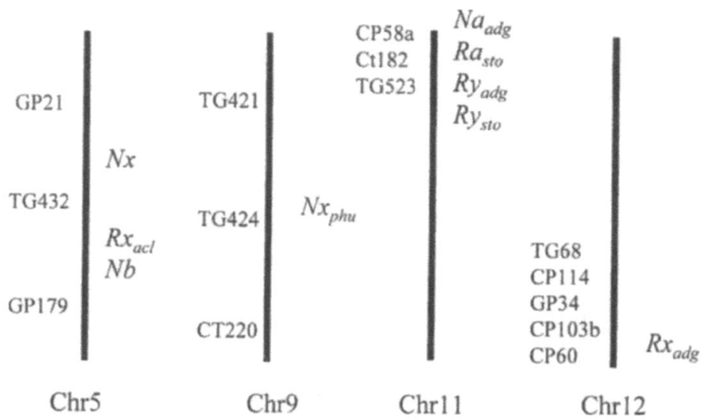

Figure 13.1: Map locations of several virus resistance genes.

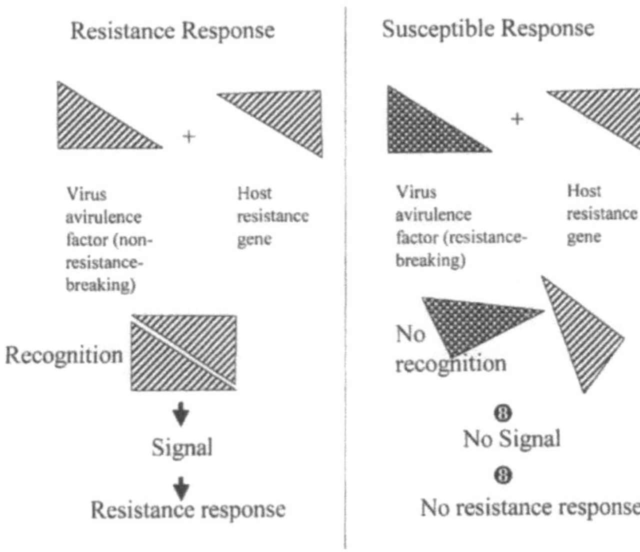

Figure 13.2. Diagrammatic representation of function of the *Rx* gene for resistance to PVX. The gene encodes a product that recognizes PVX coat protein, leading to the onset of a hypersensitive resistance response.

Breeding for Virus Resistance

Year	Stage	Virus Screening
1	Seedling population & tuber production for field planting	Exposure to virus aphid or mechanical inoculation
2	Field selection of single hills	
3	2nd year of field exposure to viruses	Graft testing for extreme resistance
4	Field exposure to viruses	Replicated plots exposed to virus by aphid and mechanical inoculation followed by winter indexing
5	2nd year selection (12-hills)	

Figure 13.3. Diagram of a sexual breeding program for potato indicating points in the scheme where selection for resistance to viruses might occur.

soils of Europe and North America where various species of the vector thrive. The disease symptom (corky ringspot, or spraing) is expressed as necrotic lesions in the tuber flesh which are sometimes visible on the tuber skin. The disease can be controlled by soil fumigation and nematicides to reduce nematode populations, but resistance is present in many *S. tuberosum* ssp *tuberosum* cultivars. A high level of resistance has been noted in particular varieties emerging from several independent breeding programs in Great Britain, Poland, Germany and the US (Robinson and Dale, 1994; Brown *et al.* 2000). Scottish researchers have pointed out that the term "resistance" refers to lack of necrotic response. Hence they have used the term "insensitivity" as a desirable (symptomless) response to TRV challenge. The inheritance of resistance is not resolved. Certain older varieties such as "Bintje" have high resistance. It appears clear, however, that resistance is not to the nematode, but rather to the virus (Brown *et al.* 2000).

13.1.8 Breeding for Virus Resistance

Potatoes are bred by use of sexual crosses between parents that complement each other for numerous traits. Sexually produced seedlings, after initial selection, are propagated vegetatively from that point on. Breeding that emphasizes virus resistance may employ selection applied to the seedlings soon after germination as shown in Figure 13.1.3. Viruliferous aphids with PLRV may be applied, or seedlings can be manually inoculated with contact-transmitted viruses by hand or by use of a pressurized spray gun. The objective is to remove susceptible seedlings and thereby expend future resources only on resistant individuals in future years. It is debatable whether selection for relative resistance to PLRV is effective at this stage, but the monogenically inherited resistance genes (e.g., Rx_{acl}, Rx_{adg}, Ry_{sto}, Ns_{adg}, Ry_{adg}, Ra_{sto}, Na_{adg}, and Gm) can be efficiently screened. Such seedling screening is rarely employed now, but field screening is often carried out after the second field selection when a sufficient number of tubers are available for replicated plot experiments. The extreme resistance of *Ry* and *Rx* genes can be effectively tested by graft inoculation at this point. One advantage of graft inoculation is that it can effectively identify bearers of the R genes versus resistant mimics that are usually difficult to infect in the field but become systemically infected upon grafting. These selections do not transmit resistance effectively to their progenies and are undesirable as parents.

Potato production worldwide is generally dominated by very old varieties and a few newer ones due to the difficulty in recovering good varieties from sexual breeding. The chances that selections with virus resistance will also be superior in enough additional characters to be commercially successful are very low. Furthermore virus resistance *per se* is often assigned a low priority. Obviously yield, appearance, and fresh market or processing characters take precedence over disease resistance as long as some method for control other than host resistance is possible. Virus-free seed is available at costs that are acceptable in the developed world. In developing countries with capital-scarce production, seed programs are nonexistent or inadequate, and the cost of importing virus-free seed is highly limiting. For these reasons, virus resistance has been emphasized at the International Potato Center (CIP) for many years. Still no variety with widespread commercial acceptance has high resistance to PVY or PLRV. Although many European cultivars are PVX and PVY immune, no economically important varieties possess PVY immunity. Exceptions to this are the varieties "Orbita" and "Belorusky 3" of Byelorussia. Important varieties in North America, such as "Atlantic" and "Ranger Russet," and a number of European varieties including "Agria" and "Saturna" harbor Rx_{adg} and are PVX immune as a result of the cosegregation of cyst nematode resistance with this trait (Swiezynski *et al.* 1998). The Ry_{sto} sources of resistance to PVY often transmit to their sexual progeny long stolons, small tubers and lateness. As a consequence wild sources of virus resistance have yet to find widespread application. Useful traits are subject to linkage drag, the inefficient separation of desirable genes from large numbers of deleterious linked genes co-segregating during the sexual process. Identification of duplex and triplex parents has been promoted as a means of accelerating breeding with the expectation that 82 %, and 100% of the progeny would be immune to particular viruses, respectively, (chromosome inheritance) when these genotypes are used as parents (Swiezynski, 1994). The main drawback to this approach is that a parent must first have good general combining ability for numerous agronomic characters for there to be a chance for the virus resistant progeny to achieve commercial success. The monogenic completely dominant inheritance of these genes means that there does not have to be an accumulation of genes for good expression of resistance. Restricting parentage to those multiplex for virus resistance monogenes would unwisely limit the choice of parents that need to be chosen primarily for a number of agronomic characters under complex polygenic control.

Bearing this in mind, it is paramount that those engaged in virus resistance breeding set reasonable goals. Combining resistance to four or five viruses derived from wild sources, although an admirable objective, will lower the probability of successful variety production to very low levels. Rather, the incorporation of one major virus resistance is an aim much more likely to combined with agronomic success. It is clear, however, that sources of wild species virus resistance require extensive pre-breeding in the fomr of an adequate number of cycles of recurrent selection and /or backcrossing to make them desirable parents sufficiently free of hitchhiking deleterious genes, in crosses directed at variety breeding.

Interestingly, Ry_{sto} and Ry_{adg} genes appear to be strong monogenes that have yet to be broken by new strains of PVY. Rx_{acl} and Rx_{adg} have been broken by PVX_{HB} and laboratory mutants, but these genes are still to be recommended due to the very limited threat that PVX_{HB} poses worldwide. Resistance to PLRV available in the most advanced breeding lines is difficult to breed into new varieties due to its polygenic nature. Alternatively, a closer look at simply inherited resistance to virus accumulation is highly warranted. However, more extreme forms of resistance from wild species and diploid cultivated germplasm appear to be simply inherited and should be the subjects of intense pre-breeding. The rapid progress in resistance gene cloning and analysis probably means that cloned natural host resistance genes will be transformable into popular varieties in the near future.

References

Akeley, R.V., Murphy, H.J. and Cetas, R.C. 1971. 'Abnaki:' a new high-yielding potato variety resistant to Verticillium wilt and leafroll. *Am. Pot. J.* 48: 230-233.

Austin, S., Baer, M.A. and Helgeson, J.P. 1985. Transfer of resistance to potato leafroll virus from *Solanum brevidens* into *Solanum tuberosum* by somatic hybridization. *Plant Science* 39: 75-82.

Baerecke, M.-L. 1958. Blattrollresistenzzuchtung. In: *Handbuch der Pflanzenzuchtung* (H. Kappert, and W. Rudorf, eds.), Verlag P. Parey, Berlin & Hamburg, Germany., pp. 97-106.

Baerecke, M.-L. 1961. Erfahrungen mit einjahrigen Kartoffelabbauversuchen unter starken Blattroll-Infectionsbedingungen. *Zeit. fur Pflanzenzuchtung* 45: 225-253.

Baerecke, M.-L. 1967a. Prufung von Saco und Saco-kreuzungen auf Resistenz gegen S-virus der Kartoffel. *Euro. J. Potato Res.* 10: 206-220.

Baerecke, M.-L. 1967b. Uberempfindlichkeit gegen das S-virus der Kartoffel in einem bolivanischen Andigena-Klon. *Zuchter* 37: 281-286.

Bagnall, R.H. and Tai, G.C.C. 1986. Potato leafroll virus: evaluation of resistance in potato cultivars. *Plant Dis.* 70: 621-623.

Barker, H. and Harrison, B.D. 1985. Restricted multiplication of potato leafroll virus in resistant potato genotypes. *Ann. appl. Biol.* 107: 205-212.

Barker, H. and Harrison, B.D. 1986. Restricted distribution of potato leafroll virus antigen in resistant potato genotypes and its effect on transmission of the virus by aphids. *Ann. appl. Biol.* 109: 595-604.

Barker, H. 1987. Multiple components of the resistance of potato to potato leafroll virus, *Ann.,appl. Biol.* 111: 641-648.

Barker, H. and Solomon, R.M. 1990. Evidence of simple genetic control in potato of ability to restrict potato leaf roll virus concentration in leaves. *Theor. Appl. Genet.* 80: 188-192.

Barker, H. and Woodford, J.A.T. 1992. Spread of potato leafroll virus is decreased from plants of potato clones in which virus accumulation is restricted. *Ann. appl. Biol.* 121: 345-354.

Barker, H., Reavy, B., Arif, M., Webster, K.D. and Mayo, M.A. 1994a. Towards immunity to potato leafroll virus and potato mop-top virus by using transgenic and host gene-mediated forms of resistance. *Aspects appl. Biol.* 39: 189-194.

Barker, H., Webster, K.D., Jolly, C.A., Reavy, B., Kumar, A. and Mayo, M.A. 1994b. Enhancement of resistance to potato leafroll virus multiplication in potato by combining the effects of host genes and transgenes. *Mol. Plant-Microbe Iinter.* 7: 528-530.

Barker, H. 1996. Inheritance of resistance to potato viruses Y and A in progeny obtained from potato cultivars containing *Ry*: evidence for the a new gene for extreme resistance to PVA. *Theor. Appl. Genet.* 93: 710-716.

Barker, H. 1997. Extreme resistance to potato virus V in clones of *Solanum tuberosum* that are also resistant to potato viruses Y and A: Evidence for a locus conferring broad-spectrum potyvirus resistance. *Theor. Appl. Genet.* 95: 1258-1262.

Bendahmane, A., Kohm, B.A., Dedi, C. and Baulcombe, D.C. 1995. The coat protein of potato virus X is a strain-specific elicitor of *Rx1*-mediated virus resistance in potato. *Plant J.* 8: 933-941.

Bendahmane, A., Kanyuka, K. and Baulcombe, D.C. 1997. High-resolution genetical and physical mapping of the *Rx* gene for extreme resistance to potato virus X tetraploid potato, *Theor. Appl. Genet.* 95: 153-162.

Bendahmane, A., Kanyuka, K. and Baulcombe, D.C. 1999. The *Rx* gene from potato controls separate virus resistance and cell death responses. *Plant Cell* 11: 781-791

Brandolini, A., Caligari, P.D.S. and Mendoza, H.A. 1992. Combining resistance to potato leafroll virus (PLRV) with immunity to potato viruses X and Y (PVX and PVY). *Euphytica* 61: 37-42.

Brigneti, G., Garcia-Mas, J. and Baulcombe, D.C. 1997. Molecular mapping of the potato virus Y resistance gene Ry_{sto} in potato. *Theor. Appl. Genet.* 94: 198-203.

Brown, C.R. 1979. General combining ability for PLRV resistance in Andigena germplasm. *Am. Potato. J.* 56: 456.

Brown, C.R. 1980. Incorporation of virus resistance and future plans. In: *Strategy for Virus Management in Potatoes.* Internatl. Potato Ctr., Lima, Peru, pp. 63-78.

Brown, C.R., Fernandez-Northcote, E.N., Jayasinghe, U. and Salazar., L. 1984a. Breeding virus-resistant potato cultivars for developing countries, *CIP Circular.* 12:1-4.

Brown, C.R., Salazar, L.F., Ochoa, C. and Chuquillanqui, C. 1984b. Strain-specific immunity to PVX_{HB} is controlled by a single dominant gene. In*: 9th Triennial Conf. Eur. Assoc. Potato Res.*, European Association for Potato Research, Interlaken, Switzerland, pp. 249-250.

Brown, C.R. 1985. Genetic studies and breeding of resistance to PLRV, PVY, and PVX. In: *Present and Future Strategies for Potato Breeding and Improvement*: Report of the XXVI Planning Conference,, International Potato Center, Lima, Peru, pp. 17-44.

Brown, C.R. and Thomas, P. E. 1994. Resistance to potato leafroll virus derived from *Solanum chacoense*: characterization and inheritance. *Euphytica* 74: 51-57.

Brown, C.R., Corsini, D., Pavek, J.J. and Thomas, P.E. 1997. Heritability of field resistance to potato leafroll virus in cultivated potato. *Plant Breeding* 116: 585-588.

Brown, C.R. Mojtahedi, H., Santo, G.S., Hamm, P., Crosslin, J.M., and Thomas, P.E. 2000. Germplasm resistant to corky ringspot disease. *Am. J. Potato Res.*, 73: 23-27.

Butkiewicz, H. 1978. Nietolerancja w stosunku do wirusa lisiozwoju wystepujaca u roslin ziemniaka Czesc I. Objawy chorobowe na rosinach zakazonych (Intolerance to potato leafroll virus (PLRV) occurring in potato plants). *Ziemniak* :5-37.

Chavez, R., Brown, C.R. and Iwanaga, M. 1988a. Application of interspecific sesquiploidy to introgression of PLRV resistance from non-tuber-bearing *Solanum etuberosum* to cultivated potato. *Theor. Appl. Genet.* 76: 497-500.

Chavez, R., Brown, C.R. and Iwanaga, M. 1988b. Transfer of resistance to PLRV titer buildup from *Solanum etuberosum* to a tuber-bearing *Solanum* gene pool. *Theor. Appl. Genet.* 76: 129-135.

Chrzanowska, M. and Muchalski, T. 1999. Potato cultivars possessing Ry_{sto} gene react to PVY with internal necroses after graft inoculation. *Abstracts of the 14thTriennial Meeting for the EAPR.* Sorrento, Italy, pp. 543-544.

Chuquillanqui, C. and Jones, R.A.C. 1980. A rapid technique for assessing the resistance of families of potato seedling to potato leaf roll virus. *Potato Res.* 23: 121-128.

Cockerham, G. 1954. Strains of potato virus X. In: *Proc. Sec. Conf. Potato Virus Diseases.* H. Veenman & Zonen- Wageningen, Lisse-Wageningen, pp. 89-92.

Cockerham, G. 1970. Genetical studies on resistance to potato viruses X and Y. *Heredity* 25: 309-348.

Corsini, D., Pavek, J.J., Martin, M.W. and Brown, C.R. 1994. Potato germplasm with combined resistance to leafroll virus and viruses X and Y. *Am. Potato J.* 71: 377-385.

Davidson, T.M.W. 1973. Assessing resistance to leafroll in potato seedlings. *Potato Res.* 16: 99-108.

Davidson, T.M.V. 1980. Breeding for resistance to virus disease of the potato (*Solanum tuberosum*) at the Scottish Plant Breeding Station. In: *Fifty-Ninth Ann. Rep. Scottish* Plant *Breeding Station 1979-1980.* pp. 100-108.

DeJong, W., Forsyth, A., Leister, D., Gebhardt, C. and Baulcombe, D.C. 1997. A potato hypersensitive resistance gene against potato virus X maps to a resistance gene cluster on chromosome 5. *Theor. Appl. Genet.* 95: 246-252.

Derrick, P.M. and Barker, H. 1997. Short and long distance spread of potato leafroll luteovirus: effects of host genes and transgenes conferring resistance to virus accumulation in potato. *J. Gen. Virol.* 78: 243-251.

Dziewonska, M.A. and Ostrowska, K. 1977. Necrotic reaction to potato virus M in *Solanum stoloniferum* and *S. megistacrolobum. Phytopathol. Z.* 88: 172-179.

Dziewonska, M.A., Czech, B., Ostrowska, K. and Was, M. 1978. Reaction to potato virus M (PVM) of hybrids with gene Rm derived from *Solanum megistacrolobum. Abstr. Seventh Triennial Conference EAPR.* Warsaw. 155-156.

Fernandez-Northcote, E.N. 1990. Variability of PVX and PVY and its relationship to genetic resistance. In: *Control of Virus and Virus-like Diseases of Potato and Sweet Potato.* International Potato Center, Lima, Peru, pp. 131-139.

Flis, B. and Wasilewicz-Flis, I. 1998. Progeny tests to identify diploid potato clones homozygous at loci controlling resistance to PLRV. *Potato Res.* 41: 219-228.

Galvez, R. and Brown, C.R. 1980. Inheritance of extreme resistance to PVY derived from *Solanum tuberosum* ssp. *andigena. Am. Potato J.* 57: 476-477.

Gibson, R.W., Jones, M.G.K. and Fish, N. 1988. Resistance to potato leaf roll virus and potato virus Y in somatic hybrids between diphaploid *Solanum tuberosum* and *S. brevidens. Theor. Appl. Genet.* 76: 113-117.

Gilbert, J., Spillane, C., Kavanaugh, T.A. and Baulcombe, D.C. 1998. Elicitation of *Rx*-mediated resistance to PVX in potato does not require new RNA synthesis and may involve a latent hypersensitive response. *Molec. Plant-Microbe Interactions* 11: 833-835.

Goulden, M.G., Kohm, B.A., Cruz, S.S., Kavanaugh, T.A. and Baulcombe, D.C. 1993. A feature of the coat protein of potato virus X affects both induced virus resistance in potato and viral fitness. *Virology* 197: 293-302.

Hamalainen, J.H., Watanabe, K.N., Valkonen, J.P.T., Arihara, A., Plaisted, R.L., Pehu, E., Miller, L. and Slack, S.A. 1997. Mapping and marker-assisted selection of a gene for extreme resistance to potato virus Y. *Theor. Appl. Genet.* 94: 192-197.

Hamalainen, J.H., Sorri, V.A., Watanabe, K.N., Gebhardt, C. and Valkonen, J.P.T. 1998. Molecular examination of a chromosome region that controls resistance to potato Y and A potyviruses in potato. *Theor. Appl. Genet.* 96: 1036-1043.

Hamann, W., Gall, H. and Moller, K.H. 1968. Erfahrungen bei der Prüfung von Kartoffelzuchtmateriel auf Blattrollvirusresistenz im Laboratorium. *Theor. Appl. Genet.* 38: 90-96.

Hinrichs, J., Berger, S. and Shaw, J.G. 1998. A hypersensitive response-like mechanism is involved in resistance of potato bearing the Ry_{sto} gene to the potyviruses potato virusY and tobacco etch virus. *J. Gen. Virol.* 79:167-176.

Hooker, W.J. 1981. *Compendium of Potato Diseases*. American Phytopathology Society, St. Paul, MN, USA.

Jones, R.A.C. 1979. Resistance to potato leaf roll virus in *Solanum brevidens*. *Potato* Res. 22: 149-152.

Jones, R.A.C. 1985. Further studies on resistance-breaking strains of potato virus X. *Plant Pathol.* 34: 182-189.

Köhm, B.A., Goulden, M.G., Gilbert, J.E., Kavanaugh, T.A. and Baulcombe, D.C. 1993. A potato virus X resistance gene mediates an induced, nonspecific resistance in protoplasts. *Plant Cell* 5: 913-920.

Makarov, P.P. 1975. Inheritance of resistance to potato virus S. *Potato Res.* 18: 326-329.

Moreira, A. and Jones, R.A.C. 1980. Properties of a resistance-breaking strain of potato virus X. *Ann. appl. Biol.* 95: 93-103.

Muñoz, F.J., Plaisted, R.L. and Thurston, H.D. 1975. Resistance to potato virus Y in *Solanum tuberosum* ssp *andigena*.. *Am. Potato J.* 52: 107-115.

Pietrak, J. 1985. Reaction of potato clones with the gene *Ns* to inoculation with PVS by grafting. *Biuletyn Inst. Ziemniaka* 32: 17-24.

Querci, M., Baulcombe, D.C., Goldbach, R.W. and Salazar, L.F. 1995. Analysis of the resistance-breaking determinants of potato virus X (PVX) strain HB on different potato genotypes expressing extreme resistance to PVX. *Mol. Plant Pathol.* 85: 1003-1010.

Ritter, E., Debener, T., Barone, A., Salamini, F. and Gebhardt, C. 1991. RFLP mapping on potato chromosomes of two genes controlling extreme resistance to potato virus X (PVX). *Mol. Gen. Genet.* 227: 81-85.

Rizvi, A. 1980. Progress in identifying resistance to PLRV. In: *Strategy for Virus Management in Potatoes.* International Potato Center, Lima, Peru, pp. 22-28

Robinson, D.J. and Dale, M.F.B. 1994. Susceptibility, resistance and tolerance of potato cultivars to tobacco rattle virus infection and spraing disease. *Aspects Appl. Biol.* 39: 61-66.

Ross, H. 1954. Die Vererbung der 'Immunität' gegen das X-Virus in tetraploidem *Solanum acaule. Caryologia* (Atti del IX Congr. Intern. Genetica) 6 Suppl.: 1128-1132.

Ross, H. 1957. Inheritance of extreme resistance to virus Y in *Solanum stoloniferum* and its hybrids with *Solanum tuberosum*. In: *Proc. Third Conf. Virus Diseases*. H. Veenman & Zonen Wageningn, Lisse- Wageningen, pp. 204-211.

Ross, H. 1960. Über die Vererbung von Eigenschaften für Resistenz gegen das Y- und A-virus in *Solanum stoloniferum* und die mögliche Bedeutung fur eine allgemeine Genetik der Virusresistenz in *Solanum* Sect. Tuberarium. In: *Proc. Fourth Conf. Potato Virus Dis.* Braunschweig, Germany, pp. 40-49.

Ross, H. 1986. Potato breeding - problems and perspectives. In: Adv. Plant Breeding. Paul Parey, Berlin and Hamburg.

Salaman, R.N. 1921. Degeneration of potatoes (sic.). In: *Rep. Intern. Potato Conf.* pp. 79-91.

Salaman, R.N. 1985. The history and the social influence of the potato (J. G. Hawkes, ed.), *Cambridge Univerisity Press,* Cambridge, U.K., pp. 685.

Santa Cruz, S. and Baulcombe, D. 1995. Analysis of potato virus X coat protein genes in relation to resistance conferred by the genes *Nx, Nb* and *Rx1* of potato. *J. Gen. Virol.* 76: 2057-2061.

Solomon-Blackburn, R.M. and Barker, H. 1993. Resistance to potato leafroll luteovirus can be greatly improved by combining two independent types of heritable resistance. *Ann. appl. Biol.* 122: 329-336.

Solomon-Blackburn, R.M., Mackay, G.R. and Brown, J. 1994. An evaluation of seedling progeny tests for resistance to potato leafroll virus in potato. *J. Agric. Sci.* 122: 231-239.

Sorri, V.A., Watanabe, K.N. and Valkonen, J.P.T. 1999. Predicted kinase-3a motif of a resistance gene analogue as a unique marker for virus resistance. *Theor. Appl. Genet.* 99: 164-170.

Swiezynski, K., Dziewonska, M. and Ostrowska, K. 1981. Inheritance of resistance to potato virus M found in *Solanum gourlayi* Haw. *Gen. Pol.* 22: 1-8.

Swiezynski, K.M., Dziewonska, M.A. and Ostrowska, K. 1988. Reaction to the potato leafroll virus (PLRV) in diploid potatoes. *Potato Res.* 31: 289-296.

Swiezynski, K.M., Dziewonska, M.A. and Ostrowska, K. 1989. Resistance to potato leafroll virus (PLRV) in diploid potatoes. *Pl. Breed.* 103: 221-227.

Swiezynski, K.M. 1994. Inheritance of resistance to viruses. In: *Potato Genetics* (J. E. Bradshaw, and G. R. Mackay, eds.), CAB International, Cambridge, UK, pp. 339-364.

Swiezynski, K.M. Sieczka, M.T., Stypa, I. and Zimnoch-Guzowska, E. 1998. Characteristics of major varieties from Europe and North America. *Plant Breeding and Seed Science.* Suppl Vol 42: 3-44

Tommiska, T.J., Hamalainen, J.H., Watanabe, K.N. and Valkonen, J.P.T. 1998. Mapping of the gene Nx_{phu} that controls hypersensitive resistance to potato virus X in *Solanum phureja* IvP35. *Theor. Appl. Genet.* 96: 840-843.

Valkonen, J.P.T. 1994. Natural genes and mechanisms for resistance to viruses in cultivated and wild potato species (*Solanum* spp.). *Pl. Breed.* 112: 1-16.

Valkonen, J.P.T., G. Brigneti, Salazar, L.F., Pehu, E. and Gibson, R.W. 1992. Interactions of the *Solanum* spp. of the Etuberosa Group and nine potato-infecting viruses and a viroid. *Ann. appl. Biol.* 120: 301-303.

Valkonen, J.P.T., Slack, S.A., Plaisted, R.L. and Watanabe, K.N. 1994a. Extreme resistance is epistatic to hypersensitive resistance to Potato Virus Y^o in a *Solanum tuberosum* subsp. *andigena*-derived potato genotype. *Plant Dis.* 78: 1177-1180.

Valkonen, J.P.T., Xu, Y.-S., Rokka, V.M., Pulli, S. and Pehu, E. 1994b. Transfer of resistance potato leafroll virus, potato virus Y and potato virus X from *Solanum brevidens* to *S. tuberosum* through symmetric and designed asymmetric somatic hybridisation. *Ann. appl. Biol.* 124: 351-362.

Vallejo, R.L., Collins, W.W. and Young, J.B. 1995. Inheritance of resistance to potato virus Y and potato virus X in hybrid *Solanum phureja* x *S. stenotomum* diploid potatoes. *J. Heredity* 86: 89-93.

Was, M., Dziewonska, M., Ostrowska, K. and Kowolska, A. 1980. Reaction of *Solanum gourlayi* and its hybrids with *S. tuberosum* to potato virus M (PVM). *Phytopathol. Z.* 97: 186-191.

Whitham, S., Dinesh-Kumar, S.P., Choi, D., Hehl, R., Corr, C. and Baker, B. 1994. The product of the tobacco mosaic virus resistance gene N: similarity to Toll and the Interleukin-1 receptor. *Cell* 78: 101-1115.

Wilson, C.R. and Jones, R.A.C. 1992. Resistance to phloem transport of potato leafroll virus in potato plants. *J. Gen. Virol.* 73: 3219-3224.

Wilson, C.R. and Jones, R.A.C. 1993a. Resistance to potato leafroll virus infection, accumulation in potato cultivars, and the effects of previous infection with other viruses on expression of resistance. *Austr. J. Agric. Res.* 44: 1891-1904.

Wilson, C.R. and Jones, R.A.C. 1993b. Evaluation of resistance to potato leafroll virus in selected potato cultivars under field conditions. *Austr. J. Exp. Agric.* 33: 83-90.

Zadina, J. and Novak, F. 1983. The inheritance of extreme intolerance to potato leaf roll virus (PLRV), Sbor. *UVTIZ – Genet. a Slecht.* 19: 189-194.

13. Resistance

13.2 Biotechnology and Resistance to Potato Viruses

Philip Berger[1] and Thomas German[2]
[1]*Dept. of Plant, Soil and Entomological Sciences, University of Idaho, Moscow, ID, 83844-2339 USA,* [2]*Dept. of Plant Pathology, University of Wisconsin, Madison, WI, 53706 USA*

13.2.1 Transformation of Potato

Potatoes are susceptible to many viruses, a situation that is exacerbated by viruses that can be transmitted through tuber seed pieces. Such vegetative propagation readily perpetuates and disperses infectious viruses and the diseases they cause. In general, most viruses are controlled by a combination of strategies that include seed certification and agronomic practices (see chapters 14, 15, 17), the most effective of which is host plant resistance. However, obtaining virus resistance is problematic. It is difficult and sometimes impossible to instill sources of resistance into cultivated potato from wild relatives, and obtain or retain agronomically useful cultivars. As a consequence, seed potato certification programs expend significant time and resources to minimize the amount of virus inoculum that enters commercial production. The economic and environmental cost of

G. Loebenstein et al. (eds.),
Virus and Virus-like Diseases of Potatoes and Production of Seed-Potatoes, 341–363.

these practices, such as the application of pesticides for vector control, is of serious concern both to potato producers and consumers.

Several inter-related technologies must be in place to genetically engineer potato. An effective transformation system requires that the target tissue(s) be relatively easily manipulated in tissue culture, and that they can be induced to form fully differentiated plants (regenerated). In addition, the laboratory must be facile with various molecular biological techniques necessary, such as vector construction and manipulation and analysis of transgenic plants. Thus, genetic engineering is a complex, relatively expensive and laborious process. Furthermore, addressing current regulatory constraints so that genetically engineered potatoes can be grown commercially, involves additional time and expense.

While transgenic potatoes can be produced by a number of methods, *Agrobacterium*-mediated plant transformation (e.g., Hoekema *et al.* 1983; Horsch *et al.* 1984) is the most efficient and rapid method, and thus is the method of choice. The first examples of Agrobacterium-mediated potato transformation date back to the 1980s (Ooms *et al.* 1987; Ooms *et al.* 1983; Ooms and Lenton, 1985; Shahin and Simpson, 1986; Sheerman and Bevan, 1988; Stiekema *et al.* 1988; Tavazza *et al.* 1988).

Several methods available for *Agrobacterium*-mediated potato transformation were demonstrated. Generally, one uses a binary vector system and leaf discs, tuber discs or stem internodes as target tissue for the *Agrobacteria*. The composition of culture media, culture conditions, choice of target tissue, potato cultivar, strain of *Agrobacterium*, etc., can all influence the efficiency of transformation. Nevertheless, genetic transformation of potato can now be performed by any plant tissue culture and molecular biology laboratory.

In the mid-1980s, partly as a result of advances in plant transformation, it also became apparent that pathogen-derived resistance could be used to protect crops against virus infection. Transgenic resistance to plant viruses has been the topic of a number of recent reviews (Baulcombe, 1996; Beachy *et al.* 1990; Candresse, 1996; Crute and Pink, 1996; Fitchen and Beachy, 1993; Hackland *et al.* 1994; Huisman *et al.* 1992; Lomonossoff, 1995; Palukaitis and Zaitlin, 1997; Prins and Goldbach, 1996; Scholthof *et al.* 1993; Wilson, 1993).

Transgenic resistance to potato viruses has been the subject of past reviews (Huisman *et al.* 1992; Martin, 1994; van den Elzen *et al.* 1989). Since the first observation of coat protein-mediated resistance (Powell Abel *et al.* 1986), numerous approaches to pathogen-derived resistance have been observed, a number of which have been applied to potatoes. While products of this research are being used commercially, the variety of mechanisms underlying pathogen-derived resistance are still under intense study, as is non-pathogen-derived resistance.

13.2.2 Transgenic Resistance to *Potato Leafroll Virus*

Potato leafroll virus genus *Polerovirus* (PLRV) is frequently the most important viral pathogen which affect potato. It is also difficult to control, using either host plant resistance or other means. Thus, attempting to instil host resistance using biotechnology seems a reasonable approach.

Coat protein (CP) mediated resistance to PLRV was first reported in 1990 (Kawchuk *et al.* 1990). Transgenic potatoes of cv. Desireé were generated containing a PLRV CP under control of an enhanced cauliflower mosaic virus (CaMV) 35S promoter. The construct was a wild type CP, with 12 nucleotides (nt) upstream of the AUG start codon and 112 nt downstream of the stop codon. Of 12 plants selected, one had an apparent abnormal phenotype. Northern blot analysis indicated that the putative transgenic lines contained detectable amounts of transcript of the expected size. However, only very low levels of CP were detected in some lines and no CP was identified in others. The titer of PLRV was followed in 5 transgenic lines for 50-70 days after inoculation with viruliferous aphids. Only low levels of virus were detected in the inoculated transgenic lines over the course of these experiments. Varying the number of viruliferous aphid from 1 to 25 did not appear to affect the resistance phenotype. Two subsequent papers (Kawchuk *et al.* 1991; van der Wilk *et al.* 1991) provided essentially similar results using PLRV CP, and Kawchuk *et al.* (1991) showed that expression of CP in an antisense orientation could also confer resistance to PLRV. Thus, while PLRV replication is not absent, the levels of virus produced in these plants is only a fraction of that produced in non-protected potato (or tobacco) plants. Such CP-mediated resistance therefore likely to be of economic value (Barker *et al.* 1992; Barker *et al.* 1993; Brown *et al.* 1995; Derrick and Barker, 1992; Kawchuk *et al.* 1990; Kawchuk *et al.* 1989; van der Wilk *et al.* 1991).

Failure to detect transgenically expressed PLRV CP or erratic detection appears to be a common feature in nearly all examples where PLRV CP plants have been generated. Thus, while levels of PLRV CP in the transgenic plants were very low, resistance was observed and apparently not correlated with CP expression levels. Modification of the CP to attempt to improve expression level by eliminating the translation initiation codon of the internal 17K open reading frame (ORF), and by altering the CP translation initiation codon context (Kozak, 1989), did not improve expression of CP in transgenic plants (Presting *et al.* 1995).

Whether a functional CP is required for resistance, whether only very small quantities are required, or whether the CP itself is not required in order to obtain resistant plants is not clear. Barker *et al.* (1993) suggest that there was a relationship between PLRV CP transcript levels in transgenic tobacco and resistance.

In retrospect, this observation is consistent with the idea that pathogen-derived resistance can be effected at the RNA level (see below). In any case, it is also possible to combine transgenic resistance with host plant resistance to further reduce levels of PLRV (Barker *et al.* 1994; Derrick and Barker, 1997).

A resistant phenotype is considered to be one that has reduced levels of virus, vs. an immunity or immunity-like phenotype where no virus can be detected. However, how these plants behave under natural conditions and natural vector pressures is a key component. PLRV can directly affect aphid transmission efficiency. Plants with lower amounts of virus contribute less to virus spread than high titer plants (Barker and Harrison, 1986; Hassan *et al.* 1985). A similar situation exists for the related *Luteovirus, Barley yellow dwarf virus* (Gray *et al.* 1994).

The effect of low titer for either can partly be compensated for by longer acquisition periods by the aphid (Gray *et al.* 1991; van den Heuvel *et al.* 1991). Thomas *et al.* (1997) demonstrated that spread was significantly reduced in CP transgenic lines, in spite of the lines tested having only moderate resistance.

The vast majority of the literature on resistance to PLRV is based on the use of CP-mediated resistance. However, plants with the PLRV replicase and not CP are being used commercially. Unfortunately, no information is available in the peer reviewed scientific literature on the use of PLRV replicase for pathogen-derived resistance.

13.2.3 Transgenic Resistance to *Potato Virus X*

While *Potato virus* genus *X Potexvirus* (PVX) is widespread, it generally is not considered to be a significant pathogen. Nevertheless, it can and does cause yield reductions, particularly in plants that are also infected with other viruses such as *Potato virus Y* genus *Potyvirus* (PVY) or, to a lesser extent, *Potato virus S* genus *Carlavirus* (PVS). The first attempts at engineering potato for resistance to any virus utilized coat protein mediated resistance to PVX. Prior to working with potato, this involved insertion of the PVX CP into tobacco plants (Hemenway *et al.* 1988), where both sense and antisense constructs were used for plant transformation. Few or no lesions were observed on inoculated leaves of PVX CP transgenic tobacco plants, and other (systemic) leaves showed no virus symptoms or a delay in the onset of symptoms and amelioration of symptoms. There did not appear to be a good correlation between CP expression levels and the level of protection observed. Hemenway *et al.* (1988) showed that protection extended to plants that had been inoculated with virus RNA, whereas, inoculation of *Tobacco mosaic virus* genus *Tobamovirus* (TMV) CP tobacco with RNA overcame coat protein-mediated resistance. Antisense PVX CP, while afforded some protection in tobacco, but did so only when low doses of inoculum were used.

Hoekema *et al.* (1989) provided the first example of genetically engineering potatoes for virus resistance. Using tuber discs and kanamycin resistance for selection, the coat protein of PVX was inserted into the susceptible cultivars, Bintje, and Escort. Although expression levels of PVX CP varied between transformation events, the protein product was easily detected by western blot analysis. On the basis of high PVX CP expression levels, these authors selected three Bintje and two Escort lines for challenge inoculations with PVX. In all cases, they found a delay in the accumulation of PVX over a two week period, and reduced PVX in these plants. While only a very few transgenic lines were evaluated, this work provided the first indication that coat protein-mediated resistance could be used for virus protection in potato. It should be noted, that in some cases potato plants displayed abnormal phenotypes and genotypes, indicating the need for thorough evaluation of regenerants for agronomic traits under field conditions. Belknap *et al.* (1994) showed that less than 10% of an original population of transgenic clones retained agronomic properties equal to the parents.

Using lines that had been generated earlier, Hoekema *et al.* (1989) evaluated how these transgenic lines behaved under field conditions. Interestingly, over 80% of the cv. Escort-derived transgenic lines had agronomic properties similar or indistinguishable from the parent, while only about 18% of the Bintje-derived transgenic lines were true-to-type (Jongedijk *et al.* 1992).

While it was presumed that differences in agronomic phenotype were due to somaclonal variation and cultivar differences, the underlying mechanisms of this phenomenon are not well understood. Many potato researchers have observed cultivar dependent abnormal phenotype transgenic potatoes, and for this reason, much of the experimental work on transgenic potato virus resistance is initially done using other hosts, particularly tobacco, before working with potatoes. A study similar to Jongedijk *et al.* (1992) was done by Xu *et al.* (1998), where the PVX CP along with the 8K cistron were inserted into cultivars Bintje, Desireé, Shepody, and Russet Burbank. While only a small number of potato clones were examined and observed resistance was variable, high level resistance was observed among all four potato cultivars. Russet Burbank clones appeared to have the most variable levels of resistance to PVX. In each case resistance could be overcome if high concentrations (50 μg/ml) of virus inoculum were used. Resistance to PVX appeared to be conferred by the CP rather than the 8K cistron, and subsequent work showed that the 8K cistron by itself did not confer resistance (Hefferon *et al.* 1997a).

While a functional 8K cistron, individually, did not confer resistance to PVX in potato, the proteins encoding the triple gene block of Potexviruses (and Carlaviruses) are involved in cell-to-cell movement of virus (Beck *et al.* 1991; Lough *et al.* 1998). By expressing a mutated version of the *White clover mosaic virus* genus *Potexvirus* (WClMV) 13K movement protein in *N. benthamiana*, Beck *et al.* (1994) showed that resistance to not only WClMV could be observed, but also resistance to other Potexviruses and to PVS. By analogy, Seppänen *et al.* (1997) showed that expression of a mutated 12K gene in the potato cultivar Pito provided not only resistance to PVX but also to *Potato aucuba mosaic virus* genus *Potexvirus* (PAMV), *Potato virus M Carlavirus* (PVM) and PVS. No resistance was observed to PVY. Presumably, the mutant 12K protein retained some functionality, but lost the ability to effect cell-to-cell movement. The fact that resistance was observed to viruses with different nucleotide sequences would also argue against RNA-mediated resistance.

Golemboski *et al.* (1990) observed that expression of a replicase gene from TMV could result in high level resistance to TMV. Assuming that a similar phenomenon would operate for PVX, Braun and Hemenway (1992) inserted into tobacco four different constructs containing all the PVX replicase, the first 2106 nts, a construct containing approximately the middle third of the gene, or another construct containing approximately the last third of the gene. Transgenic resistance mapped to the first half of PVX replicase, while plant lines expressing either of the constructs from the 3'-half of PVX replicase showed no resistance to PVX. However, while 5 of 7 lines with the full length gene were resistant, only 1 of 6 containing the truncated version was resistant. Two of the five lines with the full length gene were highly resistant with little or no virus detected up to two weeks after inoculation. One line containing the first 2106 nts was resistant and supported very low levels of virus replication. A correlation between levels of transgene expression and resistance was not observed. Interestingly, plants that were resistant after inoculation with intact virus were also resistant when inoculated with virus RNA. Since the construct containing the 3'-end of PVX replicase did not confer resistance to the virus, one can conclude that the region of the replicase containing the GDD motif is not necessary for resistance, even though this motif is highly conserved among many plant viruses.

Longstaff *et al.* (1993) subsequently showed that mutation of the GDD motif and ensuing expression of the full-length 166 kDa replicase can result in tobacco lines that are highly resistant to PVX. Protoplast experiments showed that resistance was at the level of virus replication rather than cell-to-cell movement. Thus, while expression of the native replicase did not appear to be capable of providing resistant plants, mutation of the GDD motif to ADD in particular had dramatic effects. Mutation of the GDD motif to GED or GAD provided partial resistance. However, there was no firm relationship between expression levels of the ADD mutant and the degree of resistance, While there are a number of potential explanations for this observation, Longstaff et al. (1993) suggested the possible presence of two distinct mechanisms for replicase-mediated resistance, a proposal which is now supported by other results (see below).

13.2.4 Transgenic Resistance to *Potato Virus Y*

Transgenic resistance to a potyvirus was first reported by Stark and Beachy (1989). Somewhat unexpectedly, protection to *Potato virus Y* genus *Potyvirus* was effected in tobacco by a heterologous coat protein from soybean mosaic potyvirus (SMV), thus suggesting that PVY^{O} CP might protect against different strains of PVY. Similarly, resistance to PVY can be obtained in transgenic potato expressing the coat protein of *Lettuce mosaic virus* genus *Potyvirus* (LMV) (Hassairi *et al.* 1998). The first report of resistance to PVY in potato (Lawson *et al.* 1990), did not discuss resistance to PVY^{N}. Malnoë *et al.* (1994) generated PVY coat protein transgenic potato (cv. Bintje), and showed that plants resistant to PVY^{N} could be obtained that were also resistant to PVY^{O}. These tests were done under field conditions, using aphids to transmit the virus. Work by another group confirmed resistance to heterologous strains of PVY in tobacco (McDonald *et al.* 1997).

In both cases, higher levels of resistance were observed to the homologous strain. Hefferon *et al.* (1997c) constructed a vector to obtain the highest possible levels of transgenic PVY^{O} CP, using a PVX-derived translational enhancer. Of the lines tested, and regardless of whether the inoculum was PVY^{O} or PVY^{N}, no virus could be detected in inoculated plants over 7 weeks. Whether resistance to PVY is due to the higher levels of CP produced is not clear. It is clear from the above examples as well as many others for various *Potyviridae* not cited, that coat protein-mediated resistance can be obtained to most if not all viruses in the genus *Potyvirus*. What was also becoming clear in the early 1990s, based initially on the work of Golemboski *et al.* (1990), was that other regions of the potyvirus genome could be used to instill pathogen-derived resistance, particularly the replicase domain. Resistance to PVY in tobacco using the PVY^O NIb cistron was demonstrated, where the GDD motif was shown to be essential for conferring a resistant phenotype (Audy *et al.* 1994), in contrast to the situation with PVX. With the exception of a single R_2 line, all of the other tobacco lines were susceptible to PVY^N, demonstrating that resistance is highly strain specific, consistent with what was previously reported for TMV. Indeed, the transformation of plants with many different PVY-derived constructs can provide plants with resistance to PVY, although the level of strain specificity is frequently not determined. PVY resistant plants have also been produced expressing the first three cistrons of PVY (Livneh *et al.* 1995), and NIa (Maiti *et al.* 1993; Vardi *et al.* 1993). The PVY VPg cistron and its expression will confer resistance to PVY in tobacco or potato and to *Tobacco etch virus* genus *Potyvirus* (TEV) in VPg tobacco (Berger, Himsl and Shiel, unpublished).

As stated above, there are numerous examples of coat protein-mediated resistance to viruses in the *Potyviridae*. A common occurrence is a low or even no relationship between CP expression and levels of resistance. i.e., plant lines with low or non-detectable CP expression at times had high levels of resistance to virus while those with detectable levels of the CP were susceptible (e.g., Lawson *et al.* 1990; Van der Vlugt *et al.* 1992). This phenomenon began to be clarified after the work of Dougherty and co-workers (Dougherty *et al.* 1994; Lindbo and Dougherty, 1992a; Lindbo and Dougherty, 1992b; Lindbo *et al.* 1993; Smith *et al.* 1994), and led to the concept of "RNA-mediated resistance" (reviewed in Palukaitis and Zaitlin, 1997; Prins and Goldbach, 1996). RNA-mediated resistance to PVY was demonstrated in Russet Burbank and Russet Norkotah potatoes that expressed a translatable CP, an antisense CP, and two versions of a sense CP containing one or four translational stop codons near the 5'-end of the gene (Smith *et al.* 1995). While resistant plants could be observed with any of the four constructs (albeit in only one of eight antisense CP lines), certain common features emerged. Highly resistant plants usually had multiple transgenes that were actively transcribed, but accumulated only low levels of these transcripts. These data suggest a host-mediated response rather than a direct effect of CP itself, and is now referred to as virus-induced gene silencing or post-transcriptional gene silencing (PTGS). This suggestion opened a new field of investigation, relating to many viruses in addition to viruses within the *Potyviridae*. This gene silencing phenomenon occurs in both transgenic and non-transgenic plants, and in some cases may be involved in conventional genetic virus resistance. In transgenic plants PTGS involves sequence-specific degradation of RNA wherein the targeted RNA species have similar nucleotide sequences to the transcribed part of a silencer transgene. The silencing phenomenon is derived from the fact that transcription levels in plants exhibiting gene silencing are normal, but the level of the transcript in the cytoplasm is greatly reduced (i.e., degraded). Hence, in a plant expressing viral RNA sequences, the contribution of similar sequences to the RNA pool upon virus infection triggers a sequence-specific RNA degradation process that prevents viral infection. Specific details of the mechanisms involved are still being determined, but this phenomenon will no doubt play a significant role in the development of virus resistant cultivars of the future.

13.2.5 Resistance to Other Potato Viruses

Relative to work on resistance in potato to PLRV, PVX, and PVY, little has been done on the other potato viruses, with the partial exception of multivirus resistance (below). *Nicotiana debneyii* plants expressing the PVS coat protein were resistant to PVS as well as to inoculation with PVS RNA (MacKenzie and Tremaine, 1990). PAMV CP likewise provided protection to the virus in *N. tabacum* (Leclerc and AbouHaidar, 1995). In this work it was found that deleting 46 amino acids from the amino terminus of the CP eliminated coat protein-mediated resistance, but a construct containing the conserved core of the CP provided resistance as did a CP antisense construct. Reavy *et al.* (1995) transformed *N. benthamiana* with the *Potato mop top virus* genus *Pomovirus* (PMTV) CP. Those lines with the highest CP expression levels were also highly resistant to PMTV after mechanical or graft inoculations. These same plants were also resistant to this *Spongospora subterranea*-transmitted virus, where again high level resistance was observed. In subsequent work, it was shown that a non-translatable PMTV CP construct provided at best only moderate resistance to PMTV, suggesting a requirement of the CP for resistance and little evidence of RNA-mediated resistance occurring (Barker *et al.* 1998). Resistance was maintained if RNA was used as inoculum, however, there was little correlation between CP levels in transformed plants and the level of resistance.

In some instances, researchers investigating pathogen-derived resistance to PVY have also examined resistance to other potyviruses infecting potato. There are two reports on resistance or tolerance to *Potato virus A* genus *Potyvirus* (PVA). Farinelli and Malnoë (1993) using CP or an antisense CP construct from an isolate of PVY^N, also observed tolerance to *Potato virus V* genus *Potyvirus* (PVV) as well as PVA. In another study, Hefferon *et al.* (1997b) found a delay of 1 week in the accumulation of PVA in PVY CP plants, and this level of tolerance could be overcome using high concentrations of inoculum.

Little work has been done on obtaining transgenic resistance to *Tobacco rattle virus* genus *Tobravirus* (TRV) in potato. TRV, which causes "spraing" or "corky ringspot disease", can be prevalent on a local basis, and this virus can have devastating effects under the right conditions. Van Dun and Bol (1988) showed that coat protein-mediated resistance to TRV (and *Pea early browning* genus *Tobravirus*) could be achieved in potato. However, these tests were done with virus that was mechanically transmitted, and later work showed that resistance was not effective against nematode-transmitted virus (Ploeg *et al.* 1993).

13.2.6 Multivirus Resistance

Ultimately, some sort of single gene resistance to multiple viruses would be ideal. This is particularly crucial in potato, since many commercial varieties are vegetatively propagated, and multiple transformations would be required in order to insert multiple virus resistance genes to the many viruses known to infect potato. There are, however, few examples of multivirus resistance in potato. The earliest instance is also the first case of transgenic resistance to PVY (Lawson *et al.* 1990), mentioned above. Resistance to PVX and PVY also, not unexpectedly, provides resistance to the synergistic effects of this dual virus infection, as does resistance to either individual virus. Lawson *et al.* (1990) observed instances of moderate virus replication, particularly of PVY, when plants that were resistant to each individual virus in PVX CP/PVY CP transgenic plants were simultaneously inoculated with both viruses. Nevertheless, no synergism was observed (Lawson *et al.* 1990), and at least one of these lines had effective resistance to both viruses in the field (Kaniewski *et al.* 1990).

An alternative approach to obtaining multivirus resistance is to somehow disrupt the process of cell-to-cell movement. It was already known that genes of the triple gene block (TGB) found in potex- and carlaviruses are involved in mediating cell-to-cell movement. Further it was known that expression of a modified version of the 2nd of the 3 genes of the TGB could effect resistance to the virus from which it was obtained as well as other TGB-containing viruses (Beck *et al.* 1994). Thus, Seppänen *et al.* (1997) targeted this process by inserting in potato cv. Pito two mutated 12K genes derived from PVX. Resistance was obtained to PVX and to PAMV, PVM and PVS, but not to PVY. These results suggest that expression of a viral movement protein (MP) that retains partial functionality will interfere with cell-to-cell movement of not only the virus from which the movement protein was derived but also to other viruses that presumably use similar mechanisms for cell-to-cell movement. Tacke *et al.* (1996) took a similar approach, but this time used the 17K protein of PLRV to which a 2K extension had been added to the amino terminus. It had recently become apparent that 17K was a MP or MP-like protein, albeit with phloem specificity. Presumably, these additional amino acids affected the normal protein-protein interaction function of 17K, and disrupted its MP activity in a manner similar to what was reported for TMV (Cooper *et al.* 1995; Lapidot *et al.* 1993). Thus, plants transformed with the mutant 17K gene were resistant to PLRV, and to PVY^{O}. Interestingly, plants resistant to PVX were also obtained, suggesting a common process that was inhibited by the mutated 17K protein, but the converse experiment using mutated TGB proteins had no effect. Other experimental results suggested a requirement for accumulation of the altered 17K protein, which would be consistent with the proposed mechanism of resistance (Tacke *et al.* 1996).

Multivirus resistance need not be conferred using pathogen-derived resistance, however, and one of the few examples of non-pathogen-derived resistance to plant viruses is provided by Truve *et al.* (1993). The murine 2'-5' oligoadenylate synthetase (2-5A synthetase) gene was inserted into potato cv. Pito, and transgenic plants challenge inoculated with PVX in the greenhouse and in the field. Several lines with reduced or non-detectable levels of PVX were observed. Unfortunately, in this study only PVX was used. Later Truve *et al.* (1994) reported resistance to PVY in one 2-5A synthetase transgenic potato clone that was also "tolerant" to PVX. Resistance to PVS and PVX were reported in transgenic tobacco. Why expression of 2-5A synthetase should confer resistance to plant virus in a plant is still unclear. Interferons, after induction by dsRNA, induce a cascade of activities resulting in inhibition of virus replication in mammalian cells (see, Sen and Lengyel, 1992; Sen and Ransohoff, 1993). In order for 2-5A synthetase to mimic the mammalian interferon response in a plant there must be other enzyme activities either present or also induced. There is little evidence that plants actually have homologs to their mammalian counterparts. Mitra *et al.* (1996) produced tobacco plants expressing both 2-5A synthetase and RNase L, to effect an antiviral state. When plants or detached leaves were inoculated with either TMV, *Afalfa mosaic virus* genus *Alfamovirus* or *Tobacco vein mottling virus* genus *Potyvirus*, a hypersensitive response ensued. No resistance to these viruses was observed with either individual gene. In spite of the questions surrounding the use of mammalian antiviral response genes, practical and social, the idea of reconstituting such a system in plants or perhaps obtaining plant equivalents of these various genes, if they exist, presents an intriguing approach towards generic virus resistance. Indeed, considerable work has been done to determine if plants naturally produce interferon or interferon-like molecules, and to determine if mammalian interferon can affect plant viruses. While there is evidence that interferon may have an effect on plant viruses or may activate certain genes in plant cells, much of this work has produced equivocal or negative results.

Other possibilities for non-pathogen-derived resistance also exist. These include ribosome-inhibiting proteins/antiviral proteins (Bonness *et al.* 1994; Lodge *et al.* 1993; Smirnov *et al.* 1997; Taylor *et al.* 1994; Wang *et al.* 1998; Watanabe *et al.* 1997), antisense inhibition of a host methyltransferase (Masuta *et al.* 1995), or even expression of mammalian antibodies in plants. One successful application of the latter technology resulted in resistance to a single tombusvirus (Tavladoraki *et al.* 1993). It may also be possible to target conserved virus epitopes to effect resistance to more than one virus. It may be possible to design ribozymes that either cleave regions of nucleotide sequence conserved among a number of viruses, or to insert multiple ribozymes with different specificities. A ribozyme that could cleave PLRV RNA *in vitro* has been reported (Lamb and Hay, 1990), but has not been tested *in vivo*. A ribozyme directed against *Potato spindle tuber Viroid* (PSTVd) has been described and expressed in potato, where effective resistance to the viroid was observed (Yang *et al.* 1997).

An alternative approach to pathogen- or non-pathogen-derived resistance would be to instill into potato resistance to the vector rather than to the virus. This would, of course, have the added benefit of reducing the impact of direct feeding damage by aphid. Towards this end, Gatehouse *et al.* (1996) generated cv. Desireé potato plants expressing a chitinase gene from bean, a lectin gene from snowdrop, or an α-amylase inhibitor gene from wheat. These were used either individually or in all possible pairwise combinations. Plants expressing the lectin plus either of the other two genes reduced both aphid fecundity and survival. It remains to be determined if these effects will be observed under field conditions, and whether there is any effect on transmission of aphid-borne viruses to or from these plants. Alternatively, characterizing the molecular interactions involved in the specificity of insect transmission may yield useful tools for developing strategies to prevent virus transmission. For example, the interaction between viral components and a putative receptor for *Tomato spotted wilt virus* genus *Tospovirus* [(TSWV – an emerging pathogen of potato in temperate regions (see Chapt. 8)] in its principal vector, *Franklinella occidentalis*, has bee reported (Medeiros *et al.* 2000). It should also be possible to use the components of the interaction (e.g., the viral protein that binds to vector's receptor) to produce transgenic plants from which acquisition is blocked, or greatly reduced. In turn, it may be possible to exploit these molecular interactions to direct fusion proteins in a manner specific to virus vectors and not to non-target species.

Finally, the use of a dsRNA-specific RNase may be useful and, if so, provides an elegant solution to the problem of obtaining multivirus resistance, since the majority of plant viruses use a dsRNA intermediate in their replication cycles and dsRNA is not a normal cytosolic constituent of plant cells. At least one family of enzymes possess this activity, RNase III from *E. coli* and a distantly related enzyme from the yeast *Schizosaccharomyces pombe* (Lino *et al.* 1991; Xu *et al.* 1990), known as *hcs* or *pac1*. Watanabe *et al.* (1995) demonstrated that *pac1* could confer resistance in tobacco to three different viruses, *Cucumber mosaic virus* genus *Cucumovirus*, PVY, and *Tomato mosaic virus* genus *Tobamovirus*, although few plants were tested, and the relationship between *pac1* expression and resistance was inconsistent. Recently, Sano *et al.* (1997) showed that *pac1* could also confer resistance to PSTVd in potato. While *pac1* has been shown to have dsRNA as its substrate (Rotondo and Frendewey, 1996; Rotondo *et al.* 1997), it has not been conclusively demonstrated that it is degrading plant virus dsRNA *in vivo*.

13.2.7 Future Prospects for Biotechnological Methods of Potato Virus Control

There are numerous examples of transgenic resistance to potato viruses, using many different approaches. It is the possible that biotechnological approaches could result in greatly reduced damage due to virus disease, or perhaps viruses could even be eliminated as constraints on potato production and quality. Whether this will happen is not only a matter of innovative research, but also of political and social factors that have little relationship to the science. Without doubt, combining conventional breeding with gene transfer technology has the potential to provide important and useful new germplasm and cultivars.

World-wide, more than 40 million hectares of transgenic crops were grown in 1999, but potato accounted for only about 100,000 ha. As of this writing over 40 transgenic crops have been approved for marketing and consumption in the United States (Ferber, 1999), including three types of transgenic potatoes. These are potatoes containing the δ-endotoxin gene from *Bacillus thuringiensis* (Bt) for control of the Colorado potato beetle, potatoes expressing Bt combined with the PLRV replicase for additional control of potato leafroll virus and net necrosis, and potatoes expressing Bt and PVY coat protein for control of the Colorado potato beetle and potato virus Y. Examples of the striking effect of these transgenes on resistance to PLRV or PVY are provided in Fig. 13.2.2. The number of new lines and cultivars with approved transgenes is limited, however, as additional

Figure 13.2.1. Examples of genetically engineered pathogen-derived resistance to two important potato viruses, Potato virus Y and Potato leafroll virus (PLRV). A,B – Russet Burbank potato inoculated with PLRV. The plant in panel A is transgenic with the replicase gene from PLRV, while the plant in panel B is not transgenic and shows typical leafroll symptoms. Similarly the cut tubers in panels C and D show that there is also a reduction in PLRV-caused net necrosis. The field in E was inoculated with viruliferous aphids, with plants containing the PLRV replicase transgenic Russet Burbank on the left and control plants on the right. Panel F is a side-by-side comparison of potato virus Y (PVY) coat protein transgenic Russet Burbank potato on the right and the control plant on the left, after both were inoculated with PVY. (Photos courtesy NatureMark Potatoes.)

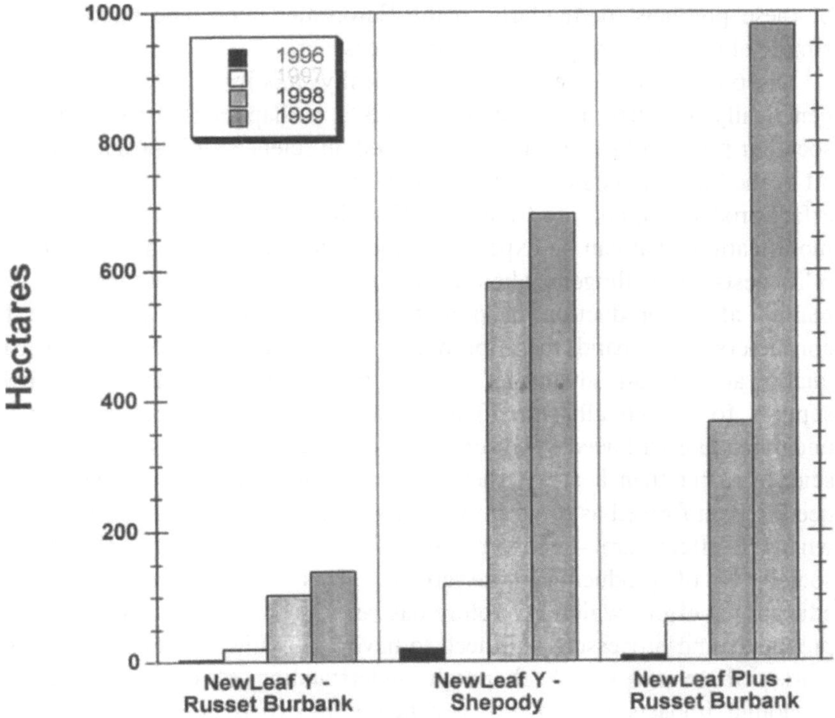

Figure 13.2.2. Production of NewLeafY and NewLeaf Plus in the US from 1996 to 1999. Hectares included are both for seed and commercial production. (Data courtesy NatureMark Potatoes.)

approvals in the US and elsewhere require significant additional molecular data.

Production of genetically engineered potatoes has increased in land area in the US for the period 1996-1999. Combined production of seed and transgenic potatoes has increased each year to meet demand for these products (Figure 13.2.1). However, it is possible that demand for genetically engineered virus resistant potatoes may plateau or decrease, as major processors eliminate genetically modified potatoes from their products. While there is no scientific evidence that any of the transgenic products currently available have had or may have detrimental effects on human health or on the environment, a number of countries will not allow import or sale of genetically modified potatoes or potato products. Several major potato processors appear to be concerned over lack of consumer acceptance of these products, particularly in the European Union, and have opted to simply eliminate these products from processing.

Discomfort over the use of genetically modified potatoes or other genetically modified agricultural products is perhaps rooted more firmly in social or political factors, rather than based on science. It is problematic that all of the biotechnological improvements that have been made to potato will affect production, and are 'transparent' to the end-user. Indeed, most other modifications that can be expected in the next few years (e.g., resistance to other pests and pathogens, alterations in processing characteristics, etc.) will mainly affect production or processing. There is little motivation for consumers to demand these products, unless they provide a significant quality and/or cost advantage. Hence, there is no groundswell of public support for genetically modified potatoes. For this to change, public understanding will need to be substantially increased, and based on facts and science rather than hearsay, rhetoric and misinformation. Consumers will need to be informed as to what these products are, how they were made, and what the effects are – economically, environmentally, agronomically and socially – of producing these crops. This will require a significant educational effort, which heretofore has been inadequate. At the same time, producers and processors will need to have compelling reasons to grow and market these crops. A thorough understanding of environmental and economic impacts of growing genetically modified potatoes is required. Without a clear quality, quantity and/or cost advantage, there is little reason to switch practices or varieties.

The developments that have occurred, representing significant and even revolutionary scientific advancements, have their roots in advances in our fundamental understanding of viruses, the genes that comprise potato viruses, and their function(s). Without the basic research that has already occurred, development of genetically engineered resistance to potato viruses would not have been possible, irrespective of the benefits or risks associated with deployment of such resistance. Research on development of virus resistant potatoes, which must continue, is one component of what is required to ultimately provide a highly desirable potato for producers, processors and consumers, which at the same time is produced in a sustainable manner.

References

Audy, P., Palukaitis, P., Slack, S. A. and Zaitlin, M. 1994. Replicase-mediated resistance to potato virus Y in transgenic tobacco plants. *Mol. Plant Microbe Interact.* 7:15-22.

Barker, H. and Harrison, B. D. 1986. Restricted multiplication of potato leafroll virus in resistant potato genotypes. *Ann. Appl. Biol.* 107:595-604.

Barker, H., Reavy, B., Kumar, A., Webster, K. and Mayo, M. 1992. Restricted virus multiplication in potatoes transformed with the coat protein gene of potato leafroll luteovirus: Similarities with a type of host gene-mediated resistance. *Ann. Appl. Biol.* 120:55-64.

Barker, H., Reavy, B., McGeachy, K. D. and Dawson, S. 1998. Transformation of *Nicotiana benthamiana* with the potato mop-top virus coat protein gene produces a novel resistance phenotype mediated by the coat protein. *Mol.Plant Microbe Interact.* 11:626-633.

Barker, H., Reavy, B., Webster, K. D., Jolly, C. A., Kumar, A. and Mayo, M. A. 1993. Relationship between transcript production and virus resistance in transgenic tobacco expressing the potato leafroll virus coat protein gene. *Plant Cell Reports* 13:54-58.

Barker, H., Webster, K. D., Jolly, C. A., Reavy, B., Kumar, A. and Mayo, M. A. 1994. Enhancement of resistance to potato leafroll virus multiplication in potato by combining the effects of host genes and transgenes. *Mol. Plant Microbe Interact.* 7:528-530.

Baulcombe, D. C. 1996. Mechanisms of pathogen-derived resistance to viruses in transgenic plants. *Plant Cell* 8:1833-1844.

Beachy, R. N., Loesch-Fries, S. and Tumer, N. E. 1990. Coat protein-mediated resistance against virus infection. *Annu. Rev. Phytopathol.* 28:451-474.

Beck, D. L., Guilford, P. J., Voot, D. M., Andersen, M. T. and Forster, R. L. S. 1991. Triple gene block proteins of white clover mosaic potexvirus are required for transport. *Virology 183:695-702.*

Beck, D. L., Van Dolleweerd, C. J., Lough, T. J., Balmori, E., Voot, D. M., Andersen, M. T., O'Brien, I. E. W. and Forster, R. L. S. 1994. Disruption of virus movement confers broad-spectrum resistance against systemic infection by plant viruses with a triple gene block. *Proc. Natl. Acad. Sci. USA* 91:10310-10314.

Belknap, W. R., Corsini, D., Pavek, J. J., Snyder, G. W., Rockhold, D. R. and Vayda, M. E. 1994. Field performance of transgenic Russet Burbank and Lemhi Russet potatoes. *Am. Pot. J.* 71:285-296.

Bonness, M. S., Ready, M. P., Irvin, J. D. and Mabry, T. J. 1994. Pokeweed antiviral protein inactivates pokeweed ribosomes; Implications for the antiviral mechanism. *Plant J.* 5:173-183.

Braun, C. and Hemenway, C. 1992. Expression of amino- terminal portions or full- length viral replicase genes in transgenic plants confers resistance to potato virus X infection. *Plant Cell* 4:735-744.

Brown, C. R., Smith, O. P., Damsteegt, V. D., Yang, C.-P., Fox, L. and Thomas, P. E. 1995. Suppression of PLRV titer in transgenic Russet Burbank and Ranger Russet. *Am. Pot. J.* 72:589-597.

Candresse, T. 1996. The use of pathogen derived resistance and other strategies to obtain transgenic plants resistant to viruses. *Acta Hort.* 424:369-374.

Cooper, B., Lapidot, M., Heick, J. A., Dodds, J. A. and Beachy, R. N. 1995. A defective movement protein of TMV in transgenic plants confers resistance to multiple viruses whereas the functional analog increases susceptibility. *Virology* 206:307-313.

Crute, I. R. and Pink, D. A. C. 1996. Genetics and utilization of pathogen resistance in plants. *Plant Cell* 8:1747-1755.

Derrick, P. M. and Barker, H. 1992. The restricted distribution of potato leafroll luteovirus antigen in potato plants with transgenic resistance resembles that in clones with one type of host gene-mediated resistance. *Ann. Appl. Biol.* 120:451-457.

Derrick, P. M. and Barker, H. 1997. Short and long distance spread of potato leafroll luteovirus: Effects of host genes and transgenes conferring resistance to virus accumulation in potato. *J. Gen. Virol.* 78:243-251.

Dougherty, W. G., Lindbo, J. A., Smith, H. A., Parks, T. D., Swaney, S. and Proebsting, W. M. 1994. RNA-mediated virus resistance in transgenic plants: Exploitation of a cellular pathway possibly involved in RNA degradation. *Mol. Plant Microbe Interact.* 7:544-552.

Farinelli, L. and Malnoë, P. 1993. Coat protein gene-mediated resistance to potato virus Y in tobacco: required for protection? *Mol. Plant Microbe Interact.* 6:284-292.

Ferber, D. 1999. GM Crops in the Cross Hairs. *Science* 286:1662-1666.

Fitchen, J. H. and Beachy, R. N. 1993. Genetically engineered protection against viruses in transgenic plants. *Annu. Rev. Microbiol.* 47:739-763.

Gatehouse, A. M. R., Down, R. E., Powell, K. S., Sauvion, N., Rahbé, Y., Newell, C. A., Merryweather, A., Hamilton, W. D. O. and Gatehouse, J. A. 1996. Transgenic potato plants with enhanced resistance to the peach-potato aphid *Myzus persicae*. *Entomol.Exp.Appl.* 79:295-307.

Golemboski, D. B., Lomonosoff, G. P. and Zaitlin, M. 1990. Plants transformed with a tobacco mosaic virus nonstructural gene sequence are resistant to the virus. *Proc. Natl. Acad. Sci. USA* 87:6311-6315.

Gray, S. M., Power, A. G., Smith, D. M., Seaman, A. J. and Altman, N. S. 1991. Aphid transmission of barley yellow dwarf virus: Acquisition access periods and virus concentration requirements. *Phytopathology* 81:539-545.

Gray, S. M., Smith, D. and Sorrells, M. 1994. Reduction of disease incidence in small field plots by isolate-specific resistance to barley yellow dwarf virus. *Phytopathology* 84:713-718.

Hackland, A. F., Rybicki, E. P. and Thomson, J. A. 1994. Coat protein-mediated resistance in transgenic plants. *Arch. Virol* 139:1-22.

Hassairi, A., Masmoudi, K., Albouy, J., Robaglia, C., Jullien, M. and Ellouz, R. 1998. Transformation of two potato cultivars 'Spunta' and 'Claustar' (*Solanum tuberosum*) with lettuce mosaic virus coat protein gene and heterologous immunity to potato virus Y. *Plant Sci.* 136:31-42.

Hassan, S., Thomas, P. E. and Mink, G. I. 1985. Tomato yellow top virus: host range, symptomatology, transmission, and variability. *Phytopathology* 75:287-291.

Hefferon, K. L., Doyle, S. and AbouHaidar, M. G. 1997a. Immunological detection of the 8K protein of potato virus X (PVX) in cell walls of PVX-infected tobacco and transgenic potato. *Arch. Virol.* 142:425-433.

Hefferon, K. L., Khalilian, H. and AbouHaidar, M. G. 1997b. Expression of the PVY^O coat protein (CP) under the control of the PVX CP gene leader sequence: protection under greenhouse and field conditions against PVY^O and PVY^N infection in three potato cultivars. *Theor. Appl. Genet.* 94:287-292.

Hefferon, K. L., Khalilian, H., Xu, H. M. and Abouhaidar, M. G. 1997c. Expression of the coat protein of potato virus X from a dicistronic mRNA in transgenic potato plants. *J. Gen. Virol.* 78:3051-3059.

Hemenway, C., Fang, R.-X., Kaniewski, W.K., Chua, N.-H. and Tumer., N.E. 1988. Analysis of the mechanism of protection in transgenic plants expressing the potato virus X coat protein or its antisense RNA. *EMBO J.* 7:1273-1280.

Hoekema, A., Hirsh, P. R., Hooykaas, P. J. J. and Schilperoort, R. A. 1983. A binary plant vector strategy based on separation of vir- and T-region of the Agrobacterium tumefaciens Ti-plasmid. *Nature 303:179-180.*

Hoekema, A., Huisman, M. J., Molendijk, L. van dan Elzen, P. J. M., and Cornelissen, B. J. C. 1989. The genetic engineering of two commercial potato cultivars for resistance to potato virus X. *Bio/Technology 7:273-278.*

Horsch, R. B., Fraley, R. T., Rogers, S. G., Sanders, P. R., Lloyd, A. and Hoffmann, N. 1984. Inheritance of functional foreign genes in plants. *Science* 223:496-498.

Huisman, M. J., Cornelissen, B. and Jongedijk, E. 1992. Transgenic potato plants resistant to viruses. *Euphytica* 63:187-197.

Jongedijk, E., de Schutter, A. A. J. M., Stolte, T. van den Elzen, P. J. M., and Cornelissen, B. J. C. 1992. Increase resistance to potato virus X and preservation of cultivar properties in transgenic potato under field conditions. *Biotechnology* 10:422-429.

Kaniewski, W., Lawson, C., Sammons, B., Haley, L., Hart, J., Delannay, X. and Tumer, N. E. 1990. Field reistance of transgenic Russet Burbank potato to effects of infection by potato virus X and potato virus Y. *Bio/Technology* 8:750-754.

Kawchuk, L. M., Martin, R. R. and McPherson, J. 1990. Resistance in transgenic potato expressing the potato leafroll virus coat protein gene. *Molec. Plant-Micro. Interact.* 3:301-307.

Kawchuk, L. M., Martin, R. R. and McPherson, J. 1991. Sense and antisense RNA-mediated resistance to potato leafroll virus in Russet Burbank potato plants. *Mol. Plant-Microbe Interact.* 4:247-253.

Kawchuk, L. M., Martin, R. R., Rochon, D. M. and McPherson, J. 1989. Identification and characterization of the potato leafroll virus putative coat protein gene. *J. Gen. Virol. 70:783-788.*

Kozak, M. 1989. The scanning model for translation: An update. *J. Cell. Biol.* 108:229-241.

Lamb, J. W. and Hay, R. T. 1990. Ribozymes that cleave potato leafroll virus RNA within the coat protein and polymerase genes. *J. Gen. Virol. 71:2257-2264.*

Lapidot, M., Gafny, R., Ding, B., Wolf, S., Lucas, W. J. and Beachy, R. N. 1993. A dysfunctional movement protein of tobacco mosaic virus that partially modifies the plasmodesmata and limits virus spread in transgenic plants. *Plant J* 4:959-970.

Lawson, C., Kaniewski, W., Haley, L., Rozman, R., Newell, C., Sanders, P. and Tumer, N. E. 1990. Engineering resistance to mixed virus infection in a commercial potato cultivar: Resistance to potato virus X and potato virus Y in transgenic Russet Burbank. *Biotechnology* 8:127-134.

Leclerc, D. and AbouHaidar, M. G. 1995. Transgenic tobacco plants expressing a truncated form of the PAMV capsid protein (CP) gene show CP-mediated resistance to potato aucuba mosaic virus. *Mol.Plant Microbe Interact..* 8:58-65.

Lindbo, J. A. and Dougherty, W. G. 1992a. Pathogen-derived resistance to a potyvirus: Immune and resistant phenotypes in transgenic tobacco expressing altered coat protein nucleotide sequence. *Mol. Plant-Microbe Interact..* 5:144-153.

Lindbo, J. A. and Dougherty, W. G. 1992b. Untranslatable transcripts of the tobacco etch virus coat protein gene sequence can interfere with tobacco etch virus replication in transgenic plants and protoplasts. *Virology* 189:725-733.

Lindbo, J. A., Silva-Rosales, L., Proebsting, W. M. and Dougherty, W. G. 1993. Induction of a highly specific antiviral state in transgenic plants: Implications for regulation of gene expression and virus resistance. *Plant Cell* 5:1749-1759.

Lino, Y., Sugimoto, A. and Yamamoto, M. 1991. *S. pombe pac1*$^{+}$, whose overexpression inhibits sexual development, encodes a ribonuclease III-like RNase. *EMBO J.* 10:221-226.

Livneh, O., Edelbaum, O., Kuznetsova, L., Livne, B., Vardi, E., Sela, I. and Stram, Y. 1995. Plants transformed with the first (nonstructural) three cistrons of potato virus Y are resistant to potato virus infection. *Transgenics* 1:565-571.

Lodge, J. K., Kaniewski, W. K. and Tumer, N. E. 1993. Broad-spectrum virus resistance in transgenic plants expressing pokeweed antiviral protein. *Proc. Natl. Acad. Sci. USA* 90:7089-7093.

Lomonossoff, G. P. 1995. Pathogen-derived resistance to plant viruses. *Annu. Rev. Phytopathol.* 33:323-343.

Longstaff, M., Brigneti, G., Boccard, F., Chapman, S. and Baulcombe, D. 1993. Extreme resistance to potato virus X infection in plants expressing a modified component of the putative viral replicase. *EMBO J.* 12:379-386.

Lough, T. J., Shash, K., Xoconostle-Cázares, B., Hofstra, K. R., Beck, D. L., Balmori, E., Forster, R. L. S. and Lucas, W. J. 1998. Molecular dissection of the mechanism by which potexvirus triple gene block proteins mediate cell-to-cell transport of infectious RNA. *Mol.Plant Microbe Interact.* 11:801-814.

MacKenzie, D. J. and Tremaine, J. H. 1990. Transgenic *Nicotiana debneyii* expressing viral coat protein are resistant to potato virus S infection. *J. Gen. Virol. 71:2167-2170.*

Maiti, I. B., Murphy, J. F., Shaw, J. G. and Hunt, A. G. 1993. Plants that express a potyvirus proteinase gene are resistant to virus infection. *Proc. Natl. Acad. Sci. USA* 90:6110-6114.

Malnoë, P., Farinelli, L., Collet, G. F. and Reust, W. 1994. Small-scale field tests with transgenic potato, cv. Bintje, to test resistance to primary and secondary infections with potato virus Y. *Plant Mol. Biol.* 25:963-975.

Martin, R. R. 1994. Genetic engineering of potato. *Am. Pot. J.* 71:347-358.

Masuta, C., Tanaka, H., Uehara, K., Kuwata, S., Koiwai, A. and Noma, M. 1995. Broad resistance to plant viruses in transgenic plants conferred by antisense inhibition of a host gene essential in *S*-adenosylmethionine-dependent transmethylation reactions. *Proc. Natl. Acad. Sci. USA* 92:6117-6121.

McDonald, J. G., Brandle, J. E., Gleddie, S., Hermans, J. A. and Kermali, I. R. 1997. Resistance to homologous and heterologous strains of potato virus Y in transgenic tobacco carrying the PVY^N coat protein gene. *Can. J. Plant Sci.* 77:167-171.

Medeiros, R. B., Ullman, D. E., Sherwood, J. L. and German, T. L. 2000. Immunoprecipitation of a 50 kDa protein: a putative receptor component for tomato spotted wilt *tosposvirus* (*Bunyaviridae*) in its insect vector, *Franklinella occidentalis. Virus Res.* (in press).

Mitra, A., Higgins, D. W., Langenberg, W. G., Nie, H. Q., Sengupta, D. N. and Silverman, R. H. 1996. A mammalian 2-5A system functions as an antiviral pathway in transgenic plants. *Proc. Natl. Acad. Sci. USA* 93:6780-6785.

Ooms, G., Burrell, M. M., Karp, A., Bevan, M. and Hille, J. 1987. Genetic transformation in two potato cultivars with T-DNA from disarmed *Agrobacterium. Theor. Appl. Genet.* 73:744-750.

Ooms, G., Karp, A. and Roberts, J. 1983. From tumour to tuber; tumour cell characteristics and chromosome numbers in crown gall-derived tetraploid potato plants (*Solanum tuberosum* cv Maris Bard). *Theor. Appl. Genet.* 66:169-172.

Ooms, G. and Lenton, J. R. 1985. T-DNA genes to study plant development: precocious tuberisation and enhanced cytokinins in *A. tumefaciens* transformed potato. *Plant Mol. Biol.* 5:205-212.

Palukaitis, P. and Zaitlin, M. 1997. Replicase-mediated resistance to plant virus disease. *Adv.Virus Res.* 48:349-377.

Ploeg, A. T., Mathis, A., Bol, J. F., Brown, D. J. F. and Robinson, D. J. 1993. Susceptibility of transgenic tobacco plants expressing tobacco rattle virus coat protein to nematode-transmitted and mechanically inoculated tobacco rattle virus. *J. Gen. Virol.* 74:2709-2715.

Powell Abel, P., Nelson, R. S., De, B., Hoffmann, N., Rogers, S. G., Fraley, R. T. and Beachy, R. N. 1986. Delay in disease development in transgenic plants that express the tobacco mosaic virus coat protein gene. *Science* 232:738-743.

Presting, G. G., Smith, O. P. and Brown, C. R. 1995. Resistance to potato leafroll virus in potato plants transformed with the coat protein gene or with vector control constructs. *Phytopathology* 85:436-442.

Prins, M. and Goldbach, R. 1996. RNA-mediated virus resistance in transgenic plants. *Arch. Virol.* 141:2259-2276.

Reavy, B., Arif, M., Kashiwazaki, S., Webster, K. D. and Barker, H. 1995. Immunity to potato mop-top virus in *Nicotiana benthamiana* plants expressing the coat protein gene is effective against fungal inoculation of the virus. *Mol.Plant Microbe Interact..* 8:286-291.

Rotondo, G. and Frendewey, D. 1996. Purification and characterization of the Pac1 ribonuclease of *Schizosaccharomyces pombe. Nucleic Acids Res.* 24:2377-2386.

Rotondo, G., Huang, J. Y. and Frendewey, D. 1997. Substrate structure requirements of the Pac1 ribonuclease from *Schizosaccharomyces pombe. RNA Publ.RNA Soc.* 3:1182-1193.

Sano, T., Nagayama, A., Ogawa, T., Ishida, I. and Okada, Y. 1997. Transgenic potato expressing a double-stranded RNA-specific ribonuclease is resistant to potato spindle tuber viroid. *Nat. Biotechnol.* 15:1290-1294.

Scholthof, K. B. G., Scholthof, H. B. and Jackson, A. O. 1993. Control of plant virus diseases by pathogen-derived resistance in transgenic plants. *Plant Physiol* 102:7-12.

Sen, G. C. and Lengyel, P. 1992. The interferon system. A bird's eye view of its biochemistry. *J. Biol. Chem.* 267:5017-5020.

Sen, G. C. and Ransohoff, R. M. 1993. Interferon-induced antiviral actions and their regulation. *Adv. Virus Res.* 42:57-102.

Seppänen, P., Puska, R., Honkanen, J., Tyulkina, L. G., Fedorkin, O., Morozov, S. Y. and Atabekov, J. G. 1997. Movement protein-derived resistance to triple gene block-containing plant viruses. *J. Gen. Virol.* 78:1241-1246.

Shahin, E. and Simpson, R. 1986. Gene transfer system for potato. *Hort. Sci.* 21:1199-1201.

Sheerman, S. and Bevan, M. W. 1988. A rapid transformation method for Solanum tuberosum using binary Agrobacterium tumefaciens vectors. *Plant Cell Repts. 7:13-16.*

Smirnov, S., Shulaev, V. and Tumer, N. E. 1997. Expression of pokeweed antiviral protein in transgenic plants induces virus resistance in grafted wild-type plants independently of salicylic acid accumulation and pathogenesis-related protein synthesis. *Plant Physiol.* 114:1113-1121.

Smith, H. A., Powers, H., Swaney, S., Brown, C. and Dougherty, W. G. 1995. Transgenic potato virus Y resistance in potato: Evidence for an RNA-mediated cellular response. *Phytopathology* 85:864-870.

Smith, H. A., Swaney, S. L., Parks, T. D., Wernsman, E. A. and Dougherty, W. G. 1994. Transgenic plant virus resistance mediated by untranslatable sense RNAs: Expression, regulation, and the fate of nonessential RNAs. *Plant Cell* 6:1441-1453.

Stark, D. M. and Beachy, R. N. 1989. Protection against potyvirus infection in transgenic plants: Evidence for broad spectrum resistance. *Bio/Technology 7:1257-1262.*

Stiekema, W. J., Heidekamp, F., Louwerse, J. D., Verhoeven, H. A. and Dijkhuis, P. 1988. Introduction of foreign genes into potato cultivars Bintje and De´sire´e using an Agrobacterium tumefaciens binary vector. *Plant Cell Rep. 7:47-50.*

Tacke, E., Salamini, F. and Rohde, W. 1996. Genetic engineering of potato for broad-spectrum protection against virus infection. *Nature Biotechnology* 14:1597-1601.

Tavazza, R., Tavazza, M., Ordas, R. J., Ancora, G. and Benvenuto, E. 1988. Genetic transformation of potato (Solanum tuberosum): An efficient method to obtain transgenic plants. *Plant Sci.* 59:175-181.

Tavladoraki, P., Benvenuto, E., Trinca, S., De Martinis, D., Cattaneo, A. and Galeffi, P. 1993. Transgenic plants expressing a functional single-chain Fv antibody are specifically protected from virus attack. *Nature* 366:469-472.

Taylor, S., Massiah, A., Lomonossoff, G., Roberts, L. M., Lord, J. M. and Hartley, M. 1994. Correlation between the activities of five ribosome-inactivating proteins in depurination of tobacco ribosomes and inhibition of tobacco mosaic virus infection. *Plant J.* 5:827-835.

Thomas, P. E., Kaniewski, W. K. and Lawson, E. C. 1997. Reduced field spread of potato leafroll virus in potatoes transformed with the potato leafroll virus coat protein gene. *Plant Dis.* 81:1447-1453.

Truve, E., Aaspôllu, A., Honkanen, J., Puska, R., Mehto, M., Hassi, A., Teeri, T. H., Kelve, M., Seppänen, P. and Saarma, M. 1993. Transgenic potato plants expressing mammalian 2'-5' oligoadenylate synthetase are protected from potato virus X infection under field conditions. *Biotechnology* 11:1048-1052.

Truve, E., Kelve, M., Aaspôllu, A., Kuusksalu, A., Seppänen, P. and Saarma, M. 1994. Principles and background for the construction of transgenic plants displaying multiple virus resistance. *Arch. Virol.* 9:41-50.

van den Elzen, P. J. M., Huisman, M. J., Posthumus-Lutke Willink, D., Jongedijk, E., Hoekema, A. and Cornelissen, B. J. C. 1989. Engineering virus resistance in agricultural crops. *Plant Mol. Biol.* 13:337-346.

van den Heuvel, J. F. J. M., Boerma, T. M. and Peters, D. 1991. Transmission of potato leafroll virus from plants and artificial diets of *Myzus persicae*. *Phytopathology* 81:150-154.

Van der Vlugt, R., Ruiter, R. and Goldbach, R. 1992. Evidence for sense RNA-mediated protection to PVY^N in tobacco plants transformed with the viral coat protein cistron. *Plant Mol. Biol.* 20:631-639.

van der Wilk, F., Posthumus-Lutke Willink, D., Hiusman, M. J., Huttinga, H. and Goldbach, R. 1991. Expression of the potato leafroll luteovirus coat protein gene in transgenic potato plants inhibits viral infection. *Plant Mol. Biol. 17:431-439.*

Van Dun, C. M. P. and Bol, J. F. 1988. Transgenic potato plants accumulating tobacco rattle virus coat protein resist infection with tobacco rattle virus and pea early browning virus. *Virology* 167:649-652.

Vardi, E., Sela, I., Edelbaum, O., Livneh, O., Kuznetsova, L. and Stram, Y. 1993. Plants transformed with a cistron of a potato virus Y protease (NIa) are resistant to virus infection. *Proc. Natl. Acad. Sci. USA* 90:7513-7517.

Wang, P. G., Zoubenko, O. and Tumer, N. E. 1998. Reduced toxicity and broad spectrum resistance to viral and fungal infection in transgenic plants expressing pokeweed antiviral protein II. *Plant Mol.Biol.* 38:957-964.

Watanabe, K., Kawasaki, T., Sako, N. and Funatsu, G. 1997. Actions of pokeweed antiviral protein on virus-infected protoplasts. *Biosci.Biotechnol.Biochem.* 61:994-997.

Watanabe, Y., Ogawa, T., Takahashi, H., Ishida, I., Takeuchi, Y., Yamamoto, M. and Okada, Y. 1995. Resistance against multiple plant viruses in plants mediated by a double stranded RNA specific ribonuclease. *FEBS Lett.* 372:165-168.

Wilson, T. M. A. 1993. Strategies to protect crop plants against viruses: Pathogen-derived resistance blossoms. *Proc. Natl. Acad. Sci. USA* 90:3134-3141.

Xu, H., Khalilian, H., Eweida, M., Squire, S. and Abouhaidar, M. G. 1998. Genetically engineered resistance to potato virus X in four commercial potato cultivars. *Plant Cell Repts.* 15:91-96.

Xu, H.-P., Riggs, M., Rodgers, L. and Wigler, M. 1990. A gene from *S. pombe* with homology to *E. coli* RNAse III blocks conjugation and sporulation when overexpressed in wild type cells. *Nucl. Acids Res.* 18:5304.

Yang, X. C., Yie, Y., Zhu, F., Liu, Y. L., Kang, L. Y., Wang, X. F. and Tien, P. 1997. Ribozyme-mediated high resistance against potato spindle tuber viroid in transgenic potatoes. *Proc. Natl. Acad. Sci. USA* 94:4861-4865.

14. Control of Potato Viruses using Meristem and Stem-cutting Cultures, Thermotherapy and Chemotherapy

G.Faccioli
UCI S.T.A.A,Istituto di Patologia vegetale,Universitá degli Studi di Bologna, Via Filippo Re 8, 40126 Bologna,Italy

14.1 Introduction

Potato viral diseases are widespread worldwide, and often cause serious yield losses. Their control is difficult because, unlike pathogenic microorganisms, viruses are unaffected by therapeutic treatments; their spread is thus best minimized by preventative measures. As potato is vegetatively propagated, infection increases through subsequent generations, so it is essential to produce healthy from infected plants.

When studying the distribution of *Tomato mosaic virus* (ToMV) in tomato roots, White (1934) first suggested that plant meristems might be free of viruses. When culturing shoot tip meristems aseptically, Morel (1948) obtained virus-free from diseased plants, and hypothesised those meristems were immune to viruses. Healthy plants of some virus-infected potato cultivars were first produced by Morel and Martin (1955) who grew excised meristems on a sterile synthetic medium, results indicating a new way of controlling potato viruses.

Meristem-tip culture (MTC) was first used to produce virus-free plants on the assumption that viruses were unable to invade the meristematic tissues of the bud (Kassanis, 1965), although it was already known that

G. Loebenstein et al. (eds.),
Virus and Virus-like Diseases of Potatoes and Production of Seed-Potatoes, 365–390.

several viruses (e.g., *Carnation mottle* and *Carnation ringspot viruses*) occurred in meristems (Hollings and Stone, 1964).

Simple and insufficiently sensitive techniques, such as mechanical inoculation of test plants and serology, were first used to detect viruses in meristems. Later more sophisticated but time-consuming methods (e.g., electron microscopy) were employed. Precise and sensitive techniques such as autoradiography and Dot-ELISA were finally adopted and showed that hitherto undetected viruses occurred in potato meristems. The presence of viruses in meristem tips did not always result in production of infected plantlets. Virus-free plants were obtained from meristems infected by *Potato virus X* (PVX), *Potato virus M* (PVM) and *Potato virus S* (PVS), indicating that virus eradication occurred during tissue culture *in vitro* (Beemster and Rozendaal, 1972; Faccioli and Rubies-Autonell, 1982; Faccioli and Colombarini, 1996).

Various hypotheses have been proposed to explain the inability of viruses to invade meristem tips of certain species, and the elimination of viruses from meristems during tissue culture. The failure of viruses to invade meristems has been attributed to the high auxin content of actively dividing meristematic cells (Wu *et al.*1960), to competition for nutrients during synthesis of normal and viral nucleoproteins (Quak, 1965) or to the absence of vascular tissues and plasmodesmata connecting meristematic cells (Quak, 1977). This last may well be so with luteoviruses and most closteroviruses that are confined to phloem cells but none of the hypotheses is fully explanatory. Numerous factors probably influence the host cell-virus relationship, so currently it can only be stated that in meristematic cells enhanced oxidative-reductive processes may create an unsuitable environment for virus replication.

Hypotheses to explain virus eradication during *in vitro* culture include the action of growth regulators, especially cytokinins. (Quak, 1977), inhibitors such as phenolamines (Martin-Tanguy, 1985), the metabolic disruption of viral replicases, and the degradation of viral RNA resulting from cell injury during explant excision (Mellor and Stace-Smith, 1977). Studies on virus multiplication in plant tissue cultures have shown that, although some viruses are inactivated, others (e.g. PVX) persist in all or most of the cultures (Mellor and Stace-Smith, 1977). Several factors probably influence virus persistence in cultured meristems and that all events influencing host cell metabolism (e.g., temperature and composition of culture media) affect virus replication.

Although the persistence of some viruses during *in vitro* culture has not yet been satisfactorily explained, MTC has been used successfully for virus eradication, and multiplication and safe storage of valuable germplasm. The simultaneous use of thermo- and chemotherapy has increased its potential for producing material not only free of viruses but also of bacteria and fungi.

More recently, some viruses have also been eradicated more quickly and simply from virus-infected potato stem cuttings by thermo- and/or chemotherapy (Griffiths *et al.* 1990; Sanchez *et al.* 1991; Faccioli and Zoffoli, 1998).

14.2 Characteristics of Meristem Tissues

Apical meristems are domes of actively dividing cells, located at the apices of shoots and roots (Fig. 14.1), that remain in an active state of division throughout the vegetative phase of the plant, forming new tissues and organs (shoots, leaves, roots); they have, therefore, the capability of producing complete plants *in vivo* and *in vitro*. This totipotency is the basis of the MTC technique by which plantlets are produced.

MTC-derived plantlets usually retain the genetic characteristics of mother plants, whereas those derived from other tissues such as protoplasts, callus, nucellus or floral primordia may differ (D'Amato, 1977).

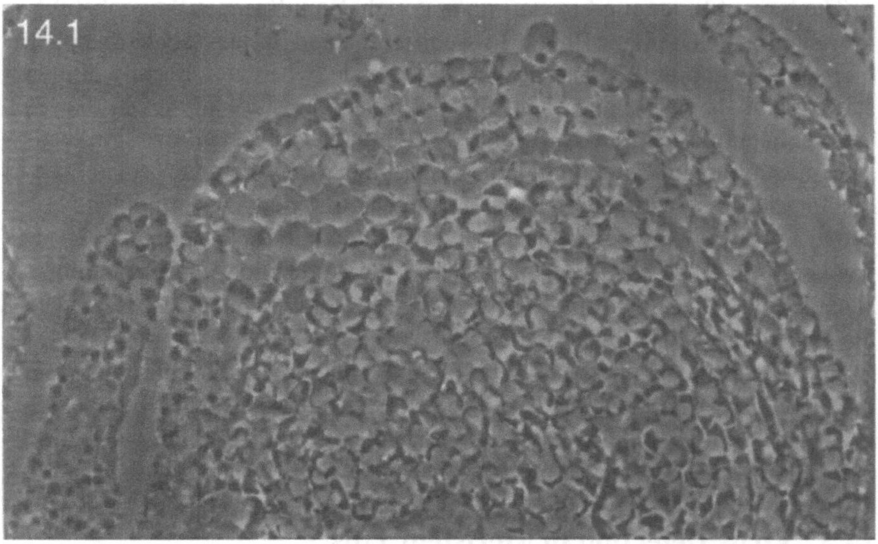

Fig.14.1. Phase-contrast micrograph of dome and part of the two first leaf primordial of cv. Majestic meristem tip.

The purpose of MTC is to eliminate viruses from a selected cultivar obtaining plants phenotypically identical to mother plants. The frequency of variants is comparable to that obtained with traditional methods of propagation (Beauchesne, 1982). Meristems are genetically stabile because they are the storage sites of genetic information, which, in seed plants, is transmitted to the germ line in the flower during the transition from vegetative to reproductive phase. Thus genetic stability of meristem cell lines is a prerequisite for genetic stability of species germ- line (D'Amato, 1978). The genetic stability of meristems is strictly controlled by DNA synthesis, which does not allow extra duplication of DNA responsible for somatic polyploidy, and the continuous cell division which eliminates any spontaneous chromosomal structural changes.

The genetic uniformity in mericlones depends on different factors, but mostly on the use of techniques that avoid the formation of callus (derived from undifferentiated tissues), which is a common source of genetic variation.

It is well known that frequency of mutation increases when adventitious shoot formation occurs in clonal propagation (micropropagation), or in callus culture (somaclonal variations) or when somatic embryogenesis occurs in callus cultures. This embryogenetic pathway, in which plants derived from single cells are solid mutants, is of particular significance in mutagenetic studies.

The type of tissue used as starting material also influences the probability of mutation (Sheridan, 1975). In undifferentiated tissues such as apical meristems, cambium, procambium and pericycle, where a uniformly diploid state of cells is maintained, the chance of mutations is usually lower than in differentiated tissues, such as pith, in which induction of cell division leads to callus cultures. Growth regulators used in growing media especially 2,4-dichlorophenoxyacetic acid (2,4-D), naphthalenacetic (NAA) and synthetic cytokinins are also responsible for inducing mutations. In particular 2-4-D has been reported as a direct inducer of polyploidization in tissue cultures (Sunderland, 1977). The induction of higher ploidy is inversely related to the concentration of growth regulators.

Nuclear changes, including polyploidy, aneuploidy and chromosomal mutations, are also associated with repeated subculturing *in vitro* and with long-term callus cultures, which can be used in mutant selection (somaclones) (Larkin and Snowcroft, 1981).

14.3 Virus Detection in Meristem Tips

The meristematic tissues of shoots and roots are plant parts with a low or, sometimes no, virus content. From the dome, virus concentration

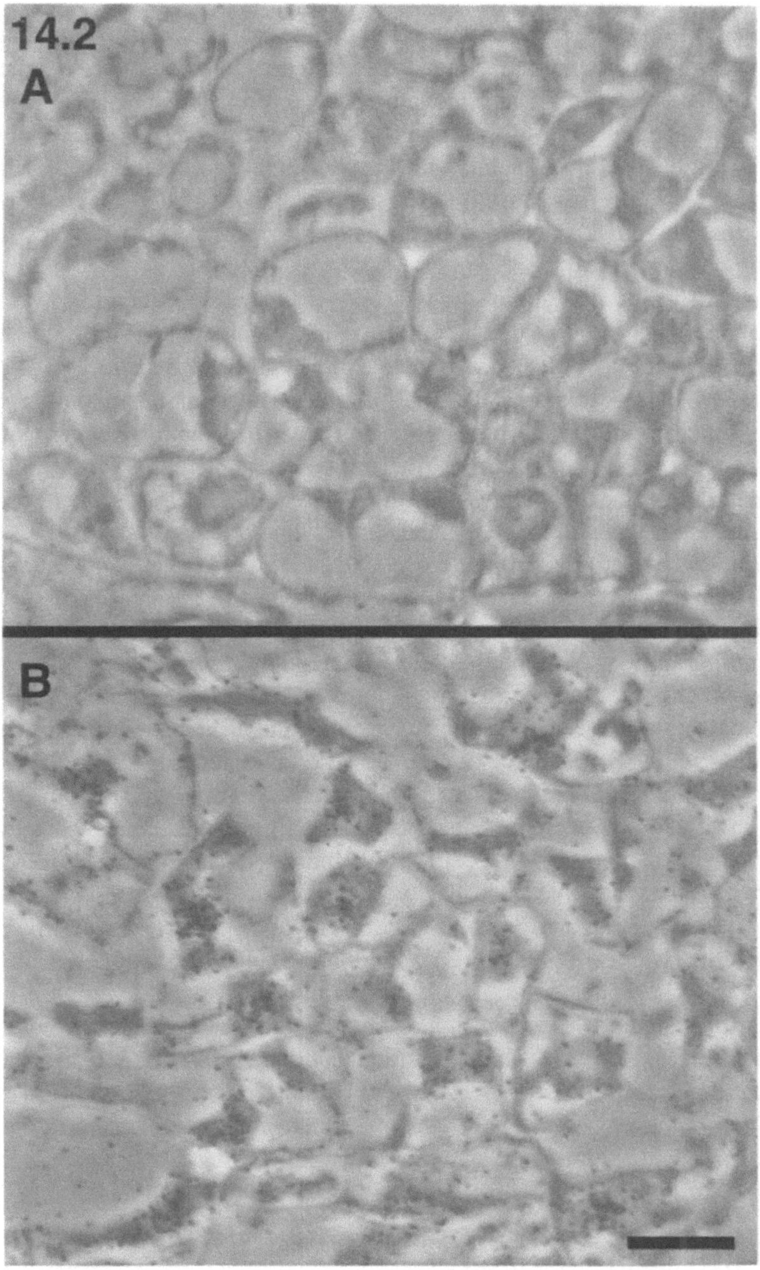

Fig.14.2. Autoradiogram of uridine-H^3 incorporation (silver grains) in cells of cv. Mjestic meristem tip (first leaf prinmordium). Samples were given 5 h pulse feeding (50 μC/ml) after a 12 h treatment with actinomycin D (250 μg/ml). A- healthy cell; B- PVX-infected cell. Bar =10μ.

increases basipetally through the primordia. The probability of obtaining virus-free plants is, therefore, inversely related to the size of the meristem used. By contrast, meristem development into a full plant is directly proportional to the size of the explant. It is necessary, therefore, to establish the maximum size of meristem, which produces the largest number of virus-free plantlets. The use of virus-free meristems obviously ensures 100% virus-free progenies but those that contain virus may also develop into healthy plants.

Virus distribution in shoot tips is mostly dependent on the virus type and host-plant. With potato, it is also cultivar- (Colombarini, 1990) and virus strain-dependent (Mellor and Stace-Smith, 1970). Before large-scale production by MTC, viral distribution for each virus-host combination should, therefore, is determined. In the past, the number of virus-free plantlets obtained by culturing meristems of different sizes was the parameter used, but nowadays the direct detection of viruses in meristem tips, and subsequent choice of the convenient explant size, is the fast approach to be used. Electron microscopy of tissues (Appiano and Pennazio, 1972) or immunofluorescence (Mori and Hosokawa, 1977) were first used for this, but were laborious and unsuitable for large-scale screening. Autoradiography with a phase-contrast microscope is a faster and more precise method for detecting virus in several meristems, but is still too complicated for mass screening (Faccioli and Rubies-Autonell, 1982; Faccioli *et al.*1988). The technique is based on the incorporation of uridine-H3 into viral RNA only, since cellular-RNA synthesis is prevented by actinomycin (AMD) treatments. "Silver grains" produced by isotopic radiation in a photographic emulsion indicate the presence of virus (Fig.14.2). The improved sensitivity of serological methods, particularly Dot-ELISA (Banttari and Godwin, 1985), now permits rapid identification in single meristems, of 1 pg of PVX and PVS (Fig. 14.3; Rubies-Autonell *et al.* 1987; 1990; Faccioli and Colombarini, 1996). By examining a high number of meristems and culturing an equivalent number of them *in vitro,* it was possible to: (i) establish that meristems containing a limited amount of virus could also produce virus-free plants (ii) determine, for particular virus-plant combinations (PVX, PVY, PVM / potato), the meristem virus level below which the progeny would be largely virus-free (Faccioli and Colombarini, 1990; 1996).

14.4 Meristem-tip Culture (MTC)

The mother plants used to provide meristems should be selected for their desirable agronomic traits. MTC involves three main operations; meristem excision, *in vitro* culture and development of meristems (i.e., shoot and root

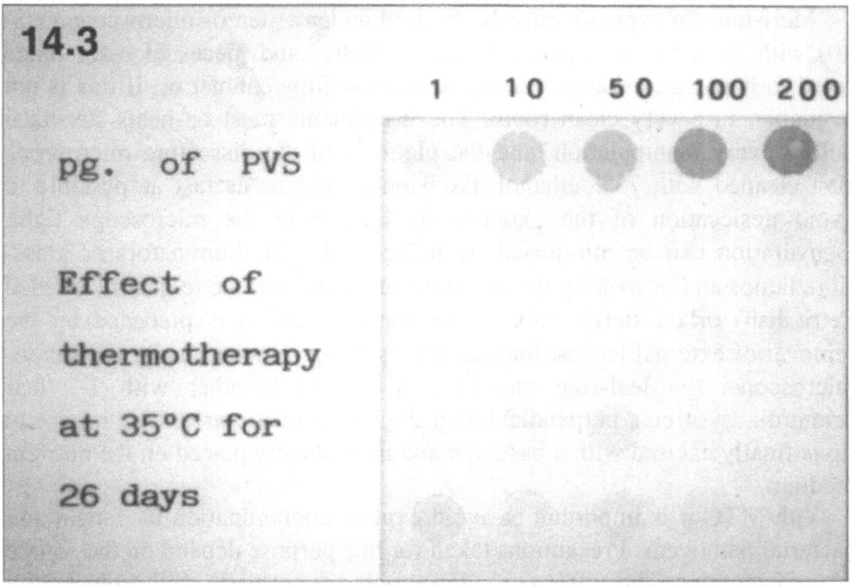

Fig.14.3. Dot-ELISA detection of PVS on nitrocellulose membrane. Top line- standard virus concentration in picograms (pg); lower lines tests for virus presence in meristems excised from heat- treated cv. Primura

induction), and transplantation of plantlets to pots and subsequent virus indexing.

14.4.1 Meristem Excision

Explants suitable for virus elimination have been variously designated "shoot tips", "tips", "meristems", and "meristem tips" (Hollings, 1965). The last two indicate explants made of meristematic dome and one or two pair of leaf primordia, measuring c. 100 μm in diameter and 250 μm in length. "Shoot tips" and "tips" is shoot apices, each including the meristem, with several leaf primordia and adjacent differentiated stem tissue.

The terms "MTC" and "clonal propagation" have often been used synonymously, probably because both have been used to obtain virus-free plants. This is erroneous, as is the designation of MTC for the culture of shoots and other tissues, which is simple clonal propagation. This distinction is important because non-meristematic tissues of virus-infected plants are infected.

Meristem-tip explants must be excised under a stereo- microscope (x 8-40) with suitable instruments (forceps, needles and pieces of razor blade fixed to handles), usually in a sterile laminar-flow cabinet or, if this is not available, in a very clean room. The instruments must be heats sterilized before every manipulation, and the platform of the dissecting microscope first cleaned with 75% ethanol. Excision should be as fast as possible to avoid desiccation of the explants by heat from the microscope light. Dehydration can be minimised by using cool-light illuminators or glass-fibre lamps and/or making the excisions on a cold surface (e.g., an ice-filled Petri dish) or on sterile moist filter paper. Excision is preceded by the removal of external leaves, internal leaves being removed under the stereo-microscope; the leaf-free meristem is excised together with 1-2 leaf primordia by 4 cuts perpendicular to the meristematic area. The meristem tip is finally excised with a basal cut and immediately placed on the nutrient medium.

With MTC it is important to avoid explant contamination by fungal and bacterial pathogens. Precautions taken for this purpose depend on the source of meristem tips. Those from growing shoots are normally well protected by several leaf primordia, so sterilization is not strictly necessary; indeed, attempted disinfection may sometimes increase the level of contamination. Normally, to grow donor plants in a greenhouse on sterilized soil, watering the soil rather than the leaves helps to reduce contamination. Periodic spraying of the plants with systemic fungicides is also useful. For field-grown plants it is advisable to take small cuttings, grow them in a nutrient solution, and then excise meristems from the new axillary buds, which are less contaminated. Meristem tips collected from underground tubers should be dipped in 75% ethanol to remove trapped air, then sterilized with sodium hypochlorite or calcium chloride for 10 min and finally rinsed three times in sterile water.

We do not normally use sterilization, but others should find the best way of avoiding contamination under their working conditions.

14.4.2 Culture and Development of Meristems

The morphogenetic potential of potato meristems to produce plantlets depends mainly on meristem size, physiological conditions of the donor plants, meristem position on the plant and also the potato cultivar.

14.4.2.1 Meristem Size

The optimum size of meristems to be used is either quoted as total meristem length or as number of leaf primordia. Tips ranging from 0.2 to 1 mm (meristematic dome plus two or few more leaf primordia) are those mostly used. To avoid confusion, it is better to specify the number of leaf primordia rather than dimensions.

The larger the size of the meristem cultured, the greater is the number of regenerated plants produced, but the number of virus-free plantlets obtainable is inversely proportional to the size of the cultured tips. The size of the explants should thus be a compromise between the two. Like that of other species, the survival rate for potato meristems without leaf primordia is very low. Low survival levels are typical of small meristems; for example, 21% of potato meristems with 2 leaf primordia (0.3 mm in length) and 69% with 3-4 leaf primordia (0.8 mm) produced plantlets (Marani and Pisi, 1977). An average development close to 80% for CVs Majestic and Kennebec and 90 - 95 % for cvs Majestic and Primura was obtained using meristems with 4 leaf primordia (Faccioli and Rubies-Autonell, 1982). Using the so-called "multimeristem" technique (Roca *et al.*1978), 7-9 times the number of the initially cultured potato meristems developed into plantlets (Faccioli *et al,* 1988; Fig.14.4).

14.4.2.2 Physiological Condition of Donor Plant

Meristems are best collected from donor plants in a very active vegetative state. It is, therefore, desirable to provide good growing conditions by using phytotrons, controlled growth cabinets or greenhouses in which light and temperature can be regulated. When plants are not kept in ideal conditions, seasonal fluctuations of light intensity and photoperiod are particularly important. For example, meristems of most potato cultivars collected in spring and early summer root more easily than those collected later (Mellor and Stace-Smith,1969; Quak,1977). For cvs Primura and Spunta, 69% of the meristems (domes with 3-4 leaf primordia, 0.8 mm in length) collected in May developed into plantlets, whereas only 37% of those taken in February did so (Marani and Pisi, 1977.

14.4.2.3 Meristem Position on the Plant

All plant meristems can be used as starting material, but the morphogenetic potential and virus content of apical or axillary meristems differ. With axillary meristems, shoot position on the stem is also important. In PVS-infected cv. Majestic shoots, for example, an increase in virus concentration from the apical meristem to the lower axillary buds was

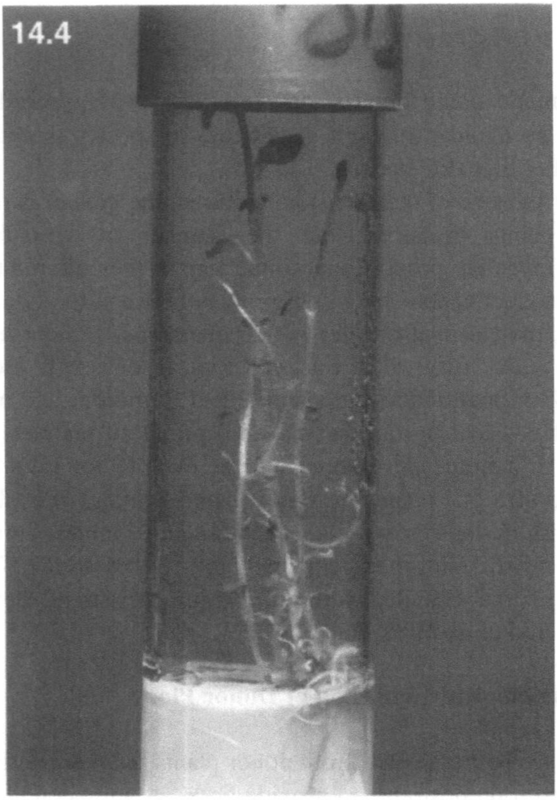

Fig. 14.4. Rooted shoots of cv. Majestic meristem tips (4 leaf primordial) in MS+1mg/l IAA and 1mg/l IBA medium of "multimeristem" culture technique after a culture for 14 weeks.

found: only the former and the two first axillary meristems had less than 10 pg of virus (Faccioli and Colombarini, 1996).

Meristem position on the plant is also important for the phenological stages of the derived plantlets but, unfortunately, studies are not available for potato.

14.4.4 Factors Affecting Meristem Development

The subsequent development of meristems *in vitro* will mainly depend on the characteristics of culture media, but also on light and temperature and, to a lesser extent, on other factors.

14.4.4.1 Culture Media

Culture media are generally balanced mixtures of mineral salts, a carbon source of energy, growth factors, vitamins, and minor components. Since meristems are unable to synthesize growth factors, their development mainly depends on the hormone composition of the culture medium. Different hormone groups have contrasting effects on meristem differentiation. It is, therefore, quite difficult to formulate a proper-balanced medium. Earlier trials with single-media culture for potato were only partially successful, and the number of developed plantlets was always much lower than that obtained using two media (Pennazio and Vecchiati, 1978). It is now common practice to use first a cytokinin-rich medium for shoot development, and a second auxin-rich medium to stimulate root formation. The use of an intermediate hormone-free medium, to drain residual cytokinins from potato shoots, facilitates increased rooting in the second medium (Faccioli, unpublished). Three media ("multimeristem" technique) were employed to stimulate the development of more shoots from a potato meristem (Roca *et al.* 1978).

Potato meristems grow well on nutrient media solidified with agar (0.6-0.8 %) but liquid media can also be used, although a paper bridge is needed to support the meristem. Concentration and quality of the agar affect meristem development; Difco Bacto agar or alternative products such as Biogel P200, alginate or Gelrite should be used (Pierrik, 1987), rather than ordinary agar which may contain toxic inorganic or organic contaminants It is advisable to use a lower concentration of growth regulators in liquid media because their uptake is faster, and the concentration *in planta* may become too high. The use of paper bridges (Heller, 1953) or other porous materials dipped in the liquid and supporting the explants is recommended (Pierick, 1987). The pH of the culture medium, normally 5.0-6.5, is an important variable, which affects meristem development of different plant species, and the control of meristem organogenesis. This probably modifies the cell uptake of growth factors, vitamins and ammonium ions and also influencing amino acid and sugar uptake by the cells (Margara, 1982). In potato, a pH of 5.7-5.8 has often been found suitable (e.g., Mellor and Stace-Smith, 1969; Pennazio and Redolfi, 1973; Gregorini and Lorenzi, 1974; Marani and Pisi, 1977; Faccioli and Rubies-Autonell, 1982). Both lower (4.8) and higher (6.2) pH inhibits root formation (Mellor and Stace-Smith, 1969).

An excess of auxins and cytokinins in media, coupled with a low agar concentration (less than 6 g/l), high relative moisture and accumulation of gases, may cause "vitrification" or "glassiness". This physiological disorder is characterized by the production of water-soaked leaves, with water-filled intercellular spaces, which may inhibit growth and/or result in transplant losses up to 60% (Maene and Deberg, 1985; Kevers *et al.*1984). A final recommendation is that only pure ingredients should be used for tissue culture.

The duration of culture on a medium before transplanting is variable depending on plant species and cultivar. With potato, rooted shoots can be transplanted after c. 3-3.5 months, the early cultivars requiring a slightly shorter time (Faccioli, unpublished).

To better understand the role of the chemical components of the growing media on meristem development, a more detailed account is given below.

Mineral salts. The major mixtures of mineral salts at different concentrations, which have been included in growing media, have been reviewed by Bhojwani and Razdan (1996) and Gamborg (1984). Although the quality of inorganic nutrients is fairly constant in all media, the quantity of salts and ions used varies. The concentrations of various inorganic nutrients (speciallyK+ and NH_4) was low in White's (1943) and Heller's (1953) media, widely used initially, and were responsible for poor development and chlorosis of potato.meristem-tips They were, therefore, later increased (Mellor and Stace-Smith, 1977). Some minor non-essential elements, like Al and Ni, were eliminated because they were unnecessary. Iron, added first in the form of $Fe_2(S0_4)_3$ was replaced by $FeCl_2$ then by Fe-EDTA (ethylendiaminetetracetic acid), which greatly improved availability of this element in the variable pH conditions of tissue culture, and also the uptake of other micronutrients probably preventing their precipitation (Gamborg *et al.*1976).

In the Murashige and Skoog (1962) medium (MS), nitrogen is supplied as nitrate and ammonium compounds. Nitrate is superior to ammonium as source of nitrogen but used alone causes a shift in the pH medium towards alkalinity, which is controlled by the presence of NH_4 compounds. Thus a correct NO_3/NH_4 balance is important for a successful development of explants. The MS medium, originally devised for tobacco meristems, is also effective for potato meristems (Stace-Smith and Mellor, 1968) and is, therefore, widely used.

Carbon and energy source. The carbon source most often utilized for plant tissue culture, is sucrose, which is rapidly hydrolized to fructose and glucose, the last of which is utilized first.

Vitamins. Vitamins such as thiamine-HCl, biotin, pyridoxin, pantothenic acid and nicotinic acid are often added to the medium at low concentration (0.1- 0.5 mg/l) even if their role is not well known. Of these only thiamine- HCl appears to be essential in the culture of most plants (Murashige, 1977)

Growth Regulators. It is widely accepted that the auxin/cytokinin ratio determines the type and the extent of organogenesis because exogeneously supplied cytokinins caused shoot proliferation of cultured tobacco pith cells whereas auxins stimulated root formation (Skoog and Miller, 1957); a similar response was found later for numerous other plant species. However, there is great variability in auxin and cytokinin requirements for morphogenesis among taxa and genotypes, thus the appropriate types and concentrations of hormones should be determined for each species, and sometimes even for cultivars. Besides these two hormone groups, other additional substances influencing growth and morphogenesis were identified. These included gibberellins, ethylene, and complex organic compounds although the biochemical mechanisms by which they alter growth responses are not fully understood.

In general, the main events or sites of auxin and cytokinin action are as yet unknown, or whether they have a direct action at the gene level or have effects through a series of intermediates.

Growth promoters are generally used at concentrations between 0.01-10 mg/l. The commonly used auxins are 3-indolacetic acid (IAA), 3-indolbutyric acid (IBA), and naphthalenacetic acid (NAA); 2,4-D should be avoided because it generally promotes callusing of the explants instead of morphogenesis.

Growth hormones are not always needed for large meristem-tip explants (500μm long or more), which sometimes may also develop into whole plants in media lacking growth regulators, but small amounts of an auxin or a cytokinin or both (0.1-0.5 mg/l) are highly beneficial and absolutely necessary for small explants.

Since the second pair of leaf primordia probably synthesizes auxins, the meristematic dome of shoot tips is not autonomous for auxins. This was demonstrated by the need for exogenous auxins of meristems lacking leaf primordia of many species to develop (Smith and Murashige, 1970).

As mentioned previously, two media are normally used. The first is an MS-based modified medium with 1 mg/l benzylaminipurine (BA), followed by a second in which BA is substituted by 1 mg/l indolbutyric acid (IBA) and 1 mg/l indolacetic acid (IAA), ensure, for example, a proper development of 3 to 4 -leaf primordia potato meristems. (Faccioli and Rubies Autonell, 1982)

The induction of more shoots per meristem depends on the species involved but mostly on the cytokinin level of the medium. Cytokinins can overcome apical dominance and induce growth of axillary shoots in the presence of the terminal apex. This stimulation of adventitious shoot formation can be repeated several times, and is the basis of clonal propagation (Bhojwani and Razdan, 1996).

The role of GA3 seems to be quite variable; it improved the growth of some, but had no effect on the meristem tips of other potato cultivars (Mellor and Stace-Smith, 1977). Rooting of potato meristems on MS medium lacking GA3 was lower (17%) than on a medium containing it (66%)(Mellor and Stace-Smith, 1977).

To induce multiple-shoot formation in potato axillary meristems (Fig.14.4), we successfully used 3 sequential media, the first containing 1mg/l BA, the second with 0.5 mg/lBA + 0.5 mg/l GA3 and the third with 1 mg/l IBA + 1mg/l IAA (Faccioli *et al.* 1988).

In addition to the production of plantlets, culture *in vitro* can be used for the production of minitubers. For this purpose, rooted shoots can be placed in a liquid medium containing high BA and sucrose concentrations (10 mg/l and 8%, respectively) as suggested by Wang and Hu (1982), or single node cuttings can be set on solid media enriched with cumarin (25 mg/l) as reported by Stallknecht and Farnsworth (1982). We found the first method to be more convenient, and the minitubers produced were viable up to 6 months when stored in the dark at 4°C on sand (Faccioli *et al.* 1988).

Antibiotics. After months of culture, contamination by endogenous bacteria causing growth inhibition may occur. This is infrequent with potato but, as a preventive measure, nontoxic antibiotics can be included in media. Gentamicin and tetracyclin (100-200 mg/l) seem to be the most effective in fruit tree culture, but they need to be tested on potato.

Minor components. Tissues cultured *in vitro* release metabolites into the media, which can be toxic, or tissue-growth inhibitors. These substances pass unnoticed, unless they cause discoloration of the medium, or browning of the tissues and adjacent medium .The latter is assumed to be due to polyphenol oxidation to quinones, which are toxic. To reduce medium browning, an initial short dark culture period may be beneficial since light promotes oxidation; antioxidant and adsorbent substances can also be used. Like ascorbic acid, which is routinely used as an antioxident in strawberry and fruit tree culture, cystine, thiourea and citric acid (Murashige, 1974) may be used in potato. As an adsorbent, activated charcoal (0.5-3%), which binds phenol but may also adsorb plant growth regulators and other organic compounds, could be employed. In potato it stimulates also meristem growth into shoots and increases rooting of shoots (Fridborg and Eriksson, 1975). Polyvinylpyrrolidone (PVP, 250-1000 mg/l) is another useful phenol absorbent (Pierik, 1987).

To free the medium of cell compounds and debris, meristems can be tansferred to fresh culture medium after one or few days, although it is unnecessary for potato and is a possible source of additional contamination.

Culture containers. Glass tubes (10 x 120 mm) closed with caps permitting air exchange, are the most commonly used containers for meristem culture and suitable for potato. Vessel dimensions should be taken into account because they can affect morphogenesis. The sealing of culture vessels to avoid contamination is not a good practice because it can inhibit gaseous exchange and then cause accumulation, for example, of carbon dioxide and ethylene which could adversely affect morphogenesis (Thomas and Murashige, 1979)

14.4.4.2 Temperature

Temperature requirements for meristem growth are assumed to be similar to those of mother plants and depending also on the interaction with light and hormone composition of the substrate. A constant temperature of c. 26°C, with a 16 hour daylight period of 3-4 Klux is appropriate for ordinary potato cultivars grown on modified MS (Faccioli et *al.* 1988).

14.4.4.3 Light

Light affects meristem organogenesis through exposure time (photoperiod), intensity and wavelength. It seems reasonable to assume that meristem requirements do not differ substantially from those of whole plants of the same species (Murashige, 1974).

For potato a photoperiod of 16 hours at 1-4 Klux has been suggested (Styler, and Chin, 1983). The quality of light which favours photosynthesis in plants, and presumably in cultured tissues, is similar for all species and preferentially occurs at those wavelengths in which chlorophyll a and b absorb light, i.e., the red (680-650 nm) and blue (440-470 nm) bands respectively. In tissue culture, these two light bands preferentially direct morphogenesis towards formation of adventitious roots (Letouze and Beauchesne, 1969) the former, and of shoots the latter (Seibert *et al.* 1975.). Various lamps have been tested to find the most effective quality of light. For potato meristems, the effects of three different light regimes have been investigated (Pennazio and Redolfi, 1973) by using either red- rich light sources, or sources approximating daylight, or a 2:1 combination of both. The most vigorous and rooted plants were obtained under mixed light conditions. Fluora 77R (Osram), with prevalent emitting bands in the red and blue regions (3 Klux), 16 hours/day stimulated a complete and satisfactory morphogenetic response from 4 leaf primordia meristems of various potato cultivars (Faccioli and Rubies- Autonell, 1982).

14.4.4.4 Plantlet Transplanting

When shoots have developed a satisfactory root system, they are transferred from culture vessels to well-drained soil, e.g. a mixture (1/1/1) of peat, soil and sand. Care is needed during transplanting to avoid root damage, and agar should be removed by gentle washing to prevent contamination by fungi and bacteria. In subsequent operations maximum care must be taken to maintain plantlets under very high humidity (90-100%) for the first 10-15 days as plants grown in test tubes have a poorly-developed cuticle and during transplantation loose water by evaporation. During the first few hours of acclimatization, plantlets undergo significant water stress: stomata do not operate properly, since in tissue culture the closure mechanism is not operative, and plantlets grown *in vitro* lack root hairs. They are, therefore, generally covered with clear plastic bags or kept under intermittent mist. During the first few days, watering with a nutrient solution, such as Hoagland's solution, aids plantlet development. The humidity is then gradually decreased to allow plant adjustment to normal conditions and, 2-4 weeks later, they can be grown under normal greenhouse conditions.

Alternatively, to overcome this critical phase potato shoots can be induced to produce microtubers in culture.

14.5 Stem-cutting Culture

Culture *in vitro* of potato stem cuttings is much simpler than that of meristems. The growth substrates utilized may be those used for meristem culture, but without growth hormones that are unnecessary. Vitamins are also probably not required, but this needs to be determined. Single-node stem cuttings, after removal of subtending leaflets, can be used but although the starting material is always infected, the smaller the explant, the larger is the chance of obtaining virus-free progenies (Faccioli and Zoffoli, 1998). This is probably because they contain lower amounts of viruses. It is better to use stem cuttings as small as possible (e.g., 0.15 cm in diameter, 0.5 cm long) since they have no growth problems, They can be taken directly from sprouting tubers, but better from smaller cuttings from sprouting tuberlets grown in the greenhouse. In this case, sterilization with 1% sodium hypochlorite in a laminar-flow cabinet is needed (Faccioli and Zoffoli, 1998). Cuttings can also be excised without sterilization from plantlets previously estabilished *in vitro* (Griffiths *et al.* 1990). All other growth requirements such as temperature and light are similar to those of meristems although, for stem cuttings, these requirements are obviously less critical.

14.6 Indexing

Plantlets transferred from culture media to soil must be tested to ensure that they are free of viruses. To ensure this, plants are tested several times during their development, because the virus concentration may initially fall below a detectable level. Mother plants kept in greenhouse, should also be tested repeatedly before they are used for multiplication.

Virus testing methods are fully reviewed in chapter 10. However, it should be mentioned that sensitivity of DAS-ELISA for some potato viruses (PVY, PVX, PVS and PVM) is in the ng range, whereas Dot-ELISA, slightly modified by trapping the virus directly on NCM, detects as little as 1 pg of potato viruses. (Rubies-Autonell *et al.*1987; Colombarini, 1990; Faccioli and Colombarini, 1996) Chlorophyll-containing plant sap, unlike meristem sap, can interfere with the staining reactions of this test, but diluting the sap and/or "bleaching" the NCM with sodium hypochlorite can overcome this. A simpler version of ELISA, the so-called "squash blot" for mass field detection, has about the same sensitivity as DAS-ELISA (Mitchell *et al.* 1990). It can be carried out by squeezing a plant leaf between two inexpensive commercial PVC strips, which are processed, as in Dot-ELISA. The method is fast and labour-saving for detecting PVY, PVX, PVS, PVM but not PLRV (Faccioli and Colombarini, 1990).

The sensitivity of molecular hybridization for detecting PLRV and PVX is comparable to that of DAS-ELISA, but less efficient than that of Dot-ELISA (Faccioli, unpublished), probably because plant sap constituents interfere with the hybridization reaction.

14.7 Virus Eradication

14.7.1 Meristem Culture

The correlation between meristem size and virus eradication in potato is well established. Early reports on potato virus eradication indicated a differential pattern among viruses: excised meristems with 2 leaf primordia produced 100% plantlets free of PLRV, PVY and PVM, but only 66% of them were PVX- and PVS-free (Accatino, 1966). High levels of eradication (95%) for PVX and PVS were obtained by culturing 0.1 mm-long meristems with or without a leaf primordium, but only 20 plantlets were regenerated from 196 meristems (Kassanis and Varma, 1967). Potato virus A (PVA) and PVY were easily eradicated (85-90 %) using meristems shorter than 0.3 mm (Morel *et al.* 1968), whereas other viruses such as PVX and PVS were not eliminated because they were able to invade meristems. With potatoes double-infected by PVY and PVA, Marani and Pisi (1977) obtained a 21% regeneration from 0.3 mm-long meristems, 90% of which developed into healthy plants, and 69% regeneration from 0.8 mm-long meristems, of which only 7% produced virus-free progeniess. By the use of autoradiography, which revealed the absence of PVY and PLRV in the dome and the first 4-leaf primordia, it was possible to obtain an almost total virus-free regeneration from this size meristems (Faccioli and Rubies-Autonell, 1982;Faccioli *et al.* 1988). A better eradication of PVX from some potato cultivars, was achieved when smaller (0.2 mm long) rather than larger (0.4 mm) meristem tips were used, (47.4% and 21.2% eradication respectively), but survival and relative growth rate of smaller meristems were lower (El-Fiki *et al.* 1992). These last eradication results are probably explained by the lower virus content of smaller meristems.

14.7.2 Thermo- and Chemotherapy

Thermo- and chemotherapy coupled with MTC improves the success of obtaining virus-free plants when dome and leaf primordia are infected.

Thermotherapy can be applied to mother plants before meristem excision, or to meristem tips during in *vitro* culture. The former is less damaging and also allows the use of larger meristems, which more easily develop into plantlets with the same possibility of being virus free, since heat treatment inhibits virus multiplication (Stace-Smith and Mellor, 1968; Wang and Hu, 1980). Meristems can also be excised from sprouting tips of heat-treated tubers. The mechanisms of virus eradication by heat are not known, but the induced changes in cell environment probably inhibit virus multiplication and/or virus transport. Heat-dependent alterations of cellular metabolism may favour the formation of o-quinones, activation of ribonucleases, inhibition of replicases, and a reduction in ribosomes. Several potato cultivars have been freed from PVX and PVS using thermotherapy (Stace-Smith and Mellor, 1968; 1970; MacDonald, 1973. Meristems taken from heath-treated tubers usually grow unsatisfactorily. Sip (1972), for example, obtained only 14% of plantlets free of PVA or PVS from 0.3-1 mm-long meristem-tips excised from sprouting tubers heat-treated for 52-124 days (33° C initially, increased up to 37° C in the fourth week).

Better and quicker yields of virus-free progenies have been obtained with plant thermotherapy. The number of PVX-free plantlets obtained by heating plants at 30°C from 14 to 42 days was increased from 12.5 to 82% using 0.3 mm-long meristems, and from 0 to 53% with 0.8 mm-long meristems (Pennazio and Vecchiati, 1978). Virus-free progenies (92-91%) were produce by culturing meristem tips with 4 leaf primordia excised from

PVX-infected cvs Majestic and Kennebec plants heat-treated at 35°C for 27 days and 93% from PVS-infected cv. Majestic similarly heat-treated 26 days; the success of eradication increased with treatment length (Faccioli and Rubies-Autonell, 1982; Faccioli and Colombarini, 1996). Moreover, for PVS-infected material a clear correlation was established between virus content of meristems and percentage of virus-free plants obtained: in most meristems (85%) the treatment reduced the virus content from 10-100 pg to less than 10 pg (Fig.14.3); thus a virus content lower than 10 pg ensured virus eradication in MTC. These results were probably influenced by the use of different cultivars, different media and various light regimes. Moreover, differences in virus strain or isolate also affected eradication (Mellor and Stace-Smith, 1970). Thus, when initiating eradication programs, all these factors need to be considered. Since prolonged constant exposure to heat may damage host tissues, alternating temperatures may be used.

Low temperature has also been used to free tissues from viruses before meristem collection. PVA and PVY for example were eradicated by growing potato tissues at 5°C (Moskovets *et al.*1973). Low temperature treatment, however, was most effective against viroids which multiply well at temperatures above 30°C. Potato spindle tuber viroid (PSTVd) was eradicated by excising meristems with 1 leaf primordium from plantlets kept at 5-6°C for 6 months (53% virus-free progenies) or from tubers kept at 8 C° for 4 months (30%) (Lizarraga *et al.*1980); by contrast, only 2.4-6% of viroid-free plantlets were obtained by two cycles of heat treatment (33-35°C) followed by low temperature (Stace-Smith and Mellor,1970).

Chemotherapy can also be applied to meristems cultured *in vitro*, or to plants before meristem collection. In the former case, the chemicals are incorporated into the culture medium, but can affect meristem growth. Numerous chemicals have been tested for antiviral activity, but few were effective (see Schuster, 1988). The most used substance is the synthetic analogue of guanosine, ribavirin (=Virazole;1-β D-ribofuranosyl-1, 2,4-triazole-3-carboxamide) which, when added to the media at concentrations of 10-50 mg/l, was effective against PVX, PVY, PVS and PVM in potato (Cassel and Long, 1982; Klein and Livingston, 1982; 1983; Cassel, 1987). However, ribavirin at active doses, which vary with the plant species and virus, is usually phytotoxic causing an increase in culture time, death of some meristems, and the need for frequent transfers to fresh media; the use of 190 mg/l for PVX eradication in potato, for example, increased culture time up to 6-8 months (Klein and Livingston, 1982). Regeneration of potato sprouts from meristematic cultures was delayed by 6 to 8 weeks, when ribavirin (10-50 mg/l) was used (Cassel, 1987).

To overcome toxicity, a low dose of ribavirin was supplied for two weeks to hydroponic cultures of potato stem cuttings, together with the non-toxic antimetabolite 2,4-dioxo-hexahydro-1, 3,5-triazine (DHT) (Rubies-Autonell et al.1987; Faccioli and Colombarini, 1996). Ninety per cent of meristems with 4 leaf primordia, taken from cuttings treated with 5 mg ribavirin + 5 mg DHT developed into plants, 74% of which produced PVS-free plants. Dot-ELISA showed that c. 90% of the meristems had less than 10 pg of virus; each of the chemicals alone (10 mg/l DHT or 5 mg/l ribavirin) had less effect. Comparable results were found for PVM, when c. 69% of virus-free plantlets was obtained in comparison to 39% of the untreated control. The simultaneous use of the two chemicals was also beneficial, because ribavirin above 5 mg/l delayed meristem development. Another base analogue, cyanoguanidin, gave similar results (61.4% virus-free plantlets vs 34.7%) with PVM (Schuster, 1982).

The three eradicating methods have been used together to obtain virus-free material from 34 *Solanum* genotypes (Sanchez *et al.* 1991). Plantlets established *in vitro* from nodal cuttings were placed on MS medium containing 20 mg/l ribavirin and, 15 days later, were thermo-treated (uninterrupted, alternating temperatures of 35°C-4 h light/ 31°C-4 h dark periods). Meristems (0.2-0.4 cm long) taken from such plantlets did not perform better than those from either thermo- or chemotherapy coupled with meristem-tip culture for PVS and PVX (43.9 and 65.8% of virus-free progenies, respectively) and much worse than those by MTC alone for PVY and PLRV (37 and 57% and 37,respectively) (Faccioli and Rubies-Autonell, 1982; Faccioli *et al.* 1988).

Chemo- or thermotherapy, either alone or in combination, must be used to obtain virus-free plantlets from virus-infected stem cuttings. Using a combined chemotherapy (50 mg/l ribavirin + 100mg/l DHT), 36-91% PVX-free plantlets of five potato cultivars (success was cultivar-dependent) were obtained from large single-node cuttings (O.3 x 2 cm), cultured *in vitro* for 4-weeks, and 100% virus-free plantlets were obtained after another 4-week period (Faccioli and Zoffoli, 1998). For PVS eradication, smaller cuttings (0.15 x 0.5 cm), a higher level of ribavirin (50 mg/l) and two 4-week treatments were needed to produce, depending on cultivar and without any phytotoxicity, 40-87% virus-free plantlets. To eradicate PVM, PLRV, PVS, PVX and PVY from 18 potato cultivars and 6 *Solanum* species, nodal cuttings of unspecified length were grown *in vitro* on MS medium containing 20 mg/l ribavirin for 14 days (Sanchez *et al.*1991). Plantlets established for a further 28 days on the same medium in a heat chamber (alternating temperature cycle 4 h light at 35°C, 4 h dark at 31°C) were mostly freed of PVS, PVM and PVX; however, no PLRV-free and only a few PVY-free plants were obtained.

14.7.3 Effects of Virus Eradication

Virus elimination by meristem-tip or stem cutting cultures can have various effects on the derived plants and their production. The effects on the increase in quantity and quality of the tubers obtained are considered in detail in Chapter 20 but besides crop yield, other characteristics of virus-infected plants change. For example they normally react differently to a second infection thus, freeing plants from one infection it may result in a different susceptibility (or resistance) to other viruses or pathogens. There is the well-known "cross protection" that strains of a virus usually confer to plants against the other strains of the same virus. Acquired resistance against virus infection is also a common phenomenon, which can protect plants from successive infections (Ross, 1961). In the above- mentioned cases, freeing the plant from viruses may, therefore, result in their increased susceptibility to virus infections. On the contrary, there are several examples which show an increased susceptibility of infected plants towards a second infection, or a greater damage caused by a mixed infection as compared to the sum of the effects of the single infections: a classical example is that of PVX-PVY infection in potato. Interactions between viruses and fungi also influence host susceptibility. Examples of increased resistance to fungus infections of virus-infected plants are numerous, and in these cases an increased susceptibility of the cured plant is found. PVX and PVY infected potatoes, for example, are less susceptible to infection by *Phytophthora infestans* (Muller and Munro, 1951). PVX-infected tubers are less susceptible to *Fusarium roseum* (dry rot) than healthy ones (Jones and Mullin, 1974). Many reports, however, indicate that virus infection increases susceptibility to fungus infection (Brunt, 1986), and in these cases eradication also increases resistance towards fungal infections. Potato plants obtained by MTC generally show few or no genetic differences from the original virus-infected sources. Minor physiological changes can, however, occur in regenerated plants following virus removal. These variations are probably frequent, but rarely considered, although somaclonal variations can also be exploited for breeding purposes.

14.8 Maintenance of Virus-free Stocks and Reinfection

Virus-free plants obtained from meristem and stem-cutting cultures can easily become reinfected. To prevent any contamination, careful procedures based on the knowledge of virus epidemiology and type of crop should be used.

Virus-free stocks (mother plants) should be grown in sterilized soil, to avoid infection by nematode and fungus-transmitted viruses, and in virus- and vector-free greenhouses. Subsequently, the material should be multiplied (two or three times) in insect-proof screenhouses, and bulk multiplication in isolated areas, where chances of reinfection are minimal due to the absence of virus sources and vectors. Mother plants should be controlled singly and frequently, while plants multiplied in nurseries or fields for commercial use should be periodically tested, in-groups or randomly. Strict hygiene should be adopted mainly to prevent infection by contact-transmitted viruses; healthy stocks can also be stored safely *in vitro.*

To reduce reinfection in the field, pesticides should be used depending on the virus vector. To reduce the spread of non-persistent aphid-transmitted viruses, mineral oils can be used although the oil coating, by causing reduced oxygen uptake in potato, reduces production (Faccioli, unpublished).

MTC-regenerated plants can be stored *in vitro* to avoid reinfection and save space. Nuclear stocks of potato shoot-tip cultures are so stored to maintain a germplasm collection (Westcott *et al.* 1977). The use of minimal nutrients in the medium, the addition of inhibitors (i.e., abscissic acid) and a low temperature reduce growth rates during conservation and the need for frequent transplanting; potatoes can be stored under these conditions for at least 1 year.

Cryopreservation using liquid nitrogen has also been used to store plant cells of different tissues and species, its success depending on the differential response of tissues to freezing. Meristems are preferred to callus cultures for long-term preservation of germplasm, because they regenerate plants more easily, are genetically more stable and withstand sudden freezing. Potato plants of three cultivars were regenerated, for example, from meristems freeze-preserved (-196°C) for 24 months (Bajaj, 1981).

Since MTC-regenerated potato virus-free stocks can be reinfected by mechanically or vector-transmitted viruses when grown in the field, sanitation programs are operative in different countries.

References

Accatino, P. 1966. Papa corahila libre de virus mediante cultivo de meristemas. *Agric.Tec. 26:* 34-39.

Appiano, A. and Pennazio, S. 1972. Electron microscopy of potato meristem tips infected with potato virus X. *J.gen.Virol.* 14: 273-276.

Bajaj, Y.P.S. 1981. Regeneration of plants from potato meristems freeze-preserved for 24 months. *Euphytica* 30: 141-145.

Banttari, E.E. and Goodwin, P.H. 1985. Detection of potato viruses S, X and Y by enzyme-linked immunosorbent assay on nitrocellulose membranes. *Plant Dis.* 69: 202-205.

Beauchesne, G. 1982. Appearance of plants not true to type during in vitro plant propagation. In: *Variability in Plants Regenerated from Tissue Culture* (eds.) E.D. Earle and Y. Demarly, pp. 268-302 Praeger Publishers, New York U.S.A.

Beemster, A.B.R. and Rozendaal, A. 1972. Potato viruses: properties and symptoms. In: *Viruses of Potatoes and Seed Potato production.* (ed.) A.de Bokx, pp. 115-143. Cent. Agric. Publ.Doc. (PUDOC), Wageningen, The Netherlands.

Bhojwani, S.S. and Razdan, M.K. 1996. *Plant Tissue Culture: Theory and Practice (a revised edition, 767 pp.* .Elsevier, Amsterdam, The Netherlands.

Brunt, A.A. 1986. Methods of producing and some benefits of growing virus-free plants. In: *Healthy Planting Material: Strategies and Technologies.* Monogr. 33, pp. 55-65. Brit. Crop Protect. Counc., Thornton Heath, U.K.

Cassel, A.C. 1987. In vitro induction of virus-free potatoes by chemotherapy. In: *Biotechnology in Agriculture and Forestry, Vol.3. Potato* .(ed.) Y.P.S. Bajaj, pp. 40-50. Springer-Verlag, Berlin, Germany.

Cassel, A.C. and Long, R.D. 1982. The elimination of potato viruses X, Y, S and M in meristem and explant cultures of potato in the presence of virazole. *Potato Res.* 25: 165-173.

Colombarini, A. 1990. Individuazione dei virus PVS e PVM in patata e risanamento mediante coltura di meristemi, chemio- e termoterapia. Dr.degree Thesis, 65 pp. University of Bologna, Italy.

D'Amato, F. 1977. Cytogenetics of differentiation in tissue and cell cultures. In: *Applied and Fundamental Aspects of Plant Cell, Tissue and Organ Culture* (eds.) J. Reinert and Y.P.S. Bajaj, pp. 343-357. Springer-Verlag, Berlin, Germany.

D'Amato, F. 1978. Chromosome number variation in cultured cells and regenerated plants. In: *Frontiers of Plant Tissue Culture* (ed.) T.A. Thorpe, pp. 287-295. Univ. Calgary Press, Canada.

El-Fiki A.I.I., El-Din T.M.N., Fawzy R.N. and Ali A.S., 1992.Tissue culture combined with thermo or chemotherapy for elimination of potato virus X (PVX). *Ann. Agric.Sci., Mosthtohor* 30: 179-194

Faccioli, G. and Colombarini, A.,1990. Risanamento della patata da seme dai principali virus a mezzo coltura di meristemi ed eventuali termoterapia e chemioterapia. *Agricoltura e Ricerca* 115: 19-22.

Faccioli, G. and Colombarini, A. 1996. Correlation of potato virus S and virus M contents of potato meristem tips with the percentage of virus-free plantlets produced in vitro. *Potato Res.* 39: 129-140.

Faccioli, G. and Rubies-Autonell, C. 1982. PVX and PVY distribution in potato meristem tips and their eradication by the use of thermotherapy and meristem-tip culture. *Phytopathol .Z.* 103: 66-76.

Faccioli, G., Rubies-Autonell, C. and Resca, R. 1984. Autoradiographic detection of PLRV, PVY and PVX in potato meristem tips and production of virus-free material by meristem-tip culture. (Abstr.) *Potato Res.* 27: 102.

Faccioli, G., Rubies-Autonell, C. and Resca, R. 1988. Potato leafroll virus distribution in potato meristem tips and production of virus-free plants. *Potato Res.* 31: 511-520.

Faccioli, G. and Zoffoli, R. 1998. Fast eradication of potato virus X (PVX) and potato virus S (PVS) by chemotherapy. *Phytopath. medit.* 37: 9-12.

Fridborg, G. and Eriksson, T. 1975. Effects of activated charcoal on growth and morphogenesis in cell cultures. *Physiol.Plant.* 34: 306-308.

Gamborg, O.L. 1984. Plant cell cultures: nutrition media. In: *Cell Culture and Somatic Cell Genetics of Plants,* Vol.1 (ed.) I.K. Vasil, pp.18-26. Academic Press, New York, USA.

Gamborg, O.L., Murashige, T., Thorpe, T.A. and Vasil, I.K. 1976. Plant tissue culture media. *In Vitro* 12: 473-478.

Gregorini, G. and Lorenzi R. 1974. Meristem tip culture of potato plants as a medium method of improving productıvity. *Potato Res.* 1 7:24-33.

Griffiths, H.M.,, Slack S.A and Dodds, J.H. 1990. Effect of chemical and heat therapy on virus concentration in vitro potato plantlets. *Can. J. Botany* 68: 1515-1521.

Heller, R. 1953. Recherches sur la nutrition minerale des tissus vegétaux cultivées in vitro. *Ann.Sci.Bot.Veg.Paris 14*: 1-223.

Hollings, M. 1965. Disease control through virus-free stock. *Annu. Rev.Phytopathol.* 3: 367-396.

Hollings, M. and Stone, O.M. 1964. Investigation of carnation viruses.1.Carnation mottle. *Ann.appl.Biol 153*: 103-118.

Jones, A.D. and Mullin, J.M. 1974. The effect of potato virus X on susceptibility of potato tubers to *Fusarium roseum "Avenacearum"*. *Amer.Potato J.* 51: 200-215.

Kassanis B. 1957.The use of tissue culture to produce virus-free clones from infected potato varieties. *Ann. appl. Biol.* 45: 422-427.

Kassanis, B. 1965. Therapy of virus infected plants. J.*R. Agric.Soc.Engl.* 126: 105-114.

Kassanis, B. and Varma, A. 1967. The production of virus-free clones of some British potato varieties. *Ann.app.Biol.* 59: 447-450.

Kevers, C., Coumans, M., Coumans-Gilles, M.F. and Gaspar, Th. 1984. Physiological and biochemical events leading to vitrification of plants in vitro. *Physiol.Plant.* 61: 69-74.

Klein, R.E. and Livingston, C.H. 1982. Eradication of potato virus X from potato by ribavirin treatment of cultured potato shoot tips. *Amer.Potato J.* 59: 359-365.

Klein, R.E. and Livingston, C.H. 1983. Eradication of potato viruses X and S from potato shoot- tip cultures with ribavirin. *Phytopathology 73*: 1049-1050.

Larkin, P.J. and Scowcroft, W.R. 1981. Somaclonal variation, a novel source of variability from cell cultures for plant improvement. *Theor.Appl.Genet.* 60: 197-214.

Letouze, R. and Beauchesne, G. 1969. Action d ' eclairements monochromatiques sur la rhizogenese de tissue topinambour. *C.R.Acad.Sci. Paris,* 269: 1528-1531.

Lizarraga, R.E., Salazar, L.F., Roca, W.M.T. and Schilde-Reutschler, L. 1980. Elimination of potato spindle tuber viroid by low temperature and meristem culture. *Phytopathology* 70: 754-755.

MacDonald, D.M. 1973. Heat treatment and meristem culture as a means of freeing potato varieties from viruses X and S. *Potato Res.* 16: 263-269.

Maene, L.J. and Debergh, P.C. 1985. Liquid medium additions to established tissue cultures to improve elongation and rooting in vivo. *Plant Cell Tissue Org. Cult.* 5: 23-33.

Marani, F. and Pisi. 1977. A meristem-tip culture and vegetative propagation in potato. *Acta Hort.* 78: 415-424.

Margara, J. 1982. Bases de la multiplication vegétative. *Les Meristèmes et l'Organogenèse, 262 pp.* INRA, Paris.

Martin-Tanguy, J. 1985. The occurrence and possible function of hydroxycinnamoyl acid amides in plants. *Plant Growth Regul.* 3: 381-399.

Mellor, F.C. and Stace-Smith, R. 1969. Development of excised buds in nutrient medium. *Can.J. Bot.* 47: 1617-1621.

Mellor, F.C. and Stace-Smith R. 1970. Virus strain differences in eradication of potato virus X and S. *Phytopathology 60*: 1587-1590.

Mellor, F.C. and Stace-Smith, R. 1977. Virus-free potatoes by tissue culture. In: *Applied and Fundamental Aspects of Plant Cell, Tissue and Organ Culture* (eds.) J. Reinert, and Y.P.S. Bajaj, pp.616-635. Springer-Verlag, Berlin, Germany.

Mitchell, D.J., Powell, P.J. and Rose, D.G. 1990. Squash blot, a novel device for sampling plant viruses. (Abstr.) *Potato Res*. 33: 144-145.

Morel, G. 1948. Recherches sur la culture associée de parasities obligatoires et de tissus végétaux. *Ann.Epiphyt.* 1: 123-234.

Morel,G.,1964. Régéneration des variétés virosée par la culture de méristémes apicaux. *Rev. Hort.* Paris, 261: 733-740.

Morel, G. and Martin, 1955 Guerison de pommes de terre atteints de maladies a virus. C.R. *Acad.Agr. Fr.* 41:472-475.

Morel, G., Martin, C. and Muller, J.F. 1968. La guerison des pommes de terre atteints de maladies a virus. *Ann.physiol.Veg.* 10: 113-119.

Mori, K. and Hosokawa, D. 1977. Localization of viruses in apical meristems and production of virus-free plants by means of meristem and tissue culture. *Acta Hort.* 78: 389-396.

Moskovets, S.N., Gorbarenko, N.I. and Zhuk, I.P. 1973. The use of the method of the culture of apical meristems in the combination with low temperature for the sanitation of potato against mosaic virus. *Sel-skokhozyaiastvennaya Biol.* 8: 271-275.

Muller, K.O.L. and Munro, J. 1951. The reaction of virus infected potato plants to *Phytophthora infestans*. *Ann.appl.Biol.* 38: 765-773.

Murashige,T. 1974. Plant propagation through tissue cultures. *Annu.Rev.Plant Physiol.* 25: 135-166.

Murashige, T. 1977. Manipulation of organ initiation in plants tissue cultures. *Bot. Bull. Acad. Sin.* 18: 1-24.

Murashige, T. and Skoog, F. 1962. A revised medium for rapid growth and bioassays with tobacco tissue cultures. *Physiol.Plant.* 15: 473-497.

Pennazio, S. and Redolfi, P. 1973. Factors affecting the culture in vitro of potato meristem tips. *Potato Res.* 16: 20-29.

Pennazio, S. and Vecchiati M. 1978. Sulla posssibilità di ottenere elevate rese di piantine per coltura in vitro di apici meristematici. *Riv. Ortoflorofrutt. Ital.*, 58: 456-460

Pierik, R.L.M. 1987. *In Vitro Culture of Higher Plants, 321 pp.* Martinus Nijhoff Publishers, Dordrecht, Netherlands.

Quak, F. 1965. Infection of tobacco-callus tissue with tobacco mosaic virus and multiplication of virus in such tissue. *Proc.Int.Conf.Plant Tissue Culture*, pp. 513-519. Berkeley, USA.

Quak, F. 1977. Meristem culture and virus-free plants. In: *Applied and Fundamental Aspects of Plant Cell, Tissue and Organ Culture* (eds.) J. Reinert and Y.P.S. Bajaj, pp. 598-615. Springer-Verlag, Berlin, Germany.

Roca, W.M., Espinoza, N.O., Roca, M.R. and Bryan, J.E. 1978. A tissue culture method for the rapid propagation of potatoes. *Amer Potato J*. 55: 691-701.

Ross, A.F. 1961. Systemic acquired resistance induced by localized virus-infections in plants. *Virology* 14: 340-358.

Rubies-Autonell, C., Faccioli, G. and Colombarini, A. 1990. Detection of PVM in potato meristem tips and preliminary results on its eradication. (Abstr.) *Potato Res.* 33: 142.

Rubies-Autonell, C., Resca, R., Faccioli, G. and Colombarini, A. 1987. PVS and PVX rapid detection in potato meristem tips and of thermo- and chemotherapy. (Abstr) *Proc.7th Congress Medit. Phytopathol. Union,* Granada, Spain: 78-80.

Sanchez,G.E, Slack, S.A. and Dodds, J.H. 1991. Response of selected Solanum species to virus eradications. *Amer. Potato J.* 68: 299-315.

Schuster, G. 1982. Antiphytovirale Verbindungen mit Guanidinstruktur. *Phytopathol.Z.* 103: 77-86.

Schuster, G. 1988. Synthetic antiphytoviral substances. *Appl.Virology Res.* 1: 265-283

Seibert, M., Wetherbee, P.J. and Job, D.D. 1975. The effects of light intensity and spectral quality on growth and shoot initiation in tobacco callus. *Plant Physiol.* 56.

Sheridan, W.F. 1975. Plant regeneration and chromosome stability in tissue culture. In: *Genetic Manipulation with Plant Material.* (ed). L. Ledoux, pp. 263-295. .Plenum Press, London, UK.

Sip, V. 1972. Eradication of potato viruses A and S. *Potato Res.* 15: 270-273.

Skoog, F. and Miller, C.0. 1957. Chemical regulation of growth and organ formation in plant tissues cultured in vitro. *Symp.Soc.Exp.Biol.* 11: 118

Smith, R.H. and Murashige, T. 1970. In vitro development of isolated shoot apical meristems of angiosperms. *Amer.J.Bot.* 57: 562-568.

Stace-Smith, R. and Mellor, F.C. 1968. Eradication of potato viruses X and S by thermotherapy and axillary bud culture. *Phytopathology* 58: 199-203.

Stace-Smith, R. and Mellor,F. 1970. Eradication of potato spindle tuber virus by thermotherapy and axillary bud culture. *Phytopathology 60*: 1857-1858.

Stallknecht,G.G. and Farnsworth, 1982. General characteristics of Cumarin-induced tuberization of axillary shoots of *Solanum tuberosum* L. cultured *in vitro. Amer. Potato J.* 59: 17-32.

Styer, D.J. and Chin, C.K. 1983. Meristem and shoot-tip culture for propagation, pathogen elimination and germplasm preservation. *Hortic. Rev.* 5: 221-277.

Sunderland, N. 1977. Nuclear cytology. In: *Plant Tissue and Cell Culture.* (ed.) H.E. Street, pp. 177-205. Blackwell Scientific Publishers, Oxford, UK.

Thomas, D.D.S. and Murashige, T. 1979. Volatile emissions of plant tissue cultures. I. Identification of the major components. *In Vitro* 15: 654-658.

Wang,P. and Hu C.,1980. Regeneration of virus-free plants through in *vitro culture.* In: *Advances in Biochemical Engineering.* (ed.) A. Fletcher) Vol. 18, Plant *Cell Culture* II, pp.61-69 .Springer-Verlag, Berlin, Germany.

Wang, P. and Hu, 1982. In vitro mass tuberization and virus-free seed potato production in Taiwan. *Amer. Potato J.* 59: 33-37.

Westcott, R.J., Henshaw, G.G. and Roca, W.M. 1977. Tissue culture storage of potato germplasm: culture initiation and plant regeneration. *Plant Sci. Letters 9*: 309-315.

White, P.R., 1934. Multiplication of the viruses of tobacco and aucuba mosaic in growing exiced tomato roots. *Phytopathology* 24: 1003-1011.

White, P.R. 1943. A Handbook of Plant Tissue Culture, 227 pp. J.Cattel Press, Lancaster, PA, USA.

Wu, J.H., Hildebrandt, A.C. and Riker, A.J. 1960. Virus-host relationships in plant tissue culture. *Phytopathology* 50: 587-594.

15. Rapid Propagation of Virus-tested Potatoes

A.A.Watad[1]*, C. Sluis[2], A. Nachmias[3] and R. Levin[4]

[1] *Department of Ornamental Horticulture, ARO, The Volcani Center. P.O.B. Bet-Dagan 50250, Israel,* [2] *Tissue Grown Corp. Woodland, CA, USA,* [3] *Potato R&D, P.O.B. 515, Ofaquim, Israel,* [4]*Osmotek Ltd. P.O.B. 550 Rehovot Israel; * Dr. A. Watad passed away on March 15, 2000, after completing this chapter.*

15.1 Introduction

High quality potato seed production is typically begun with the selection of true to type tubers that have been tested and found to be free of specific viral, fungal and bacterial pathogens. In most current seed production programs, plant tissue culture methods are employed for the initial increase of these tubers (Bryan, 1988; Jones, 1988). The second increase is generally in the greenhouse after which the tubers are multiplied in the field over successive generations under conditions that control infectious diseases that can be transmitted through the progeny tubers.

Traditionally, high quality potato seed has been produced in north temperate climates where the weather conditions preclude the emergence of virus vectors during the period of seed potato growth. Since the rate of

G. Loebenstein et al. (eds.),
***Virus and Virus-like Diseases of Potatoes and Production of Seed-Potatoes*, 391–406.**

potato multiplication in the field is only about ten fold per year (Beukema and Van der Zaag, 1990), potato seed production is a lengthy process requiring large production areas. For example, in 1990 North American certified seed programs required 6-7 years of field multiplication to produce seed for ware potato production and utilized almost 270.000 acres of land per year (Slack, 1993).

In North temperate climates, this traditional system is cost effective, but relatively inflexible and can suffer from problems with bacterial and fungal pathogens. For example, in northwest Europe *Erwinia* spp. infections have caused difficulties in seed production. In areas where there are no vector free production areas high, quality potato seed can not be produced by traditional means. In these areas quality potato seed can only be produced by a reduced generation scheme (Ranalli *et al.* 1994b).

In a reduced generation scheme, seed tubers are produced after as little as one and three field multiplications, (Kim *et al.* 1999; Maene, 1994). The reduction in field generations is made possible by increasing true to type, specific pathogen free potato tubers under controlled conditions until the number of propagules is sufficient to produce the seed required in the desired number of field generations. Typically, the controlled environment increase begins with the *in vitro* production of plantlets or microtubers. Plantlets (Levy, 1988) and microtubers (Haverkort *et al.* 1991) can be transferred directly to the field, however, it is more common for these propagules to be transferred to a screenhouse, greenhouse or other controlled growth environment for production of minitubers, which are then transferred to the field.

This chapter will review aspects of the technology of potato propagule production and the incorporation of these propagules into seed production schemes.

15.2 Production of *in vitro* Derived Plantlets

Over 90% of the traditional seed production schemes use *in* vitro derived transplants for the initial increase of pathogen tested clonal selections (Jones, 1988). The most common method of production of these transplants is single node cuttings (Dodds *et al.* 1992) although other methods have been reported.

15.2.1 Single Node Multiplication

Ranalli (1997) reported a typical method of multiplication of potato plantlets by single node cuttings. In this method explants derived from virus tested tubers are sterilized, cut into 5 – 10 mm nodal segments and placed in vessels on agar based medium. Plantlets grow upright from an axillary bud. These plantlets are multiplied by cutting 4-mm segments including a node, placing the segments on new medium, and incubating the cultures for 4 weeks. The rate of multiplication is x3 to x 5 every four weeks (Plate 1).

The basic medium for nodal culture is typically based on MS salts (Murashige and Skoog, 1962) with sucrose, myo-inositol, thiamine HCl, and agar. Growth regulators are generally omitted from the medium out of concern that their use may produce off types in the propagules. The pH of the medium is usually adjusted to 5.7, but can range from 5.2 to 6.5 (Jones, 1988).

A large range of types and sizes of containers and closures are used for *in vitro* culture. Growth of plantlets in large vessels have may be weak, but can be improved by the addition of silver salts to the medium (Dodds *et al.*, 1992) or venting of the vessels (Levin and Watad, unpublished).

15.2.2 Mechanized Production of Single Node Cuttings

In developed countries, labor costs are approximately 70% of the cost of *in vitro* production (George, 1993). Consequently, a number of efforts have been made to reduce this cost by mechanization or the use of a vision based robot system. For example, Dodds *et al.* (1992) reported the use of a modified blender as well as a grid of blades to produce nodal cuttings from plantlets. The use of a color machine vision system to identify transplantable precut nodal segments from morphologically regularized potato plantlets has also been reported (Alchanatis *et al.* 1994). Further, Gorst and Teasdale (1999) describe a robot capable of removing a morphologically regularized plant from a vessel, cutting it into internodes and transplanting the cut internodes to a new vessel. This robot which is capable of transplanting 4,000,000 units per year has been tested on potato plantlets, but is currently being employed primarily in the forestry industry (Teasdale, personal communication).

15.2.2 Multiple Shoot Multiplication

Multiplication of multiple shoots from stem cuttings of *in vitro* plantlets has been reported by Dodds *et al.* (1992). In this procedure large leaves are removed from a stem with three or four nodes and the stem placed in a flask with 15ml of liquid medium. The flasks produce 60 –70 nodes after incubation on a shaker at 80 RPM for two to three weeks. Similar results were reported by Leclerc *et al.* (1994) where vegetative growth was produced prior to microtuber induction. In this study, shoots of six nodes each were placed in plastic containers with 50 ml of liquid medium and incubated on shelves in a growth room for four weeks.

15.2.4 Bud Cluster Multiplication

Bud cluster culture has been used for multiplication of potato meristems in liquid medium (Ziv and Shemesh, 1996). Bud cluster culture is the proliferation of masses of meristems having reduced leaf and stem development (Plate 15.1). The clusters can be proliferated in airlift bioreactors such as the LifeReactor™ or on floating membrane rafts such as the LifeRaft™ both manufactured by Osmotek Ltd., Rehovot, Israel (Watad, *et al.* 1995, 1999).

Bud clusters are produced from single node explants cultured on liquid medium containing growth retardants and cytokinins (Watad *et al.* 1999; Ziv and Shemesh, 1996). Compact bud clusters are produced in flasks incubated on a shaker (Ziv and Shemesh, 1996) and more leafy bud clusters are produced on floating membrane rafts (Watad *et al.* 1999).

Bud clusters produced in flasks can be chopped, washed, and inoculated into bioreactors containing liquid medium. The bud clusters can then be chopped again and used to inoculate other bioreactors for further multiplication or be spread on solid medium or rafts for plantlet regeneration (Plate 15.1).

Bud clusters produced on rafts are compact shoot clusters with very small leaves that can be mechanically chopped with a device such as the VitroCut™ manufactured by Osmotek Ltd., Rehovot, Israel.

15.2.5 Genetic Stability of Propagules

Shoot formation (Roest and Bokelmann, 1980; Karp *et al.* 1982) show methods that employ existing buds for multiplication appear to be more stable than multiplication by adventitious bud production. Plants produced by adventitious evidence of somoclonal variation. However, plants produced by nodal culture appear to be stable (Ahloowalia 1994).

In bud cluster culture, most of the buds observed appear to develop from pre-existing axillary buds, although some adventitious regeneration was noted (Ziv and Shemesh, 1996). However, whether bud cluster culture will provide a level of phenotypic fidelity needed for a reduced generation seed production scheme still needs to be established.

Plate 15.1. *In vitro* Potato Plants Multiplication and Microtuber Production.

a-plantlets from single node cuttings.
b-bud cluster multiplied in bioreactor.
c-clusters mechanically separated by VitroCut™.
d-plantlets derived from mechanically separated bud clusters and grown on membrane raft.
e-microtubers developed on plantlets from single node cuttings.
f-microtubers developed on plantlets regenerated from bud clusters.

15.2.6 Utilization of Propagules

Plantlets can be transferred to the field either as transplants after hardening in the greenhouse or directly from tissue culture (Levy, 1988). However, under typical growing conditions plantlets are not hardy enough for direct transfer to the field (Ranalli, 1997). Plantlets are often used to produce minitubers in the greenhouse, although minituber yield is better when microtubers are used as starting material (Rolot and Seutin, 1998; Wiersema *et al.* 1987). However, plantlet production cost is well below that of microtubers. Consequently, it is often more cost effective to use plantlets rather than microtubers for minituber production.

15.3 Production of Microtubers and Minitubers

Microtubers are small tubers that are produced *in vitro*. Minitubers are small tubers produced *ex vitro* in controlled environments from *in vitro* propagules. Although microtubers are generally smaller than minitubers, microtubers and minitubers are not defined by their size, but whether they are produced *in vitro* or *ex vitro*. Microtubers have been reported to be transferred to the field in sizes ranging from 0.024g (Ranalli *et al.* 1994b) to 10g (Wang and Hu, 1982). Minitubers are typically larger than microtubers, however, both microtubers and minitubers are smaller than traditional seed tubers, which typically weight about 50 – 70 grams

15.3.1 Production of Microtubers

Microtubers can be developed on buds of single node cuttings (Seabrook *et al.* 1993), shoots (Leclerc, 1994) or bud clusters (Ziv and Shemesh, 1996), and can be produced for a broad range (Plate 15.1) of potato genotypes (Estrada *et al.* 1986). Although a number of methods for producing microtubers have been reported, a simple and effective method was developed by Estrada *et al.* (1986). In this method, stem pieces of *in vitro* plantlets are layered in 250 ml flasks with 20 ml of liquid propagation medium including MS salts, 0.4 mg/l thiamine HCl, 2.0 mg/l Ca-pantothenic acid, 0.4 mg/l GA, 0.5 mg/l BA, 0.01 mg/l NAA, 100mg/l myo-inositol, and 2% sucrose. Flasks are incubated on an orbital shaker at 60 RPM with illumination 16 hours per day. After 2 – 3 weeks of plantlet development the media is replaced with a tuber induction medium including MS salts, 0.4 mg/l thiamine HCl, 5mg/l BA, 500 mg/l CCC (chlorochlorine chloride), 100mg/l myo-inositol and 8% sucrose. The flasks are then incubated on shelves in the dark at 22°C for 40 days. Factors that influence tuber initiation and development include genotype, environmental factors such as illumination and temperature and media components such as sucrose, nitrogen and growth regulators.

15.3.1.1 Genotype

Microtuber production has been reported in over 50 varieties, but the number and size of the microtubers produced vary between genotypes when a single procedure is used (Estrada *et al.* 1986). Other differences in tuberization between genotypes have also been noted. For example, photoperiod (Lentini and Earle, 1991; Seabrook *et al.* 1993) and sucrose levels and photoperiod (Garner and Blake, 1989) can influence tuberization. Microtuberization protocols may therefore need to be optimized for mass production of different genotypes.

15.3.1.2 Illumination

Most microtuber production protocols employ low light intensities and short days or dark for tuber induction and development. However, the effects of illumination on *in vitro* tuberization, either alone or together with other factors, are complex. For example Dobranszki *et al.* (1998) report an interaction between the affects of light intensity and photoperiod on the development of microtubers. Further, Oka and Sluis (1996) reported that incubating plants on tuberization medium for 1 – 4 weeks in the light before transfer to the dark promotes tuberization.

In the whole plant, tuberization is promoted by short days, however, for *in vitro* tuberization, the effects of photoperiod are not clear-cut. Some studies report that a long photoperiod during tuberization promotes tuber formation (Hussey and Stacey, 1981), other studies have found no photoperiod effects (Hussey and Stacey, 1984) while yet other reports indicate that short photoperiods promote early tuber formation (Garner and Blake 1989) or larger microtubers (Slimmon *et al.* 1989).

15.3.1.3 Temperature

Cool temperatures favour the induction of tuberization in whole plants and *in vitro*. Microtubers are produced at temperatures ranging from 15°C - 25°C (McCown and Joyce, 1991) with 18°C - 20°C being preferable (Ranalli, 1997).

15.3.1.4 Sucrose

Sugar is a critical component of the tuberization medium, and a typical medium includes 8% sucrose. Sucrose may effect tuberization in a number of ways. It may act as an energy source or as an osmoticum, and it may initiate tuberization. The level of sucrose in the medium is critical. At low levels (1% - 3%) shoots develop. At high (8%) levels, sucrose induces tuber formation and starch accumulation (Vreugdenhil *et al.* 1998). However, at very high (12%) levels, sucrose inhibits both shoot growth and tuber formation (Garner and Blake, 1989). In the tuberization medium, sucrose is rapidly hydrolyzed to glucose and fructose and replacement of the medium is needed to support continued growth of the microtubers (Yu et al. 1999).

15.3.1.5 Nitrogen

While limiting nitrogen to potato plants in the field reduces growth and increases tuberization, there is little evidence that nitrogen inhibits *in vitro* tuberization. Under typical *in vitro* tuberization conditions and at concentrations of up to the 60mM N of the Murashige and Skoog (1962) medium, tuberization is not inhibited provided that the ratio of ammonium to nitrate is near 1(Ewing and Struik, 1992).

15.3.1.6 Growth regulators

Both cytokinins and antigibberellins have been reported to facilitate *in vitro* tuber production. Gibberellins appear to play a significant negative role in tuberization under field conditions by shifting growth away from tubers and towards stolons (Ewing and Struik, 1992). Gibberellins also seem to inhibit tuberization *in vitro* (Hussey and Stacey, 1984). It therefore seems likely that the addition of antigibberellins would increase *in vitro* tuberization. This assumption has been confirmed by several authors with reports of facilitation of microtuber production with compounds such as CCC (chlorochlorine chloride), coumarin, ancymidol, paclobutrazol, and compounds related to jasmonic acid (Dodds *et al.* 1992; Hussey and Stacey, 1984; Koda *et al.* 1991; Opatrna *et al.* 1998; Stallknecht and Farnsworth, 1982; Ziv and Shemesh, 1996).

Similarly there is evidence to indicate that cytokinins can promote *in vitro* tuberization (Palmer and Smith, 1969) and cytokinins have been incorporated in the tuberization medium in a number of studies (Dodds *et al.* 1992: Ziv and Shemesh, 1996). However, it should be noted that tuberization can also be obtained with media that does not include growth regulating substances.

15.3.1.7 Mechanized Production of Microtubers

The integration of microtubers into a seed production scheme will require large-scale methods of microtuber production. Consequently, there have been several reports of scale up methods using bioreactors. For example, McCown and Joyce (1991) report a method for multiplication of microtubers in which microtubers are induced to form multiple shoots that are then elongated and subsequently tuberized. Multiple microtubers are developed from each initial microtuber in a form of temporary immersion bioreactor for large-scale propagation using this process.

Microtuber production in temporary immersion bioreactors have also been reported by Akita and Takayama (1994), Teisson and Alvard, 1998 and Hulscher *et al.* 1998. Further, Watad *et al.* (1999) reported the tuberization of mechanically chopped bud clusters which were regenerated and tuberized on floating rafts. In this study a raft produced about 140 microtubers with 30% being larger than 0.5g.

15.3.1.7 Dormancy and Storage

Microtubers have the essential characteristics of potato tubers. Therefore, like normal tubers they exhibit dormancy, i.e. a period of time between production and normal sprouting. The dormant period is affected by a number of factors including genotype, production and storage conditions, tuber size and a number of other factors.

Dodds *et al.* (1992) reported that when tubers were produced in the dark and stored at 4°C sprouting occurred after 210 days but when they were produced under short days only 60 days was required. Ranalli *et al.* (1994a) found that the length of dormancy is inversely related to tuber size and length of storage at 3°C. Leclerc *et al.* (1994) found a significant genotype effect on dormancy that correlated with the dormancy characteristic of the cultivar and that smaller microtubers have longer dormancy periods. Microtubers can easily be stored in cold (4° – 10° C) and humid conditions (McCown and Joyce, 1991). However, shortening the dormancy period requires chemical treatment (Kim *et al.* 1999; Leclerc *et al.* 1994)

15.3.1.8 Utilization of Microtubers

Microtubers are typically planted in nursery bed in a protected environment and used to produce minitubers. When planted in a nursery bed microtubers yield more larger size minitubers than plantlets (Rolot and Seutin, 1998; Wiersema *et al.* 1987), however, plantlets are less expensive to produce. Consequently, in North America and Europe, plantlets rather than microtubers have been typically used to produce minitubers (Jones, 1988).

Microtubers can be planted directly in the field. However, since microtubers generally weigh less than 0.2 grams, they typically perform poorly in the field. For example, when Ranalli *et al.* (1994b) compared the performance of field planted microtubers (average size 0.09g – 0.12g), mini tubers (average size 1.5g – 3.4g) and normal seed they found that microtubers emerged later than minitubers that they, in turn, were later than normal seed with ground cover developing more slowly in the smaller tubers. Planting microtubers at high densities produced a large number of small tubers, but the yield in terms of tons per hectare was low. Further, the use of improved husbandry techniques do not show any significant practical yield increase for field planted microtubers (Struik and Lommen, 1998).

Recent studies have reported the production of microtubers weighing between 0.5 grams and more than one gram (Oka and Sluis, 1996: Yu *et al.* 1999), and larger microtubers do appear to show better field performance. In an unpublished study, a commercial company (Small Potatoes Inc., Madison, USA) compared the field performance of large microtubers (average weight 0.73g, weight range (0.4g – 2.3g) and minitubers (average weight 5.41g, weight range 1.93g – 7.83g). Both minitubers and microtubers reached full ground cover in 12 weeks and there was no significant difference in the yield between the two groups. Further, the results of Struik and Lommen (1990) and Lommen and Struik, (1994) with minitubers suggest that small tubers larger than 0.5g can produce acceptable results in field planting. However, farmers can have difficulties even with microtubers weighing up to one gram.

15.3.2 Production of Minitubers

Minitubers are small tubers produced from either microtubers or plantlets that are planted at high-densities in beds in protected environments. The size and number of minitubers produced are influenced by a number of factors including cultivar, the type of protected environment, season, light and growing media, planting density and husbandry practices (Jones, 1988). Two of the most important factors are planting density and the husbandry practices (Lommen and Struik, 1992a, 1992b).

Increasing planting density increases the number of minitubers produced per m^2 but decrease the size of the minitubers produced. Repeated non-destructive harvest also increases the number of minitubers produced per m^2, but again at the expense of the size of the minitubers. For example Lommen and Struik (1992a) report that using three harvests 1800 minitubers per m^2 but weighing an average between 1 and 2 grams and up to 3,500 minitubers per m^2 for tubers larger than 5 grams. Tuber yield can also be increased to some extent by a continuous supply of low-dose nutrient starting with the first harvest.

15.3.3 Storage and Dormancy

Freshly harvested minitubers are usually cured i.e. left undisturbed at a high temperature and humidity for two weeks. They are then stored at a low temperature for between several months and a year and then presprouted by transferring them to a higher temperature for several weeks to a month. Storage and curing affect minituber performance. For example, the length of storage can affect the field behavior of minitubers and this effect is cultivar dependent (Lommen and Struik, 1993).

15.3.4 Utilization of Minitubers

Minitubers are planted in the field after storage as part of a seed production scheme. The major factor in field performance of minitubers is the size of the tuber. Smaller minitubers have a longer dormancy (Lommen, 1993) and emerge later. They have thinner stems which are more susceptible to stress and have fewer reserves to resume growth after damage (Lommen, 1994). They have slower foliar ground cover and higher plant death after emergence (Lommen and Struik, 1994). Planting smaller minitubers also leads to a more variable crop (Lommen and Struik, 1995).

The performance of small minitubers, however, can be improved by agrotechnical means. For example small minitubers larger than 0.5g, can produce acceptable results in field planting at proper planting densities (Lommen and Struik, 1994, 1995). Field performance of small minitubers can also be improved by techniques such as encapsulation in moist peat-lite mix (Melching *et al.* 1993).

Minitubers play a central role in reduced cycle potato production. For example, in Israel, Potato R&D Ltd. is producing seed for ware potatoes in two years (see Table 15.1). In the first year plantlets or microtubers with 2 stems are planted in a greenhouse at a density of 50 to 200 units per m^2 depending on the cultivar. Plantlets yield 1,000 – 2,000 tubers per m^2 sized between 20mm – 40mm, and microtuber yield exceeds this by about 50%. In the second year, minitubers are planted in soil under insect proof mesh. The crop is harvested, stored and replanted in the field in the in the same year. Seed for ware production is then sold the third year at $250 -$300 per ton, which is below the cost of imported seed.

Other instances of low cost seed production using small minitubers include methods used by Technituber and American Ag-Tech. Technituber produces an approximately 1 gram minituber under controlled environment conditions. American Ag-Tech employs a process that produces 1,000, 1 – 5 gram minitubers per m^2 at low cost by increasing the plantlets in the controlled environment before tuberization.

15.4 Progress towards Reduced Generation Production

The typical seed production scheme used in north temperate climates requires at least 8 years of multiplication to reach seed for ware production

(Table 15.1) In South Africa seed production takes four years. The laboratory and greenhouse increase takes place in the first year, there are two increases a year the second and third years and seed sales take place the fourth year. In Israel, The first year is devoted to laboratory and greenhouse production and there are two field increase in the second year. The first is in soil, but under protected condition and the second is in the open field. Seeds are sold in the third year.

The 2nd year (Israel) and 3rd year (South Africa) production schemes are made possible by weather conditions which permit two crops a year. It would be desirable if this kind of a reduction could also be achieved in north temperate climates. However, two conditions need to be met to achieve this goal. The first is a reduction in the price of the propagule to the field to a point where it is cost effective to produce seed potatoes in one or two generations. The second is to be able to produce a propagule at that price which will also perform well in the field.

Table 15.1

Year	Typical Long Cycle Scheme		South Africa		Israel	
	Site	Increase Gen.	Site	Increase Gen.	Site	Increase Gen.
1	tissue culture	- 0	tissue culture	-	tissue culture	–
1			greenhouse	0	greenhouse	0
2	greenhouse	0	open field	1	screenhouse	1
2			open field	2	open field	2
3	open field	1	open field	3	ware seed	5
3			open field	4		
4	open field	2	ware seed			
5.	open field	3				
6.	open field	4				
7	open field	5				
8.	open field	6				
9.	ware seed					

For production of seed potatoes in one or two generations a propagule would have to be sold for about $0.10 or less. Unfortunately, a low cost propagule whether a microtuber or a minituber would need to be small. Although the results of Lommen and Struik (1994, 1995) suggest that satisfactory results can be obtained with tubers as small as 0.5 grams, farmers have a great deal of difficulty dealing with propagules of less than 2 grams. Moreover, small propagules are still being sold for more than $0.20.

Consequently, in many cases there is no clear economic basis for only one or two field generations. Propagators may therefor opt for larger propagules that are less problematic in the field even if they are somewhat more expensive.

As the cost of a propagule that performs well in the field approaches $0.10, field multiplications will be reduced to one or two generations.

Acknowledgement

Contribution from the Agricultural Research Organization, The Volcani Center, Institute of Horticulture, Bet Dagan, Israel. No. 220, 1999 series.

References

Ahloowalia, B.B. 1982. Plant regeneration from callus culture in potato.*Euphytica.* 3: 755-759.

Akita, M. and Takyama, S. 1994. Stimulation of potato (*Solanum tuberosum* L.) tuberization by semicontinious liquid medium surface level control. *Plant Cell Rep..* 13: 184 – 187.

Alchanatis, V., Peleg, K. and Ziv, M. 1994. Morphological control and mensuration of potatoplantlets from tissue cultures for automated micropropagation. *Plant Cell Tssue and Organ Culture.* 36: 331-338.

Beukema, H.P. and van der Zaag, D.E.1990. *Introduction to Potato Production.* Pudoc, Wageningen.

Bryan, J.E. 1988. Implementation of rapid multiplication and tissue culture methods in third
world countries. *Am. Potato J.* 6: 199-207.

Dobranszki, J., Taboriand, K.M. and Ferenczy, A. 1998. Environmental effects of *in vitro*tuberization of potato plants. *Development of Integrated systems For Large Scale Propagation of Elite Plants Using In Vitro Techniques.* Conference on Potato Seed Production by Tissue Culture, Brussels, Abs.

Dodds, J.H., Silva-Rodriguez, D. and Tovar P. 1992. Micropropagation of potato (*Solanum tuberosum L.*) In: *Biotechnology in Agriculture and Forestry* (ed) Y.P.S. Bajaj. Vol. 19: 91-106, Springer-Verlag, Berlin.

Ewing, E.E. and Struik, P.C. 1992. Tuber formation in potato: Induction, initiation and growth. *Hort. Rev.* 14: 89-198.

Estrada, R. Povar, P. and Dodds, J.H. 1986. Induction of *in vitro* tubers in a broad range of genotypes. *Plant Cell, Tissue and Organ Culture* 7: 3-10.

Garner, N. and Blake, J. 1989. The induction and development of potato microtubers *in vitro* on media free of growth regulating substances. *Annals of Bot.* 63: 663-674.

George E.F. 1993. *Plant Propagation by Tissue Culture Part 1 The Technology.* Exegetics Ltd. Edington, England.

Gorst J.R. and Teasdale, R.D. 1999. Robotic Micropropagation for Commercial Forestry. In: *Plant Biotechnology and In Vitro Biology in the 21stCentury.* pp 633- 636. Kluwer Academic Publisher, The Netherlands.

Haverkort, A.J., Waart, M. vande, Marinus, J. 1991. Field performance of Potato microtubers as propagation material. *Potato Res.* 34: 353-356.

Hulscher, M. Krijgsheld, H., van Vark, A. and van de Haar, J. 1998. Large scale production of potato microtubers in biorectors. *In vitro* production of potato microtubers in liquid medium using temporary immersion. *Development of Integrated systems For Large Scale Propagation of Elite Plants Using In Vitro Techniques.* Conference on Potato Seed Production by Tissue Culture, Brussels, Abst.

Hussey, G. and Stacey, N.J. 1981. *In vitro* propagation of potato (*Solanum tuberosum* L). *Ann. of Bot.* 48: 787-796.

Hussey, G. and Stacey, N.J. 1984. Factors affecting the formation of *in vitro* tubers of potato. (*Solanum tuberosum* L). *Ann. of Bot.* 53: 565-578.

Jones, E.D. 1988. A current assessment of *in vitro culture* and other rapid multiplication methods in North America and Europe. *Am. Potato J.* 65: 209-221.

Karp, A., Nelson, R.S. and Thomas, E. 1982. Chromosome variation in protoplast-derived potato plants. *Theor. Appl. Gent.* 63: 265- 272.

Kim, H.S., Jeon, J.H., Choi, K.W., Joung, Y.H. and Joung, H. 1999. Effects of Rindite on breaking dormancy of potato microtubers. *Amer. J. of Potato Res.* 76: 5-8.

Koda, Y. Kikuta, Y., Tazaki, H., Tsujino, Y. Sakamura, S. and Yoshihara, T. 1991. Potato tuber inducing activities of Jasmonic acid and related compounds. *Phytochemistry.* 30: 1435-1438.

Leclerc, Y., Donnelly, D.J and Seabrook, J.E.A. 1994. Microtuberization of layered shoots and nodal cuttings of potato: The influence of growth regulators and incubation period. *Plant Cell Tissue and Organ Culture.* 37: 113- 120.

Lentini, Z. and Earl, E.D. 1991. *In vitro* tuberization of potato clones from different maturity groups. *Plant Cell Rep.* 9: 691-695.

Levy, D. 1988. Propagation of potato by the transfer of transplants of *in vitro* proliferated shoot cuttings into the field. *Scientia Hort.* 36: 165-171

Lommen, W.J.M. and Struik, P.C. 1992a. Production of potato minitubers by repeated harvesting: plant productivity and initiation, growth and resorption of tubers. *Neth. J. Ag, Sci.* 40: 341–358

Lommen, W.J.M. and Struik, P.C. 1992b. . Production of potato minitubers by repeated harvesting: Effects of crop husbandry on yield parameters. *Potato Res.* 35: 419-432.

Lommen, W.J.M. 1993. Post harvest characteristics of potato minitubers with different fresh weights and from different harvests. I. Dry-matter concentration and dormancy. *Potato Res.* 36: 265- 272.

Lommen, W.J.M. and Struik, P.C. 1993. Performance of potato minitubers in a controlled environment after different storage periods. *Potato Res.* 36: 283- 292.

Lommen, W.J.M. 1994. Efect of weight of potato minitubers on sprout growth, emergence, and plant characteristics at emergence. *Potato Res.* 37: 315–322.

Lommen, W.J.M. and Struik, P.C. 1994. Field performance of minitubers with different fresh weights and conventional seed tubers: Crop estabishment and yield formation. *Potato Res.* 37: 301–313.

Lommen, W.J.M. and Struik, P.C. 1995. Field performance of potato minitubers with different fresh weights and conventional seed tubers: Multiplication factors and progeny yield variation. *Potato Res.* 38: 159-169.

Maene, L. 1994. Minitubers for seed production. *Acta Hort.* 355: 233-242.

McCown, B.H. and Joyce, P.J. 1991. Automated propagation of microtubers of potato. In: *Scale Up and Automation in Plant Propagation. Cell Culture and Somatic Cell Genetics of Plants*, Vol. 8. pp. 95- 109. (ed) Vasil, I.K., Academic Press, New York.

Melching, J.B., Slack, S.A. and Jones, E.D. 1993. Field Performance of peat-lite mix encapsulated small minitubers. *Amer. Potato. J.* 70: 285- 299.

Murashige, T. and Skoog, F. 1962. A revised medium for rapid growth and bioassay with tobacco tissue cultures. *Physiol. Plantarum* 15: 473-497.

Oka ,I. and Sluis, C. 1996. Method for producing potato micotubers. US Patent # 5, 498,541.

Opatrna,J., Novak, P., Grospietsch, M. and. Opatrny, Z. 1998. The effect of growth retardants on potato micropropagation and *in vitro* regeneration. *Development of Integrated systems For Large Scale Propagation of Elite Plants Using In Vitro Techniques.* Conference on Potato seed production by tissue culture, Brussels, Abstr..

Palmer, C.E. and Smith, O.E. 1969. Cytokinin and tuber initiation in the potato *Solanum tuberosum* L. *Nature* 221: 270–280.

Ranalli, P., Bizarri, M., Borghi, L. and Mari, M. 1994a Genotypic influence on *in vitro* induction, dormancy length, advancing age and agronomical performance of potato microtubers (*Solanum tuberosum* L.). *Ann. appl. Biol.* 12: 161-172.

Ranalli, P., Bassi, F., Ruaro G., Del Re, P., Di Candilo, M. and Mandolino, G. 1994b. Microtuber and minituber production and field performance compared with normal tubers. *Potato Res.* 37: 383-391.

Ranalli, P. 1997. Inovative propagation methods in seed tuber multiplication programes. *Potato Res.* 40: 439-453.

Roest, S. and Bokelmann, G.S. 1980. *In vitro* adventitious bud technique for vegetative propagation and mutation breeding of potato (*Solanum tuberosum L.)* I. Vegetative propagation in vitro through adventitious shoot formation. *Potato Res.* 23: 167-181.

Rolot J.L. and Seutin H. 1998. Soil-less production of potato minitubers using an hydroponic technique. *Development of Integrated systems For Large Scale Propagation of Elite Plants Using In Vitro Techniques.* Conference on Potato Seed Production by Tissue Culture, Brussels, (Abstr.)

Seabrook, J.E.A., Coleman, S. and Levy, D. 1993. Effect of photoperiod on *in vitro* tuberization of potato (*Solamum tuberosum* L.). *Plant Cell, Tissue and Organ Culture.I* 34: 43- 51.

Slack, S.A. 1993. Seed certification and seed improvement programs. In: *Potato Health Management* (ed) R.C. Rowe. pp 61- 65. APS Press, St. Paul, Minnesota, USA.

Slimmon, T., Machado, V.S. and Coffin, R. 1989. The effects of light on *in vitro* microtuberization of potato cultivars. *Am. Potato J.* 66: 843-848.

Stallknecht, G.F. and Farnsworth, S. 1982. General characteristics of coumarin-induced tuberization of axillary shoots of *Solanum tuberosum* L. cultured in vitro. *Am. Potato J.* 59: 69-75.

Struik, P.C. and Lommen, W.J.M. 1998. Improving the field performance of micro-and minitubers. *Development of Integrated systems For Large Scale Propagation of Elite Plants Using In Vitro Techniques.* Conference on Potato Seed Production by Tissue Culture, Brussels, (abstr.)

Struik P.C. and Lommen,W.J.M. 1990. Production Storage and use of micro and minitubers. *Proc. 11th Triennial Conf., Eur. Assoc. Potato Res.*, Endinburgh UK. pp. 122 –133.

Teisson , C. and Alvard, D. 1998. In vitro production of potato microtubers in liquid medium using temporary immersion. *Development of Integrated systems For Large Scale Propagation of Elite Plants Using In Vitro Techniques.* Conference on Potato Seed Production by Tissue Culture, Brussels, (abstr.)

Vreugdenhil, D. Boogaad, Y. Visser, R.G.F. and de Bruijn, S.M. 1998. Comparison of tuber and shoot formation from *in vitro* cultured potato explants. *Plant Cell, Tissue and Organ Culture.* 53: 197-204.

Wang, P. and Hu, C. 1982. *In vitro* mass tuberization and virus-free seed-potato production in Taiwan. *Amer. Potato. J.* 59: 33-37.

Watad, A.A., Kochba, M., Nissim, A. and Gaba, V. 1995. Improvement of *Aconitum napellus* micropropagation by liquid culture on floating membrane rafts. *Plant Cell Reports* 14: 345-348.

Watad, A.A., Alper, Y., Stav, R. and Levin, R. 1999. Mechanization of micropropagation. In: *Plant Biotechnology and In Vitro Biology in the 21stCentury*. pp. 663–666. Kluwer Academic Publishers, The Netherlands.

Wiersema, S.G., Cabello, R., Tovar, P. and Dodds, J.H. 1987. Rapid seed muliplication by planting into beds microtubers and in vitro plants. *Potato Res.* 30: 117-120.

Yu, W.C., Joyce, P.J., Cameron, C. and McCown, B.H. 1999. Sucrose utilisation during microtuber growth in bioreactors. *Plant Cell Reports*. (In Press).

Ziv, M. and Shemish, D. 1996. Propagation and tuberization of bud clusters from bioreactor culture. *In Vitro Cell. Dev. Biol. – Plant*. 32: 31-36.

16. Seed Certification as a Virus Management Tool

Gary D. Franc
University of Wyoming, Department of Plant Sciences, P.O. Box 3354, Laramie, WY 82071-3354 (francg@uwyo.edu)

16.1 Introduction

Vegetative or clonal propagation of the potato plant enables tuber borne pathogens to persist from one growing season to the next. Systemic pathogens such as viruses are especially likely to be present in tubers held back for seed. The so-called "running out" or declining productivity of seed tubers that was described over one hundred years ago was eventually found to result from the presence of one or several viruses. Potato seed improvement programs reversed the trend of declining seed productivity and are examples of our earliest attempts at virus management in the seed tuber crop, even before the nature of infectious viruses was known or understood.

During the late 1800's, methods to increase the productivity of seed tubers were in their infancy. The early seed improvement programs by Dutch and German scientists demonstrated that seed productivity could be improved by selective propagation of certain tubers (Shepard and Claflin, 1975; Jones, 1991). The benefit of these seed improvement efforts was quickly recognized by growers who realized increased yields and improved tuber quality. As seed selection methods improved and disease tolerances and regulations were adopted, seed improvement programs became more complicated and evolved into seed certification programs (Rieman, 1956). Dr. Otto Appel is credited with providing the impetus for development of potato seed certification in Europe (Shepard and Claflin, 1975), and these

G. Loebenstein et al. (eds.),
Virus and Virus-like Diseases of Potatoes and Production of Seed-Potatoes, 407–420.

European programs served as models for development of programs in North America (Rieman, 1956).

Seed certification programs are characterized as regulatory programs developed to deal with pathogens, varietal mix, and other seed-performance related issues (Shepard and Claflin, 1975; Jones, 1991). Although varying degrees of complexity exist, the basic tenet of seed certification remains the same; seed certification strives to improve and maintain the health of the seed tuber crop so that it performs in a predictable manner. Important technological advances for virus eradication from potato tissue, virus detection, and the rapid clonal propagation of potato under controlled conditions, all have contributed greatly to the success of modern seed certification programs worldwide (Jones, 1988).

This chapter focuses on the evolution of virus disease management practices from the earliest days of seed improvement. Although steps in this evolution are categorized below, the development and application of various virus management practices overlapped considerably (Darling, 1967). The production of basic (nuclear) seedstocks is emphasized because the principles of virus management hold true for all classes of certified seed.

16.2 Evolution of Virus Management Practices

Tuber increase coupled to an intensive program of identifying and eliminating undesirable plants, whether infected by virus or possessing some other undesirable feature, is a tried and proven method for improving and maintaining seed productivity. Early seed improvement programs selectively propagated the best hill or line selections in an effort to improve the performance of basic seedstocks. Tuber-unit and tuber indexing procedures further improved the quality of basic seedstocks because these procedures aided in the detection and elimination of tubers that carried virus and other pathogens. The development of various virus detection methods described elsewhere in this book enabled great strides in the detection and elimination of infected propagation stock, decreasing reliance on visible symptom expression as a selection criterion. The key to success of all seed certification programs is, and always has been, the detection and elimination of primary inoculum sources during the production of basic seedstocks.

16.2.1 Hill and Line Selections

During the late 1800's, it was commonly accepted that the potato plant had become weakened from continued and repeated clonal (asexual) propagation from seed tubers (Stevenson, 1956). It was widely believed that the only way to restore vigor to the potato was through the planting of true (botanical) seed harvested from the fruit of the potato plant (Stevenson, 1956). However, in 1885 E.S. Goff demonstrated that the propagation of tubers harvested from productive plants resulted in greater yields than did the propagation of tubers harvested from non-productive plants (Rieman, 1956).

Goff proved that vigor in the potato plant could be maintained and increased by repeated selection of the best tubers for propagation and, from 1900 to 1912, the transmission of "productivity" in tuber lines also was demonstrated by numerous other workers (Rieman, 1956). These early workers often used the equivalent terms "hill" or "line" selection to describe the process where single tubers from the most productive plants were identified and then selectively increased by clonal propagation. By 1920, clonal propagation was practiced by several prominent growers in the Netherlands (Hiddema, 1972) and, in the United States, yield increases of 400 percent and greater were routinely achieved by this simple selection process (Werner and Howard, 1923). Thus, the earliest seed improvement programs merely relied upon selecting tubers from the most productive plants for clonal increase to provide basic seedstocks. Hill or line selections served as a continuous source of basic seedstocks because desirable tubers were annually re-selected for their physical appearance or some other characteristic deemed important (Trank, 1967).

McConnell (1983) emphasized the continued importance of grower participation in selecting hills or lines with the most desirable characteristics for propagation. Growers are able to visit seedplots with greater frequency compared to seed certification officials and, therefore, are able to make selections based on plant characteristics observed throughout the growing season (Schutter, 1983). McConnell (1983) also stated that mechanization has decreased the opportunity for the modern seedsperson to scrutinize the performance of individual hills, as the harvest is now a continual blur of potatoes from the field to the storage. The modern trend to decrease emphasis on hill or line selections for basic seedstocks production was considered undesirable (McConnell, 1983; Schutter, 1983).

Dykstra (1948) reported that hill or line selections reduced virus incidence somewhat, but concluded that this type of selection was unsatisfactory as a primary means of virus management. In one experiment, it was revealed that 15 percent of the tubers selected from healthy-appearing hills actually carried rugose mosaic caused by *Potato virus Y* (PVY; Dykstra, 1948). Virus infection that occurred late in the season was assumed to account for tuber infection that occurred in the absence of foliar symptoms. Although virus was not reliably eliminated by hill or line selections, Dykstra (1948) conceded that this type of selection was better than no selection at all.

16.2.2 Tuber-unit Methods

H.J. Webber is credited with introducing the tuber-unit method in 1908 (Rieman, 1956). A tuber-unit planting of hill or line selections aided in the identification and roguing of undesirable plants (Werner and Howard, 1923). Thus, not only are the best tubers retained for propagation via line or hill selections (positive selection), tubers considered infected or weak also are being eliminated from the seed plot (negative selection).

The tuber-unit method starts by selecting tubers ideal in appearance from a given seedlot. All seedpieces from a single tuber are then planted consecutively within the row as a "tuber-unit" and each unit is separated from neighboring units to enable identification of all progeny from a single seed tuber (Dykstra, 1948). Plants within a unit were typically spaced at 30 to 38 cm and a blank space of 60 to 76 cm within the row remained to separate consecutive units. Although tuber-units commonly consisted of four hills, the actual number of hills within each unit varied slightly according to the size of the mother tuber.

Tuber-unit plantings enabled growers and inspectors to scrutinize all hills derived from a single seed tuber (Dykstra, 1948). If any plant within the unit was symptomatic, all plants from that unit were immediately eliminated by roguing. During roguing, seedpieces, daughter tubers and foliage were carefully removed from the field plot area. Allowing the rogued plant to remain in the field posed considerable risk as an inoculum source. Recommendations were to carry the plant out of the plot within a closed container and to dispose the plant at least 150 m outside of the seedplot area. Covering the plant while exiting the seedplot prevented the escape of viruliferous aphids, and rogued plants also were covered before wilting to prevent the movement of aphids to more attractive plants.

Because virus is not uniformly distributed among all eyes of the mother tuber (Beemster, 1961), the ability to detect non-uniformity or other undesirable characteristics associated with the original seed tuber was maximized by tuber-unit seedplots (Dykstra, 1948). Diseased plants (infected units) were rogued earlier from tuber-unit plantings compared to mass plantings of a given seedlot. The tuber-unit method was quickly adopted as the most favorable method for control of virus diseases through field-plot selection, and mass roguing was found to be less efficient for virus elimination (Dykstra, 1948). The importance of grower participation in the inspection and roguing of tuber-unit plantings cannot be over-emphasized. Frequent inspections by growers ensured the early detection and removal of undesirable plants, reducing the potential for secondary virus spread within the seedplot. Tuber-unit plantings were typically conducted under conditions of reduced fertility to enhance the onset of virus symptomatology or the detection of other perceived weaknesses (Schutter, 1983; K.W. Knutson, personal communication). As improved virus detection methods became available, foliar assays played an increasingly important role for virus detection and the elimination of infected units that appeared healthy during visual inspection (Darling, 1967).

16.2.3 Tuber Indexing

Following the development of practical virus assays, tuber indexing became one of the most effective methods for eliminating infected tubers from basic seedstocks (Koehnke and Shaughnessy, 1942). Tuber indexing aids in the identification and removal of primary inoculum sources from a seedlot before planting, thus reducing the risk of secondary spread. By the 1920's, tuber and hill indexing became standard procedures for the production of basic seedstocks (Jones, 1991).

A typical tuber indexing scheme is initiated by sequentially numbering all tubers from a seedlot (Dykstra, 1948). Seedpieces consisting of a single eye are removed from each tuber and planted. The numbering scheme is such that each plant or "index" grown from the seedpiece corresponds to the original seed tuber. Each index is examined to determine the presence of virus or any other reason for eliminating the original seed tuber. Early tuber indexing programs relied upon visual examination of the index to detect virus infection and gradually became more sophisticated as various assays for virus and other pathogens were developed (Darling, 1967; Trank, 1967). Tuber indexing is commonly done between growing seasons, or well in advance of the growing season during which the indexed tubers are to be planted. In the United States, indexing typically was performed in the greenhouse, although seedpieces were sometimes shipped to southern states for grow-out of the index plants in field plots (Koehnke and Shaughnessy, 1942).

Trank (1967) and others (Koehnke and Shaughnessy, 1942) described tuber indexing and its incorporation into the Nebraska seed certification program during its early development. During tuber indexing, one eye from each tuber was planted in the greenhouse and resultant plants were scrutinized for any indication of disease or undesirable phenotype. Any "off-type" plants and their corresponding tubers were immediately discarded. If plants appeared healthy, extensive testing via serology, indicator plants and electron microscopy was performed on foliage to detect and eliminate tubers that carried viruses (Trank, 1967). Tubers that passed all tests were planted in isolated field plots for tuber-unit increase. All progeny of each indexed tuber, called a line selection, were handled separately during increase. Foliage from all line selections was tested by serology and undesirable lines were rogued. Following a second year of field increase and testing, line selections derived from indexed stocks were sold to growers that met production restrictions for isolation and disease management. In order to provide a continuous source of indexed tubers, 10 to 20 tubers were selected each year from the best index seedstocks for each variety and re-tested via tuber-indexing.

This production scheme worked well for early attempts aimed at the management of *Potato virus X* (PVX) (Trank, 1967). During the growing seasons from 1961 to 1966, the PVX-tested acreage increased from 0.5 ha to 479 ha and PVX-tested seed availability increased from 16.9 t to 11,249 t. However, this program was labor intensive and required five technicians and three inspectors to process leaf samples during 1966. Regardless, growers felt the program was worthwhile and experienced up to an estimated 5.7 t per ha increase in yield compared to unimproved seedlots. Tuber indexing is proving to be of increasing importance in modern seed certification programs because it is the only means for detecting pathogens in the absence of overt symptoms (De Boer *et al.* 1996). The need for indexing is largely driven by the international market place which permits only the importation of pathogen-tested seed tubers (De Boer *et al.* 1996). The elimination of tuber borne pathogens is becoming a more realistic goal following the development of practical testing methods for viruses and the widespread adoption of *in vitro* methods for rapid increase of pathogen-tested materials (described below).

16.2.4 Early Harvest

Late season infection of the potato plant may result in virus translocation to tubers harvested from otherwise healthy-appearing plants. Experiments conducted in Maine revealed that seed tubers harvested near the end of July had less than 0.1 percent virus incidence compared to 35 percent mosaic-infected and 33 percent *Potato leaf roll virus* (PLRV) -infected when tuber harvest was delayed by two months (Dykstra, 1948). A second year of this experiment revealed an increase from 0 percent for tubers harvested July 31, to 0.9 percent PLRV incidence for tubers harvested on August 8, to 20 percent mosaic-infected and 30 percent PLRV-infected for tubers harvested on September 24. Vine (haulm) killing by herbicide or vine removal by mechanical means before late season infection occurs is important for preventing the movement of virus into seed tubers (Dykstra, 1948; Hiddema, 1972). The terms "early harvest" or "early lifting" are often used to describe this method for avoiding virus infection.

In the Netherlands, the date of vine-kill is fixed by the regulatory agency (Hiddema, 1972). This date is based on aphid-flight monitoring for each seed class grade and cultivar, and becomes more relaxed for lower classes of seed. Since basic seedstocks are to remain free of virus, the date selected is so early for this seed class that the chance of virus infection via aphid vectors is considered to be negligible. Inspectors also determine whether or not the entire stem is dead, as required, by the compulsory vine-kill date. Vine-kill is done by flame, chemical desiccation, vine-pulling, or green harvest is sometimes practiced. North American growers also practice early vine-kill to avoid late season tuber infection by viruses. Although early harvest is most often used to manage viruses vectored by insects (Hiddema, 1972), it also may limit late season spread of mechanically transmitted viruses (Franc and Banttari, 1996).

16.2.5 Winter-tests

Because it is difficult to detect late season virus infection during field inspection and by tuber symptomatology, post-harvest winter-tests were developed. Winter-tests attempt to measure disease incidence in a seedlot and also predict seedlot performance before it is planted the following season (Slack, 1993). Winter-tests are sometimes referred to as a winter index, since it is a variation of the tuber index process (Shepard and Claflin, 1975).

A standard subsample of each seedlot (e.g., 400 tubers) is collected shortly after harvest and delivered to the seed certification agency (Slack, 1993). After breaking dormancy by chemical means, tuber subsamples from all seedlots are shipped to a central location and planted in the field. After plant emergence, each seedlot subsample is visually inspected and plants are tested for the presence of various pathogens, according to the rules of the seed certification agency. Certification tags are issued if tolerances are not exceeded. The first winter-test programs in North America were initiated in 1925 and most winter test programs continue to be practiced today (Jones, 1991; Slack, 1993). Winter-test programs are costly and, with the development of improved virus detection methods, there is an increasing reliance being placed on laboratory-based tuber tests for virus detection (De Boer *et al.* 1996). Thus, tuber indexing for pathogen detection in the subsample is supplementing or replacing seedlot inspections to detect disease symptomatology.

16.3 Evolution of Clonal Propagation Methods for Basic Seedstocks Production

Early seed improvement programs relied upon the planting of whole or cut seed for clonal increase of hill or line selections. This method remained relatively unchanged until the 1960's and 1970's when meristem-tip culture, stem-cutting, and leaf-bud cutting propagation methods were adopted for basic seedstocks production. *In-vitro* production of pathogen-tested plantlets quickly replaced earlier methods for rapid propagation of desirable clones in North America and Europe (Jones, 1988). These clonal increase methods are described in greater detail elsewhere in this book.

Although some increase methods can be used to eliminate virus from vegetatively propagated material, most methods for rapid clonal increase do not eliminate viruses. Instead, these methods enabled the rapid increase of virus-tested materials under controlled conditions, thus, providing growers with a greater volume of basic seedstocks. The greater volume of basic seedstocks, in turn, enabled the widespread adoption of limited generation or true "flush out" programs that restricted the number of years that seedlots remained eligible for certification (Slack, 1993). Additional improvements to our existing technology is expected as scientists and seed producers strive to produce high quality seed tubers with greater efficiency.

16.3.1 Shoot-tip and Meristem-tip Culture

Clonal propagation methods listed above relied greatly upon visual inspection for elimination of viruses. However, most strains of PVX and *Potato virus S* (PVS) typically do not produce symptoms readily detectable in the potato plant and are considered latent under most field conditions (Wright, 1988). Early field selection methods did little to eliminate these viruses and most important cultivars were universally infected (Mellor and Stace-Smith, 1967; Wright, 1988). Various propagation methods in combination with therapy to decrease virus titer, such as thermotherapy and chemotherapy, were developed in attempts to generate virus-tested clones from infected tissue.

Rooted cuttings taken from PVX-infected potato plants remained infected in the absence of thermotherapy (Mellor and Stace-Smith, 1967). Thermotherapy at an average air temperature of 35° C (33 to 37° C) increased plant survival compared to exposure at a constant air temperature of 37° C, and enabled thermotherapy to be continued for an extended period of time. Shoot tips (1 to 5 mm) were excised from stems and grafted onto tomato every 1 to 2 weeks during thermotherapy treatment. The first PVX-free scion was obtained after 15 weeks of thermotherapy and the proportion of PVX-free scions increased with longer treatment. Approximately 46 percent of the scions survived and 63 percent were freed of PVX. However, this method failed to eradicate PVS (Stace-Smith and Mellor, 1968). An improvement of this method for virus elimination was attempted through meristem-tip culture.

The meristematic region is a dome of actively growing cells at the apex of shoots and roots and in axillary buds. Depending on the virus-host relationship, virus titer is reduced or virus may be absent in meristematic and underlying tissue of infected mother plants (Hollings, 1965). Therefore, the regeneration of plantlets or "explants" from excised meristem-tips is a method of vegetative propagation that potentially avoids virus infection. Application of meristem-tip culture for the large-scale production of basic seedstocks did not occur until the 1960's (Wright, 1988).

Meristem-tip culture in the absence of therapy proved to be relatively inefficient for elimination of PVX and PVS (Stace-Smith and Mellor, 1968). For example, in the absence of thermotherapy, the frequency of virus-free explants was only 4 percent and 2 percent for PVX and PVS, respectively. However, thermotherapy of mother plants for 18 to 20 weeks prior to excision and culture of meristem-tips (0.3 to 1.0 mm) increased the frequency for recovery of PVX-free explants to almost 100 percent and, of these, 23 percent also were freed of PVS. Chemotherapy with antiviral compounds was subsequently developed as a means to eliminate viruses and was used in conjunction with meristem-tip culture (Cassels and Long, 1982). However, due to extensive time requirements for re-generation of explants from meristem-tips (at least 6 months; Stace-Smith and Mellor, 1968), larger shoot tips were excised and these cultures were treated with antiviral compounds for virus elimination (Klein and Livingston, 1982; Wambugu *et al.* 1985). Ribavirin (Virazole), a synthetic riboside with antiviral properties, increased the recovery of PVX-free plantlets by at least 80 percent compared to shoot tip cultures that lacked ribavirin (Klein and Livingston, 1982). Although ribavirin could substitute for the time-consuming thermotherapy treatment, ribavirin is phytotoxic and slowed plantlet regeneration. Therefore, the time required for virus-free plantlet regeneration from shoot tips ultimately proved similar to that required for explant regeneration from meristem-tips.

Slack (1980) described a typical application of meristem-tip culture to Wisconsin's certified seed program. Once meristem-tip culture plantlets were established in soil, extensive testing was done to eliminate those infected by virus or other pathogens. Repeated tests were conducted in an effort to detect low virus titer that may have gone undetected during earlier testing. After plantlets were determined to be virus-free, efforts were concentrated on propagation and increase without recontamination. Meristem-tip culture derived material was increased in isolated field plots. Tubers from selected plants were harvested and increased according to the leaf-bud cutting method described below. The meristem-tip culture method was typically not repeated unless a cultivar became reinfected and healthy material was no longer available.

16.3.2 Stem Cuttings

Propagation of potato seed stocks by stem cuttings was developed in the 1960's as a means for rapidly increasing virus-tested material (Cole and Wright, 1967). Stem cuttings also aided in the elimination of bacterial and fungal pathogens normally carried over by tuber propagation.

Plants are grown in the greenhouse from seedpieces or eyes until approximately 20 to 25 cm in height (Cole and Wright, 1967). The growing tips are removed to stimulate axillary stem growth. When axillary stems reach approximately 13 cm in length, their tips (approximately 5 cm) are removed and set into washed sand with the cut surface downward. A 33- by 46-cm greenhouse flat will accommodate 60 cuttings. Intermittent mist is applied and, after approximately 2 weeks, cuttings that rooted are transplanted into greenhouse soil. The cycle is repeated after rooted cuttings grow sufficiently large, and multiple generations of stem cuttings are subsequently produced during a single season. For example, eight plants produced from a single tuber yielded approximately 5000 rooted cuttings 6 months later. Cuttings are transplanted to field plots for the production of a tuber crop. Transplanting is done by setting cuttings at the bottom of a shallow trench and covering roots with soil. Soil also is periodically added to the base of stems, burying them to greater depths (15 to 20 cm final depth) as growth and stem elongation occurred. Plants produced from cuttings matured approximately 2 weeks earlier than plants grown from seed tubers, likely reducing the potential for reinfection by PVS and PVY (McDonald, 1987; Franc and Banttari, 1996). Seed tubers produced by stem cuttings had a tendency to be rougher in appearance (Cole and Wright, 1967).

Virus-tested clones were developed from meristem-tips of five clonal selections of Russet Burbank (Netted Gem) which had been grown separately for at least 40 years (Wright and Mellor, 1976). Three years of comparative tests under uniform field conditions revealed that none of the clones were intrinsically superior, even after 40 years of independent field selection by seed growers. Therefore, meristem-tip culture and propagation by stem cuttings did not appear to alter the growth characteristics of Russet Burbank (Wright and Mellor, 1976), other than to increase yield potential due to freedom from virus (Wright, 1970). Similar conclusions were made following a four-year comparison of ten virus-free potato cultivars; no deleterious effects resulted from virus eradication by thermotherapy and meristem-tip culture, followed by stem cutting propagation (Wright, 1983). However, Schutter (1983) observed that virus-free clones may appear less uniform, at least during the earliest generations of field increase.

In North America, the Vancouver Agriculture Research Station was instrumental in applying stem cutting technology, and since 1968 released virus-free seedstock to growers and potato improvement agencies in Canada, United States, Australia, New Zealand, and several other countries in Europe, North Africa, and Central and South America (Wright, 1988). The increased volume of virus-tested seedstock enabled revision in 1969 of Canadian seed potato certification regulations to establish a "flush-through" system for seed production (Wright, 1988) similar to the limited generation increase programs described below. The flush-through system facilitated control of PVS and PVX provided that proper sanitation was practiced to prevent re-introduction of virus. In 1970, Scotland became the first country to require that all seed lots entered for certification be derived from virus-tested stem cuttings (Jones, 1991). The introduction of stem cutting virus-tested seedstocks into Oregon in 1972 contributed to a 17 percent decreased incidence (5 year average) of seedlots with detectable PLRV, presumably because inoculum sources from which vectors acquired virus were systematically eliminated (O. Gutbrod, personal communication).

16.3.3 Leaf-bud Cuttings

The use of leaf-bud cuttings as a tool for rapid clonal increase was first proposed for seed certification and breeding programs by Lauer (1977). Propagation is done by cuttings that consist of a leaf, axillary buds, and a small stem segment. Cuttings from the entire haulm of the plant are prepared at one time. For the six cultivars studied, the number of cuttings ranged from 37 to 288 with an average of 114. The stem segment for each cutting is buried in water-saturated sand so that only the leaflets are exposed. After 2 to 3 weeks, tubers begin to form on the axillary buds on the stem segments. After 7 weeks most cuttings formed tubers and leaves started to senesce. Leaf-bud cuttings from the apex tended to form both roots and tubers while cuttings from the basal portion of the plant produced the smallest tubers. Tubers were small and ranged in size up to 3 cm in diameter. After dormancy, tubers from leaf-bud cuttings were planted in the field or greenhouse to start a second cycle of production. Lauer (1977) demonstrated that each mother plant propagated by leaf-bud cuttings and followed by one season of field increase yielded an average of 43 kg of tubers. If each mother plant is first increased by stem-cuttings and then by leaf-bud cuttings, the yearly increase is proportionately greater.

16.3.4 *In-vitro* Plantlet Propagation

Micropropagation of plantlets greatly improved our ability to increase basic seedstocks in a manner that protects clones from reinfection (Roca *et al*. 1978). Plantlets derived from healthy material, usually by meristem-tip culture, are then maintained, or propagated, in tissue-culture to serve as a healthy clone-bank (Jones, 1988). This method also has been used in conjunction with therapy and serology to select plantlets freed of viruses, reducing or eliminating the need for virus elimination by meristem-tip culture (Sanchez *et al*. 1991). Periodic transfers and repeated pathogen-testing ensures that a continuous supply of healthy *in vitro* material is maintained and available for increase.

For plantlet increase, nodal cuttings are prepared from *in vitro* plantlets and sub-cultured to fresh growth medium (Roca *et al*. 1978). A seven-fold increase in the number of plantlets may be achieved for each generation (2 to 3 weeks) for most cultivars. When the number of plantlets produced is sufficient, they are carefully transplanted to the greenhouse or field to produce a seed tuber crop (Jones, 1988). Tubers are harvested and replanted during subsequent field seasons to increase the volume of basic seedstocks. This method, or variations of this method quickly gained acceptance throughout North America and Europe and was incorporated into seed certification programs for basic seedstocks production (Jones, 1988).

16.3.5 Limited Generation Increase

The availability of healthy tissue-culture derived seedstocks and the ability to rapidly increase the number of plantlets enabled the adoption of limited generation programs by most seed certification agencies (Slack, 1993). Limited generation programs in North America are characterized as those that limit the number of generations that a certified seedlot may be increased in the field. The premise is that seedlots planted in the field are at greater risk of becoming infected by pathogens. The majority of programs limit the number of field generations to 7 years, and ranged from 4 to 8 years (Slack, 1993). Once seed has been increased the maximum number of generations, it is no longer eligible for re-certification. The adoption of limited-generation programs for increase of seedstocks from pathogen-tested *in vitro* sources enabled virus testing to become progressively more systematic and comprehensive (De Boer *et al*. 1996).

16.4 Summary

During the last 100 years, common sense, experience and scientific study have been brought to bear on the problems associated with the production of sufficient quantities of true-to-type, agronomically superior, disease-free seed potato tubers. Seed potato certification organizations have taken the leadership role in developing and applying simple and practical virus management practices, including methods such as visual inspection to remove diseased or off-type plants, various means to vegetatively increase pathogen-free materials, and the isolation of seed production areas. More recently, the use of tissue culture for clonal propagation and application of laboratory tests such as ELISA and PCR for the detection of pathogens has improved both the quality and quantity of material available to the industry.

To maintain their effectiveness, seed potato certification programs will have to respond to ever-changing social, scientific and environmental issues. The potential emergence of transgenic cultivars and the changing patterns of end-product use will require constant vigilance and adaptation to meet changing demands. The concentration of farms into fewer and larger units will change both the economics and politics of seed potato production. Increasing concerns about the pesticide use in the presence of increased pest pressures and heightened environmental awareness will create new and challenging problems for the producer of high quality seed tubers.

Acknowledgment

The author is grateful to T.L. German and K.W. Knutson for critical review of this chapter and suggestions for its improvement.

References

Beemster, A.B.R. 1961. Translocation of leaf roll and virus Y in the potato. *Proc. Fourth Conf. Potato Virus Diseases*. Braunschweig, 1960: 60-67.

Cassels, A.C. and Long, R.D. 1982. The elimination of potato viruses X, Y, S, and M in meristem and explant cultures of potato in the presence of Virazole. *Potato Res.* 25: 165-173 .

Cole, C.F. and Wright, N.S. 1967. Propagation of potato by stem cuttings. *Am. Potato J* 44:301-304.

Darling, H.M. 1967. Present Seed Potato Improvement Programs. Pp. 1-8. In: *Proc. Seed Potato Improvement Symp.*, Potato Assoc. of America, Orono, ME.

De Boer, S.H., Slack, S.A., van den Bovenkamp, G.W. and Mastenbroek, I. 1996. A role for pathogen indexing procedures in potato certification. *Adv. Botanical Res.* 23: 217-242.

Dykstra, T.P. 1948. *Production of disease-free potatoes.* USDA Circ. No. 764, Beltsville, MD.

Franc, G. D. and Banttari, E.E. 1996. Translocation and mechanical spread of a Minnesota isolate of potato virus S in potatoes. *Am. Potato J.* 73: 123-133.

Hiddema, J. 1972. Inspection and grading of seed potatoes. Pp. 206-215. In: *Viruses of Potatoes and Seed-Potato Production.* J.A. De Bokx (ed.). PUDOC Wageningen, Netherlands.

Hollings, M. 1965. Disease control through virus-free stock. *Ann. Rev. Phytopathol* 3: 367-396.

Jones, E.D. 1988. A current assessment of *in vitro* culture and other rapid multiplication methods in North America and Europe. *Am. Potato J.* 65: 209-220.

Jones, E.D. 1991. Progress in seed production technology. *Am. Potato J.* 68: 247-248.

Klein, R.E. and Livingston, C.H. 1982. Eradication of potato virus X from potato by ribavirin treatment of cultured potato shoot tips. *Am. Potato J.* 59: 359-365.

Koehnke, M. and Shaughnessy, J. 1942. Nebraska seed improvement program for Nebraska potatoes. *Am. Potato J.* 19: 161-166.

Lauer, F.L. 1977. Tubers from leaf-bud cuttings: a tool for potato seed certification and breeding programs. *Am. Potato J.* 54: 457-464.

McConnell, R. 1983. Advanced Techniques for the Modern Seedsman. Pp. 143-147. In: *2nd Ann. North American Seed Potato Seminar Proc.* National Potato Council, Denver, CO.

McDonald, J.G. 1987. Comparative levels of potato viruses S and Y infection of microplants and tuber-propagated plants in the field. *Am. Potato J.* 64: 517-521.

Mellor, F.C. and Stace-Smith, R. 1967. Eradication of potato virus X by thermotherapy. *Phytopathology* 57: 674-678.

Rieman, G.H. 1956. Early History of Seed Certification in North America, 1913-1922. Pp. 6-10. In: *Potato Handbook Seed Certification Issue*, Potato Assoc. America, Orono, ME.

Roca, W.M., Espinoza, N.O., Roca, M.R. and Bryan, J.E. 1978. A tissue culture method for the rapid propagation of potatoes. *Am. Potato J.* 55: 691-701.

Sanchez, G.E., Slack, S.A. and Dodds, J.H. 1991. Response of selected *Solanum* species to virus eradication therapy. *Am. Potato J.* 68: 299-315.

Schutter, J. 1983. Advanced Techniques for the Modern Seedsman. Pp. 148-151. In: *2nd Ann. North American Seed Potato Seminar Proc.* National Potato Council, Denver, CO.

Shepard, J.F. and Claflin, L.E. 1975. Critical analyses of the principles of seed potato certification. *Ann. Rev. Phytopathol.* 13: 271-293.

Slack, S.A. 1980. Pathogen-free plants by meristem-tip culture. *Plant Dis.* 64: 15-17.

Slack, S.A. 1993. Seed Certification and Seed Improvement Programs. Pp. 61-65. In: *Potato Health Management*, R.C. Rowe, (ed.). APS Press, St. Paul, MN.

Stace-Smith, R. and Mellor, F.C. 1968. Eradication of potato viruses X and S by thermotherapy and axillary bud culture. *Phytopathology* 58: 199-203.

Stevenson, F.J. 1956. Potato Breeding in the United States. Pp. 25-30. In: *Potato Handbook Seed Certification Issue*, Potato Assoc. America, Orono, ME.

Trank, W.E. 1967. The Nebraska "Virus X-Free" Foundation Seed Program. Pp. 40-46. In: *Proc. Seed Potato Improvement Symp.*, Potato Assoc. America, Orono, ME.

Wambugu, F.M., Secor, G.A. and Gudmestad, N.C. 1985. Eradication of potato virus Y and S from potato by chemotherapy of cultured axillary bud tips. *Am. Potato J.* 62: 667-672.

Werner, H.O. and Howard, R.F. 1923. Seed potato investigations. *University of Nebraska Agric. Exp. Sta. Bull.* 24, Lincoln, NE.

Wright, N.S. 1970. Combined effects of potato viruses X and S on yield of Netted Gem and White Rose potatoes. *Am. Potato J.* 51: 202-205.

Wright, N.S. 1983. Uniformity among virus-free clones of ten potato cultivars. *Am. Potato J.* 60: 381-388.

Wright, N.S. 1988. Assembly, quality control and use of a potato cultivar collection rendered virus-free by heat therapy and tissue culture. *Am. Potato J.* 65: 181-198.

Wright, N.S. and Mellor, F.C. 1976. A comparison of five clones of virus-free Netted Gem potato from British Columbia and Idaho. *Am. Potato J* 53: 99-103.

17. Common Seed Potato Certification Schemes

O. A. Gutbrod and A. R. Mosley
Certification Specialist and Extension Potato Specialist, respectively, Department of Crop and Soil Science, Oregon State University, Corvallis, Oregon 97331-3001, U.S.A.

17.1 Introduction

Superior potato planting stock is becoming ever more important as world food needs increase. With current technology, maximum production per unit area of land can only be achieved by planting vigorous, disease-free tubers derived through well-designed and carefully monitored production and certification schemes. Such tubers are asexually propagated, or cloned, by dividing tubers of preceding generations. While cloning provides excellent genetic uniformity, yields, and quality, it also presents serious pest control problems. Cloning effectively spreads diseases, nematodes and other organisms from generation to generation.

17.2 Characteristics of Good Seed

Good seed, properly used, produces acceptable yields of high quality tubers with minimal production inputs. It typically satisfies the following requirements (Mosley and Chase, 1993).

Good seed:
is clonally pure, or true to type;
is inspected and approved by an official certification agency;
meets established certification tolerances for diseases and pests;
is free of diseases and pests quarantined by an importing country;

G. Loebenstein et al. (eds.),
Virus and Virus-like Diseases of Potatoes and Production of Seed-Potatoes, 421–438.

is post-harvest tested;
has a desirable physiological age;
meets tuber quality and size requirements; and
contains minimal amounts of soil, stones, and other debris.

Because the potato is asexually propagated or cloned, all seed potato stocks eventually become infected by a number of disease organisms, especially viruses; furthermore, infection levels increase with time in the field. These two familiar principles dictate important attributes of all good seed production schemes.

A successful seed production scheme must:

Create/identify and increase seed stocks free of known potato pathogens; minimize field exposure to pathogens through a limited-generation or flushout increase program; and monitor crop health, especially during the field years, through an accredited certification agency responsible for enforcing established tolerances for disease and other quality factors.

Earlier chapters in this volume addressed the elimination of diseases, sophisticated seed testing, and propagation of healthy plant materials in protected environments. This chapter will review present day seed certification schemes, or protocols, and outline important minimum requirements for field production of healthy seed crops.

17.2.1 Importance of Seed Quality

The use of disease-free planting stocks is considered to be the single most important integrated pest management practice available to potato growers. Most diseases of potato can be seedborne. Late blight, for example, can easily move in seed potatoes and be introduced to areas formerly free of the disease. The negative impact of some diseases, such as PVY on yield, is directly related to growing conditions and the length of the season where the crop is grown. Longer-season production regions, especially those experiencing high temperature and moisture stress, appear to be especially susceptible to yield loss from seedborne viruses (Hane and Hamm, 1999; Rykbost *et al.* 1998).

17.2.2 The Role of Certification

The establishment and application of strict disease and quality tolerances is an essential component of all good seed production schemes. Seed certification agencies monitor seed quality using a combination of traditional and high technology methods ranging from visual inspection in the field and greenhouse to sophisticated serological techniques to electron microscopy. Broadly speaking, potato seed certification agencies function primarily by monitoring and characterizing disease levels, clonal/varietal purity, and tuber physical condition. Visual inspections during the growing season, and especially during post harvest evaluations, including winter growouts, may also identify chemical damage, freezing injury, and other non-pathogenic problems.

The certification process addresses not only seed health and quality but also certain required production inputs including: field history; isolation; correct labelling; seed source and generation status; planting methods; record keeping; reporting and labelling; and a number of other considerations.

17.3 Major Certification Schemes or Protocols

Most seed potatoes are produced under either the UN/ECE (European) or North American (Canada and the United States) seed production schemes (Table 17.1).

17.3.1 UN/ECE (European) Scheme

Holland and Germany are credited with initiating the certified seed potato industry. The first European certification schemes were introduced in the early 1900's (Slack, 1993). They utilized a clonal selection process for introducing and maintaining seed stocks. Today, in addition to clonal selection, disease testing and tissue culture are major ingredients of the European protocol.

Because Europeans have extensively produced and marketed seed tubers worldwide, their protocol is more familiar than the North American Scheme in most countries. Holland has long been, and still remains, the major exporter of seed potatoes (Consolidated Exports Highlights, 1999).

The UN/ECE scheme uses categories and subdivisions rather than crop generations to define the flush out concept (Table 17.1). It appears somewhat complex because it allows individual countries to establish subdivisions within the three major categories. The number of field generations following tissue culture can extend to 10 or even 12 years. It uses generic small seed (non-potato) certification terminology as primary categories (pre-basic, basic, and certified). This system still emphasizes clonal selection and permits non-meristem, pathogen-tested production. Clonal selection insures varietal purity and reduces grower costs. This scheme has merit in countries without laboratories, greenhouses, or technical expertise for the production of disease-free planting stock.

The United Nations Economic Commission for Europe (UN/ECE) Working Party on Standardization of Perishable Produce and Quality Development (United Nationals/Economic Commission for Europe Standards S-1, 1998) is currently attempting to promote standardization of seed certification programs internationally. The UN/ECE consists of 55 member countries including all of Europe, Canada, and the United States. The Working Party of the UN/ECE has established a "Specialized Section for the Standardization of Seed Potatoes". This section is not limited to member countries. Invitations to participate have been extended to all seed importing and exporting countries. The UN/ECE Standards established by these countries serve as a model certification scheme. The standards facilitate fair trade and help standardize certification methods internationally. Adherence to UN/ECE standards is voluntary. These standards may also be used to settle disagreements between countries when seed potatoes do not meet all expectations.

17.3.2 The North American Scheme

Seed Potato Certification began in Canada and the United States around 1914 (Slack, 1993). The current North American scheme, (Table 17.1) or protocol, follows very rigid production regimes (Canada Food Inspection Agency (CFIA), 1996; U.S. Export Standards for Seed Potatoes, 1999). All certified seed originates from meristem culture and is first increased under greenhouse conditions. Certification is limited to seven years of field increase under a protocol commonly referred to as a "Limited Generation" or "Flush Out" scheme. U.S. Seed Export Standards (U.S. Export Standards for Seed Potatoes, 1999) permit only five field increases. These standards list specific requirements and disease tolerances for each generation and essentially simulate the category concept associated with UN/ECE standards.

The Limited Generation scheme defined in the U.S. Export Standards could easily provide for the certification of non-meristemmed cultured (clonal selection) seed. If necessary, tissue culture origin could be verified by adding the postfix TC to the generation number (ex., G1TC).

17.4 Certification Methods

The preceding paragraphs describe the two major international schemes for producing and certifying seed potatoes. Both schemes stipulate limited-generation increase. Major differences between the European and North American schemes involve terminology, source of original planting material (clonal or meristemmed/tissue culture), and the number of field generations permitted. While the European scheme uses categories and subdivisions, the North American Scheme uses actual generation numbers to identify the various years or stages of increase. The term "generation" actually refers to a level of quality based on both years from Pre-Basic or Nuclear status and the disease level and varietal purity. Diseased seed lots may advance more than one generation per season, or be rejected for certification outright.

Certification schemes have changed dramatically in recent years because of the widespread availability of greenhouse minitubers grown from tissue-cultured plants free of known potato pathogens. Both the European and North American schemes have been redesigned to regulate production from the greenhouse through successive generations of field increase in order to produce tubers relatively free of degenerating diseases.

Large-scale seed production requires strict attention to recommended management practices and procedures. "Pre-Basic" or "Nuclear" (depending on the country of origin) minitubers or transplants grown in a protected greenhouse are propagated through open field increases according to strict protocols (Table 17.1). The ability to detect and eliminate diseases, pests, and off types is clearly essential for success with either production scheme. Both depend absolutely on the development and application of effective, independent, professionally run seed certification programs.

The following concepts must be addressed by all certification agencies, whether operating under the UN/ECE or North American production schemes.

17.4.1 Legal Recognition

Legal recognition of the certification organization is essential. The certification agency is often referred to as the "Designated Authority". Once a seed law has been enacted and a certification agency identified, no other body or agency can officially carry out the certification process or assign labels, certificates, or any other documents indicating the seed has met official certification standards. The existence of a single, legally-recognized certification agency is particularly important in preventing fraudulent practices (Hiddema, 1972).

Once the designated authority (i.e., certification agency) has been established, advisory committees and boards can be appointed for the purpose of setting standards and regulations. The certification agency then assigns trained personnel to monitor and verify all standards and guidelines to ensure satisfactory final product quality.

17.4.2 Officially Recognized Varieties

Only officially approved varieties should be certified. Such varieties are approved and recognized by an agricultural experiment station, seed certification agency, breeding program, or plant variety protection authority. The sponsor for each variety provides the certification agency with information relative to its origin, characteristics, disease reaction, and ownership. The "approved variety" requirement is common to all major seed certification schemes.

17.4.3 Approved Seed Source

Seed "source" strongly impacts the certification process. Seed may originate from either a clonal selection process or from a meristem program depending on the seed protocol being followed.

Earlier chapters in this volume referenced the production of "Basic Seed Stock", often referred to as "Pre-basic" or "Nuclear" depending on the country of production. "Limited Generation" systems are most commonly used in today's certification schemes. *In vitro* plant material vigorously evaluated for known potato pathogens is increased through a meristem/tissue culture program. These *in vitro* plants are maintained in good health and provide the "nucleus" or "basis" of the generation system. This seed source is essentially free of all known potato pathogens.

The seed certification agency applies standards that require growers to plant only approved seed of approved varieties. The "category", "generation" or "source" of the seed stock must be documented. In most cases, the seed originates from a previously certified source. Exceptions may occur, but only infrequently, and should be carefully considered. With clonal selection (UN/ECE scheme), the seed source and quality must be known even though the seed may be many generations removed from the original Pre-Basic category.

Clonal selection under the UN/ECE scheme requires planting back selected tubers from the previous crop each year. There is no limit to the number of reselections, provided greenhouse and field generation requirements are met. Lack of varietal purity and excess disease can, however, restrict further increases from a given seed lot. In some cases, lab-tested tubers may be used to produce the clonal type of the seed source for further multiplication. Some clonal material may originate from a tissue culture source.

17.4.4 Limited Generation Production

The limited generation concept is fundamental to any reliable seed certification scheme (Table 17.1). The term "generation" was introduced by U.S. potato certification programs in the mid 1970's. It typically refers to a given production season or number of years from meristem culture. However, because of strict disease tolerances, infected seed lots may increase several "generations" in a single year or be rejected outright for certification purposes.

Plants or tubers that have been tested and found free of all potato pathogens are planted in the first field generation to produce Pre-Basic or Generation 1 stocks. These stocks are then planted in the second field season to produce Basic or Generation 2, provided all standards are met, and so on. Successive categories or generations follow within prescribed limits. The North American Scheme typically ends with the production of Generation 5 while the European Scheme allows up to 10 field seasons if all tolerances are satisfied (European and Mediterranean Plant Protection Organization, 1999). As noted, the U.S. flush-out system is limited to five field generations to prevent excessive disease (U.S. Export Standards for Seed Potatoes, 1999). The "Limited-Generation" production system is also referred to as a "flush-out" program. The limited-generation concept is based on the assumption that disease levels increase with time in the field. Seed is "flushed out" of the program when it no longer meets limited-generation criteria either based on years of increase or actual disease levels.

As mentioned earlier, excessive disease can cause seed to move through the limited-generation scheme at an accelerated rate in either the European or North American schemes. Lots are downgraded one or more generations or totally eliminated from the certification process in a single season if diseases exceed established tolerances. This potential downgrading is the most important aspect of the flush-out, limited-generation scheme. The number of field generations is as important as the actual disease levels in determining generational status for a given seed lot. The most important aspect of the limited-generation scheme is the continuous addition of pathogen-tested, disease-free planting stock at the laboratory and greenhouse levels, thus older seed must be "flushed out" to make room for the new.

The limited-generation scheme has nearly eradicated some potato diseases and reduced the frequency of others. For example, blackleg has been dramatically reduced in Canada and the U.S. since the advent of limited-generation programs in the 1970's (personal observation).

17.4.5 Land History

Standards for land history are essential to prevent varietal mixtures as well as reduce potential disease and pest problems. A long crop rotation cycle is an integral part of any good seed production scheme. The minimal time allowed between seed crops depends on the local conditions. Very cold climates, periods of standing water, and the use of resistant varieties can reduce the non-potato period. Ordinarily fields should not be cropped to potatoes more often than every third or fourth year. Early seed generations may require much longer crop rotations. The use of fumigation or other approved practices to destroy pathogens and volunteers may shorten the rotation period.

The potato tuber is very hardy and readily survives when buried in the soil, even under very cold conditions with a deep snow load. Potatoes surviving from the preceding season are often called “volunteers” or “ground keepers”. Field flooding normally eliminates this problem, but flooding is usually not an option for most growers. In the absence of flooding, volunteers are typically controlled through long potato-free rotations, which typically involve other crops, tillage practices and herbicides unfavorable to volunteer potatoes. Each certification agency determines the number of non-potato seasons required to eliminate volunteers and lists requirements in official seed production standards.

Certification standards may require pre-plant soil testing to determine the presence of certain nematode pests, including *Globodera* and *Meloidogyne* spp. Many soil pests are internationally prohibited by import quarantines. Each country has established a specific list of quarantine pests; thus, requirements vary from country to country. The presence of these pests should eliminate a field from seed production.

17.4.6 Isolation

Pest management for ware potatoes is typically less intensive than for seed potatoes. Therefore, seed potatoes should ideally be planted in areas well separated from ware crops. The distance between a seed field and an infection source is commonly referred to as the “isolation” distance. Seed potatoes are often grown in isolated, cool, high altitude, or high latitude areas which are usually not conducive to high potato yields expected from a ware crop. Therefore, isolation from ware potatoes may occur naturally.

Official isolation distance requirements may range from a single vacant row to several kilometres. Fungal spores and aphids travel great distances by wind and storms. Thus, complete isolation may be impossible to achieve for some potato pathogens or pests. The isolation distance depends somewhat on the class of seed being produced. When producing several lots of the same generation and variety, some type of clear identification of the rows is usually sufficient. Skip rows between varieties of the same generation will help prevent mixture at harvest. Disease research has shown the importance of planting crop borders adjacent to the seed field rather than leaving the soil bare (DiFonzo *et al.* 1996).

Legislative seed control areas also help preserve seed potato quality. Several countries and/or states use legally defined “quarantine” or “controlled seed” areas to protect seed from the introduction of unwanted pests and diseases. Thus, seed potatoes not grown locally can be prohibited from entering a seed area if the seed potentially contains pests or diseases not common to the control area.

17.4.7 Record Keeping

Record keeping is a significant component of the seed certification process. Accurate long-term production records are especially critical with an asexually propagated seed crop. Normally the record keeping process begins when the lab produces *invitro* plants or the first tubers from *invitro* plants. The certification "paper trail" begins with the certification application form. This is the official record submitted by the applicant to substantiate seed source and lot history. The prospective seed grower fills out the application when requesting a lab, greenhouse, or field inspection.

The term "seed lot" is commonly used to identify a specific quantity of seed or seed from a specific production or field area (United Nations/Economic Commission for Europe Standards S-1, 1998). A seed lot is clearly identified by number and/or letter codes. Fields may contain numerous "lots". The importance of assigning letters and/or numbers to each particular increase so that the lot can be tracked back through its entire production history cannot be overemphasized.

17.4.8 Field Inspection

Evaluating the growing crop for variety mixtures, off types, and diseases is an essential part of the certification process. Trained personnel experienced in identifying potato varieties and diseases are essential for these inspections. Trial plantings with many potato varieties and various diseases are highly useful for training inspectors. Plantings should be repeated each season, well ahead of the first inspection. The plots should include bacterial, fungal, viral and off-type examples common to the producing region. Disease training plots should be extremely well isolated from seed fields.

Field crops should be inspected at least two or three times while the plants are still vigorous. Plants are normally first inspected when six or eight inches tall. The last inspection should occur shortly before kill down but before plants lose disease symptoms due to senescence, frost, or other causes.

Certification inspectors typically recommend the following practices, along with many others:

1. Always evaluate plants with the sun behind you.
2. Use a hat to shade your eyes and anti-glare sunglasses which provide a better view of mild disease symptoms.
3. Inspect a constant number of plants in every row, keeping a record of diseased and off type plants within the observed population.
4. Determine the plant population observed.

5. Stake or flag examples of off-type or diseased plants for grower education.

6. Obtain samples of diseased plants for possible laboratory verification.

7. Always stake or flag plants with diseases considered unacceptable in any seed lot, such as *Clavibacter michiganensis* subsp. *sepedonicus*, or *Ralstonia solanacearum.*

To determine the total number of plants examined during inspections, the inspector typically examines five or six plants per row, while walking diagonally across the rows, counts the number of rows, and multiplies that number by the number of plants observed in each row. Other methods may also work well. Regardless of the method used, inspectors should be certain to observe at least a few plants in every row. Normally a minimum of 1,000 plants per lot or 300 plants per hectare are observed. Larger numbers of plants increase the accuracy of the inspection estimate. The number of off type and diseased plants is divided by the total number of plants observed to determine infection percentages.

17.4.9 Roguing

Roguing is the common term for removing undesirable plants from a seed field. Trained personnel able to identify off-types and diseased plants should do the roguing or mark plants to be rogued. Training should also include information regarding plants not to be removed (*Clavibacter michiganensis* subsp. *sepedonicus*, or *Ralstonia solanacearum*) as mentioned above.

All rogued plants and associated plant parts should be removed from the field and destroyed. If left in the field, they could survive and provide a source of inoculum.

17.4.10 Tolerances

Each seed category or generation is assigned an allowable percentage tolerance for diseased and off-type plants Tables (17.2;17.3;17.4). Percentages range from 0 to 10% depending on the particular factor and certification program. The official seed certification agency establishes these local tolerances. Tolerances are normally based on the impact the disease exerts on yield and quality in ware potato fields.

Pre-basic and Nuclear seed potatoes grown in a protected environment must meet a "0" tolerance for all pathogens. Depending on the particular disease, some allowance/tolerance can be made for succeeding generations and categories.

Certain diseases show a "0" tolerance for all seed generations. "0" tolerance means "none found" in an official inspection of the plants and tubers. However, none found does not guarantee that the seed is free of a particular disease or even of other varieties. Countries have independently established "0" tolerances for specific pests and diseases. However, some are identified with a "0" tolerance by all countries. Examples include *Clavibacter michiganensis* subsp. *sepedonicus, Ralstonia solanacearum, Globodera rostochiensis, Globodera pallida, Meloidogyne* spp., and *Synachytricum endobioticum.*

17.4.11 Early Kill Down

Many certification programs recommend early haulm destruction in certain instances. Normally these kill down dates are determined by the category/generation of seed being produced and the influx of virus-vectoring aphids, particularly the Green Peach Aphid (*Myzus persicae*). Aphid trapping devices are sometimes used to help monitor aphid populations and establish kill down dates.

17.4.12 Harvest Inspection

Some diseases and pests can only be observed during seed harvest. Certification standards normally require tubers to be inspected during harvesting and storage. If "0" tolerance pests or diseases are identified, the crop will not be eligible for certification and must be sold for ware or other non-seed uses. Seed growers choosing to sell rejected seed as ware potatoes can avoid the long storage season normally required for seed.

The harvest inspection permits certification personnel to assess the condition of the seed storage facility, including the degree of separation of varieties and lots. Maintaining "lot" identity from planting through storage ensures the integrity of the system. The harvest inspection can also determine the quantity of seed available for ware planting.

17.4.13 Post-harvest Testing

Certification schemes normally recommend that harvested tubers be evaluated for diseases for final certification. So-called "Post Harvest" or "Direct Progeny" testing helps quantify disease infection which may have occurred late in the growing season after the last field inspection. Post-harvest testing should always be required for recertification.

Post-harvest testing may involve sampling a determined number of tubers per lot and serologically testing them in a lab or growing them out for visual inspection in greenhouses or fields in warm climates. The number of tubers selected for testing is normally based on the size of the area being sampled.

A minimum sample size is generally two to four hundred tubers per seed lot. This could be increased based on lot size. The confidence level increases with increasing sample size.

Post-harvest grow-out testing is accomplished by breaking dormancy soon after harvest and planting the seed tubers either in a greenhouse or field in a warmer climate (European and Mediterranean Plant Protection Organization,1999). Post-harvest disease tolerances must be established for the various classes of seed. The amount of disease (mainly viruses) permitted for re-certification normally ranges from zero to two percent infection.

17.4.14 Marketing/Dispatch Inspection

Before seed tubers are shipped they may be evaluated for diseases, variety mixtures, off types, and general quality defects according to certification standards. This inspection is often referred to as a "Shipping Point Inspection". The primary concerns at this stage are soil and extraneous matter, dry and wet rots, external defects (e.g. misshapen or damaged tubers), common scab, rhizoctonia, powdery scab, prohibited nematodes, prohibited bacteria and tuber size (minimum/maximum). Other quality constraints may be imposed by individual countries/states.

Internationally, six diseases and/or pests have been classed as unacceptable in any seed potato lot. They are *Clavibacter michiganensis* subsp. *sepedonicus, Ralstonia solanacearum, Globodera rostochiensis, Globodera pallida, Meloidogyne* spp., and *Synachytricum endobioticum* (United Nations/Economic Commission for Europe Standards S-1, 1998).

Two sets of tuber quality standards are observed in seed potato marketing. First are those minimum standards established by legal authority and, secondly, those agreed upon by the buyer and seller. Prohibited diseases and/or pests are not acceptable in either instance. Thus, some minimum quality requirements cannot be arbitrated.

17.4.15 Seed Labelling

When seed is marketed, a certification label is attached to each container by the designated authority/certification agency or their representative. The label should be attached with a sealing device. This must be done in such a manner that any tampering would be readily evident.

Large bulk containers or trucks, which transport seed, can also be sealed with appropriate devices. A special certificate of certification, replacing certification labels, is issued. The identification number used on the seal must also appear on the certificate to maintain the integrity of the seed lot.

Seed labels or certificates should contain the following minimal information: seed lot identification, variety, country of origin, certification authority, category or class of seed, and quality of tubers (grade). Other information could be added such as weight, closing date, and size of tubers.

17.4.16 Financing Certification Programs

Record keeping, inspections, testing and other administrative functions are paid for by the applicant. Certification agencies must devise fee schedules for their services. Programs are funded in various ways. The grower is sometimes required to cover the entire expense. Some countries or agencies subsidize certification programs from general funds or provide office space, administration, and inspections, thereby eliminating some charges to the grower. Charges for certification services may include a membership fee, an acreage fee, a weight fee, or all three.

17.5 Future Trends

The basic principles of seed potato certification schemes will probably remain stable for the foreseeable future. However, changes will likely occur in three areas as technology evolves.

The number of field generations required to produce adequate supplies of clean, vigorous seed for the production of ware and processed potatoes will decline. With newer techniques and an improved propagation system evolving, more "disease free" plant material will be available for the first field increase, thereby reducing the number of additional generations required. The shorter the duration between the introduction of disease free seed stocks and the last seed increase, the greater will be the impact on many potato pests. The "flush out" system will help maintain sustainable potato production practices for many years to come.

Secondly, "true seed" or "potato seeds" could become the preferred planting material for producing seed potatoes in some countries. Newer breeding/genetic techniques will permit the production of hybrid potato clones, which will allow for increases of two or three tuber generations while still maintaining the tuber quality uniformity expected in asexually propagated varieties (Almekinders, 1995; Pallais *et al.* 1985). Seed certification programs will need to incorporate appropriate genetic plant "type" standards for varieties derived from true seed, since not all plant types are phenotypically the same. A limited amount of true seed increase is already evident in established seed certification programs (Oregon Directory of Certified Seed Growers, 1999). The number of increases directly from true seed most likely will be limited to two or three field generations. True seed has the advantage of providing large volumes of "virus free" planting material. Very few potato pathogens are transmitted through botanical potato seeds. Acceptance of true seed varieties will depend on their quality uniformity, particularly for processing.

Thirdly, as new or improved quality factors are incorporated into existing varieties, phenotypic inspections will become more difficult and less meaningful. Maintaining true identity of varieties developed through biotechnology, from their basic seed source through the end user, will require certification agencies to institute "Identity Preserved" schemes. Such schemes are already in existence for true seed programs (General Standards for the Identity Preserved, IP Program, 1999). These schemes follow the final field increase from certified seed to the consumer. Lab testing for genetic markers may be necessary to maintain the integrity of these new varieties. The "Identity Preserved" systems may be used for genetically modified cultivars and for clonal selections from established varieties developed by traditional methods. Much has been accomplished in the area of clonal selection to improve varietal qualities. Many of these selections are now recognized as new varieties. Some have been submitted through plant variety protection laws. In most cases, these new varieties are phenotypically almost identical to the parent. Certification agencies will need to refine their system of variety identification, which uses written plant descriptors, to place greater emphasis on maintaining good cropping history records and developing testing protocols for reliable variety identification.

Acknowledgement

The authors are grateful to R.L. Cook and R.L. Henderson for critical review of this chapter and suggestions for its improvement.

References

Almekinders, C. J.M. 1995. On flowering and botanical seed production in potato (*Solanum tuberosum* L.). PhD. Thesis Landbouwuniversiteit, Wageningen, The Netherlands

Canada Food Inspection Agency (CFIA). 1996. Seed Regulations, Seed Act SOR/96-252.s.l Part II Seed Potatoes.

Consolidated Exports Highlights. 1999. World Statistical Export Summary for Seed Potatoes. United States Potato Board. pp. 8.

DiFonzo, C.D., D.W. Ragsdale and E.B. Radcliffe. 1996. Crop borders reduce potato virus Y Potyvirus (PVY) incidence in seed potato. *Ann. appl Bio.*. 129: 289-302.

European and Mediterranean Plant Protection Organization, Certification Scheme, Seed Potatoes. 1999. 99/7479.

General Standards for the Identity Preserved (IP) Program. 1999. Operational Procedures of the Association of Official Seed Certifying Agencies. pp 118-122.

Hane, D.C. and Hamm, P.B. 1999. Effects of seedborne potato virus Y infection in two potato culivars expressing mild disease symptoms. *Plant Dis.* 83: 43-45.

Hiddema, J. 1972. Inspection and quality grading of seed potatoes. *Viruses of Potatoes and Seed Potato Production* Ed. J.A de Bokx. pp. 206-215.

Mosley, A. R. and Chase, R. W. 1993. Selecting cultivars and obtaining healthy seed lots. In: *Potato Health Management*. Ed. R. C. Rowe. pp 23-24.
Oregon Directory of Certified Seed Growers. 1999. Varieties Catalina and 88.EX.2..
Pallais, N., N. Fong and D. Berrios. 1985. Research on the physiology of potato sexual seed production. In: *Innovative Methods for Propagation of Potatoes*, CIP, Lima, Peru. pp 149-168.
Rykbost, K.A., Hane, D.C., Hamm, P.B., Voss, R. and Kirby, D. 1998. Effects of seedborne potato virus Y on Russet Norkotah performance. *Amer. J. Potato Res.* 75: 91-96.
Slack, S. A. 1993. Seed certification and seed improvement programs. In: *Potato Health Management.* Ed. R.C. Rowe. pp 61-65.
United Nations/Economic Commission for Europe Standards S-1. 1998. Concerning the Certification and Commercial Quality Control of Seed Potatoes Moving in International Trade Between and to UN/ECE Member Countries.
U.S. Export Standards for Seed Potatoes. 1999. Appendix P Export Certifica
tion Manual. USDA/APHIS.

Table 17.1. Generalized Certification Procedures, UN/ECE and North American Schemes

UN/ECE Categories	**U.S. Export Standards, Limited Generation [b]**	**Comments**
Pre-Basic	Pre-Nuclear	UN/ECE – Tissue culture optional U.S. – In-vitro production only
Pre-Basic[a]	Nuclear	UN/ECE – Add TC postfix to Pre-Basic if from Tissue Culture Canada/U.S. – Greenhouse or controlled environment production
Pre-Basic or Basic[a]	Generation 1 Pre-Elite[b]	First field increase. UN/ECE classification determined by individual countries
Basic[a]	Generation 2 Elite 1[b]	Second field increase
Basic[a]	Generation 3 Elite 2[b]	Third field increase
Basic or Certified[a]	Generation 4 Elite 3[b]	Fourth field increase
Certified[a]	Generation 5 Elite 4[bc]	UN/ECE, Canada -- Eligible for further re-certification[c] U.S. – Final field increase.
Certified[a]	Foundation[b]	UN/ECE – Re-certification may continue depending on disease and other factors
Certified[a]	Certified[b]	UN/ECE -- May produce up to 10 field generations from tissue cultured seed source. Canada – Final field increase

[a]Individual countries may establish subdivisions within categories.
[b]Canada follows a very similar Limited Generation Certification Scheme.
[c]Canada permits seven field increases.

Table 17. 2. Tolerance Comparisons (% of Plants Affected), Final Inspection, First Field Increase

	Category		
Factor	**UN/ECE Pre-Basic**	**Canada Pre-Elite (G1)**	**U.S. Export Standards Generation 1**
Growing Crop			
Potato Spindle Tuber	Not specified	0.00	0.00
Virus	0.10	0.00	0.00
Blackleg (*Erwinia*)	0.00	0.00	0.00
Off-types/Other varieties	0.01	0.00	0.00
Direct Progeny [a]			
Virus	0.10	Not specified	Not specified[c]
Off-types	0.01[b]	Not specified	Not specified[c]
Major Disease/Pest			
Globodera spp.	0.00	0.00	0.00
Synchytrium endobioticum	0.00	0.00	0.00[d]
Clavibacter michiganensis, subsp. *sepedonicus*	0.00	0.00	0.00
Phoma exigua pv.foveata	Not specified	0.00[d]	0.00[d]
Rolstonia solanacearum	0.00	0.00[d]	0.00
Meloidogyne spp.	Not specified	0.00	0.00
Ditylenchus destructor	Not specified	0.00	0.00

[a]Post-harvest testing.
[b]Includes other varieties
[c]Determined by individual states.
[d]Does not occur in Canada or the United States.

Table 17..3. Disease Tolerances (% Infected Plants), Final Inspection, Third Field Increase

	Category		
Factor	**UN/ECE Basic**[a]	**Canada Elite 2 (G3)**	**U.S. Export Standards Generation 3**
Growing Crop			
Potato Spindle Tuber	Not specified	0.00	0.00
Virus	Not specified	0.00	0.10
Blackleg(*Erwinia*)	1.00	0.00	0.10
Off-types	Not specified	0.00	0.10
Direct Progeny[b]			
Virus	4.00	Not specified	Not specified[c]
Off-types	0.25	Not specified	Not specified[c]
Other varieties	0.10	Not specified	Not specified[c]
Major Diseases/Pests			
Globodera spp.	0.00	0.00	0.00
Synchytrium endobioticum	0.00	0.00	0.00[d]
Clavibacter michiganensis Subsp. sepedonicus	0.00	0.00	0.00
Phoma exigua pv.foveata	Not specified	0.00[d]	0.00[d]
Rolstonia solanacearum	0.00	0.00[d]	0.00
Meloidogyne spp.	Not specified	0.00	0.00
Ditylenchus destructor	Not specified	0.00	0.00

[a] May exceed three field years; eligible for recertification.

[b] Post-harvest testing.

[c] Determined by individual states.

[d] Does not occur in Canada or the United States.

Table 17.4. Tolerance Comparisons (% Affected Plants), Final Inspection, Fifth Field Increase.

	Category		
Factor	**UN/ECE Certified**	**Canada Elite 4 (G5)**	**U.S. Export Standards Generation 5**
Growing Crop			
Potato Spindle tuber	Not specified	0.00	0.00
Virus	Not specified	0.10	0.50
Blackleg (*Erwinia*)	2.00	0.25	0.50
Off-types	Not specified	0.00	0.50
Direct Progeny[a]			
Virus	10.00[b]	Not specified	Not specified[c]
Off-types	0.50	Not specified	Not specified[c]
Other varieties	0.20	Not specified	Not specified[c]
Major Diseases/Pests			
Globodera SPP.	0.00	0.00	0.00
Synchytrium endobioticum	0.00	0.00	0.00[d]
Clavibacter michiganensis subsp. sepedonicus	0.00	0.00	0.00
Phoma exigua pv.foveata	Not specified	0.00[d]	0.00[d]
Rolstonia stolanacearum	0.00	0.00[d]	0.00
Meloidogyne spp.	Not specified	0.00	0.00
Ditylenchus destructor	Not specified	0.00	0.00

[a]Post-harvest testing.
[b]Severe virus only.
[c]Determined by individual states.
[d]Does not occur in Canada or the United States

18. Emerging Potato Viruses That Confront Regulations

L. G. Brown
USDA/APHIS/PPQ-Center for Plant Health Science and Technology, Raleigh, North Carolina 27606 USA.

18.1 Introduction

There have been numerous examples where, despite regulatory policies intended to prevent introduction of non-indigenous pathogens, introduction has still occurred. As fewer governmental resources are devoted to regulation of international movement of plant materials, germplasm, etc., and as international trade increases, it is imperative that we understand potato virus biology and dynamics as they relate to the regulatory framework. Case studies of how the regulatory process has functioned can be used to illustrate the scientific and organizational factors that comprise regulatory decisions and actions.

18.2 Background

The cases of the tobacco necrosis pathotype of *Potato virus Y* (PVY^N) and the tuber necrosis pathotype of PVY (PVY^{NTN}), *Potato yellow mosaic virus* (PYMV), *Potato virus A* (PVA), and *Tomato spotted wilt virus* (TSWV) can be used to study and clarify regulatory concepts.

G. Loebenstein et al. (eds.),
Virus and Virus-like Diseases of Potatoes and Production of Seed-Potatoes, 439–450.

Pest risk assessment is the process of evaluating biological or other scientific and economic evidence to determine whether a pest should be regulated and the strength of any phytosanitary measures to be taken against the pest. Pest risk assessment has to answer and ask many questions based on science. It identifies and communicates information about pest pathways and determines the likelihood of pest entry, establishment, and spread. Risk assessment is practiced by seed improvement organizations. These organizations turn to state, federal and ultimately international organizations to develop common rules for the safe trade in seed potatoes.

Pest Risk Analysis = Pest Risk Assessment + Pest Management + Communication

A number of national and international organizations are involved in various aspects of pest risk analysis. The North American Plant Protection Organization (NAPPO) is the regional plant protection organization of the International Plant Protection Convention that coordinates efforts among Canada, the United States and Mexico to protect their plant resources from the entry, establishment and spread of regulated plant pests, while facilitating trade (http://www.nappo.org).

The European Plant Protection Organization (EPPO) is an intergovernmental organization responsible for international cooperation in plant protection in the European and Mediterranean region. Under the International Plant Protection Convention, EPPO is the regional organization for Europe. Founded in 1951, EPPO objectives are to develop an international strategy against the introduction and spread of pests that damage crops and forests, encourage harmonization of phytosanitary regulations and all other areas of official plant protection action, promote the use of modern, safe, and effective pest control methods, and provide a documentation service on plant protection (http://www.eppo.org).

The International Plant Protection Convention (IPPC) is based upon a multilateral treaty deposited with the Director-General of the Food and Agriculture Organization of the United Nations (FAO) and administered through the IPPC Secretariat located in FAO's Plant Protection Service. One hundred and eleven (111) governments are currently contracting parties to the IPPC (http://www.fao.org).

An A2 Pest is a quarantine pest present in an area but not widely distributed and also is under an official control program. An A1 pest is a quarantine pest that is not present in an area (Anon, 1999). An area is a geographic or geopolitical zone that has been determined by agreement. PVY^N is an A2 pest in North America, while PVY^{NTN} is an A1 pest in North America. No PVY strain is an A1 or A2 pest in Europe.

18.3 Taxonomic Issues

Before a pathogen is considered to be emerging there should be a correct diagnosis. Pest risk management begins with a correct diagnosis of the pest. Van Regenmortel *et al.* (1995) states that virus identification, that is unambiguous, is a virtue and it is of particular value when scientists require precise naming of viruses for citation. Precise naming avoids problems or confusion caused by synonyms, transliterated vernacular names, and local laboratory jargon that can be problems confronting regulators. The ICTV operates under the auspices of the Virology Division of the International Union of Microbiological Societies, and fosters standardized names by defining and approving virus families, genera and species. Study groups within the ICTV recommend virus subgroups, strains, and isolates. For the case studies here is important to first discuss the taxonomy of the *Potyviridae*, *Geminiviridae*, and Tospoviruses.

18.3.1 Potyvirus Taxonomy

Barnett (1992) noted that virologists have not reached a consensus about taxonomic placement because of the genetic and biological diversity of viruses in the family *Potyviridae* and an inability to separate taxonomic and diagnostic characteristics. He states that the need for a useful and acceptable taxonomy requires that some conclusions be drawn; otherwise a situation of disjointed names and classification will occur. The taxonomy of the *Potyviridae* has, in recent years, become clearer (Berger *et al.* 1997; 1999).

Defining terms is required before conclusions can be drawn and articulated. According to Barnett (1992), an <u>isolate</u> is a single isolation from an infected plant and the term does not have a taxonomic implication. A <u>strain</u> is a collection of naturally occurring isolates of a virus species having the same recognizable properties. A <u>pathotype</u> is a collection of isolates that cause a recognizable host response. A <u>serotype</u> is a collection of isolates, which differ from another collection of isolates by antigenic properties. A <u>variant</u> is a novel or distinctive isolate occurring naturally or one that is man-made. A <u>subgroup</u> is a collection of virus species, between the genus and species level that are closely related and have many properties in common, making them difficult to separate (Barnett, 1992).

PVY is the type species of the genus *Potyvirus*, which consists of the aphid-transmitted members of the family *Potyviridae*. PVY is in a subgroup of the genus *Potyvirus*

Another potato-infecting member of the genus *Potyvirus* is PVA, which has not been studied extensively because it causes mild symptoms in most cultivars. However, de Bokx and van der Want (1987) believe this aphid transmitted virus deserves more attention because it is wide spread, and in combination with either *Potato virus X* or PVY it causes severe symptoms and yield loss.

18.3.2 Tospovirus Taxonomy

Tomato spotted wilt virus (TSWV) is the type species of the genus *Tospovirus* in the family *Bunyaviridae* (Murphy *et al.* 1995). This thrips transmitted virus is found in subtropical and temperate climate zones. Many economical hosts are impacted and the host list continues to expand (Cho *et al.* 1988). TSWV had a period of low economic impact after World War II until the spread of the Western Flower Thrips, *Frankliniella occidentalis* increased crop losses due to the virus. Potatoes should be included as a crop that is economically impacted (Weingartner, D.P., personal communication and Bhat, *et al.* 1999).

18.3.3 Geminiviridae Taxonomy

Geminiviridae includes the genus *Begomovirus*, with Bean golden mosaic virus as the type species. Members of this genus are transmitted by the whitefly *Bemisia tabaci*. The virus is retained when the vector molts. It does not multiply in the vector nor is it transmitted congenitally to the progeny of the vector (Murphy *et al.* 1995).

Just as TSWV increased after the emergence of western flower thrips, a potential impact on potatoes from *Begomovirus* species may follow the emergence of the biotype B of *Bemisia tabaci* (Faria *et al* 1997; Polston and Anderson 1997). A strain of Potato yellow mosaic virus (PYMV) has moved into tomatoes in Venezuela (Guzman *et al.* 1997). Whiteflies transmit Tomato yellow vein streak virus (ToYVSV) to potatoes in laboratory studies, but the virus has not been detected in field potatoes (Faria *et al.* 1997). ToYVSV is an emerging virus in Brazilian tomatoes (Polston and Anderson 1997; Ribeiro *et al.* 1998). Therefore, since the growing seasons for potatoes and tomatoes overlap these two geminiviruses could also emerge as a production problem in potatoes.

18.4 Regulatory Case Studies

The case studies focus on the regulation of two strains of PVY, PVA, and TSWV. PVA occurs worldwide in most potato growing areas. The mild mosaic and crinkle symptoms in the field are difficult to distinguish from PVX and PVY either when hosts have a mixed or single virus infection.

This makes serology a practical tool for diagnosing these viruses (Murphy *et al.*1995).

Host plant reactions, serology, and possibly viral inclusion bodies must be used together to discriminate between strains of PVY. Therefore, a combination of methods is necessary to make regulatory decisions when new observations outside regulated areas are involved. One or a combination of two methods would suffice for confirming the presence of PVY strains in an existing regulated area. This point is stressed because regulatory decisions must be made about a virus strain even though the research community has not yet reached a consensus about taxonomic placement of the strain.

18.4.1 Classification of PVY

The conventional classification of necrotic isolates of PVY is based on symptoms in primary hosts (potato and tomato) and on those induced in indicator plants (Chachulska *et al.* 1997). According to De Bokx and Huttinga (1981) PVY^{O}, PVY^{N}, and PVY^{C} are the three main strains in the PVY subgroup. It may be more useful to consider PVY^{NTN}, PVY^{N}, PVY^{O}, and PVY^{C} as strains of PVY. Some scientists include PVY^{Z}, as a strain of PVY (Blanco-Urgoiti *et al.* 1998), but its relationships to the other strains is not yet clear. The correlation between tests using primary hosts, indicator plants, and serology to characterize a PVY strain is helpful but not absolute.

When symptoms on potatoes are used to describe PVY strains, usually two general pathotypes are observed. The acronym PVY^{N} is used for convenience and encompasses those pathotypes that causes veinal necrosis on tobacco. Those, which cause mild symptoms in most potato varieties, are called PVY^{O} (Chrzaniwska, 1991;Weidemann, 1988) and those that cause necrotic rings are called potato tuber necrotic disease or PVY^{NTN} (Le Romancer *et al.* 1994;Chachulska *et al.*1997).

There are significant risks that these viruses will be moved long distances in potato seed tubers or through the exchange of infected plant material. Boonham *et al.* (1999), for example, report that PVY^{NTN} causes a disease of trailing petunias, and these ornamental hosts could be a source of inoculum.

Classification of PVY isolates by serology is a matter of interpretation. Coat protein directed polyclonal antibodies do not differentiate between PVY. A simple and accurate method to diagnose closely related viruses is needed for prediction and control to enable effective regulatory actions. Monoclonal antibodies (MABs) are specific to most necrotic strains of PVY (Gugerli and Fries, 1983). MABs specific to PVY^{O} and PVY^{N} have been used to characterize selected PVY isolates, allowing determination of possible strain-specific epitopes in the N-terminal part of the coat protein (Hatatam *et al.* 1994). In Poland the majority of necrotic PVY isolates from potatoes (pathogenic to many tobacco cultivars and less pathogenic to potatoes) gave no reaction when BIOREBA MABs were used against the VAM-B pathotype (Chranowska and Doroszcwska, 1997). PVY^{N-NY} and PVY^{N-WI} phenotypically go with the necrotic PVY pathotype (Chranowska and Doroszcwska, 1997 but the latter did not react with MABs, that recognize PVY^{N-NY}. Chachulska *et al.* (1997) phenotypically separated Polish strains of PVY^{N-NY} and PVY^{N-WI} into the necrotic pathotype PVY^{NTN}. They also showed that PVY^{O-LW} is a PVY^{O} strain. Garbaczewska *et al.* (1996) report cytological characters that may be useful to distinguish members of PVY that cause necrosis but this method is useful for characterizing isolates and not for large-scale testing.

Chachulska *et al.* 1997 believed that molecular sequence analysis based clustering of isolates generally follows the phenotypic classification of strains. The notable exceptions are the isolates that induce tobacco veinal necrosis. Therefore, the ability to induce tobacco veinal necrosis does not follow the molecular classification based on the analysis of the CP and 3'NTR sequences. To characterize the isolates, coat protein genes were sequenced and compared with sequences from databases (Chachulska *et al.* 1977). PVY^{N-WI} isolate showed 99% amino acid homology with PVY^{O-LW} and differed from the PVY^{N-NY} isolate. A sequence based homology matrix and phylogenetic analysis classified PVY^{N-NY} into group 1 which encompasses solely necrotic pathotype isolates. Group I is monophyletic whereas PVY^{N-WI} is in a phenotypically heterogeneous group II.

The potato tuber necrotic ringspot strain of PVY (PVY^{NTN}) was distinguished from other members of the PVY necrotic isolate as well as PVY^{O} and PVY^{C} using RT-PCR. Studies indicated that PVY^{NTN} infections do not necessarily lead to potato tuber necrotic ringspot disease (PTNRD) symptoms and that tomatoes can act as a virus reservoir (Weidemann and Maiss, 1996). Three California isolates which induced PTNRD symptoms in potato did not show the expected 835 bp bands, indicating that these isolates differed from those collected in Europe (Weidemann and Maiss, 1996). Reliable detection of PVY^{NTN} is critical because of the fast dispersal of PVY^{NTN} in Europe, its transmission by *Mzyzus persicae*, and the fact that this virus goes to both potatoes and tomatoes, which are produced in adjoining fields.

18.4.2 The Regulatory Story of PVY^{N}

PVY^{N} has minimal effect on potatoes but has a devastating effect on field tobacco. This was the cause of the A1 quarantines in the USA and Canada. PVY^{N} was detected in Canada in 1990 in Prince Edward Island (PEI) and in a tobacco field near a potato field in New Brunswick (NB) (Coffin *et al.* 1991). PVY^N had also been detected in Ontario (Singh 1991). These observations were associated with the planting of illegally imported potato cultivars; PVY^N is prevalent in Europe and South America (Brunt *et al.* 1996). States in the USA were concerned with tobacco transplants being exposed to PVY^N and becoming a means for spread of the strain into the large tobacco growing areas. The USDA regulations (Code of Federal Regulations 319.37-2) prohibited the movement of potatoes into the USA from everywhere except Canada because of PVY^N. When PVY^N was confirmed in Canada, many questions about eradication and management arose. Had it already been shipped to the USA in seed potatoes and had it become established in USA fields and seed system. The USDA prohibited certain potatoes from entry into the US and placed restrictions on others in an effort to keep PVY^N out of the country.

The USDA revised import regulations for Canadian potatoes several times in an attempt to prevent the entry of PVY^N. Although quarantine regulations provided a relatively high degree of protection against PVY^N introduction, the regulations did not prevent its entry into the US (Curlett, 1995). The Canadian Food Inspection Agency was conducting an eradication program as suggested by Singh (1991) in certified seed potato systems of PEI and NB. Canada's definition of eradication is similar to that of The North American Plant Protection Organization, which is based on that of the United Nations Food and Agriculture Organization (FAO) that define an eradication program as the application of phytosanitary measures to eliminate a pest from an area (Anon. 1999). Canada detected PVY^N in a few fields in NB, Nova Scotia, Ontario, and PEI and the seed potatoes plant used in all fields was traced back to PEI (Singh, 1992). At issue was whether PVY^N was found only in the seed lots in question or whether other seed lots in Canada and the USA had become infected from earlier shipments of potatoes from Canada. Another issue was whether PVY^N was in the USA. Canada detected PVY^N in a shipment(s) of table stock potatoes from the USA but the origin of these potatoes was never established.

18.4.3 The US Survey for PVY^N

Initial testing in the USA of Canadian winter grow-out tests of seed production was negative for PVY^N. Canadian potato seed lots from NB and PEI were grown in Florida during the winter for general virus testing (Singh, 1992). The Florida Department of Agriculture / Division of Plant Industry and USDA / PPQ survey of winter seed tests in Florida conducted by Canada was also negative for PVY^N.

A key question for the US was how to prove that the virus was not already present in the seed system. A survey of potato fields in Florida planted from US seed potatoes produced outside of Florida was key to determining the extent of virus spread in the US. Also, Florida and the other states would directly benefit from any incidence data collected in Florida because these data would be critical to permitting trace-back of virus-infected seed sources and for designing eradication plans in the US if necessary. Field surveys in Florida during 1992 and 1993 for PVY^N represented a good cross-section of seed potatoes produced in Eastern Canada and in the US from the Mississippi drainage basin to the east. A virus detection level of 0.1% at a 95% confidence level was set for the 29,290 acres of potato production that was surveyed each year. This was a timely and cost effective method to determine if PVY^N was in the seed production programs of Midwestern and Eastern North America. PVY^N was confirmed in one Florida potato field in 1992 and in 5 in 1993 (Brown and Weingartner, 1993). Confirmation was by ELISA using $MabY^N$ 295.5 and Burley 21 tobacco bioassay with independent confirmation by USDA/APHIS/PPQ.

All trace-backs of seed-potatoes planted in infected fields indicated that the source was Canada. These tests were conducted with the cooperation the Florida Department of Agriculture and Consumer Services, USDA Plant Protection and Quarantine, and Agriculture Canada. PVY^N was not detected in any fields planted with USA produced seed potatoes. Re-sampling of fields at the level of 3,000-samples/field indicated that the level of PVY^N infection never exceeded 0.1%, and that PVY^N was randomly distributed within fields. The virus probably came in on seed potatoes and there was no detectable infield spread. No evidence in Florida was found that PVY^N became established in volunteer potatoes or in susceptible weeds (L. Brown, unpublished results). PVY^N was also detected in Pennsylvania but did not become established there either.

18.4.4 Lessons Learned from PVY^N.

An agreement between Canada and the US was reached that could be used in lieu of regulations. The Canadian / US PVY^N Management Plan was developed in 1993 to facilitate trade of potatoes within and between Canada and the US, and was a reasonable and responsible response a trade situation. If the countries had continued with existing regulations, then the required testing would have made seed potatoes too expensive to trade between Canada and the US. The plan emphasized control of PVY^N in Canada and the US while protecting the tobacco industry from harm caused by PVY^N. As a result, all quarantine regulations concerning PVY^N were revoked in

Canada and the US for potatoes. The lesson was that a reasonable and responsible response to the situation was preferable to regulations.

Another lesson was that a protocol using just serology is useful as an early step when dealing with a regulatory issue. However, both the bioassay on tobacco and the testing for PVX that confuses symptoms on PVY^O infected Burley 21 is necessary as a final step for all quarantine actions. In the Florida survey Brown and Weingartner (1993) used (a polyclonal to PVY^O and MAB for PVY^N) using 10 to 15 leaf composites followed by a bioassay on Burley 21 tobacco for confirmation. The Canadian survey of potato fields for PVY^N (McDonald *et al.* 1994) used serology for confirmation. Suspects that were positive for PVY^O were tested as 100-leaf composite samples using $MabY^N$ 295.5. A suspect was any sample that had a signal to noise ratio (S: N) above 4:1. A second ELISA screen used 10 leaf sub-composites with a panel of monoclonal antibodies.

18.4.5 Lessons Learned from Relying Solely on Field Observations to Detect PVA, PVY, PVY^N, and PVY^{NTN}.

Potato improvement programs were formed in many countries in part to eliminate or reduce the impact of virus diseases on the quality of seed potatoes. High quality means the reduction or absence of pests, preservation of genetic identity and purity. Most programs start with clonal material and sell the subsequent field increases that are produced over the next six or seven years. These improvement programs use a one-way production system that always begins with indexed clonal material that has not been field increased.

Improvement programs are labour intensive and must balance benefits and costs throughout the process of field increases. Each increase involves larger size lots and laboratory testing becomes increasingly expensive. To contain costs most programs until recently relied on field observations for symptoms to remove virus infected material from the system. A common model was that both a growing season and winter grow-out inspection was conducted. The winter grow-out was conducted on a sample of the lot that passed the growing season inspection. Winter tests were performed in tropical areas so that a second season of observation could be conducted in a timely manner. It was felt that trained observers could reliably detect viruses. However, visual assessment has limitations. In the past the observation method missed the increase of PVY^{NTN} as it spread from Eastern to Western Europe after World War II. The spread occurred not because inspectors and researchers were inattentive but because this virus only showed symptoms on the underground tubers. Tomassouli *et al.* (1998) summarized the occurrence of potato tuber necrosis ringspot disease

(PTNRD) in Italy. PVY^{NTN} is not known to occur in North America by lack of symptom expression. Thus, a walk-through field inspection had no chance of detecting PVY because PVY tolerant varieties, like Jemseg (Young *et al.* 1979) were introduced which did not show leaf symptoms when infected. These tolerant varieties became virus reservoirs that allowed PVY to build-up undetected and spread to susceptible varieties during field increase. Serological testing of field generations of seed potatoes was necessary to reduce PVY below tolerances for improved seed potatoes. Serological testing coupled with appropriate sampling methods needs to be implemented on a cost effective basis.

18.5 *Potato virus* A

PVA is causing increasing problems in seed programs in the Northwestern US due to the unobserved increase of PVA in seed stocks in symptomless infected potato varieties. PVA spreads to fields of susceptible varieties that may carry PVX and PVY, resulting in more severe disease. The late 1990's witnessed the re-emergence of PVA as a threat to certain varieties of potatoes in Idaho. Spread of PVA occurred concurrently with an increase of PVY. The large increase in aphid populations coupled with a virus tolerance level of 15% all attributed to the increase.

18.6 Conclusions

When regulators are confronted with pests that have ambiguous taxonomy they must arrive at a consensus on how to identify the pest. This process starts with using common terms and accepting scientific definitions. This chapter has outlined an approach to do this with the *Potyviridae,* Tospoviruses, and *Geminiviridae.* Application of this approach is given in the regulatory case study of PVY^{N}. Host plant reactions, serology and other methods must be used together to identify a pest for regulatory purposes.

The concept of a reasonable and responsible response to trade is illustrated by the lessons learned from the necrotic strains of PVY.

Observation without testing is a practice that is not viable anymore. Case studies with PVY and PVA support this contention.

Acknowledgements

I acknowledge O.W. Barnett, S.R. Poe, S C. Redlin, and R.A. Sequeira for many helpful discussions and comments

References

Anon. 1999. Glossary of phytosanitary terms. NAPPO, Ottawa.

Barnett, O.W. 1992. A summary of potyvirus taxonomy and definitions. In: *Potyvirus Taxonomy*. Springer, Wien New York, pp 435-444. Arch. Virol. [Suppl] 5).

Berger, P.H., Barnett, O.W., Brunt, A.A., Colinet, D., Edwardson, J.R., Hammond, J., Hill, J.H., Jordan, R.L., Kashiwazaki, S., Makkouk, K., Morales, F.J., Rybicki, E., Spence, N., Ohki, S.T., Uyeda, I., van Zaayen, A. and Vetten, H.J. 1999. *The Potyviridae.* in: Virus Taxonomy. 7th Report of the International Committee on Taxonomy of Viruses. (van Regenmortel, M.H.V., Fauquest, C.M., Bishop, D.H.L., Carstens, E.B., Estes, M.K., Lemon, S.M., Maniloff, J., Mayo, M.A., McGeoch, D.J., Pringle, C.R. and Wickner, R.B., eds.) Academic Press, NY.

Berger, P.H., Wyatt, S.D., Silbernagel, M., Shiel, P.J. and Mink, G.I. 1997. Phylogenetic analysis of the *Potyviridae* with emphasis on the legume-infecting potyviruses. *Arch. Virol.* 142:1979-1999.

Bhat, A.I., Pappu, S.S., Pappu, H.R., Deom, C..M. and Culberath, A.K. 1999. Analysis of the intergenic region of tomato spotted wilt Tospovirus medium RNA segment. *Virus Res.* 61:161-170.

Blanco-Urgoiti, B., Tribodet, M., Leclere, S., Ponz, F., de Roman, C.P.S., Legorburu, F.J. and Kerlan, C. 1998. Characterization of potato potyvirus Y (PVY) isolates from seed potato batches. Situation of the NTN, Wilga, and Z isolates. *Eur. J. Plant Pathol.* 104:811-819.

Boonham, N., Hims, M., Barker, I. and Spence, N. 1999. Potato virus Y form petunia can cause symptoms of potato tuber necrotic ringspot disease. *Eur. J. Plant Pathol.* 105:617-621.

Brown, L.G. and Weingartner, D.P. 1993. Survey for potato virus Y, necrotic strain, in potato production areas of Florida for 1992 and 1993. *Phytopathology* 83:1400A. (abstr.).

Brunt, A.A., Crabtree, K., Dallwitz, M.J., Gibbs, A.J., Watson, L. and Zurcher, E.J. (eds.), (1996 onwards). '*Plant Viruses Online*: Descriptions and Lists from the VIDE Database.

Chachulska, A.M., Chrzanowska, M., Robagilia, C. and Zagorski, W. 1997. Tobacco veinal necrosis determinants are unlikely to be located within the 5' and 3' terminal sequences of the potato virus Y genome. *Arch. Virol.* 142:765-779.

Cho, J.J., Mau. R.F.L., Hammasaki, R.T. and Gonsalves, D. 1988. Detection of tomato spotted wilt virus in individual thrips by ELISA. *Phytopathology* 78:1348-1352.

Chrzanowska, M. 1991. New isolates of the necrotic strain of potato virus Y found recently in Poland. *Pot. Res.* 34:179-182.

Chrzanowska, M. and Doroszcwska, T. 1997. Comparison between PVY isolates obtained from potato and tobacco plants grown in Poland. *Phytopathol. Pol.* 13:63-71.

Coffin, R. Singh, R., DeHann, T.L., Parker, A., Keenan, M.K. and MacQuarrie, J. 1991. Eradication measures for a new occurrence of tobacco veinal necrosis virus (PVY^{N}) in potatoes. *Potato J.* 68: 601 (abstr.).

Curlett, E. C. 1995. Potato Virus Y (PVY^{N}). *USDA/APHIS Plant Prot. Quar. Fact Sheet.*

De Bokx, J.A. and Huttinga, H. 1981. Potato virus Y. *CMI/ABB Desc. Pl. Vir.* No. 242,

De Bokx, J.A. and van der Want, J.P.H. (eds.). 1987. *Viruses of potatoes and seed-potato production.* Sec. ed. Pudoc Wageningen. 259 pp.

Faria, J.C., Souza-Dias, J.A.C., Slack, S.A. and Maxwell, D.P. 1997. A new Geminivirus associated with tomato in the state of Sao Paulo, Brazil. *Plant Dis.* 81:423.

Garbaczewska, G., Chrzanowska, M. and Kerlan, C. 1996. Ultrastructural studies of Nicotiana clevelandii and potato plants infected with PVYN-NTN. *Ann. Warsaw Agric. University,* 31:15-22.

Glais, L., Tribodet, M., Gauthier, J.P., Astier-Manifacier, S., Robaglia, C. and Kerlan, C. 1998. RFLP mapping of the whole genome of ten viral isolates representative of different biological groups of potato virus Y. *Arch. Virol.* 143: 2077-2091.

Gugerli, P. and Fries, A. 1983. Characterization of monoclonal antibodies to potato virus Y and their use for virus detection. *J. Gen. Virol.* 64: 2471-2477.

Guzman, P., Arredondo, C.R., Emmatty, D., Portillo, R.J. and Gilbertson, R.L. 1997. Partial characterization of two whitefly-transmitted Geminiviruses infecting tomatoes in Venezuela. *Plant Dis.* 81:312.

Hatatam, T., Inouc, A.K., Oshima, K. and Shikata, E. 1994. Characterization and strain identification of a potato virus Y isolate non-reactive with monoclonal antibodies specific to the ordinary and necrotic strains. *Intervirology* 37: 12-19.

Le Romancer, M., Kerlan, C. and Nedellec, M. 1994. Biological characterization of various geographical isolates of potato virus Y inducing superficial necrosis on potato tubers. *Plant Pathology* 43: 138-144.

McDonald, J., MacLatchy, I.A., Sheidow, N.W., Watson, M.C. and Reeleder, R.D. 1991. Tobacco veinal necrosis strain of potato virus Y (PVY^N). *Ontario Min. Agric. Food (FAM)* Fact Sheet, January.

McDonald, J.G., Kristjansson, G.T., Singh, R.P., Ellis, P.J. and McNab, W.B. 1994. Consecutive ELISA screening with monoclonal antibodies to detect potato virus Y^N. *Amer. Potato J.* 71:175-183.

McDonald, J.G. and Singh, R.P. 1996. Host range, symptomatology, and serology of isolates of potato virus Y (PVY) that share properties with both PVY^N and PVYO strain groups. *Amer. Potato J.* 73: 309-315.

McDonald, J.G. and Kristjansson, J.G. 1993. Properties of strains of potato virus Y-N in North America. *Plant Dis.* 77:87-89.

Polston, J.E. and Anderson, P.K. 1997. The emergence of whitefly-transmitted geminiviruses in tomato in the Western Hemisphere. *Plant Dis.* 81: 1358-1369.

Ribeiro, S.G., de Ávila, A.C., Bezerra, I.C., Fernandes, J.J., Faria, J.C., Lima, M.F., Gilbertson, R.L., Maciel-Zambolim, E. and Zerbini, F.M. 1998. Widespread occurrence of tomato Geminiviruses in Brazil, associated with the new biotype of the whitefly vector. *Plant Dis.* 82: 830.

Singh, R.P. 1991. *A suggestion for eradicating PVY^N from Canada*: A discussion paper. distributed to Agriculture Canada, PVY^N group in February 1991, Agriculture Canada 3 pp.

Singh, R.P. 1992. Incidence of the tobacco veinal necrotic strain of potato virus Y (PVY^N) in Canada in 1990 and 1991 and scientific basis for eradication of the disease. *Canad. Plant Dis. Survey.* 72: 113-119.

Tomassoli. L. and Lumia, V. 1998 Occurrence of potato tuber necrotic ringspot disease (PTNRD) in Italy. *Plant Dis.* 82: 350.

Weidemann, H.L. 1988. Importance and control of potato virus YN in seed potato production. *Potato Res.* 31: 85-94.

Weidemann, HL. and Maiss, E. 1996. Detection of the potato tuber necrotic ringspot strain of potato virus Y (PVY^{NTN}) by reverse transcription and immunocapture polymerase chain reaction. *J. Plant Dis. Prot.* 103: 337-345.

Weilguny, H. and Singh, R.P. 1998. Separation of Slovenian isolates of PVY^{NTN} from the North American isolates of PVY^N by a 3-primer PCR. *J. Virol. Meth.* 71: 57-68.

Young, D.A., Davies, H.T. and Johnston, G.R. 1979. Jemseg: A new, early high-yielding potato variety with high resistance to virus Y and immune to virus X. *Am. Pot. J.* 56:325-328.

Index

MIX
Papier aus verantwortungsvollen Quellen
Paper from responsible sources
FSC® C105338

If you have any concerns about our products,
you can contact us on
ProductSafety@springernature.com

In case Publisher is established outside the EU,
the EU authorized representative is:
Springer Nature Customer Service Center GmbH
Europaplatz 3, 69115 Heidelberg, Germany

Printed by Libri Plureos GmbH
in Hamburg, Germany